PREDICTION OF POLYMER PROPERTIES

PLASTICS ENGINEERING

Founding Editor

Donald E. Hudgin

Professor
Clemson University
Clemson, South Carolina

1. Plastics Waste: Recovery of Economic Value, *Jacob Leidner*
2. Polyester Molding Compounds, *Robert Burns*
3. Carbon Black-Polymer Composites: The Physics of Electrically Conducting Composites, *edited by Enid Keil Sichel*
4. The Strength and Stiffness of Polymers, *edited by Anagnostis E. Zachariades and Roger S. Porter*
5. Selecting Thermoplastics for Engineering Applications, *Charles P. MacDermott*
6. Engineering with Rigid PVC: Processability and Applications, *edited by I. Luis Gomez*
7. Computer-Aided Design of Polymers and Composites, *D. H. Kaelble*
8. Engineering Thermoplastics: Properties and Applications, *edited by James M. Margolis*
9. Structural Foam: A Purchasing and Design Guide, *Bruce C. Wendle*
10. Plastics in Architecture: A Guide to Acrylic and Polycarbonate, *Ralph Montella*
11. Metal-Filled Polymers: Properties and Applications, *edited by Swapan K. Bhattacharya*
12. Plastics Technology Handbook, *Manas Chanda and Salil K. Roy*
13. Reaction Injection Molding Machinery and Processes, *F. Melvin Sweeney*
14. Practical Thermoforming: Principles and Applications, *John Florian*
15. Injection and Compression Molding Fundamentals, *edited by Avraam I. Isayev*
16. Polymer Mixing and Extrusion Technology, *Nicholas P. Cheremisinoff*
17. High Modulus Polymers: Approaches to Design and Development, *edited by Anagnostis E. Zachariades and Roger S. Porter*
18. Corrosion-Resistant Plastic Composites in Chemical Plant Design, *John H. Mallinson*
19. Handbook of Elastomers: New Developments and Technology, *edited by Anil K. Bhowmick and Howard L. Stephens*
20. Rubber Compounding: Principles, Materials, and Techniques, *Fred W. Barlow*
21. Thermoplastic Polymer Additives: Theory and Practice, *edited by John T. Lutz, Jr.*
22. Emulsion Polymer Technology, *Robert D. Athey, Jr.*
23. Mixing in Polymer Processing, *edited by Chris Rauwendaal*
24. Handbook of Polymer Synthesis, Parts A and B, *edited by Hans R. Kricheldorf*
25. Computational Modeling of Polymers, *edited by Jozef Bicerano*

Additional Volumes in Preparation

PREDICTION OF POLYMER PROPERTIES

SECOND EDITION, REVISED AND EXPANDED

JOZEF BICERANO

The Dow Chemical Company
Midland, Michigan

Marcel Dekker, Inc.　　　　　New York • Basel • Hong Kong

Library of Congress Cataloging-in-Publication Data

Bicerano, Jozef.
    Prediction of polymer properties / Jozef Bicerano. — 2nd ed., Revised and Expanded.
      p. cm. — (Plastics engineering ; 38)
    Includes bibliographical references and index.
    ISBN 0-8247-9781-7 (hardcover : alk. paper)
    1. Polymers. I. Title. II. Series: Plastics engineering (Marcel Dekker, Inc.) ; 38.
TA455.P58B46 1996
668.9—dc20   96-25977
                                                             CIP

The publisher offers discounts on this book when ordered in bulk quantities. For more information, write to Special Sales/Professional Marketing at the address below.

This book is printed on acid-free paper.

MARCEL DEKKER, INC.
270 Madison Avenue, New York, New York 10016

Current printing (last digit):
10 9 8 7 6 5 4 3 2 1

PRINTED IN THE UNITED STATES OF AMERICA

To my wife, Cynthia, whose patience, support and companionship have facilitated my work and made my life enjoyable; and to the memory of my father, Salamon, who would have enjoyed seeing the publication of this book.

PREFACE TO THE SECOND EDITION

The commercial successes of both the first edition of this book and the software package (SYNTHIA) based on the method described therein, as well as the positive feedback which I have received directly from many readers of the book and users of the software encouraged me to develop this revised and expanded second edition. (It should be noted that, after a recent merger, the vendor of SYNTHIA has been renamed Molecular Simulations, Inc.)

Some of the revisions are improvements and/or extensions of the methods developed earlier to predict the physical properties of polymers. For example, the discussion of the glass transition temperature in Chapter 6 has been extended to include new quantitative structure-property relationships for the effects of the average molecular weight and crosslinking on this property. The discussion of the melt viscosity in Chapter 13 has been extended to include equations for the shear rate dependence, while the first edition only dealt with the zero-shear viscosity.

Many other revisions consist of more detailed background information and discussion on the topics discussed in the book, including extensive tabulations of additional experimental data and literature references. To mention just two of many examples, Chapter 7 now contains tables listing experimental values of the surface tensions of polymers and interfacial tensions between pairs of polymers, as well as a discussion and references on the interfacial tension; and Chapter 11 contains a much broader discussion of the behavior of polymers under large deformations.

Finally, some of the revisions involve the reorganization of the material discussed in a given chapter in a manner which may facilitate comprehension.

Revisions and extensions of the book were made for two major reasons. One goal was to increase the utility of the book as a research monograph presenting a new method to calculate polymer properties. Another important goal was to make the book much more self-contained, thus encouraging its more extensive use as both a general reference and a textbook. It is hoped that readers will find this edition useful in their work.

Jozef Bicerano

PREFACE TO THE FIRST EDITION

This book was written primarily to help scientists and engineers working on applied problems in polymer science and technology in the chemical and plastics industries. An empirical approach is therefore used. No new theories or fundamental concepts are presented. Instead, a set of quantitative structure-property relationships is offered. The reader can use these relationships to predict many important physical properties of polymers without being limited by the lack of group contributions. Some properties are predicted with great accuracy, others more approximately.

The reasons for the selection of the types of structural descriptors used in the quantitative structure-property relationships, the nature of these relationships, the relationships for specific properties, and their strengths and limitations are all discussed in great detail. Fundamental considerations are often deliberately not addressed in detail, to avoid lengthy digressions from the practical focus. It is hoped, however, that the information provided by the many new relationships developed for the physical properties, and the extensive tables of data used in developing them and listed here, will serve as starting points for fundamental research.

This manuscript is not a comprehensive textbook but a research monograph based on the author's work at The Dow Chemical Company. Nonetheless, basic background information is provided in the first section of each chapter, to refresh the reader's memory, to enable the reading and comprehension of the remainder of the chapter without having to refer to other sources, and to provide perspective and context for the later sections in which new techniques and relationships are presented. The references listed at the end of each chapter should provide a good starting point for the reader who wishes to explore the literature in greater depth. Sufficient procedural details are presented and examples are given to enable the reader to apply all of the techniques to polymers of interest in a simple and straightforward manner. The book can therefore also be used as a textbook to teach students how to calculate the industrially important properties of polymers.

The efficient design of new polymers for many different technological applications almost always requires that two major challenges be overcome. First, the properties of many candidate polymers intended for the application of interest must be predicted prior to synthesis. Then, the predictions must be used to evaluate, screen, and help prioritize the synthesis of these candidates.

The solution of these problems often requires significant extensions of existing quantitative structure-property relationships. In particular, the candidate polymers for advanced "high-tech" applications requiring outstanding performance characteristics very often contain exotic structural units for which the standard simple additive (group contribution) techniques cannot be applied. Some of the required group contributions to the physical properties are often not available, and there are no previously obtained experimental data to use in estimating these missing group contributions. This limitation is inherent to group contribution methods and unavoidable in applying such methods to truly novel types of structures.

This difficulty was overcome by developing a new method in which many physical properties are expressed in terms of topological variables (connectivity indices). In addition, geometrical variables and/or other structural correction terms are used to obtain more refined correlations. The remaining properties are calculated from relationships expressing them in terms of the properties calculated by using the topological variables. This method enables the prediction of the properties of all polymers of interest, without being limited by the absence of the group contributions for the structural units from which a polymeric repeat unit is built. It is equivalent to the prediction of the properties of polymers by the summation of additive contributions mainly over atoms and bonds instead of groups. The values of these atom and bond contributions are dependent on the nature of the environment of each atom and bond in a particularly simple relationship.

The quantitative structure-property relationships developed in this book therefore enable their users to transcend the limitations of traditional group contribution techniques in predicting the properties of polymers. The work described in this book owes much, however, to the solid foundation of earlier quantitative structure-property relationships in polymers, developed over many decades by the meticulous efforts of many workers. In particular, much of the information provided in the second and third editions of D. W. van Krevelen's classic textbook, *Properties of Polymers* [Elsevier, Amsterdam (1976 and 1990)], was extremely valuable in our work.

The new methodology was tested extensively in practical work at Dow. It was found to be able to predict the properties of novel polymers as accurately and reliably as can be reasonably expected from any scheme based on simple quantitative structure-property relationships.

The only computational hardware required to perform the calculations described in this book is a good hand calculator. The method was nonetheless automated by implementation in a simple interactive computer program, SYNTHIA, enabling its easier use by non-specialists, resulting in much greater efficiency and significantly reducing the possibility of human error. The use of this computer program is simply a matter of drawing the structure(s) of the repeat unit(s), specifying the calculation temperature (also the mole fractions or weight fractions of the repeat units for copolymers), and asking the program for the predicted values of the properties. In addition, this program allows the user to obtain graphs of many of the predicted properties as a function of the temperature and, for copolymers, also as a function of the composition. This computer program is commercially available from Biosym Technologies, Inc., in San Diego, California, USA, to which it has been licensed for commercialization by The Dow Chemical Company.

Jozef Bicerano

ACKNOWLEDGEMENTS

Jerry Seitz provided many stimulating discussions, strong support and encouragement, and valuable insights on structure-property relationships in polymers based on his previous work.

David Porter and Warren Knox critically read the numerous internal reports issued at earlier stages of this work, and a draft of the first edition of this book, and made helpful suggestions.

Ingrid Knox, Angela Buske, and Sharilyn Smith performed literature searches on many of the topics discussed in this book, thus helping to gather necessary data and background information.

Mike Winkelman prepared a patent application covering the methodology.

Eric Eidsmoe wrote a computer program which implemented the relationships presented in this book. This computer program and many of the relationships incorporated in it were improved and their usefulness in solving practical problems was demonstrated as a result of extensive use by many researchers at Dow. Steve Mumby, Steve Harris, Fiona Case and David Klipstein (all from Biosym Technologies, Inc.), made various contributions which resulted in the successful commercialization of the software package.

Cathy Wedelstaedt, Andy Ahn, Laura Beecroft, Chuck Broomall, Jim Curphy, Eric Fouch, Kurt Kuhlmann, Jean McPeak, Pam Percha, Tom Tomczak, and Tanisha Sangster, ran a series of experiments to obtain data which supplemented the experimental data found in the literature, and thus helped in the development of the quantitative structure-property relationships.

The following is a list of persons who provided helpful technical discussions, information, and/or collaboration, in areas covered in this book, during my years at Dow: Patricia Andreozzi, Edgar Andrews, Chuck Arends, David Babb, Steve Bales, Eric Bancroft, Charlie Berglund, Matt Bishop, Clive Bosnyak, Gary Billovits, Clive Bosnyak, Jim Brewbaker, Bob Bubeck, Al Burmester, Bruce Burton, Craig Carriere, Hayden Clark, Katherine Clement, Phil DeLassus, Tad DeVilbiss, Jake Eichhorn, Rich Fibiger, David Goldwasser, Bob Harris, Tony Hopfinger, Steve Humphrey, Susan Jordan, Jason Kinney, Bob Kirchhoff, Tom Kling, Brian Landes, Charlie Langhoff, David Laycock, Werner Lidy, Mike Mang, Maurice Marks, Mike Mazor, Mark McAdon, Dave Moll, Eric Moyer, Mike Mullins, Andy Pasztor, David Porter, Duane Priddy, Mike Radler, Jim Rieke, Nelson Rondan, James Ruiz, Bob Sammler, Ed Sanders, Nitis Sarkar, Dan Scheck, Jerry Seitz, Deepak Sharma, Craig Silvis, Dick Skochdopole, Pat Smith, Var StJeor, Ulrich Suter, Oomman Thomas, Bob Turner, John Warakomski, Cathy Wedelstaedt, Ritchie Wessling, Jerry White, Larry Whiting, and Ed Woo. I have tried to make this list as complete as possible, and regret any unintentional omissions.

Patricia Andreozzi, Warren Knox, Don Dix, and Bob Nowak offered management support to this work.

My wife Cynthia patiently and cheerfully supported and encouraged the writing of this book.

CONTENTS

Contents XIII

Contents   XV

CHAPTER 1

INTRODUCTION

1.A. Properties of Polymers

The ability to *predict* the key physical and chemical *properties of polymers* from their molecular structures *prior to synthesis* is of great value in the design of polymers. Performance criteria which must be satisfied for the technological applications of polymers have become increasingly more stringent with the recent rapid advances in many areas of technology. At the same time, the applications of most of the familiar polymers with relatively simple repeat unit structures have reached their limits. The chemical structures of polymers suitable for advanced applications have therefore increased in complexity. Consequently, development of predictive computational schemes to evaluate candidates for specific applications has gained urgency.

The fact that a *systems approach* must be used in predicting the properties of the types of polymers required to satisfy the performance requirements of most applications further complicates the task of the polymer scientist. In a systems approach, individual physical and chemical properties must not be viewed by themselves, but instead in relation to all of the other properties which might be of interest in a certain application. In order to be useful as the principal predictive tool for a wide variety of applications in a systems approach, a computational scheme must be able to predict many different properties of polymers.

The following important factors render a systems approach to the prediction of the properties of polymers indispensable:

1. The usefulness of a polymer in an application generally depends on the balance between several complementary properties, any one of which can often only be improved upon by reducing the performance in the others. Two simple examples of the complementarity of certain properties of polymers are as follows:

    (a) Polymers with very low thermal expansion coefficients are desirable for applications in electronics. Such polymers are usually brittle. Brittleness is an undesirable attribute for a polymer, since a brittle specimen will break easily.

    (b) Polymers with very high heat distortion temperatures are desirable for applications requiring the retention of physical properties at very high use temperatures. Such polymers are usually very difficult to process in the melt. Difficulty in melt processing is an undesirable attribute for a polymer.

2. The recognition of such interrelationships between properties has two benefits:

    (a) It can significantly reduce the number of independent variables which have to be considered, and thus the amount of effort required, for the design and development of a new polymer or a new fabrication process.

(b) It can provide a more realistic set of expectations concerning the maximum attainable performance and the tradeoffs which often have to be made between mutually conflicting "ideal" performance requirements.

3. The recognition of interrelationships between polymeric properties by using a systems approach can also guide fundamental research towards understanding the physical phenomena involved in the behavior of materials.

The properties of a polymer fall into two general classes:

1. *Material properties* are mainly related to the nature of the polymer itself.
2. *Specimen properties* are primarily consequences of the size, shape and layout of the finished specimens prepared from that polymer, and the process used to prepare these specimens.

The separation of the properties of polymers into these two classes is unfortunately not always straightforward. Material properties affect the specimen properties, and fabrication conditions affect the observed material properties. Two examples are given below:

1. The density is essentially a material property. The density of a given specimen, however, can be affected by the nature of the specimen. For example, many polymers can be prepared with different percent crystallinities, and hence with different densities, by changing the preparation method or annealing the specimens after preparation.
2. The amount of directional variation ("anisotropy") induced in various physical properties by uniaxial or biaxial orientation ("drawing") is a process-related specimen property. The response of a specimen to a given set of drawing conditions, however, depends on the material properties of the polymer from which it is made.

This book will mainly focus on the material properties of polymers. In dealing with material properties whose observed values are significantly affected by the preparation conditions or testing methods used, significant effort will be exerted to compare polymers fabricated and/or tested under equivalent sets of conditions.

The material properties of polymers can be roughly divided into two general types:

1. *Fundamental properties*, such as the van der Waals volume, cohesive energy, heat capacity, molar refraction and molar dielectric polarization, are directly related to the following very basic physical factors:

   (a) Materials are constructed from assemblies of atoms with certain sizes and electronic structures.

   (b) These atoms are subject to the laws of quantum mechanics. They interact with each other via electrical forces arising from their electronic structures.

   (c) The sizes, electronic structures and interactions of atoms determine the spatial arrangement of the atoms.

(d) The interatomic interactions and the resulting spatial arrangements determine the quantity and the modes of absorption of thermal energy.

2. *Derived properties*, such as the glass transition temperature, density, solubility parameter and modulus, are more complex manifestations of the fundamental properties, and can be expressed in terms of combinations of them.

The separation of polymeric properties into the two classes of "fundamental" and "derived" properties facilitates the development of new computational schemes. If the fundamental properties can be calculated with reasonable accuracy, the derived properties can be expressed in terms of these fundamental properties, greatly reducing the number of independent correlations which must be developed. Direct correlations can often also be independently developed for the derived properties, providing more than one alternative method for predicting these properties. It is sometimes useful both to express the derived properties in terms of the fundamental properties, and to develop alternative direct correlations for them.

Additive or *group contribution techniques* are commonly used for the prediction of the properties of polymers from their molecular structures. These techniques provide many extremely useful simple correlations for predicting the properties of polymers. Group contribution techniques will be briefly reviewed in Section 1.B.

The main approach used in the scheme of correlations developed in this book is based on *topological techniques*. The philosophy of this approach, and the scope of the work presented in this book, will be discussed in Section 1.C. The detailed technical implementation of the scheme of correlations will be postponed to later chapters.

The subjects which will be covered in the later chapters of this monograph will be summarized in Section 1.D.

1.B. Group Contribution Techniques

1.B.1. Basic Technique

Additive (group contribution) techniques have a long tradition of successful use in predicting the properties of both ordinary molecules and macromolecules (polymers). They have formed the backbone of the *quantitative structure-activity relationships* [1,2] (QSAR) used to predict the chemical reactivity and the biological activity of molecules in medicinal and agricultural chemistry. They have also been used extensively in many *quantitative structure-property relationships* (QSPR) developed for the physical and chemical properties of polymers.

Van Krevelen [3] has published the classic textbook on group contribution techniques in the QSPR of polymers. This book contains a compendium of useful QSPR relationships for polymers, as well as tables of large amounts of experimental data. The information contained in this book has been extremely valuable in the development of many of our correlations, and we will consequently refer very frequently to this book. We will also often refer to a later review article by van Krevelen [4] containing a significant amount of additional or revised information.

A structure is conceptually broken down into small *fragments*, i.e., *groups*, in additive (i.e., group contribution) techniques. Properties are then expressed as sums of "group contributions" from all of the fragments constituting the structure. The values of the group contributions themselves are estimated by fitting the observed values of the properties of interest, which are expressed as regression relationships in terms of these group contributions, to experimental data on other polymers containing the same types of fragments. Group contribution techniques are therefore essentially empirical. Their development involves the statistical analysis of data.

As a simple example of the use of correlations based on group contributions, consider the calculation of the glass transition temperature T_g, and the density ρ at room temperature, of polystyrene. As shown in Figure 1.1, the structure of the repeat unit of this polymer can be broken down into two groups. The contribution m_i of each group to the total molecular weight M of the repeat unit is, of course, exactly determined. On the other hand, the contribution Y_{gi} of each group to the "molar glass transition function" Y_g, and the contribution V_i of each group to the amorphous molar volume V at room temperature [3], have been determined by fitting regression equations to extensive tables of experimental values of T_g and V, respectively. M, Y_g and V are each equal to the sum of the contributions of the two groups. T_g is then estimated by Equation 1.1, while ρ is defined by Equation 1.2:

$$T_g \approx \frac{Y_g}{M} \tag{1.1}$$

$$\rho \equiv \frac{M}{V} \tag{1.2}$$

More generally, the total value of a molar property in a group contribution scheme is given by Equation 1.3 for a polymer whose repeat unit can be formally broken down into "groups":

$$\text{(Additive Property)} \approx \Sigma(\text{"additive" group contributions}) + \Sigma(\text{"constitutive" structural terms}) \tag{1.3}$$

In Equation 1.3, Σ denotes summation. The summation over structural terms only has to be included for certain properties where the assumption of the strict additivity of group contributions results in large errors because the environment of a given group also plays an important role in the contribution of the group to that particular property. For example, structural terms must be used in calculating the Y_g of certain types of polymers [3].

Figure 1.1. Calculation of glass transition temperature T_g and density ρ at room temperature, of polystyrene, as an example of the application of group contribution techniques. (a) The structure of a polystyrene repeat unit. (b) Formal breakdown of this structure into two "groups". (c) Each group makes a contribution to the molecular weight M of the repeat unit, to the "molar glass transition function" Y_g, and to the amorphous molar volume V [3a]. M, Y_g and V are sums of these contributions:

$$M = m_1 + m_2 = 14.03 + 90.12 = 104.15 \text{ grams/mole.}$$
$$Y_g = Y_{g1} + Y_{g2} = 2700 + 35000 = 37700 \quad \text{(degrees Kelvin)·(grams/mole).}$$
$$V = V_1 + V_2 = 15.85 + 82.15 = 98.0 \text{ cc/mole.}$$

(d) T_g is expressed as the quotient of Y_g and M, while ρ is expressed as the quotient of M and V: $T_g \approx Y_g/M = 362K$, in good agreement with the accepted experimental value of 373K.
$\rho \equiv M/V = 1.06$ grams/cc, in good agreement with the experimental value of 1.05 grams/cc.

In spite of its usefulness, the group contribution approach has some inherent limitations. The most important limitation is the need for a substantial body of experimental data for a given property in order to derive reliable values for the contributions of a large number of molecular fragments to that property. Attempts to bypass the need for experimental data result, instead, in the necessity to perform extensive amounts of time-consuming force field or quantum mechanical calculations [5]. If a polymer contains a structural unit whose additive contribution to a certain property cannot be estimated, the value of that property cannot be predicted for that polymer.

1.B.2. An Extension

A set of correlations developed over many years by Seitz [6] resulted in a considerable simplification in the use of group contributions for the prediction of the properties of polymers. This procedure allows the prediction of a large number of the "derived" physical properties of polymers from just five "fundamental" properties:

1. Molecular weight of the repeat unit. This quantity is determined exactly from the structure.
2. Length (i.e., end-to-end distance) of the repeat unit in its "fully extended" conformation. This quantity can generally be estimated in a simple and accurate manner by using an interactive molecular modeling software package, building a rough structure of the repeat unit, and asking the modeling program to list the distance between the two specified end atoms of the repeat unit. If greater accuracy is required, the geometry of the rough initial structure can be rapidly optimized by using force field techniques prior to measuring the length of the repeat unit.
3. Van der Waals volume of the repeat unit. This quantity is usually estimated by using group contributions. If necessary, it can also be calculated, in a very tedious manner, from the assumed bond lengths and bond angles of all atoms of the repeat unit.
4. Cohesive energy. This quantity is usually estimated by using group contributions.
5. A parameter related to the number of rotational degrees of freedom of the backbone of a polymer chain. This quantity can be exactly determined from the structure of the repeat unit.

If these five fundamental properties can all be estimated with reasonable accuracy, then the correlations developed by Seitz [6] allow the prediction of many additional physical properties without the necessity to derive a separate group contribution table for each property.

The group contributions to the cohesive energy are unknown for many structural units present in novel polymers. The inability to estimate the cohesive energy was thus the major difficulty in attempting to use the correlations of Seitz. The work presented in this book was initiated as an attempt to develop a new method for estimating the cohesive energies of polymers of arbitrary structure in a simple manner without having to rely on the availability of group contributions.

This problem was solved by developing a topological approach whose conceptual framework will be outlined in the next section. It soon became apparent that the method developed for calculating the cohesive energy could be extended to calculate many other properties. These correlations were gradually combined into a scheme which could be used to predict the physical properties of polymers with reasonable accuracy even in the absence of group contributions. This scheme was successfully tested and refined by application to many practical problems. This book represents the first publication of the new scheme of correlations in the open literature.

1.C. Topological Technique

1.C.1. Topology and Geometry

The spatial arrangement of the atoms constituting a material is completely specified by the *topology* and the *geometry*. The *topology* is simply the pattern of interconnections between atoms. It is often expressed in the form of a *connectivity table*. The *geometry* also encompasses the values of the coordinates of the atoms, usually in Cartesian (x, y and z) coordinates, but sometimes in alternative coordinate systems such as spherical, cylindrical or internal coordinates.

The topology and the geometry of a system provide complementary types of information. For example, carbon in the diamond form and crystalline silicon have the same topology. This topology is defined by tetrahedral bonding, in which each atom in the lattice has four nearest neighbors. On the other hand, diamond and crystalline silicon have different geometries. The bond lengths in silicon are longer than the bond lengths in diamond, so that the coordinates of the atoms in silicon are different from the coordinates of the atoms in diamond.

It is also possible for the same geometry of a given system to result in the need to consider different types of topologies depending on the nature of the physical phenomenon of interest. For example, the connectivity table expressed in terms of the pattern of covalent bonds in a material will include much fewer terms than a tabulation of all non-negligible long range nonbonded interactions such as the electrostatic (Coulombic) interactions between atoms of different partial charges in a material with many highly polar bonds.

The mathematical discipline of *geometry* involves the complete study of the arrangement of the components of a system, including their coordinates. On the other hand, the mathematical discipline of *topology* consists of the study of the interconnections of the components of a system, without considering the detailed coordinates of these components.

1.C.2. Graph Theory and Connectivity Indices

Graph theory is a subdiscipline of *topology*. Graph theory has been extensively used [7-23] to study the chemical physics of polymers, and especially those properties of polymer chains that can be studied in solutions, including viscoelastic behavior and chain configurations. This work has provided many valuable physical insights, but no simple predictive methods and correlations that could be used routinely in industrial research and development.

The formalism of *connectivity indices* is an embodiment of graph theory. Connectivity indices are intuitively appealing because each index can be calculated exactly from valence bond (Lewis) diagrams familiar to organic chemists, which depict molecular structure in terms of atoms, inner shell and valence shell electrons, valence shell hybridization, σ and π electrons, bonds and lone pairs. These indices can then be correlated with the physical or chemical properties of interest. Connectivity indices have, in the past, been very useful in treating molecular systems with well-defined chemical formulae and fixed numbers of atoms [24,25].

Early attempts to use simple topological indices to predict polymeric properties met with very limited success. For example, in the "topological extrapolation method" [26], the prediction of the properties of a polymer requires the availability of experimental data (*for each property to be predicted)* on a series of oligomers of the same polymer with increasing degrees of polymerization, in order to perform the extrapolation. Such a requirement makes it impossible to apply the topological extrapolation method as a routine tool for polymer design.

The QSPR techniques developed for polymers in this monograph utilize connectivity indices as their principal set of descriptors of the topology of the repeat unit of a polymer. An extensive literature search has resulted in the finding of only two papers (both recent) on the application of connectivity indices in a manner simple enough to be of practical utility in predicting industrially important properties (the dielectric constant [27] and the oxygen permeability [28]) of polymers. It is hoped that the present monograph will stimulate further work in this area.

1.C.3. Nature and Scope of the New Approach

The familiar formalism of connectivity indices [24,25] was used as the starting point of this work. Several extensions were made in the methods for calculating connectivity indices, to be able to correlate the properties of polymers in a simple, efficient, and reasonably accurate manner with connectivity indices:

1. A simple treatment was developed for calculating the needed subset of the connectivity indices of polymer chains of arbitrary lengths and unknown numbers of atoms.

2. A simple graph theoretical distinction was made between "extensive" (molar) properties which depend on the amount of material present and "intensive" properties which do not. (The molecular weight per repeat unit, the molar volume, and the molar glass transition function, are examples of extensive properties. The density and the glass transition temperature are examples of intensive properties.)

3. The contributions made to "locally anisotropic" physical properties by the "backbone" and "side group" portions of the connectivity indices, and by the shortest path through the backbone, were separated from each other. (The thermal conductivity is an example of locally anisotropic properties.)

4. Specialized topological indices were defined to calculate certain conformational properties of polymer chains.

The extended method was then used as the foundation of a scheme of predictive techniques and correlations for the calculation of the physical properties of polymers. Connectivity indices were correlated either with experimental data or with tables of group contributions. For example, there is a sufficient amount of reliable experimental data for the molar volume or the heat capacity at room temperature to correlate the connectivity indices directly with experimental data. On the other hand, the cohesive energy of a polymer cannot be directly measured, but can only be inferred indirectly from the results of various types of experiments. It was, therefore, found to be more effective to correlate connectivity indices with group contribution tables for cohesive energy.

Although the correlations were of necessity developed by utilizing data on specific polymers, they appear to be quite robust with respect to extrapolation to different types of polymers. Available data are used to estimate the dependence of the properties of polymers on the topological indices. The correlations developed in this manner then allow the prediction of these properties without being limited by the availability of group contributions. Group contribution methods can only be used if the contribution of each structural building block in the polymeric repeat unit is known for each property of interest. On the other hand, a strong correlation in terms of connectivity indices, if it was obtained by using a large and diverse dataset of polymers, can be utilized to predict properties with a reasonable degree of confidence for polymers containing completely novel types of structural units.

It will be shown that the method developed in this book amounts to the prediction of the properties of polymers by the summation of additive contributions over atoms and bonds, instead of groups. The values of these atom and bond contributions are dependent on the nature of the environment of each atom and bond in a simple manner.

Easier recognition and quantification of relationships between properties is an added benefit of the approach used in this work. It is easier to identify relationships between physical properties from simple correlations with a few indices than from separate group contribution tables for each property. Examples of such interconnections will be given in later chapters.

Furthermore, when it is possible to make comparisons, the new correlations are often of higher quality than the correlations [24,25] for similar properties of simple molecules. Many properties of simple molecules can only be predicted reliably by using connectivity indices if different correlation equations are used for molecules containing different types of structural units. By contrast, similar properties of polymers often turn out to be predictable for all polymers by using a single correlation equation containing a few parameters.

It may be worthwhile to speculate that this improved quality of many of the correlations for polymers relative to correlations for simple molecules is due to the fact that the constraints imposed by the macromolecular structure reduce the degrees of freedom and somewhat smear out the complete spectrum of differences of various structural units relative to their possible range of behavior in small molecules. These constraints involve the facts that the structural units are attached to long polymer chains, which are themselves surrounded by a dense medium of other polymer chains. For example:

1. The heat capacity of a polymer is normally smaller than the heat capacity of a simple molecule of the same stoichiometry and similar structure, at the same temperature.

2. Polymers have much smaller effective dipole moments in the solid phase than in dilute solutions. The dipole moments of polar and nonpolar solid polymers thus differ much less

from each other than the dipole moments of polar and nonpolar polymers in dilute solution or of liquids of polar and nonpolar simple molecules.

An extensive amount of work on ordinary molecules [24,25] has shown that correlations in terms of connectivity indices decrease in accuracy and reliability with increasing complexity of the properties being correlated. For example, physical properties such as the molar volume are easier to correlate than chemical properties such as reactivity. On the other hand, chemical properties such as reactivity are easier to correlate than biological properties such as toxicity.

Our work on polymers using the formalism of connectivity indices has mainly focused on the physical properties. Material properties predicted from the correlations presented in this book can both be used to evaluate the inherent performance characteristics of polymers, and to specify the input parameters for continuum mechanical simulations of the performance of finished parts made from novel polymeric structures which have not yet been synthesized.

The outcome of our work is, hence, a *scheme of correlations* which allows the prediction of the physical properties of polymers constructed from the following nine elements which are most commonly incorporated into polymers with major technological applications: *carbon, nitrogen, oxygen, hydrogen, fluorine, silicon, sulfur, chlorine and bromine.* Whenever it is stated in this manuscript that a correlation is "applicable to all polymers", this statement refers only to polymers constructed from these nine elements, which are highlighted in Figure 1.2. Furthermore, except when explicitly stated otherwise, the correlations are also subject to the limitations, which are summarized below, concerning the general types of polymers to which they are applicable.

				H	He
B	C	N	O	F	Ne
Al	Si	P	S	Cl	Ar
Ga	Ge	As	Se	Br	Kr
In	Sn	Sb	Te	I	Xe

Figure 1.2. A portion of the periodic table, identifying the nine elements included within the scope of our work (C, N, O, H, F, Si, S, Cl and Br) by large bold symbols in shaded boxes.

No attempt was made to generalize the correlations by including data for more exotic organic, organometallic and inorganic polymers containing phosphorus [29], boron [30,31], or other types of elements [32-34]. The use of the correlations in their present form is thus only recommended for polymers constructed from the nine elements listed above. Readers interested in polymers containing other elements are encouraged to incorporate data on polymers containing such elements into the extensive databases which will be provided in later chapters for each correlation, and extend the correlations to include polymers containing such elements by re-fitting the data.

Except where explicitly indicated otherwise, the methods in this book were mainly developed for isotropic (unoriented) amorphous linear polymers, i.e., for isotropic amorphous polymers not containing a significant crosslink density or long chain branching. They are also applicable to amorphous phases of semicrystalline polymers, and to properties not very sensitive to the percent crystallinity for highly crystalline polymers. Correlations and models are available in the literature to predict the effects of crystallinity, orientation, long chain branching and crosslinking on many physical properties of polymers. The correlations presented in this book can be combined with these correlations and models to predict the effects of such structural features on the properties of polymers. Extensive reviews of crystallinity, orientation, branching and crosslinking, however, are outside the scope of this research monograph whose aim is to present the new tools developed for predicting the basic material properties of polymers via topological methods.

The highly specialized "ladder polymers", where each repeat unit has more than one bond connecting it to the repeat units on both sides, are also outside the scope of this work. The amount of available data on such polymers is, at present, insufficient to extend the new correlations to these polymers in a statistically significant manner.

The unique properties of polymers such as polyacetylene, whose backbones consist of an alternating succession of single and double bonds, and most of which show extraordinary electrical, optical and magnetic properties including electrical conductivity when "doped" with electron donors or acceptors [35], are also outside the scope of this work. Sophisticated quantum mechanical treatments are required to model these properties of such polymers adequately.

Similarly, the properties of polymers in the ionized state [36], where the charges on the polymer chains are balanced by counterions, are outside the scope of this book.

Molecules consisting of hollow cages of carbon atoms, named "buckministerfullerenes", are a very active area of current research. A copolymer of the C_{60} "buckyball" with p-xylylene was recently synthesized [37]. The correlations presented in this book are not parameterized to treat structures containing hollow cages of atoms, whose incorporation may be expected to endow a polymer with many unique properties. These correlations can, however, be extended to such polymers in the future, as data become available on the physical properties of such polymers.

Finally, the study of biopolymers is a vast area of research with different a emphasis, on key problems related to the functioning of such polymers in the environment of a living organism. Biopolymers are, hence, outside the scope of our work which is focused on synthetic polymers.

Connectivity indices, and therefore also correlations utilizing them, are not derived from first principles. They are obtained empirically, by considering the correlation of the topological features of molecules (polymeric repeat units in this work) with the properties of interest. Such empirical correlations serve two purposes:

1. They enable the prediction of many physical properties and are therefore practically useful.
2. They reveal trends and patterns in the physical properties. Discovery of these trends and patterns can suggest directions for future efforts to provide a more rigorous understanding of the physical phenomena involved.

The fact that our correlations for the properties of polymers involve fitting graph theoretical connectivity indices to quantities which can be expressed as sums of additive quantities provides a clue as to the source of their success. Group contribution techniques are based on the realization that many properties of a material arise from the intrinsic properties and mutual interactions of small "localized" moieties, namely "groups". It has been shown [38-40] from fundamental considerations that all additive properties of a chemical system, namely all properties which can be calculated by summing over its individual constituents, belong to the so-called *graph-like state of matter*. In other words, all additive properties can be expressed in terms of linear combinations of graph theoretical invariants. There is, thus, an underlying theoretical basis for correlations utilizing connectivity indices, despite their empirical derivation. The success of linear combinations of a small number of members of a *specific set* of graph theoretical invariants in providing a good approximation to additive properties is a result of the fact that certain graph theoretical invariants have been developed, through decades of experience on small molecules, to quantify many of the essential physical effects which determine the values of the properties as a function of the molecular structure.

1.D. Outline of the Remaining Chapters of This Book

The basic formalism of connectivity indices will be reviewed, and the extensions made in order to apply this formalism to polymers will be presented, in Chapter 2.

Chapters 3 to 16 will be devoted to the development of new correlations for the properties of polymers, based on the general formalism outlined in Chapter 2.

Correlations for the key volumetric properties (van der Waals volume, molar volume, density and coefficient of thermal expansion) will be developed in Chapter 3, followed by a general discussion of pressure-volume-temperature relationships.

A new type of rotational degrees of freedom parameter will be defined for the backbones and side groups of polymers, and correlations for the heat capacity and related thermodynamic functions (enthalpy, entropy and Gibbs free energy) will be developed utilizing both the connectivity indices and the rotational degrees of freedom, in Chapter 4.

Correlations for the cohesive energy and the solubility parameter will be presented in Chapter 5, to allow the calculation of these properties at the same level of accuracy as can be attained by group contributions but for much wider classes of polymers.

A new correlation which utilizes both the cohesive energy density and the degrees of freedom of a polymer chain will be presented for the glass transition temperature in Chapter 6. This will be followed by the development of a new correlation for the effects of the average molecular weight, the discussion of the effects of plasticization, and the development of a new correlation for the effects of crosslinking, on the glass transition temperature. Finally, the effects of tacticity on the glass transition temperature, the relationships between secondary relaxations and the glass transition temperature, and the crystalline melting temperature, will each be discussed.

The surface tension will be discussed, and a new correlation for the molar parachor will be presented, in Chapter 7. The calculation of the interfacial tension from the surface tensions of the components will also be discussed, and shown to be in need for significant improvements.

Various optical properties will be discussed, a new correlation for the refractive index at room temperature will be presented, and the molar refraction of many polymers and the specific refractive increment of many polymer-solvent combinations will be calculated, in Chapter 8.

Electrical properties will be discussed, and a new correlation will be presented for the dielectric constant at room temperature, in Chapter 9. The molar polarization, dipole moment, electrical losses and dielectric strength, will also be considered in Chapter 9.

Magnetic properties will be discussed, and a new correlation for the molar diamagnetic susceptibility will be presented, in Chapter 10.

It will be shown in Chapter 11 that the correlations developed in this monograph can be combined with other correlations found in the literature, to predict many of the key mechanical properties of polymers. In addition, new correlations in terms of connectivity indices will be developed for the molar Rao function and the molar Hartmann function. A large amount of the most reliable literature data on the mechanical properties of polymers will also be listed.

Some important properties of polymer chains in dilute solutions (steric hindrance parameter, characteristic ratio, intrinsic viscosity and viscosity at small but finite concentrations) will be discussed, and new correlations will be presented for the steric hindrance parameter and the molar stiffness function, in Chapter 12.

The average molecular weight, polydispersity, temperature, hydrostatic pressure, and shear rate dependences of polymer melt viscosity will be discussed in Chapter 13, resulting in a set of correlations which can be used to obtain a rough estimate of melt viscosity as a function of all of these variables. A new correlation will be presented for the molar viscosity-temperature function. The dependences of the zero-shear viscosity of concentrated polymer solutions on the average molecular weight and on the temperature will also be discussed.

Thermal conductivity and thermal diffusivity will be discussed in Chapter 14, where a new correlation will be presented for the thermal conductivities observed at room temperature.

The permeability of polymers to small gas molecules will be discussed, and a new correlation will be presented for the oxygen permeabilities observed at room temperature, in Chapter 15.

Thermal and thermooxidative stability will be discussed, and a correlation for the molar thermal decomposition function will be presented, in Chapter 16.

Several simple extensions, generalizations and shortcuts, which can be made to facilitate the calculations, increase their accuracy, and/or widen the range of problems which can be solved by using the methods developed in this book, will be presented in Chapter 17. These include the development of "designer correlations" for the more accurate prediction of the properties of special families of polymers; and the use of the new method to generate group contributions, to perform calculations on alternating and random copolymers, and to provide input parameters for composite models which can be used to predict the thermoelastic and transport properties of multiphase materials. A commercial software package implementing the key correlations from this book will be described briefly. Promising recent research on quantitative structure-property relationships by other workers, which is based on or dependent upon the methods developed in this book, will be discussed. Possible directions for future work will also be summarized.

Detailed examples of the use of the key correlations to calculate the properties of specific polymers (polystyrene, and random copolymers of styrene and oxytrimethylene), will be provided in Chapter 18. Chapter 18 therefore complements chapters 3 to 16, where each physical property was considered individually, and calculated for many different polymers.

Basic background information will be provided, mainly in the first section of each chapter (but also in later sections when more appropriate), to refresh the reader's memory, to facilitate the reading and comprehension of the remainder of the chapter without having to refer to other sources, and to provide the perspective and context for the new techniques and correlations.

A large number of references will be listed in at the end of each chapter. Many of these references are mainly intended to provide a good starting point for the reader who wishes to explore the literature on the subject matter of that chapter in greater depth.

Sufficient procedural details will be presented and examples will be given in each chapter to enable the reader to apply all of the techniques and correlations to polymers of interest. The only computational hardware required to perform any of these calculations is a hand calculator. We have, however, developed an interactive computer program which has resulted in greater efficiency and reduced human error.

A glossary of the symbols, acronyms and other abbreviations used, and an extensive subject index, will both be provided at the end of the book.

References and Notes for Chapter 1

1. Y. C. Martin, *J. Medicinal Chemistry*, *24*, 229-237 (1981).

2. A. J. Hopfinger, *J. Medicinal Chemistry*, *28*, 1133-1139 (1985).

3. D. W. van Krevelen, *Properties of Polymers*. (a) Second edition, Elsevier, Amsterdam (1976). (b) Third edition, Elsevier, Amsterdam (1990). (Most of our work was completed before the third edition of van Krevelen's book became available, so that we will refer to the second edition more often than the third edition.)

4. D. W. van Krevelen, in *Computational Modeling of Polymers*, edited by J. Bicerano, Marcel Dekker, New York (1992), Chapter 1.

5. A. J. Hopfinger, M. G. Koehler, R. A. Pearlstein and S. K. Tripathy, *J. Polym. Sci., Polym. Phys. Ed.*, *26*, 2007-2028 (1988).

6. J. T. Seitz, *J. Appl. Polym. Sci.*, *49*, 1331-1351 (1993).

7. S. I. Kuchanov, S. V. Korolev and S. V. Panyukov, *Advances in Chemical Physics*, *72*, 115-326 (1988).

8. K. Kajiwara and M. Gordon, *J. Chem. Phys.*, *59*, 3623-3632 (1973).

9. K. Kajiwara and C. A. M. Ribeiro, *Macromolecules*, *7*, 121-128 (1974).

10. B. E. Eichinger, *J. Polym. Sci., Symp. Ser.*, *54*, 127-134 (1976).

11. W. C. Forsman, *J. Chem. Phys.*, *65*, 4111-4115 (1976).

12. M. Gordon, *Polymer*, *20*, 1349-1356 (1979).

13. J. E. Martin, *Macromolecules*, *17*, 1263-1275 (1984).

14. D. J. Klein and W. A. Seitz, *Stud. Phys. Theor. Chem.*, *28*, 430-445 (1983).

15. H. Galina, *Macromolecules*, *19*, 1222-1226 (1986).

16. S. V. Korolev, S. G. Alekseyeva, S. I. Kuchanov, I. Ya. Slonim, G. S. Mathelashvili and M. G. Slin'ko, *Polymer Science USSR*, *29*, 2618-2628 (1987).

17. H. Galina and M. M. Syslo, *Discrete Applied Mathematics*, *19*, 167-176 (1988).

18. Y. Yang and T. Yu, *Makromol. Chem., Rapid Commun.*, *5*, 1-9 (1984).

19. Y. Yang and T. Yu, *Makromol. Chem.*, *186*, 513-525 (1985).

20. Y. Yang and T. Yu, *Makromol. Chem.*, *186*, 609-631 (1985).

21. Y. Yang and T. Yu, *Makromol. Chem.*, *187*, 441-454 (1986).

22. Y. Yang and T. Yu, *Makromol. Chem.*, *188*, 401-420 (1987).

23. Y. Yang, *Makromol. Chem.*, *190*, 2833-2845 (1989).

24. L. B. Kier and L. H. Hall, *Molecular Connectivity in Chemistry and Drug Research*, Academic Press, New York (1976).

25. L. B. Kier and L. H. Hall, *Molecular Connectivity in Structure-Activity Analysis*, John Wiley & Sons, New York (1986).

26. Ov. Mekenyan, S. Dimitrov and D. Bonchev, *European Polymer Journal*, *19*, 1185-1193 (1983).

27. A. J. Polak and R. C. Sundahl, *Polymer*, *30*, 1314-1318 (1989).

28. M. R. Surgi, A. J. Polak and R. C. Sundahl, *J. Polym. Sci., Polym. Chem. Ed.*, *27*, 2761-2776 (1989).

29. H. R. Allcock, *Phosphorus-Nitrogen Compounds*, Academic Press, New York (1972), Part III.

30. K. O. Knollmueller, R. N. Scott, H. Kwasnik and J. F. Sieckhaus, *J. Polym. Sci., A-1*, *9*, 1071-1088 (1971).

31. E. N. Peters, *Ind. Eng. Chem. Prod. Res. Dev.*, *23*, 28-32 (1984).

32. C. E. Carraher, *J. Macromol. Sci.-Chem.*, *A17*, 1293-1356 (1982).

33. J. Bicerano and S. R. Ovshinsky, in *Applied Quantum Chemistry*, edited by V. H. Smith, H. F. Schaefer III and K. Morokuma, D. Reidel Publishing Company, Holland (1986), 325-345.

34. J. Bicerano and D. Adler, *Pure & Applied Chemistry*, *59*, 101-144 (1987).

35. *Conjugated Conducting Polymers*, edited by H. G. Kiess, Springer-Verlag, Heidelberg (1992).

36. R. D. Lundberg, article titled "*Ionic Polymers*", *Encyclopedia of Polymer Science and Engineering*, *8*, Wiley-Interscience, New York (1987).

37. D. A. Loy and R. A. Assink, *J. Am. Chem. Soc.*, *114*, 3977-3978 (1992).

38. M. Gordon and W. B. Temple, from *Chemical Applications of Graph Theory*, edited by A. T. Balaban, Academic Press, New York (1976), Chapter 10.

39. J. W. Kennedy, from *Computer Applications in Chemistry*, edited by S. R. Heller and R. Potenzone, Jr., *Analytical Chemistry Symposium Series*, *15*, 151-178 (1983).

40. D. H. Rouvray, *Report No. TR-41, Order No. AD-A177812/5/GAR*, available from NTIS, US Government Report Announcement Index, *87 (12)*, Abstract No. 723796 (1987).

TOPOLOGICAL METHOD FOR STRUCTURE-PROPERTY CORRELATIONS

2.A. Review of Connectivity Index Calculations for Simple Molecules

We will use a topological formalism in developing most of our structure-property correlations for polymers. This formalism utilizes *connectivity indices* defined via graph theoretical concepts as its main structural and topological descriptors. Connectivity indices have been widely used for simple molecules. The review provided in this section is intended to familiarize the reader with connectivity indices before discussing their extension to polymers. The information given in this section is summarized from two books by Kier and Hall [1,2], to which the reader is referred for additional details. The first book [1] is more detailed, while the second book [2] includes the results of research performed over the decade after the publication of the first book.

The graph theoretical treatment of molecular properties starts by the construction of the *hydrogen-suppressed graph* of the molecule. Starting with the valence bond (Lewis) structure of the molecule, the hydrogen atoms are omitted. Each remaining *atom* becomes a *vertex* in the *graph*, while each remaining *bond* becomes an *edge*. As will be shown below, the omission of the hydrogen atoms is compensated by the manner in which the atomic indices for each vertex in the hydrogen-suppressed graph are defined to include information concerning the number of omitted hydrogen atoms attached to the atom corresponding to that vertex. The construction of a hydrogen-suppressed graph is shown in Figure 2.1, using vinyl fluoride as an example.

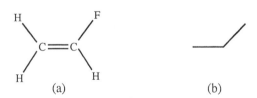

Figure 2.1. Construction of the hydrogen-suppressed graph of a molecule, using vinyl fluoride as an example. (a) Valence bond (Lewis) structure. (b) Hydrogen-suppressed graph, obtained by omitting the hydrogen atoms, and connecting all of the remaining atoms ("vertices" of the graph) with bonds ("edges" of the graph).

The values of two indices (δ and δ^v, see Table 2.1 and the subsequent discussion), which describe the electronic environment and the bonding configuration of each non-hydrogen atom in the molecule, are next assigned, and listed at the vertices of the hydrogen-suppressed graph.

Table 2.1. The δ and δ^v values used in the calculations. Hyb denotes the state of hybridization (only specified if indicated below). N_H denotes the number of hydrogen atoms bonded to the atom in question. Note that the δ^v values of the atoms from the second and lower rows of the periodic table are fractional.

Atom	Hyb	N_H	δ	δ^v	Atom	Hyb	N_H	δ	δ^v
C	sp^3	3	1	1	O	sp^3	1	1	5
		2	2	2			0	2	6
		1	3	3		sp^2	0	1	6
		0	4	4	F	---	0	1	7
	sp^2	2	1	2	Si^{b,c}	sp^3	1	3	1/3
		1	2	3			0	4	4/9
		0	3	4	P^d	sp^3	2	1	1/3
	sp	1	1	3			1	2	4/9
		0	2	4			0	3	5/9
N	sp^3	2	1	3	S	sp^3	1	1	5/9
		1	2	4			0	2	2/3
		0	3	5		note^e	0	4	8/3
	sp^2	1	1	4	Se^d	sp^3	0	2	2/9
		0	2	5	Cl	---	0	1	7/9
	sp	0	1	5	Br	---	0	1	7/27
	note^a	0	3	6	I^d	---	0	1	7/47

a This set of numbers refers to nitrogen in nitro (-NO$_2$) groups, where one oxygen atom is assumed to have $\delta=1$ and $\delta^v=5$, while the other oxygen atom is assumed to have $\delta=1$ and $\delta^v=6$.

b The use of $\delta^v=1/3$ or $4/9$, obtained from the definition given by Equation 2.1, results in the overestimation of the effect of the extra inner shell of electrons in silicon atoms on certain physical properties. Whenever this happens, the replacement Si→C (i.e., $\delta^v=3$ or 4) will be made in calculating the valence connectivity indices to correlate the property in question. For such properties, differences between Si and C atoms will be taken into account by introducing a correction term related to the chemical composition of the polymeric repeat unit.

c The correlations developed in this book are only applicable for the two bonding configurations of silicon listed in this table, since property data were not available for polymers containing silicon in any other configurations.

d The correlations developed in this book are not applicable to polymers containing phosphorus, selenium, and/or iodine, since data for polymers containing these elements were not considered. The values of δ and δ^v for these elements are listed for readers who may wish to analyze data on such polymers, and extend the correlations to include them.

e This set of numbers refers to sulfur in its highest oxidation state, in the bonding configuration R-SO$_2$-R' found in the "sulfone" (-SO$_2$-) group.

The first atomic index [Figure 2.2(a)] is the simple connectivity index δ, equal to the number of non-hydrogen atoms to which a given non-hydrogen atom is bonded. Equivalently, the δ of each vertex in the hydrogen-suppressed graph is the number of edges emanating from that vertex.

The second atomic index [Figure 2.2(b)] is the valence connectivity index δ^v, incorporating information on details of the electronic configuration of each non-hydrogen atom. Its value for the lowest oxidation states of the elements will generally be assigned by Equation 2.1 [2], where Z^v is the number of valence electrons of an atom, N_H is the number of hydrogen atoms bonded to it, and Z is its atomic number (i.e., Z equals Z^v plus the number of inner shell electrons).

$$\delta^v \equiv \frac{(Z^v - N_H)}{(Z - Z^v - 1)} \tag{2.1}$$

The first-row atoms C, N, O and F only have two inner shell electrons. Consequently, their δ^v values are $(Z^v - N_H)$, since for these four atoms $(Z - Z^v - 1) = [(Z^v + 2) - Z^v - 1] = 1$. The only exception is nitrogen atoms in nitro ($-NO_2$) groups, where we have assigned $\delta^v = 6$.

The number of inner shell electrons increases, so that the denominator in Equation 2.1 increases and δ^v decreases, in going down any column of the periodic table. The assignment of δ^v values for the higher oxidation states of the elements (such as sulfur in its highest oxidation state, in the bonding configuration $R-SO_2-R'$) is less straightforward but also feasible [2].

An alternative procedure to using a single basic set of δ^v values is optimizing the value of δ^v for each electronic configuration of each atom, for each property being correlated. This alternative approach is very cumbersome. Its use can also result in the loss of some useful information. Physical insights can be often be gained by using a single set of values of δ^v, and observing which atomic configurations and/or properties are best described by this set of values and when alternative values of δ^v and/or correction terms are required.

Bond indices β and β^v can be defined (Equations 2.2 and 2.3) for each bond not involving a hydrogen atom, as products of the atomic indices (δ and δ^v, respectively) at the two vertices (i and j) which define a given edge or bond. The assignment of atomic and bond indices for vinyl fluoride is depicted in Figure 2.2(a) and (b).

$$\beta_{ij} \equiv \delta_i \cdot \delta_j \tag{2.2}$$

$$\beta^v{}_{ij} \equiv \delta^v{}_i \cdot \delta^v{}_j \tag{2.3}$$

The *zeroth-order* (atomic) *connectivity indices* $^0\chi$ and $^0\chi^v$ *for the entire molecule* are defined in terms of the following summations over the *vertices* of the hydrogen-suppressed graph:

$$^0\chi \equiv \sum_{\text{vertices}} \left(\frac{1}{\sqrt{\delta}} \right) \tag{2.4}$$

$$^0\chi^{\text{v}} \equiv \sum_{\text{vertices}} \left(\frac{1}{\sqrt{\delta^{\text{v}}}} \right) \tag{2.5}$$

The *first-order* (bond) *connectivity indices* $^1\chi$ and $^1\chi^{\text{v}}$ *for the entire molecule* are defined in terms of the following summations over the *edges* of the hydrogen-suppressed graph:

$$^1\chi \equiv \sum_{\text{edges}} \left(\frac{1}{\sqrt{\beta}} \right) \tag{2.6}$$

$$^1\chi^{\text{v}} \equiv \sum_{\text{edges}} \left(\frac{1}{\sqrt{\beta^{\text{v}}}} \right) \tag{2.7}$$

The application of equations 2.4-2.7 to vinyl fluoride is shown in Figure 2.2(c).

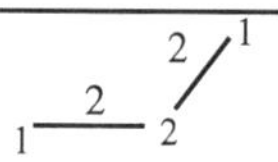

$$^0\chi = \frac{1}{\sqrt{1}} + \frac{1}{\sqrt{2}} + \frac{1}{\sqrt{1}} \approx 1 + 0.7071 + 1 = 2.7071$$

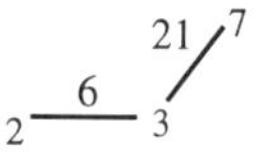

$$^0\chi^{\text{v}} = \frac{1}{\sqrt{2}} + \frac{1}{\sqrt{3}} + \frac{1}{\sqrt{7}} \approx 0.7071 + 0.5774 + 0.3780 = 1.6625$$

$$^1\chi = \frac{1}{\sqrt{2}} + \frac{1}{\sqrt{2}} \approx 0.7071 + 0.7071 = 1.4142$$

$$^1\chi^{\text{v}} = \frac{1}{\sqrt{6}} + \frac{1}{\sqrt{21}} \approx 0.4082 + 0.2182 = 0.6264$$

Figure 2.2. Calculation of the zeroth-order and first-order connectivity indices, using vinyl fluoride as an example. Hydrogen-suppressed graph with (a) δ values at the vertices and β values along the edges, and (b) δ^{v} values at the vertices and β^{v} values along the edges. (c) Summation of the reciprocal square roots of the δ values to calculate $^0\chi$, of the δ^{v} values to calculate $^0\chi^{\text{v}}$, of the β values to calculate $^1\chi$, and of the β^{v} values to calculate $^1\chi^{\text{v}}$.

Similarly, a *finite number* of higher-order connectivity indices can be defined in terms of the sums of the reciprocal square roots of products of larger numbers of δ and δ^{v} values [1,2]. These higher-order indices incorporate such structural and topological information as the "paths", branches, "clusters" of non-hydrogen atoms, rings, and types of ring substitution in the molecule. *For the sake of simplicity, the use of higher-order indices will be avoided in the present work. Various types of structural parameters, which will be discussed below, will be utilized instead, to supplement the zeroth-order and first-order connectivity indices, thus keeping the calculations simple enough for all of the correlations to be applied by using a hand calculator.*

The complete set of connectivity indices contains a vast amount of information about the molecule, including the numbers of hydrogen and non-hydrogen atoms bonded to each non-hydrogen atom, the details of the electronic structure of each atom (its inner shells, valence shell hybridization, σ and π electrons, and lone pairs), and larger-scale structural features (paths, branches, clusters and rings). Connectivity indices owe their success in correlating and predicting molecular properties to the fact that they incorporate such vast amounts of information.

Linear regression equations are usually used, in terms of a few indices most likely to be important for a given property. The general forms of the two types of correlation equations used for most of the structure-property relationships for polymers in this manuscript will be presented in Section 2.C. The total number and the types of the required connectivity indices are determined by the extent to which a property depends on certain types of interactions. It should be emphasized that the values of the connectivity indices are exactly determined by the structure of a molecule or a polymeric repeat unit. The only adjustable parameters are the coefficients multiplying the connectivity indices in the regression equations.

2.B.   Extension of Connectivity Index Calculations to Polymers

It is not possible to use the method described above for ordinary molecules without making any modifications and calculate the χ values of polymer chains correctly. A simple formal extension of the technique for calculating connectivity indices will therefore be developed. This extension will enable the consistent and uniform definition and calculation of zeroth-order and first-order χ values, which are the only ordinary connectivity indices used in our correlations, for polymer chains of indeterminate length, containing arbitrary numbers of atoms.

Two difficulties make the "brute force" application of connectivity indices as developed for ordinary molecules [1,2] and reviewed in Section 2.A to polymer chains a hopeless task unless some modifications are made:

1. The number of atoms in a typical polymer chain is very large (usually many thousands).
2. Polymer samples often manifest polydispersity. In other words, they contain chains of different molecular weights and hence different numbers of atoms.

The necessity of taking advantage of the fact that the physical properties of polymers mainly depend on specific short range structural features and interactions, and thus reducing the problem to the calculation of the χ values of an appropriate finite molecular system, is clearly seen from the existence of these two very major difficulties.

On the other hand, arbitrarily designating a finite molecule as a representation of the polymer chain by tying up those valences that would lead to chain continuation in the polymer with hydrogen atoms is not acceptable either. Such a procedure can introduce large truncation errors in the calculated χ values. The relative magnitudes of these truncation errors are largest for "standard" polymers built from small and relatively simple monomers. The relative truncation errors in the calculation of χ values are smaller for the larger and more complex monomers from which most of the novel high-performance polymers are built; however, since correlations developed by mainly using the polymers with smaller repeat units would be unreliable, the predictions for the more complex polymers would also be of poor quality.

Fortunately, it is possible to develop a very simple procedure to calculate connectivity indices up to the first order, which utilizes a small molecular unit to calculate the χ values, which takes chain continuation into account in a consistent and unambiguous manner, and which does not introduce any truncation errors.

This procedure is illustrated and described in detail in Figure 2.3, using poly(vinyl fluoride) (PVF) as an example. Note that the values calculated for the $^0\chi^v$ and $^1\chi^v$ of PVF in this manner (1.6624 and 1.0347, respectively) are quite different from the values (2.0851 and 0.9744) calculated for the fluoroethane (C_2H_5F) molecule (which represents a repeat unit of PVF terminated by hydrogen atoms) by using the procedure for simple molecules that was illustrated in Figure 2.2. In fact, one of the indices is larger and the other one is smaller in C_2H_5F, so that the truncation error does not even have the same sign for both indices.

The number N of vertices in the hydrogen-suppressed graph, the zeroth-order connectivity indices $^0\chi$ and $^0\chi^v$, and the first-order connectivity indices $^1\chi$ and $^1\chi^v$, are listed in Table 2.2 for a diverse set of 357 polymers. All of the common polymers are included in this table. Many specialized and/or exotic polymers are also included. The polymers are listed in the order of increasing N. The easiest way to look up the connectivity indices of a polymer in Table 2.2 is to draw the structure of the repeat unit of the polymer, count the number of non-hydrogen atoms in this repeat unit, and search among the polymers with that value of N. For example, poly(vinyl fluoride), which is shown in Figure 2.3, has N=3. Its connectivity indices can be easily found by searching for it among the small number of polymers with N=3. Polymers with equal values of N are ordered according to increasing $^1\chi$. Polymers with equal values of both N and $^1\chi$ are ordered according to increasing $^1\chi^v$.

Some effort has been made to use a consistent nomenclature scheme for the polymers. On the other hand, the consistency of nomenclature has been deliberately sacrificed for the sake of simplicity or the use of a more familiar name in many cases. The formal names of a few polymers are far too long to fit into the space available in the table, and shorter labels were therefore assigned to them. The structures of these polymers are shown in Figure 2.4. The structures of

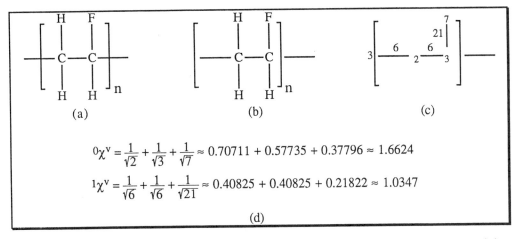

$$0\chi^V = \frac{1}{\sqrt{2}} + \frac{1}{\sqrt{3}} + \frac{1}{\sqrt{7}} \approx 0.70711 + 0.57735 + 0.37796 \approx 1.6624$$

$$1\chi^V = \frac{1}{\sqrt{6}} + \frac{1}{\sqrt{6}} + \frac{1}{\sqrt{21}} \approx 0.40825 + 0.40825 + 0.21822 \approx 1.0347$$

(d)

Figure 2.3. Example of the calculation of the zeroth-order and first-order valence connectivity indices for polymer chains via a simple procedure which avoids truncation errors.
(a) Schematic representation of the poly(vinyl fluoride) (PVF) chain in terms of its repeat unit.
(b) Alternative representation. Repetition of the unit inside the square brackets reproduces the PVF chain. The square brackets demarcating the repeat unit do not intersect backbone bonds.
(c) Hydrogen-suppressed graph of the schematic representation in (b). The δ^V values are shown at the vertices and the β^V values are shown along the edges.
(d) Summation of reciprocal square roots of the indices at the vertices of the unit enclosed within the square brackets in (c) to obtain $^0\chi^V$, and along the edges of the unit enclosed within the square brackets in (c) to obtain $^1\chi^V$. The δ^V value of the atom outside the square brackets in (c) is *not* included in the summation. It is indicated merely to take chain continuation into account, and to allow the assignment of the correct δ^V and β^V values for all of the atoms and bonds enclosed *within* the square brackets.

several other polymers are not apparent from their names, and are also shown in Figure 2.4. The structures of many polymers are shown in other figures throughout this book, and these figures are also listed. Commonly used alternative names, abbreviations and/or tradenames are also listed for many of the polymers in Table 2.2. In later chapters of this book, some polymers whose full names are very long and cumbersome, or which are better known by an abbreviated name or tradename, will often be referred to by these alternative names.

As shown in figures 2.5-2.8, the zeroth-order and first-order connectivity indices correlate quite strongly with N, and thus also with each other. The correlations of N with the valence connectivity indices $^0\chi^V$ and $^1\chi^V$ are both much weaker than the correlations of N with $^1\chi$ and $^0\chi$. The strongest correlation with N is found for the first-order simple connectivity index $^1\chi$, the second strongest correlation is found for $^0\chi$, the third strongest correlation is found for $^0\chi^V$, and the weakest correlation is found for $^1\chi^V$.

As a result of the correlations shown in figures 2.5-2.8, the four zeroth-order and first-order connectivity indices are *not* a set of linearly independent descriptors of the topology. Very often, after one or two of these four indices have been used in a correlation, the addition of extra terms proportional to the remaining zeroth-order and first-order connectivity indices does not result in statistically significant improvements.

In spite of the existence of these correlations between N and the zeroth-order and first-order connectivity indices, expressing the physical properties of polymers in terms of the connectivity indices is generally preferable to expressing them in terms of N, for the following two reasons:

1. The connectivity indices contain a vast amount of structural and electronic information omitted in a simple count of the number of non-hydrogen atoms in a molecule or a polymeric repeat unit. For example, the -CH_3, -CH_2-, trivalent carbon or silicon, tetravalent carbon or silicon, -NH_2, -NH-, trivalent nitrogen, -OH, -O-, -SH, -S-, -F, -Cl and -Br units all contribute one vertex to the hydrogen-suppressed graph, and therefore all increase the value of N by 1. On the other hand, each of these units makes a different contribution to the connectivity indices.

2. The connectivity indices therefore have different values for many polymers which have the same value of N. They thus enable distinctions to be made between polymers which have equal values of N but different structures, and different physical and chemical properties. Polyisobutylene, polybutadiene, polyacrylonitrile, poly(vinylidene chloride) and poly(dimethyl siloxane), all of which have N=4, are examples of such polymers.

As mentioned in a footnote to Table 2.1, the use of $\delta^v=1/3$ or $\delta^v=4/9$ for silicon atoms, as obtained from the definition of δ^v (Equation 2.1), causes the overestimation of the effect of the extra inner shell of electrons in silicon atoms on certain physical properties. Whenever this happens, the replacement Si→C (i.e., $\delta^v=3$ or 4) will be made in calculating the valence connectivity indices to correlate that property. For such properties, the differences between Si and C atoms will be taken into account by introducing an atomic correction term for the number of silicon atoms in the repeat unit. The alternative sets of $^0\chi^v$ and $^1\chi^v$ values obtained for silicon-containing polymers by making the replacement Si→C in the hydrogen-suppressed graph of the polymeric repeat unit, are listed in Table 2.3.

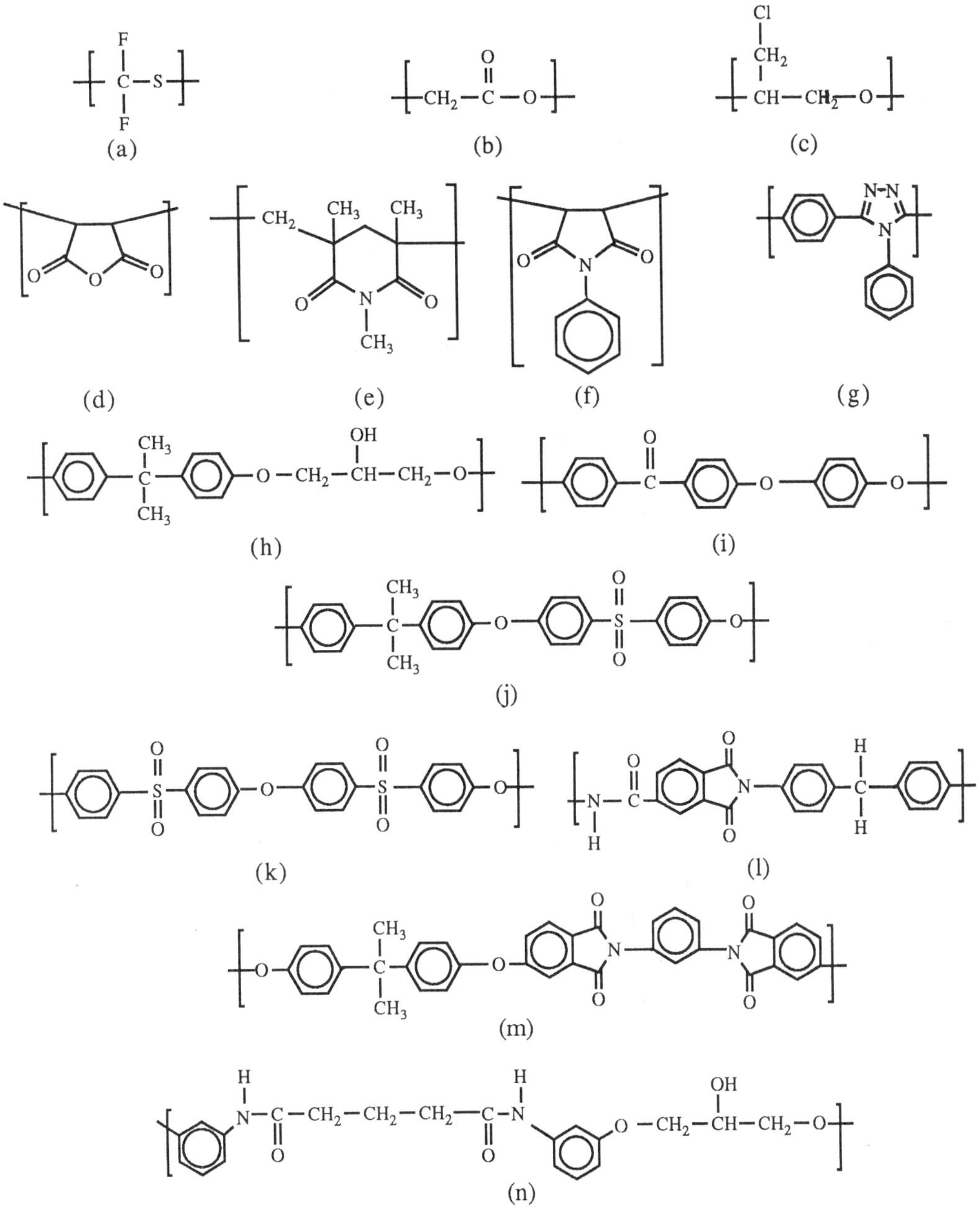

Figure 2.4. Schematic illustrations of repeat units of several of the polymers listed in Table 2.2. (a) Poly(thiocarbonyl fluoride). (b) Poly(glycolic acid). (c) Polyepichlorohydrin. (d) Poly(maleic anhydride). (e) Poly(N-methyl glutarimide). (f) Poly(N-phenyl maleimide). (g) Poly[3,5-(4-phenyl-1,2,4-triazole)-1,4-phenylene]. (h) Phenoxy resin. (i) Poly(oxy-1,4-phenylene-oxy-1,4-phenylene-carbonyl-1,4-phenylene). (j) Udel. (k) Victrex. (l) Torlon. (m) Ultem. (n) Resin F.

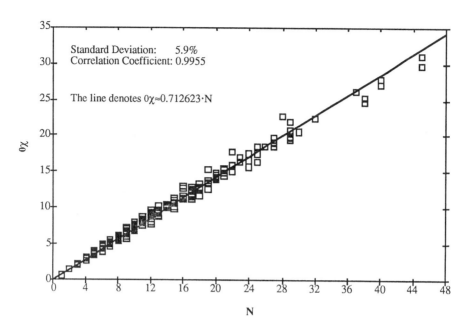

Figure 2.5. Correlation between N and $^0\chi$, for the 357 polymers listed in Table 2.2.

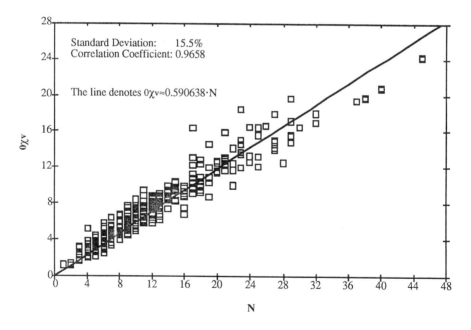

Figure 2.6. Correlation between N and $^0\chi^v$, for the 357 polymers listed in Table 2.2.

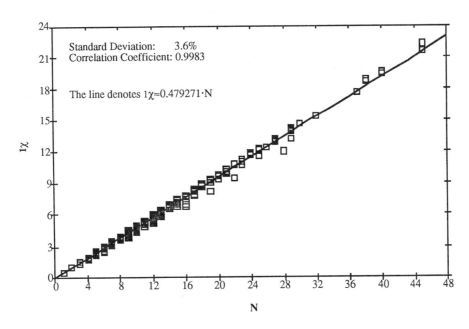

Figure 2.7. Correlation between N and $^1\chi$, for the 357 polymers listed in Table 2.2.

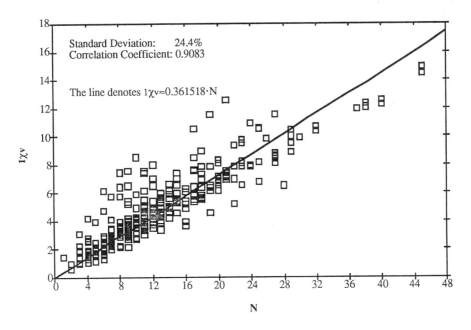

Figure 2.8. Correlation between N and $^1\chi^v$, for the 357 polymers listed in Table 2.2.

Table 2.2. The number N of vertices in the hydrogen-suppressed graph, the zeroth-order connectivity indices $^0\chi$ and $^0\chi^v$, and the first-order connectivity indices $^1\chi$ and $^1\chi^v$, for a diverse set of 357 polymers. Common abbreviations and/or tradenames are also listed for many of the polymers. The polymers are listed in the order of increasing N. Polymers with equal values of N are ordered according to increasing $^1\chi$. Polymers with equal values of both N and $^1\chi$ are ordered according to increasing $^1\chi^v$. For the silicon-containing polymers in this table, alternative sets of $^0\chi^v$ and $^1\chi^v$ values, which are listed in Table 2.3, are used in some of the correlations. All five topological quantities are identical among polymers which differ only by having *meta* rather than *para* bonding at phenyl rings. For example, the values of N, $^0\chi$, $^0\chi^v$, $^1\chi$ and $^1\chi^v$ for poly(ethylene isophthalate) and poly(ethylene terephthalate) are identical.

Polymer		N	$^0\chi$	$^0\chi^v$	$^1\chi$	$^1\chi^v$
Sulfur		1	0.7071	1.2247	0.5000	1.5000
Polyoxymethylene	*(See Fig. 17.6)*	2	1.4142	1.1154	1.0000	0.5774
Polyethylene	*(See Fig. 17.6)*	2	1.4142	1.4142	1.0000	1.0000
Poly(vinyl fluoride)	*(See Fig. 2.3)*	3	2.2845	1.6624	1.3938	1.0347
Poly(vinyl alcohol)	*(See Fig. 4.1)*	3	2.2845	1.7317	1.3938	1.0747
Polypropylene		3	2.2845	2.2845	1.3938	1.3938
Poly[oxy(methylhydrosilylene)]		3	2.2845	3.1403	1.3938	3.1463
Poly(vinyl chloride)	*(See Fig. 9.1)*	3	2.2845	2.4183	1.3938	1.4711
Poly(vinyl bromide)		3	2.2845	3.2484	1.3938	1.9504
Polyoxyethylene		3	2.1213	1.8225	1.5000	1.0774
Poly(ethylene sulfide)		3	2.1213	2.6390	1.5000	2.2320
Poly(vinylidene fluoride)	*(See Fig. 9.3)*	4	3.2071	1.9630	1.7071	1.0851
Poly(thiocarbonyl fluoride)	*(See Fig. 2.4)*	4	3.2071	2.4807	1.7071	1.6027
Polyisobutylene		4	3.2071	3.2071	1.7071	1.7071
Poly(vinylidene chloride)	*(See Fig. 9.1)*	4	3.2071	3.4749	1.7071	1.8410
Poly(vinylidene bromide)		4	3.2071	5.1350	1.7071	2.6711
Poly(dimethyl siloxane)	*(See Fig. 5.1)*	4	3.2071	3.9083	1.7071	4.2248
Poly(glycolic acid)	*(See Fig. 2.4)*	4	2.9916	2.0236	1.8938	1.0505
Poly(propylene oxide)		4	2.9916	2.6927	1.8938	1.5100
Poly(propylene sulfide)		4	2.9916	3.5092	1.8938	2.5587
Polyacrylonitrile		4	2.9916	2.2317	1.9319	1.3288
Poly(vinyl methyl ether)		4	2.9916	2.6927	1.9319	1.4604
Poly(1,2-butadiene)	*(See Fig. 6.13)*	4	2.9916	2.5689	1.9319	1.5581
Poly(1-butene)		4	2.9916	2.9916	1.9319	1.9319
Poly(vinyl methyl sulfide)		4	2.9916	3.5092	1.9319	2.7483
Polyoxytrimethylene	*(See Fig. 17.6)*	4	2.8284	2.5296	2.0000	1.5774

☞ ☞ ☞   **TABLE 2.2 IS CONTINUED IN THE NEXT PAGE.** ☞ ☞ ☞

Table 2.2. CONTINUED FROM THE PREVIOUS PAGE.

Polymer		N	$^0\chi$	$^0\chi^v$	$^1\chi$	$^1\chi^v$
Poly(1,4-butadiene)	*(See Fig. 5.1 and Fig. 6.13)*	4	2.8284	2.5689	2.0000	1.6498
Polytrifluoroethylene		5	4.0774	2.2112	2.1547	1.1735
Polymethacrylonitrile	*(See Fig. 4.1)*	5	3.9142	3.1543	2.2678	1.6807
Poly[oxy(vinylmethylsilylene))]		5	3.9142	4.1927	2.2678	3.9990
Poly(acrylic acid)	*(See Fig. 4.1)*	5	3.8618	2.6399	2.3045	1.5329
Polyacrylamide		5	3.8618	2.7701	2.3045	1.5980
Poly(vinyl methyl ketone)		5	3.8618	3.1927	2.3045	1.8903
Poly(β-propiolactone)		5	3.6987	2.7307	2.3938	1.5505
Poly(β-alanine)		5	3.6987	2.8225	2.3938	1.6612
Polyisoprene	*(See Fig. 5.1)*	5	3.6987	3.4916	2.3938	2.0505
Polychloroprene	*(See Fig. 5.1)*	5	3.6987	3.6255	2.3938	2.1174
Poly(2-bromo-1,4-butadiene)		5	3.6987	4.4555	2.3938	2.5325
Poly(vinyl formate)		5	3.6987	2.6783	2.4319	1.5236
Poly(vinyl ethyl ether)		5	3.6987	3.3998	2.4319	2.0480
Polyepichlorohydrin						
[poly(3-chloropropylene oxide), see Fig. 2.4]		5	3.6987	3.5337	2.4319	2.1427
Poly(1-pentene)		5	3.6987	3.6987	2.4319	2.4319
Poly(oxymethyleneoxyethylene)		5	3.5355	2.9378	2.5000	1.6547
Polyoxytetramethylene		5	3.5355	3.2367	2.5000	2.0774
Poly(1,4-pentadiene)		5	3.5355	3.2760	2.5000	2.1498
Polytetrafluoroethylene	*(Teflon)*	6	5.0000	2.5119	2.5000	1.2559
Polychlorotrifluoroethylene		6	5.0000	3.2678	2.5000	1.6339
Poly(3,3,3-trifluoropropylene)		6	4.7845	2.9184	2.6052	1.6721
Poly(vinyl sulfonic acid)		6	4.7845	3.1605	2.6052	1.9439
Poly(vinyl trimethylsilane)		6	4.7845	5.7845	2.6052	6.1825
Poly(propylene sulfone)		6	4.7845	3.7133	2.6278	2.2722
Poly(methacrylic acid)		6	4.7845	3.5626	2.6505	1.8848
Poly(methyl isopropenyl ketone)		6	4.7845	4.1154	2.6505	2.1612
Poly(α-methyl acrylamide)		6	4.7845	3.6927	2.6505	1.9499
Poly(vinyl acetate)	*(See Fig. 16.1)*	6	4.5689	3.6010	2.7877	1.9604
Poly(vinyl isopropyl ether)		6	4.5689	4.2701	2.7877	2.4426
Poly(4-methyl-1-pentene)		6	4.5689	4.5689	2.7877	2.7877
Poly[oxy(diethylsilylene)]		6	4.6213	5.3225	2.8284	4.7603

☞ ☞ ☞ **TABLE 2.2 IS CONTINUED IN THE NEXT PAGE.** ☞ ☞ ☞

Table 2.2. CONTINUED FROM THE PREVIOUS PAGE.

Polymer		N	$^0\chi$	$^0\chi^v$	$^1\chi$	$^1\chi^v$
Poly(methyl acrylate)		6	4.5689	3.6010	2.8425	1.9217
Poly(N-methyl acrylamide)		6	4.5689	3.6927	2.8425	2.0593
Poly(γ-butyrolactone)		6	4.4058	3.4378	2.8938	2.0505
Poly(2-methyl-1,4-pentadiene)		6	4.4058	4.1987	2.8938	2.5505
Polyfumaronitrile		6	4.5689	3.0491	2.8974	1.6912
Poly(3-methoxypropylene oxide)		6	4.4058	3.8081	2.9319	2.0378
Poly(1-hexene)		6	4.4058	4.4058	2.9319	2.9319
Poly(p-phenylene)	(See Fig. 17.6)	6	3.9831	3.3094	2.9663	2.0714
Poly(α,α-dimethylpropiolactone)		7	5.4916	4.5236	3.1278	2.3005
Poly(1-butene sulfone)		7	5.4916	4.4204	3.1659	2.8103
Poly(methyl methacrylate)	(PMMA, see Fig. 11.1)	7	5.4916	4.5236	3.1885	2.2736
Poly(methyl α-chloroacrylate)		7	5.4916	4.6575	3.1885	2.3405
Poly(N-methyl methacrylamide)		7	5.4916	4.6154	3.1885	2.4112
Poly(methyl α-bromoacrylate)		7	5.4916	5.4876	3.1885	2.7556
Poly[oxy(mercaptopropylmethylsilylene)]		7	5.3284	6.3712	3.2678	5.7341
Poly(vinyl isobutyl ether)		7	5.2760	4.9772	3.2877	2.9038
Poly(maleic anhydride)	(See Fig. 2.4)	7	5.0165	3.3794	3.3045	2.0605
Poly(2-isopropyl-1,4-butadiene)		7	5.2760	5.0689	3.3045	2.9939
Poly(vinyl propionate)		7	5.2760	4.3081	3.3257	2.5211
Poly(vinyl chloroacetate)		7	5.2760	4.4420	3.3257	2.6158
Poly(vinyl sec-butyl ether)		7	5.2760	4.9772	3.3257	2.9806
Poly(ethyl acrylate)		7	5.2760	4.3081	3.3425	2.5092
Poly(δ-valerolactone)		7	5.1129	4.1449	3.3938	2.5505
Poly(vinyl n-butyl ether)		7	5.1129	4.8140	3.4319	3.0480
Poly(vinyl n-butyl sulfide)		7	5.1129	5.6305	3.4319	4.0967
Poly[oxy(p-phenylene)]		7	4.6902	3.7176	3.4495	2.2296
Poly(vinyl formal)		7	4.6902	4.0925	3.4495	2.6817
Poly[thio(p-phenylene)]		7	4.6902	4.5341	3.4495	3.0461
Poly(2-vinylthiophene)		7	4.6902	4.7413	3.4663	3.3800
Poly(vinyl cyclopentane)		7	4.6902	4.6902	3.4663	3.4663
Poly(oxymethyleneoxytetramethylene)		7	4.9497	4.3520	3.5000	2.6547
Polyoxyhexamethylene		7	4.9497	4.6509	3.5000	3.0774

☞ ☞ ☞   **TABLE 2.2 IS CONTINUED IN THE NEXT PAGE.** ☞ ☞ ☞

Table 2.2. CONTINUED FROM THE PREVIOUS PAGE.

Polymer		N	$^0\chi$	$^0\chi^v$	$^1\chi$	$^1\chi^v$
Poly(2-*t*-butyl-1,4-butadiene)		8	6.1987	5.9916	3.6052	3.3005
Poly(ethyl methacrylate)		8	6.1987	5.2307	3.6885	2.8611
Poly(ethyl α-chloroacrylate)		8	6.1987	5.3646	3.6885	2.9281
Poly(methyl α-cyanoacrylate)		8	6.1987	4.4708	3.7491	2.2472
Poly(methyl ethacrylate)		8	6.1987	5.2307	3.7491	2.8343
Poly(ethylene oxalate)		8	5.9831	4.0472	3.8045	2.1438
Poly(oxy-2,2-dichloromethyltrimethylene)						
	(Penton™, see Fig. 9.1)	8	6.0355	6.0045	3.8284	3.5951
Poly(oxy-1,1-dichloromethyltrimethylene)		8	6.0355	6.0045	3.8284	3.6570
Poly[oxy(dipropylsilylene)]		8	6.0355	6.7367	3.8284	5.7603
Poly(vinyl acetal)		8	5.5605	4.9628	3.8433	3.1532
Poly(phenylmethylsilane)		8	5.6129	5.8868	3.8552	6.4107
Poly(cyclohexylmethylsilane)		8	5.6129	6.6129	3.8552	7.4325
Poly(N-vinyl pyrrolidone)	*(See Fig. 4.1)*	8	5.5605	4.7612	3.8770	3.1722
Poly(ε-caprolactone)		8	5.8200	4.8520	3.8938	3.0505
Poly(ε-caprolactam)	*(Nylon 6)*	8	5.8200	4.9438	3.8938	3.1612
Poly(vinyl *n*-pentyl ether)		8	5.8200	5.5211	3.9319	3.5480
Poly(*p*-xylylene)	*(See Fig. 17.6)*	8	5.3973	4.7236	3.9495	3.0285
Poly(*p*-vinyl pyridine)	*(See Fig. 5.1)*	8	5.3973	4.5411	3.9663	2.8656
Poly(*o*-vinyl pyridine)		8	5.3973	4.5411	3.9663	2.8757
Polystyrene	*(See Fig. 1.1)*	8	5.3973	4.6712	3.9663	3.0159
Poly(vinyl cyclohexane)		8	5.3973	5.3973	3.9663	3.9663
Poly[oxy(methyl γ-trifluoropropylsilylene)]	*(See Fig 9.3)*	9	7.1213	5.9564	3.9142	5.2059
Poly(trifluorovinyl acetate)		9	7.2845	4.4504	3.9165	2.1793
Poly(*tert*-butyl acrylate)		9	7.0689	6.1020	3.9889	3.2176
Poly(vinyl pivalate)		9	7.0689	6.1010	3.9990	3.2104
Poly(isopropyl methacrylate)		9	7.0689	6.1010	4.0443	3.2558
Poly(isopropyl α-chloroacrylate)		9	7.0689	6.2349	4.0443	3.3227
Poly[ethylene-N-(β-trimethylsilyl ethyl)imine]		9	6.9058	7.7756	4.0783	7.6014
Poly(1-hexene sulfone)		9	6.9058	5.8346	4.1659	3.8103
Poly(2-fluoroethyl methacrylate)		9	6.9058	5.3158	4.1885	2.9213
Poly(2-hydroxyethyl methacrylate)		9	6.9058	5.3850	4.1885	2.9703
Poly(*n*-propyl methacrylate)		9	6.9058	5.9378	4.1885	3.3611

☞ ☞ ☞   **TABLE 2.2 IS CONTINUED IN THE NEXT PAGE.** ☞ ☞ ☞

Table 2.2. CONTINUED FROM THE PREVIOUS PAGE.

Polymer		N	$^0\chi$	$^0\chi^v$	$^1\chi$	$^1\chi^v$
Poly(n-propyl α-chloroacrylate)		9	6.9058	6.0717	4.1885	3.4280
Poly(2-chloroethyl methacrylate)		9	6.9058	6.0717	4.1885	3.4559
Poly(ethylmercaptyl methacrylate)		9	6.9058	6.2794	4.1885	3.6028
Poly(2-bromoethyl methacrylate)		9	6.9058	6.9018	4.1885	4.0428
Poly(isobutyl acrylate)		9	6.8533	5.8865	4.1983	3.3651
Poly[oxy(methyl n-hexylsilylene)]		9	6.7426	7.4438	4.2678	6.4925
Poly[oxy(2,6-dimethyl-1,4-phenylene)]	*(PPO)*	9	6.4307	5.5629	4.2709	3.0629
Poly(α-methyl styrene)		9	6.3200	5.5939	4.3123	3.3678
Poly[oxy(methylphenylsilylene)]	*(See Fig. 6.11)*	9	6.3200	6.2950	4.3123	5.3854
Poly(n-butyl acrylate)		9	6.6902	5.7233	4.3425	3.5092
Poly(p-hydroxybenzoate)	*(Ekonol™)*	9	6.2676	4.6259	4.3602	2.6837
Poly(p-fluoro styrene)		9	6.2676	4.9718	4.3602	3.1155
Poly(5-vinyl-2-methylpyridine)		9	6.2676	5.4637	4.3602	3.2863
Poly(p-methyl styrene)		9	6.2676	5.5939	4.3602	3.4265
Poly(p-chloro styrene)	*(See Fig. 5.1)*	9	6.2676	5.7277	4.3602	3.4935
Poly(chloro-p-xylylene)	*(See Fig. 5.1)*	9	6.2676	5.7802	4.3602	3.5121
Poly(p-bromo styrene)		9	6.2676	6.5578	4.3602	3.9085
Poly(o-methyl styrene)		9	6.2676	5.5939	4.3770	3.4325
Poly(o-chloro styrene)		9	6.2676	5.7277	4.3770	3.4995
Poly(vinyl propional)		9	6.2676	5.6699	4.3813	3.6912
Poly(3-butoxypropylene oxide)		9	6.5271	5.9294	4.4319	3.6253
Poly(vinyl n-hexyl ether)		9	6.5271	6.2282	4.4319	4.0480
Poly(2,5-benzoxazole)	*(AB-PBO, see Fig. 11.2)*	9	5.8449	4.5875	4.4327	2.8435
Poly(2,5-benzothiazole)		9	5.8449	5.4040	4.4327	3.6600
Poly(3-phenyl-1-propene)		9	6.1044	5.3783	4.4495	3.4890
Poly(styrene oxide)		9	6.1044	5.0795	4.4663	3.1320
Poly(styrene sulfide)		9	6.1044	5.8960	4.4663	4.1807
Polyoxyoctamethylene		9	6.3640	6.0651	4.5000	4.0774
Poly(t-butyl methacrylate)		10	7.9916	7.0236	4.3349	3.5695
Poly(trifluoroethyl acrylate)		10	7.7760	4.9420	4.4890	2.7226
Poly(isobutyl methacrylate)		10	7.7760	6.8081	4.5443	3.7170
Poly(sec-butyl methacrylate)		10	7.7760	6.8081	4.5823	3.7938
Poly(sec-butyl α-chloroacrylate)		10	7.7760	6.9420	4.5823	3.8607

☞ ☞ ☞   **TABLE 2.2 IS CONTINUED IN THE NEXT PAGE.** ☞ ☞ ☞

Table 2.2. CONTINUED FROM THE PREVIOUS PAGE.

Polymer		N	$^0\chi$	$^0\chi^v$	$^1\chi$	$^1\chi^v$
Poly(*sec*-butyl α-bromoacrylate)		10	7.7760	7.7721	4.5823	4.2758
Poly(*n*-butyl methacrylate)		10	7.6129	6.6449	4.6885	3.8611
Poly(*n*-butyl α-chloroacrylate)		10	7.6129	6.7788	4.6885	3.9280
Poly(N-butyl methacrylamide)		10	7.6129	6.7367	4.6885	3.9719
Poly[oxy(2,6-dimethyl-5-bromo-1,4-phenylene)]		10	7.3010	7.4496	4.6984	3.9676
Poly[oxy(methyl *m*-chlorophenylsilylene)]		10	7.1902	7.3515	4.7061	5.8631
Poly(3,4-dichlorostyrene)		10	7.1378	6.7843	4.7709	3.9771
Poly(ethylene maleate)		10	7.3973	5.2019	4.7877	2.8045
Poly(ethylene succinate)		10	7.3973	5.4614	4.7877	3.1010
Poly(2,6-dichlorostyrene)		10	7.1378	6.7843	4.7877	3.9831
Poly(octylmethylsilane)		10	7.4497	8.4497	4.8107	8.5178
Poly(vinyl butyral)	*(See Fig. 2.9)*	10	6.9747	6.3770	4.8813	4.1912
Poly(8-aminocaprylic acid)	*(Nylon 8)*	10	7.2342	6.3580	4.8938	4.1612
Poly(*p*-methoxy styrene)		10	6.9747	6.0021	4.8981	3.5389
Poly(*o*-methoxy styrene)		10	6.9747	6.0021	4.9149	3.5449
Poly(1-decene)		10	7.2342	7.2342	4.9319	4.9319
Poly(4-phenyl-1-butene)		10	6.8116	6.0854	4.9495	3.9890
Poly(trifluoroethyl methacrylate)		11	8.6987	5.8646	4.8349	3.0745
Poly(neopentyl methacrylate)		11	8.6987	7.7307	4.8349	4.0076
Poly(isopentyl methacrylate)		11	8.4831	7.5152	5.0443	4.2170
Poly(1-methylbutyl methacrylate)		11	8.4831	7.5152	5.0823	4.2938
Poly(1,3-dichloropropyl methacrylate)		11	8.4831	7.7830	5.0823	4.4657
Poly(2,3-dibromopropyl methacrylate)		11	8.4831	9.4431	5.0823	5.4932
Poly(dimethyl itaconate)		11	8.4831	6.5472	5.1430	3.2972
Poly[oxy(acryloxypropylmethylsilylene)]		11	8.3200	7.6305	5.1616	6.1793
Poly(vinyl dimethylphenylsilane)		11	7.8973	8.1712	5.2103	7.3432
Cellulose		11	8.0081	5.7520	5.2364	3.5170
Poly[oxy(methyl *n*-octylsilylene)]		11	8.1569	8.8580	5.2678	7.4925
Poly(*p*-isopropyl styrene)		11	7.8449	7.1712	5.2709	4.3699
Poly(trimethylene succinate)		11	8.1044	6.1685	5.2877	3.6010
Poly(*o*-chloro-*p*-methoxystyrene)		11	7.8449	7.0586	5.3089	4.0225
Poly(vinyl benzoate)		11	7.6818	5.9877	5.3601	3.6211
Poly(vinyl 2-ethylhexyl ether)		11	8.1044	7.8056	5.3637	4.9798

☞ ☞ ☞   **TABLE 2.2 IS CONTINUED IN THE NEXT PAGE.** ☞ ☞ ☞

Table 2.2. CONTINUED FROM THE PREVIOUS PAGE.

Polymer		N	$^0\chi$	$^0\chi^v$	$^1\chi$	$^1\chi^v$
Poly(3-hexoxypropylene oxide)		11	7.9413	7.3436	5.4319	4.6253
Poly(vinyl n-octyl ether)		11	7.9413	7.6424	5.4319	5.0480
Poly(2-heptyl-1,4-butadiene)		11	7.9413	7.7342	5.4319	5.1112
Poly(p-phenylene-2,5-oxadiazole)		11	7.2591	5.6121	5.4327	3.3768
Poly(p-phenylene-2,5-thiodiazole)		11	7.2591	6.4286	5.4327	4.1933
Poly(5-phenyl-1-pentene)		11	7.5187	6.7925	5.4495	4.4890
Poly(2,2,2-trifluoro-1-methylethyl methacrylate)		12	9.5689	6.7348	5.2557	3.5340
Poly(1,2,2-trimethylpropyl methacrylate)		12	9.5689	8.6010	5.2557	4.4671
Poly(3,3-dimethylbutyl methacrylate)		12	9.4058	8.4378	5.3349	4.5076
Poly(N-methyl glutarimide)	*(See Fig. 2.4)*	12	9.1463	7.6779	5.3903	4.2169
Poly(1,3-dimethylbutyl methacrylate)		12	9.3534	8.3854	5.4382	4.6496
Poly($\alpha,\alpha,\alpha',\alpha'$-tetrafluoro-$p$-xylylene)		12	8.9831	5.8213	5.4603	3.7273
Poly(m-trifluoromethylstyrene)		12	8.7676	6.2278	5.5715	3.7435
Poly(p-t-butyl styrene)		12	8.7676	8.0939	5.5715	4.6765
Poly(1-methylpentyl methacrylate)		12	9.1902	8.2223	5.5823	4.7938
Poly(2-ethylbutyl methacrylate)		12	9.1902	8.2223	5.6203	4.7930
Poly[oxy(methyl m-chlorophenylethylsilylene)]		12	8.6044	8.7658	5.6793	7.0273
Poly(vinyl dimethylbenzylsilane)		12	8.6044	8.8783	5.6834	8.0074
Poly(n-hexyl methacrylate)		12	9.0271	8.0591	5.6885	4.8611
Poly(phenyl α-bromoacrylate)		12	8.6045	7.8743	5.7061	4.4621
Poly(cyclohexyl α-bromoacrylate)		12	8.6044	8.6005	5.7061	5.3996
Poly(phenyl methacrylate)		12	8.6045	6.9104	5.7061	3.9802
Poly(cyclohexyl methacrylate)		12	8.6044	7.6365	5.7061	4.9176
Poly(cyclohexyl α-chloroacrylate)		12	8.6044	7.7704	5.7061	4.9845
Poly(ethylene adipate)		12	8.8116	6.8756	5.7876	4.1010
Poly[oxy(dicyanopropylsilylene)]		12	8.8640	8.0453	5.8284	6.5004
Poly(β-vinyl naphthalene)	*(See Fig. 17.3)*	12	7.9663	6.8259	5.9327	4.4206
Poly(α-vinyl naphthalene)	*(See Fig. 17.3)*	12	7.9663	6.8259	5.9495	4.4265
Poly(acenaphthylene)	*(See Fig. 17.3)*	12	7.7067	6.6188	5.9663	4.4821
Poly(pentafluoropropyl acrylate)		13	10.2760	6.1979	5.7390	3.3506
Poly(2-nitro-2-methylpropyl methacrylate)		13	10.2760	7.9944	5.7783	4.0609
Poly(4-fluoro-2-trifluoromethylstyrene)		13	9.6378	6.5284	5.9822	3.8491
Poly(2-bromo-4-trifluoromethyl styrene)		13	9.6378	8.1144	5.9822	4.6421

☞ ☞ ☞   **TABLE 2.2 IS CONTINUED IN THE NEXT PAGE.** ☞ ☞ ☞

Table 2.2. CONTINUED FROM THE PREVIOUS PAGE.

Polymer		N	$^0\chi$	$^0\chi^v$	$^1\chi$	$^1\chi^v$
Poly(1-methylcyclohexyl methacrylate)		13	9.4747	8.5067	6.0420	5.2766
Poly[oxy(2,6-isopropyl-1,4-phenylene)]		13	9.5854	8.7176	6.0922	4.9497
Poly(m-cresyl methacrylate)		13	9.4747	7.8330	6.1000	4.3908
Poly(p-bromophenyl methacrylate)		13	9.4747	8.7970	6.1000	4.8728
Poly(4-methylcyclohexyl methacrylate)		13	9.4747	8.5067	6.1000	5.3114
Poly(3-methylcyclohexyl methacrylate)		13	9.4747	8.5067	6.1000	5.3114
Poly(o-cresyl methacrylate)		13	9.4747	7.8330	6.1168	4.3968
Poly(o-chlorobenzyl methacrylate)		13	9.4747	7.9669	6.1168	4.4637
Poly(2-methylcyclohexyl methacrylate)		13	9.4747	8.5067	6.1168	5.3282
Poly(2-chlorocyclohexyl methacrylate)		13	9.4747	8.6406	6.1168	5.4055
Poly(benzyl methacrylate)		13	9.3116	7.6175	6.2061	4.4183
Poly(N-benzyl methacrylamide)		13	9.3116	7.7092	6.2061	4.5290
Poly(N-vinyl phthalimide)		13	8.9996	6.8576	6.2877	4.2575
Poly(trimethylene adipate)		13	9.5187	7.5827	6.2877	4.6010
Poly(vinyl p-ethylbenzoate)		13	9.2591	7.6175	6.2920	4.5925
Poly(N-phenyl maleimide)	*(See Fig. 2.4)*	13	8.9996	6.8052	6.3045	4.2338
Polyoxynaphthoate	*(See Fig. 17.3)*	13	8.8365	6.7806	6.3265	4.0884
Polyundecanolactone		13	9.3555	8.3876	6.3938	5.5505
Poly(11-aminoundecanoic acid)	*(Nylon 11)*	13	9.3555	8.4793	6.3938	5.6612
Poly(vinyl n-decyl ether)		13	9.3555	9.0566	6.4319	6.0480
Poly(1-phenylethyl methacrylate)		14	10.1818	8.4877	6.6168	4.8778
Poly(vinyl p-isopropyl benzoate)		14	10.1294	8.4878	6.6647	4.9752
Poly(n-octyl methacrylate)		14	10.4413	9.4733	6.6885	5.8611
Poly[N-(2-phenylethyl)methacrylamide]		14	10.0187	8.4163	6.7061	5.0290
Poly(cyclohexyl α-ethoxyacrylate)		14	10.0187	8.7518	6.7668	5.6175
Poly(ethylene terephthalate)		14	9.9663	7.3566	6.7709	4.2152
Poly(ethylene phthalate)		14	9.9663	7.3566	6.7876	4.2212
Poly(ethylene suberate)		14	10.2258	8.2898	6.7876	5.1010
Poly(tetramethylene adipate)		14	10.2258	8.2898	6.7876	5.1010
Methyl cellulose		14	10.1294	8.6351	6.8124	4.6692
Poly(12-aminododecanoic acid)	*(Nylon 12)*	14	10.0626	9.1864	6.8938	6.1612
Poly(2-decyl-1,4-butadiene)		14	10.0626	9.8555	6.9318	6.6112
Poly(3,3,5-trimethylcyclohexyl methacrylate)		15	11.2676	10.2996	6.8071	6.0185

☞ ☞ ☞   TABLE 2.2 IS CONTINUED IN THE NEXT PAGE. ☞ ☞ ☞

Table 2.2. CONTINUED FROM THE PREVIOUS PAGE.

Polymer	N	$^0\chi$	$^0\chi^v$	$^1\chi$	$^1\chi^v$
Poly(vinyl *p-t*-butyl benzoate)	15	11.0520	9.4104	6.9653	5.2818
Poly(p-methacryloxy benzoic acid)	15	11.0520	8.1885	7.0107	4.5686
Poly[(1-(*o*-chlorophenyl)ethyl methacrylate]	15	11.0520	9.5442	7.0275	5.3614
Poly[di(*n*-propyl) itaconate]	15	11.3115	9.3756	7.1430	5.4723
Polytridecanolactone	15	10.7698	9.8018	7.3938	6.5505
Poly(vinylene diphenylsilylene)	15	10.1400	9.4282	7.4175	7.3868
Poly(vinyl *n*-dodecyl ether)	15	10.7698	10.4709	7.4319	7.0480
Poly(N-vinyl carbazole)	15	9.8281	8.3505	7.4495	5.4266
Perfluoropolymer 3　　　*(See Fig. 6.1)*	16	13.0000	6.7796	6.7500	3.6398
Poly(heptafluorobutyl acrylate)	16	12.7760	7.4538	6.9890	3.9785
Poly(*m*-nitrobenzyl methacrylate)	16	11.7591	8.8038	7.5107	4.8823
Poly(β-naphthyl methacrylate)　　*(See Fig. 17.3)*	16	11.1734	9.0651	7.6725	5.3849
Poly(α-naphthyl methacrylate)　　*(See Fig. 17.3)*	16	11.1734	9.0651	7.6893	5.3908
Poly(tetramethylene terephthalate)	16	11.3805	8.7708	7.7709	5.2152
Poly(ethylene sebacate)	16	11.6400	9.7040	7.7877	6.1010
Poly(hexamethylene adipamide)　　*(Nylon 6,6)*	16	11.6400	9.8876	7.7877	6.3225
Poly(pentachlorophenyl methacrylate)	17	12.9557	12.1931	7.7764	6.4042
Poly(pentabromophenyl methacrylate)	17	12.9557	16.3434	7.7764	8.4794
Poly[2-(phenylsulfonyl)ethyl methacrylate]	17	12.5187	9.7534	7.9401	5.8039
Poly(α-naphthyl carbinyl methacrylate)　*(See Fig. 17.3)*	17	11.8805	9.7722	8.1893	5.8290
Poly[oxy(methyl *n*-tetradecyl silylene)]	17	12.3995	13.1006	8.2678	10.4925
Poly[methane *bis*(4-phenyl)carbonate]	17	11.6649	9.0506	8.2928	5.3705
Poly[thio *bis*(4-phenyl)carbonate]	17	11.6649	9.5683	8.2928	5.8881
Ethyl cellulose	17	12.2507	10.7564	8.3124	6.4318
Polypentadecanolactone	17	12.1840	11.2160	8.3938	7.6505
Poly(diethnyl diphenylsilylene)	17	11.5542	10.2735	8.4175	7.5714
Poly[3,5-(4-phenyl-1,2,4-triazole)-1,4-phenylene] *(See Fig. 2.4)*	17	11.2423	9.0378	8.4327	5.5501
Poly(*p*-cyclohexylphenyl methacrylate)	18	12.5876	10.9459	8.6725	6.9960
Poly(dodecyl methacrylate)	18	13.2697	12.3017	8.6885	7.8611
Poly(*p*-phenylene terephthalamide)　　*(Kevlar™)*	18	12.5352	9.4353	8.7203	5.5510
Poly[1,1-ethane *bis*(4-phenyl)carbonate]	18	12.5352	9.9209	8.7203	5.8181
Poly(*o*-phenylene terephthalamide)	18	12.5352	9.4353	8.7372	5.5570
Poly(ethylene-2,6-naphthalenedicarboxylate)	18	12.5352	9.5113	8.7372	5.6199

☞ ☞ ☞　**TABLE 2.2 IS CONTINUED IN THE NEXT PAGE.** ☞ ☞ ☞

Table 2.2. CONTINUED FROM THE PREVIOUS PAGE.

Polymer		N	$^0\chi$	$^0\chi^v$	$^1\chi$	$^1\chi^v$
Poly(hexamethylene isophthalamide)	*(Selar™ PA)*	18	12.7947	10.3685	8.7709	6.4367
Poly(*p*-phenylene-2,6-benzo[1,2-*d*:5,4-*d'*]bisoxazole)						
	(Polybenzobisoxazole, cis-PBO, see Fig. 6.10)	18	11.6899	9.1750	8.8653	5.6870
Poly(*p*-phenylene-diimidazobenzene)		18	11.6899	9.3585	8.8653	5.8705
Poly(*p*-phenylene-2,6-benzo[1,2-*d*:4,5-*d'*]bisthiazole)						
	(Polybenzobisthiazole, trans-PBT)	18	11.6899	10.8080	8.8653	7.3200
Poly(1-octadecene)		18	12.8911	12.8911	8.9319	8.9319
Poly(nonafluoropentyl acrylate)		19	15.2760	8.7097	8.2390	4.6065
Poly[2,2-propane *bis*(4-phenyl)carbonate]						
	(Bisphenol-A polycarbonate, see Fig. 5.1)	19	13.4579	10.8435	9.0537	6.1634
Poly(diphenylmethyl methacrylate)		19	13.2947	10.8745	9.1893	6.4998
Poly[oxy(methyl *n*-hexadecylsilylene)]		19	13.8137	14.5149	9.2678	11.4925
Poly(hexamethylene azelamide))	*(Nylon 6,9)*	19	13.7613	12.0089	9.2877	7.8225
Poly[oxy(2,6-diphenyl-1,4-phenylene)]		19	12.6565	10.3365	9.4158	6.3843
Cellulose triacetate		20	14.8614	11.3598	9.3799	6.1692
Poly[2,2-butane *bis*(4-phenyl)carbonate]		20	14.1650	11.5506	9.6144	6.7241
Poly[1,1-dichloroethylene *bis*(4-phenyl)carbonate] *(See Fig. 6.13)*		20	14.1125	11.6113	9.6310	6.5472
Poly[1,1-(2-methyl propane) *bis*(4-phenyl)carbonate]		20	14.1125	11.4982	9.6310	6.7287
Poly(1,2-diphenylethyl methacrylate)		20	14.0018	11.5816	9.6725	6.9729
Poly(1,4-cyclohexylidene dimethylene terephthalate) *(Kodel™)*		20	13.9494	11.3397	9.7204	7.1647
Poly[1,1-butane *bis*(4-phenyl)carbonate]		20	13.9494	11.3351	9.7583	6.8561
Poly(hexamethylene sebacate)		20	14.4684	12.5325	9.7877	8.1010
Poly(hexamethylene sebacamide)	*(Nylon 6,10)*	20	14.4684	12.7160	9.7877	8.3225
Poly(2,2,2'-trimethylhexamethylene terephthalamide)		21	15.4578	13.0317	9.8718	7.5377
Poly[2,2-propane *bis*-4-(2-methylphenyl)carbonate]		21	15.1983	12.6888	9.8750	6.9967
Poly[2,2-propane *bis*-4-(2-chlorophenyl)carbonate]		21	15.1983	12.9566	9.8750	7.1306
Phenoxy resin	*(See Fig. 2.4)*	21	14.8721	12.3741	10.0537	7.2030
Poly[2,2-pentane *bis*(4-phenyl)carbonate]		21	14.8721	12.2577	10.1144	7.2241
Poly[1,1-cyclopentane *bis*(4-phenyl)carbonate]		21	14.2863	11.6720	10.2608	7.3705
Poly[oxy(methyl *n*-octadecylsilylene)]		21	15.2279	15.9291	10.2678	12.4925
Poly(undecafluorohexyl acrylate)		22	17.7760	9.9657	9.4890	5.2345
Poly[1,1-cyclohexane *bis*(4-phenyl)carbonate]	*(See Fig. 6.2)*	22	14.9934	12.3791	10.7608	7.8705

☞ ☞ ☞   **TABLE 2.2 IS CONTINUED IN THE NEXT PAGE.** ☞ ☞ ☞

Table 2.2. CONTINUED FROM THE PREVIOUS PAGE.

Polymer	N	$^0\chi$	$^0\chi^v$	$^1\chi$	$^1\chi^v$
Poly(oxy-1,4-phenylene-oxy-1,4-phenylene-carbonyl-1,4-phenylene)					
[Poly(ether ether ketone) or PEEK, see Fig. 2.4]	22	14.9409	11.6529	10.7760	6.9847
Nylon 6,12	22	15.8826	14.1302	10.7877	9.3225
Poly[2,2-propane *bis*{4-(2,6-difluorophenyl)}carbonate]	23	16.9389	12.0460	10.6964	6.5860
Poly[2,2-propane *bis*{4-(2,6-dimethylphenyl)}carbonate]	23	16.9389	14.5341	10.6964	7.8301
Poly[2,2-propane *bis*{4-(2,6-dichlorophenyl)}carbonate]	23	16.9389	15.0697	10.6964	8.0979
Poly[2,2-propane *bis*{4-(2,6-dibromophenyl)}carbonate]					
(Tetrabromobisphenol-A polycarbonate)	23	16.9389	18.3899	10.6964	9.7580
Cellulose tripropionate	23	16.9828	13.4812	10.9939	7.8512
Poly[di(*n*-heptyl) itaconate]	23	16.9684	15.0325	11.1430	9.4723
Poly[4,4-heptane *bis*(4-phenyl)carbonate]	23	16.2863	13.6719	11.1750	8.2847
Poly[1,1-(1-phenylethane) *bis*(4-phenyl)carbonate]	24	16.5708	13.2303	11.6588	7.8241
Poly(octadecyl methacrylate)	24	17.5124	16.5444	11.6885	10.8611
Poly[2,2'-(*m*-phenylene)-5,5'-bibenzimidazole] *(Celazole™,*					
see Fig. 5.1)	24	15.6730	12.6679	11.8316	7.9419
Poly[2,2-hexafluoropropane *bis*(4-phenyl)carbonate]	25	18.4579	12.1114	11.5537	6.7973
Poly[di(*n*-octyl) itaconate]	25	18.3826	16.4467	12.1430	10.4723
Poly(dicyclooctyl itaconate)	25	17.5373	15.6014	12.1783	10.5851
Poly(2,5-benzoxazolediylmethylene-5,2-benzoxazolediyl-1,3-phenylene)					
(Polyheteroarylene X)	25	16.3801	13.1915	12.3148	8.2155
Poly[1,1-cyclohexane *bis*{4-(2,6-dichlorophenyl)}carbonate]	26	18.4744	16.6053	12.4035	9.8050
Poly(quinoxaline-quinoxaline-*p*-phenylene)	26	17.0872	13.7171	12.8316	8.4828
Poly[1,1-(1-phenyltrifluoroethane) *bis*(4-phenyl)carbonate]	27	19.0708	13.8642	12.9089	8.1410
Poly(bisphenol-A terephthalate)	27	19.0183	15.0612	12.9307	8.6888
Poly[carbonylimino-(6-hydroxy-1,3-phenylene)-methylene-(4-hydroxy-1,3-phenylene)iminocarbonyl-1,3-phenylene]					
(Polyphenylenediamide)	27	18.9659	14.1915	12.9912	8.3600
Resin F *(See Fig. 2.4)*	27	19.0623	14.8119	13.0805	8.8184
Torlon™ *(See Fig. 2.4)*	27	18.5432	14.2299	13.0974	8.6211
Poly[di(*n*-nonyl) itaconate]	27	19.7968	17.8609	13.1430	11.4723
Poly(pentadecafluorooctyl acrylate)	28	22.7760	12.4775	11.9890	6.4904
Resin G *(Proprietary phenoxy-type thermoplastic)*	29	20.8552	17.1215	13.8794	10.0426
Poly[2,2-hexafluoropropane *bis*{4-(2,6-dibromophenyl)}carbonate]	29	21.9388	19.6578	13.1964	10.3918

☞ ☞ ☞   **TABLE 2.2 IS CONTINUED IN THE NEXT PAGE.** ☞ ☞ ☞

Table 2.2. CONTINUED FROM THE PREVIOUS PAGE.

Polymer	N	$^0\chi$	$^0\chi^v$	$^1\chi$	$^1\chi^v$
Poly[N,N'-(*p,p*'-oxydiphenylene)pyromellitimide]					
(*Kapton™, see Fig. 6.10*)	29	19.8611	14.7092	14.0249	8.8638
Poly[4,4'-diphenoxy di(4-phenylene)sulfone] (*Radel™*)	29	19.8467	15.4830	14.0757	9.4643
Poly[diphenylmethane *bis*(4-phenyl)carbonate]	29	19.6837	15.6171	14.2640	9.4848
Poly[*o*-biphenylenemethane *bis*(4-phenyl)carbonate] (*See Fig. 2.10*)	29	19.4241	15.4623	14.2640	9.5681
Poly[1,1-biphenylethane *bis*(4-phenyl)carbonate]	30	20.5539	16.5397	14.6252	9.8955
Poly[4,4'-sulfone diphenoxy di(4-phenylene)sulfone] (*Victrex™,*					
see Fig. 2.4, where two repeat units are used as a repeat unit, to					
make the difference between Radel, Udel and Victrex obvious)	32	22.3467	16.9119	15.3197	10.3267
Poly[4,4'-isopropylidene diphenoxy di(4-phenylene)sulfone] (*Udel™,*					
see Fig. 2.4)	32	22.3467	17.9830	15.3197	10.7143
Poly[oxy-(4-amino-1,3-phenylene)iminocarbonyl(4,9-dicarboxytricyclo[4.2.2.0^{2,5}]-7-decane-3,10-diyl)-carbonylimino-					
(6-amino-1,3-phenylene)] (*Polyheteroarylene XII*)	37	26.1705	19.3280	17.6508	11.8801
Poly(1,4-phenylene-1,3,4-oxadiazole-2,5-diyl-1,4-phenylene-2,1-benzo[c]furone-3,3-diyl-1,4-phenylene-					
1,3,4-oxadiazole-2,5-diyl) (*Polyheteroarylene XIV*)	38	25.2690	19.6595	18.6866	12.0147
Poly(pyroazolino[3,4-c]quinazoline-3,5-diyl-1,3-phenylenepyroazoline[3,4-c]quinazoline-5,3-diyl-1,4-phenylene)					
(*Polyheteroarylene VIII*)	38	24.7940	19.8153	18.8316	12.3279
Poly(hydrazocarbonyl-1,4-phenylene-2,1-benzo[c]furone-3,3-diyl-1,4-phenylenecarbonylhydrazoterephthaloyl)					
(*Polyheteroarylene XIII*)	40	27.8548	20.6871	19.3630	12.2203
Poly(imino-1,2-phenylene-4H-1,2,4-triazole-3,5-diyl-1,3-phenylene-4H-1,2,4-triazole-3,5-diyl-1,2-					
phenyleneiminoterepthaloyl) (*Polyheteroarylene IX*)	40	27.0535	20.8430	19.6193	12.5001
Ultem™ (*See Fig. 2.4*)	45	31.0344	24.2362	21.6847	14.4148
Poly[6,6'-*bis*(3,3'-diphenylquinoxaline)-2,2'-diyl-1,4-phenyleneoxy-1,4-phenylene]					
(*Polyheteroarylene XV*)	45	29.7437	24.0535	22.2474	14.8873

Table 2.3. Alternative $^0\chi^v$ and $^1\chi^v$ values obtained by replacement Si→C in hydrogen-suppressed graph of polymeric repeat unit and used in some correlations for silicon-containing polymers.

Polymer	$^0\chi^v$	$^1\chi^v$
Poly[oxy(methylhydrosilylene)]	1.9856	1.0488
Poly(dimethyl siloxane)	2.9082	1.4082
Poly[oxy(vinylmethylsilylene)]	3.1927	1.6052
Poly[oxy(diethylsilylene)]	4.3225	2.5296
Poly(vinyl trimethylsilane)	4.7845	2.6052
Poly(phenylmethylsilane)	4.8868	2.9107
Poly[oxy(methyl γ-trifluoropropylsilylene)]	4.9564	2.6823
Poly[oxy(methylphenylsilylene)]	5.2950	3.0689
Poly[oxy(mercaptopropylmethylsilylene)]	5.3712	3.2105
Poly(cyclohexylmethylsilane)	5.6129	3.8552
Poly[oxy(dipropylsilylene)]	5.7367	3.5296
Poly[oxy(methyl m-chlorophenylsilylene)]	6.3515	3.5466
Poly[oxy(methyl n-hexylsilylene)]	6.4438	3.9689
Poly[oxy(acryloxypropylmethylsilylene)]	6.6305	3.6557
Poly[ethylene-N-(β-trimethylsilyl ethyl)imine]	6.7756	3.8943
Poly[oxy(dicyanopropylsilylene)]	7.0453	4.2697
Poly(vinyl dimethylphenylsilane)	7.1712	4.2659
Poly(octylmethylsilane)	7.4497	4.8107
Poly[oxy(methyl m-chlorophenylethylsilylene)]	7.7658	4.7108
Poly[oxy(methyl n-octylsilylene)]	7.8580	4.9689
Poly(vinyl dimethylbenzylsilane)	7.8783	4.7230
Poly(vinylene diphenylsilylene)	8.4282	5.2321
Poly(diethnyl diphenylsilylene)	9.2735	5.5714
Poly[oxy(methyl n-tetradecyl silylene)]	12.1006	7.9689
Poly[oxy(methyl n-hexadecylsilylene)]	13.5149	8.9689
Poly[oxy(methyl n-octadecylsilylene)]	14.9291	9.9689

2.C. General Forms of the Correlations in Terms of Connectivity Indices

The general forms of the correlations will be presented in this section. Correlations having these general forms will be developed for the properties in later chapters. These correlations can be applied to all polymeric repeat units. It should be kept in mind, however, that a simple scheme for predicting the physical properties of polymers from their structures cannot obviate the user's need to have an understanding of materials. Simple correlation schemes are most useful if the user has some knowledge of polymer properties, and exercises good judgment in deciding which properties are relevant or even meaningful for a particular polymer or problem.

Certain physical properties of materials, such as the cohesive energy, molar volume, molecular weight per repeat unit, molar heat capacity, molar enthalpy and molar entropy, are *extensive properties*. An extensive property depends upon the size of the system. Its value increases in direct proportion to the amount of material present. For example, the molar volume increases with increasing total number of atoms in the repeat unit of the polymer, since the number of atoms per mole increases with the number of atoms per repeat unit. The molar volume correlates even more strongly with the number of non-hydrogen atoms in the repeat unit, or equivalently with the number of vertices (N) in its hydrogen-suppressed graph, than with the total number of atoms.

As can be seen from equations 2.4-2.7, from Table 2.2, and most dramatically from figures 2.5-2.8, the χ values are also extensive properties. They are sums over all vertices or edges of the hydrogen-suppressed graph. The number of terms in each summation increases in direct proportion to the size of the molecule or the polymeric repeat unit. This is the reason why the χ values are proportional to N to a good approximation. They are, therefore, logical choices of topological descriptors to correlate with extensive properties.

On the other hand, many important properties of materials are *intensive properties*. The values of intensive properties are essentially independent of the amount of material present, provided of course that this amount is not zero. An intensive property can usually be expressed in terms of the quotient of a pair of extensive properties. For example, the density equals the molecular weight per repeat unit divided by the molar volume. The solubility parameter equals the square root of the cohesive energy density (defined as the cohesive energy divided by the molar volume). As shown in Chapter 1, the glass transition temperature (an intensive property) can often be estimated in terms of the molar glass transition function divided by the molecular weight of a repeat unit of the polymer.

Intensive properties should not be directly correlated with an extensive property such as χ. It is much more reasonable to *scale* the values of the χ indices in an appropriate manner and convert them to a set of intensive indices to be used in developing correlations for intensive properties. *Scaling* is a familiar concept in polymer physics, and is used in many different contexts [3]. The very simple manner in which we will use a version of scaling in this manuscript is one more context for the application of this very useful concept.

The Greek letter ξ will be used for intensive values of the connectivity indices. It is desirable, for consistency, to remain within the framework of graph theory, by only using quantities defined in terms of the hydrogen-suppressed graph. The number N of non-hydrogen atoms in the system (i.e., the number of vertices in the hydrogen-suppressed graph of the system) will therefore be used as the scaling factor in defining the ξ indices in terms of the χ indices:

$$\xi \equiv \frac{\chi}{N} \tag{2.8}$$

Each ξ index has the same subscripts and superscripts as the corresponding χ index. For example, $^0\xi^v$ denotes $^0\chi^v/N$, and $^1\xi$ denotes $^1\chi/N$.

Equation 2.9 is the general form of the linear regression for correlating an extensive property with χ-type indices:

(Extensive property) $\approx \Sigma(a\chi)$

$\quad$ + (extensive structural parameters, and atomic or group correction terms) $\tag{2.9}$

Equation 2.10 is the general form of the linear regression for correlating an intensive property with ξ-type indices:

(Intensive property) $\approx \Sigma(b\xi)$

$\quad$ + (intensive structural parameters, and atomic or group correction terms) + c $\tag{2.10}$

The set of linear regression coefficients (a or b), the correction terms (if needed), and the constant c in Equation 2.10, are adjustable parameters. The χ and ξ values are exactly determined from the hydrogen-suppressed graph of the repeat unit. There is no additive constant term in Equation 2.9 because the value of a constant does not change as a function of the amount of material present. A constant is therefore an intensive property which does not belong in correlations for extensive properties. For example, so long as there is *some* material present, the density (an intensive property) has the same nonzero and constant value. On the other hand, the total volume becomes infinitesimally small (approaches zero) in the limit of an exceedingly small amount of material. The omission of the constant in Equation 2.9 is therefore essential to prevent the introduction of a computational artifact into the correlations for extensive properties.

The values of the properties which will be fitted by using equations 2.9 and 2.10 will be selected from available and apparently reliable experimental data whenever there are sufficient amounts of such data. Some important properties of polymers, such as the van der Waals volume (Chapter 3) and the cohesive energy (Chapter 5), are not directly observable. They are inferred indirectly, and often with poor accuracy, from directly observable properties such as the molar volume (or equivalently the density) and the solubility behavior. When experimental data are unavailable or unreliable, the values of the properties to be fitted will be estimated by using group contributions. The predictive power of such correlations developed as direct extensions and generalizations of group contribution techniques will then be demonstrated by using them to predict the value of the property of interest for polymers for which some of the required group contributions are not available to predict the value of the property.

Some properties (such as the heat capacity, which is discussed in Chapter 4) are best correlated with a combination of connectivity indices and geometrical parameters such as the "number of rotational degrees of freedom" N_{rot}. The systematic variation of the heat capacity with the polymeric structure contains both a "global" component which is accounted for by the connectivity indices, and a "local" component strongly dependent upon local conformational degrees of freedom which is accounted for by the geometrical parameters. N_{rot} is an example of a *structural parameter*, i.e., a parameter which describes the presence of certain structural features of great generality, independent of the atoms or groups which constitute that structural feature in a particular repeat unit. Another example of a structural parameter is the total number of fused rings in the repeat unit, regardless of the types or atomic compositions of these fused rings.

The *correction terms* mentioned in equations 2.9 and 2.10 depend on the property being correlated. Each correction term can be of either one of two general types:

1. An *atomic correction term* refers to the number of atoms of a given electronic configuration in a repeat unit. For example, the term $5N_{Si}$, i.e., five times the total number of silicon atoms in the repeat unit, is an atomic correction term. The use of atomic correction terms can only be avoided by following the very cumbersome procedure (see Section 3.B.2) of treating the value of δ^v for each electronic configuration of each atom as a separate fitting variable in the correlation for each property instead of using the δ^v values listed in Table 2.1.

2. A *group correction term* refers to the number of functional groups of a given type in the repeat unit. For example, the term $5N_{amide}$, i.e., five times the total number of amide groups in the repeat unit, is a group correction index. Treating the value of δ^v as a separate fitting variable while using only zeroth-order and first-order connectivity indices is usually insufficient to correct for structural features requiring group correction terms. Some such structural features could be corrected for by introducing higher-order connectivity indices. On the other hand, the effects of certain structural features on some properties are too strong and/or unique to be fully taken into account by using connectivity indices of any order.

Structural parameters, atomic correction terms and group correction terms are used whenever neeeded, in order to correct for or to account for any combination of the following circumstances:

1. Structural features, such as the difference between the effects of *meta* and *para* substitution on a phenyl ring, or the effect of torsional degrees of freedom on the heat capacity, which cannot be accounted for in terms of zeroth-order and first-order connectivity indices alone. Such features require higher-order indices, and especially higher-order path indices [1,2] whose calculation is far more complicated for polymer chains than for simple molecules.

2. Structural features (such as *cis* versus *trans* isomerization around a double bond, and different tacticities) which cannot be accounted for at all, by the standard types of connectivity indices.

3. Structural features whose effects in determining a given physical property are overestimated or underestimated, either as a result of the omission of higher-order indices, or because of the intrinsic limitations of a purely topological approach. An example (Chapter 5) is the gross underestimation of the effect of the presence of hydrogen-bonding structural units (such as hydroxyl, amide and urea groups) in the repeat unit on the cohesive energy.

    This technique is equivalent to the use of a "hybrid approach" in which the properties of polymers are predicted by the summation of additive contributions over atoms and bonds, supplemented by other structural, atomic or group contributions when appropriate. The values of the atom and bond contributions are dependent on the nature of the environment of each atom and bond in a particularly simple manner. Group contributions are not required; however, a small number of group contributions is sometimes used to improve a correlation.

    Since the use of more generalized types of variables will improve the quality of the predictions made for polymers with a greater diversity of structural features, structural parameters are preferred over both atomic and group correction terms, and atomic correction terms are in turn preferred over group correction terms, when used to supplement the zeroth-order and first-order connectivity indices. The correlations developed in this work are of reasonable to excellent quality (i.e., have acceptable to very high correlation coefficient and low standard deviation) *before including group correction terms*. Any group correction terms used in these correlations improve the quality of correlations which are already quite accurate with the use of only much more general variables. This approach is one of the reasons for the robustness of our method.

    The use of χ or ξ values in correlating polymeric properties via equations 2.9 and 2.10 is a *direct* method of developing correlations. An equally valid alternative method is to correlate properties with connectivity indices *indirectly*, by utilizing correlations in terms of connectivity indices to estimate certain "fundamental" properties, and then substituting the values of these properties into equations defining additional "derived" properties in terms of the fundamental properties. The choice of method for a given property will be based on the consideration of three factors, namely expediency, expected degree of accuracy, and the amount of fundamental insight obtained from each approach.

    A few examples of the use of the different methods are listed below:

1. The van der Waals volume and the cohesive energy will be correlated directly with χ-type connectivity indices.

2. The solubility parameter will be estimated indirectly, by combining the correlations for the cohesive energy and the molar volume.

3. The molar heat capacity will be correlated directly with a combination of χ-type connectivity indices and the rotational degrees of freedom parameters of the backbones and the side groups of the polymer chains.

4. The refractive index measured under "standard" conditions (with commonly used wavelengths of light, at room temperature) will be correlated directly with ξ-type connectivity indices.

5. The refractive index as a function of the temperature or percent crystallinity will be correlated indirectly with connectivity indices, in terms of an equation utilizing the molar refraction along with the molar volume.

Extensions made in the formalism of connectivity indices in order to treat certain properties of polymers which depend upon well-defined structural and/or topological features not accounted for either by the conventional formalism [1,2] or by the extensions we have outlined thus far will be described in the next three subsections.

2.D. Backbone and Side Group Portions of the Connectivity Indices

Define a *locally anisotropic* property as a property whose value is highly sensitive to whether each atom or bond is in the chain backbone or in a side group. By contrast, while the value of a *globally anisotropic* property may be very sensitive to the overall orientation of the polymer chains, it is somewhat less sensitive to the precise location of any given atom or bond in the structure. For example, the refractive index is a globally anisotropic property. The stress-optic coefficient, whose value depends both on the refractive index and on the difference in the polarizability of a polymer chain segment parallel and perpendicular to the chain, is a locally anisotropic property. Similarly, the thermal conductivity is a locally anisotropic property, since the coupling of the vibrational modes which results in heat transfer is much more effective along the chain backbone than perpendicular to it. (The refractive index and the thermal conductivity will both be discussed in greater detail later in this monograph.)

In order to treat locally anisotropic properties by using connectivity indices, it is necessary to distinguish between the contributions made to the connectivity indices by the *chain backbone* (BB), and by the *side groups* (SG) such as side chains (SC), branches and/or dead ends. The backbone of a polymeric repeat unit can be rigorously identified from the structure of the repeat unit, by using concepts and techniques borrowed and adapted from percolation theory and fractal mathematics [4,5]. If the repeat unit is viewed as a percolation path between its own two "ends", then the portions of the repeat unit that are connected to the percolation path by only a single bond are the side groups. The contribution of the bond connecting the backbone to a side group, as well as the contributions of all of the bonds and atoms in the side groups, are added to the SG component of the appropriate connectivity index. The contributions of backbone atoms and of all bonds connecting two backbone atoms to one another are added to the BB component of the appropriate connectivity index.

See Figure 2.9 for an illustration, using poly(vinyl butyral) as an example. Note the similarity to the familiar example of the daily route of a school bus. Imagine that each vertex in the hydrogen-suppressed graph is a stop at which the bus picks up children. Provided that the value of δ^v is not fractional, δ^v may be thought of as the number of children to be picked up at the stop corresponding to a given vertex. (All analogies have their limits!) Each edge is the portion of the road between two successive stops. The bus can proceed from its point of departure (the garage, at the left end of the graph) to its destination (the school, at the right end) without retracing its path only if it remains on the backbone. On the other hand, if it goes into the "dead end" representing the aliphatic chain segment dangling from the ring in the repeat unit to pick up children, it has to retrace its path to reach the school. Note that a single atom attached to the backbone is a "side group" with a single link as defined here. For example, the fluorine atom in poly(vinyl fluoride) (Figure 2.3) contributes to the SG components of the zeroth-order indices, and the C-F bond contributes to the SG components of the first-order indices.

The definition given above for the BB and SG components of a polymeric repeat unit is rigorous and unambiguous, but it is *not* all-inclusive. There is a small number of borderline cases, namely polymers whose repeat unit backbone includes a disubstituted atom of a ring. See Figure 2.10 for an example. Our calculations suggest that such ring structures (of course, with the exception of the disubstituted atom) should be considered as side groups.

2.E. Shortest Path Across the Backbone of a Polymeric Repeat Unit

When we travel between two cities, we are often interested in finding out about all of the possible routes to our destination. Some people might prefer to drive the shortest distance possible, others to take the most scenic route, or the most fuel-efficient route, or the road with the fewest number of detours and turns, or the road with the least restrictive speed limits. The number of cars arriving in Los Angeles starting from New York on any given day is the sum of the numbers of cars which were driven over *all* of these possible routes and their combinations. On the other hand, certain routes are far more likely to be followed than some others.

Similarly, in the electrical circuit analogy [4,5] which forms the basis of our definition of the backbone and the side groups of a polymeric repeat unit, all paths are not equally effective in the transport of electrical current between the two ends (A and B) of a complicated electrical circuit. For example, if the resistances of all of the wires making up the circuit are equal, a larger fraction of the current (starting at End A and reaching End B) at any instant is then likely to have taken the

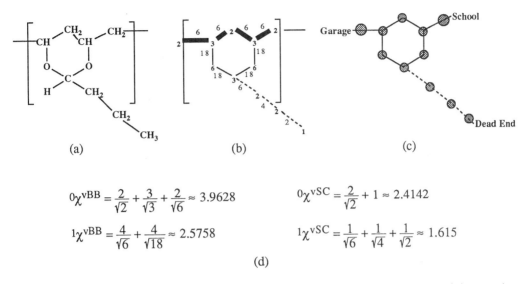

$$0\chi^{vBB} = \frac{2}{\sqrt{2}} + \frac{3}{\sqrt{3}} + \frac{2}{\sqrt{6}} \approx 3.9628 \qquad 0\chi^{vSC} = \frac{2}{\sqrt{2}} + 1 \approx 2.4142$$

$$1\chi^{vBB} = \frac{4}{\sqrt{6}} + \frac{4}{\sqrt{18}} \approx 2.5758 \qquad 1\chi^{vSC} = \frac{1}{\sqrt{6}} + \frac{1}{\sqrt{4}} + \frac{1}{\sqrt{2}} \approx 1.615$$

(d)

Figure 2.9. Calculation of the backbone (BB) and side group (SG) components of the zeroth-order and first-order valence connectivity indices, using poly(vinyl butyral) as an example.
(a) Structure of the repeat unit.
(b) Hydrogen-suppressed graph with δ^v values of the atoms at the vertices and β^v values of the bonds along the edges. All solid lines lie on paths which traverse the backbone. The shortest path across the backbone is indicated by thick lines. The side group is indicated by dashed lines. (Simple connectivity indices can be calculated by using δ and β instead of δ^v and β^v.)
(c) Analogy to the route of a school bus.
(d) Calculation of the BB and SG components of $0\chi^v$ and $1\chi^v$. The side group in this polymer is a "side chain", i.e., a series of divalent "links" which is attached to the backbone at one end and terminated by a univalent (dangling) "link" at the other end. It is therefore represented by the abbreviation SC instead of SG below.

shortest path than the longest path. It is therefore important to know not only the structure of the complete backbone but also the length of the shortest path between its two ends.

This situation also holds in assigning quantitative descriptors to polymeric repeat units for the purpose of predicting certain locally anisotropic physical properties. Some locally anisotropic properties of polymers are more sensitive to the total number N_{BB} of atoms on the chain backbone. Some other locally anisotropic properties are more sensitive to the length of the shortest path, namely the number of atoms N_{SP} between the two ends. The following inequality is a direct consequence of the definitions of N_{BB} and N_{SP}:

$$N_{SP} \le N_{BB} \qquad\qquad\qquad (2.11)$$

$N_{SP}=N_{BB}$ for all polymers with a single path across their backbones. The largest general class of polymers with $N_{SP}=N_{BB}$ are polymers with a vinylic backbone (see Figure 2.3 for an example), which have only one path containing two atoms across their backbones, so that $N_{SP}=N_{BB}=2$. $N_{SP}<N_{BB}$ for polymers with more than one possible path across their backbones. For example, $N_{SP}=4$ and $N_{BB}=7$ for poly(vinyl butyral) (Figure 2.9).

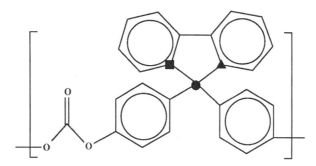

Figure 2.10. The repeat unit of poly[o-biphenylenemethane bis(4-phenyl)carbonate] contains a structural unit which is a borderline case between a true backbone component and a side group. The carbon atom indicated by a dark circle (●) occupies a unique location. Assume that, during the journey from one end of the repeat unit to the other end, we enter the large ring structure above this atom by taking a step either towards the atom marked by a dark square (■) or towards the atom marked by a dark triangle (▲). Since there are *two* bonds linking this large ring structure to the rest of the repeat unit, we could get back to the atom marked by ● without retracing our steps, for example by proceeding clockwise or counterclockwise through the large ring structure. On the other hand, there would be no way to get back on track, heading towards the opposite end of the repeat unit, without returning to the "traffic circle" represented by the atom labeled ●. Is the large ring structure in this repeat unit a part of the backbone, or a side group? It is connected to the rest of the repeat unit by two bonds (as in the backbone), but we can only exit from it by re-visiting a location we have been to before (as in a side group).

2.F. Extensions for the Calculation of Some Conformation-Related Properties

In the past, connectivity indices have found almost no use in calculations of molecular properties whose values are significantly affected by rotations of molecular fragments around bonds [1,2]. They have thus been commonly believed to be useless in predicting such properties.

Our work has shown, however, that it is possible to correlate certain conformation-related properties of polymer chains either directly or indirectly with appropriately chosen or defined connectivity indices. Such conformation-related properties include the steric hindrance parameter σ and the characteristic ratio C_∞, which are best measured in dilute solutions of the polymer in weak solvents under "theta conditions".

For example, σ and C_∞ are very sensitive functions of the ease of rotational or oscillational motions around backbone atoms (i.e., the "stiffness" of the backbone bonds), and the typical bond lengths and bond angles along the backbone. Specialized indices were defined and used to correlate the values of σ and C_∞ with the structure of the repeat unit of the polymer. The detailed discussion of these specialized indices will be postponed to Chapter 12, where the correlation for σ will be developed.

It must be emphasized that the standard connectivity indices are indeed incapable of correlating or predicting these conformational properties. It is the flexibility of not restricting ourselves just to these standard indices that has allowed us to develop the correlations for σ and C_∞.

While we have thus been able to correlate certain conformation-related properties of polymers with connectivity indices, we have not been able to do so for many other conformation-related properties. For example, the very complex nature of the stress-optic coefficient defies a simple treatment. Such complicated conformation-related properties are still best predicted via more sophisticated calculations [6-9], such as those using rotational isomeric state theory.

References and Notes for Chapter 2

1. L. B. Kier and L. H. Hall, *Molecular Connectivity in Chemistry and Drug Research*, Academic Press, New York (1976).

2. L. B. Kier and L. H. Hall, *Molecular Connectivity in Structure-Activity Analysis*, John Wiley & Sons, New York (1986).

3. P.-G. deGennes, *Scaling Concepts in Polymer Physics*, Cornell University Press, Ithaca, New York (1979).

4. R. Zallen, *The Physics of Amorphous Solids*, John Wiley & Sons, New York (1983).

5. J. Feder, *Fractals*, Plenum Press, New York (1988).

6. P. J. Flory, *The Principles of Polymer Chemistry*, Cornell University Press, Ithaca, New York (1953).

7. P. J. Flory, *Statistical Mechanics of Chain Molecules*, Interscience Publishers, New York (1969).

8. A. E. Tonelli, article titled "*Conformation and Configuration*", *Encyclopedia of Polymer Science and Engineering*, *4*, Wiley-Interscience, New York (1986).

9. P. J. Flory, *Macromolecules*, *7*, 381-392 (1974).

CHAPTER 3

VOLUMETRIC PROPERTIES

3.A. Background Information

The space occupied by one mole of a material is the *molar volume* V of that material. One mole of a polymer contains Avogadro's Number N_A (i.e., $6.022169 \cdot 10^{23}$) of repeat units of the polymer. It is sometimes more useful to consider the *specific volume* v, defined as the volume per unit weight, or the *density* ρ, defined as the weight per unit volume. If M denotes the molecular weight of one mole of repeat units of the polymer, then the specific volume v and the density ρ are defined as follows in terms of V and M:

$$v \equiv \frac{V}{M} \tag{3.1}$$

$$\rho \equiv \frac{M}{V} = \frac{1}{v} \tag{3.2}$$

The molecular weight per polymeric repeat unit is determined exactly from the structure of the polymeric repeat unit. The specific volume and the density are therefore both known if the molar volume is known.

The molar volume is a function of the temperature T. It normally increases with increasing temperature as a result of the increasing atomic motions resulting from the added energy. Exceptions to this rule do exist. For example, when ice melts, the liquid water immediately above the melting temperature occupies less space than the more "open" structure imposed on ice by the highly directional hydrogen bonds frozen into a crystalline lattice. Thermal expansion is therefore not a fundamental law of nature, but merely the most commonly observed net result of the combined effects of the many fundamental physical processes which take place when the temperature is increased.

The *coefficient of volumetric thermal expansion* α is defined as the fractional rate of change of V(T) as a function of T, and can be best estimated from the dependence of V on T if the functional form of V(T) is known:

$$\alpha(T) \equiv \frac{1}{V(T)} \left[\frac{\partial V(T)}{\partial T} \right] \tag{3.3}$$

In Equation 3.3, ∂ denotes a partial derivative. V is also a function of the pressure. The pressure dependence of V is usually taken into account by using a "thermodynamic equation of state" which describes the behavior of V as a function of the temperature and the pressure simultaneously. This topic will be addressed briefly in Section 3.E.

The *coefficient of linear thermal expansion* β, which is another useful quantity commonly quoted in the literature, simply equals one third of the coefficient of volumetric thermal expansion for an isotropic (unoriented) polymer:

$$\beta \equiv \frac{\alpha}{3} \tag{3.4}$$

The molar volume of a material is the sum of three components:
1. Space truly occupied by its atoms, often called the *van der Waals volume* V_w. More formally, the V_w of a molecule is defined as the space occupied by this molecule, which is impenetrable to other molecules with normal thermal energies corresponding to ordinary temperatures [1].
2. The amount of additional "empty" space, i.e. the *packing volume*, taken up due to packing constraints imposed by the sizes and shapes of the atoms or molecules which constitute the material. The packing volume is equal to the difference between the molar volume at absolute zero temperature and the van der Waals volume.
3. The *expansion volume* resulting from thermal motions of atoms is the difference between the molar volume at the temperature of interest and the molar volume at absolute zero temperature.

According to classical thermodynamics, the degrees of motional freedom resulting in thermal expansion are all frozen at absolute zero temperature (T=0K). They gradually become available with increasing T, so that α slowly increases with T. At extremely low temperatures (very close to absolute zero), quantum mechanical fluctuations and the resulting "zero-point vibrations" attributed to the validity of the Heisenberg uncertainty principle modify this simple classical physical picture of thermal expansion. The most important physical factors determining the dependence of α and V on T in various temperature regimes [2-10] are as follows:
1. Quantum mechanical fluctuations play the key role in the immediate vicinity of T=0K.
2. Harmonic vibrational modes dominate at low temperatures above this immediate vicinity.
3. Three factors generally become increasingly important with increasing temperature:

(a) "Hard sphere" repulsions prevent atoms from occupying the same space, while atoms can move arbitrarily far away from each other. Consequently, there is an inherent asymmetry (anharmonicity) in the vibrational modes. This anharmonicity increases with increasing temperature. Anharmonicities of the vibrational modes cause thermal expansion by increasing the average interchain distance with increasing temperature.

(b) "Frozen-in" and "dynamic" components of the entropy. According to the third law of thermodynamics, the entropy of a perfect crystal at absolute zero temperature equals zero. On the other hand, because of the disordered spatial arrangement of the atoms, some entropy (disorder) is "frozen into" an amorphous material even at T=0K. Atomic motions increase with increasing temperature, and the entropy increases as a result of these motions.

(c) Secondary (sub-T_g) relaxations, where T_g denotes the glass transition temperature. These relaxations signal the inception of certain relatively localized motions of chain segments or of side groups, resulting from an increase of the thermal energy sufficient to overcome the activation energies for such motions.

4. The amount of "free volume". Free volume can be "frozen into" a material as a result of the slowing down of the molecular-level relaxation processes as T decreases significantly below T_g. In addition, the free volume increases with increasing T as a result of thermal expansion. The free volume can therefore play a role in thermal expansion processes both at low and at high temperatures. The detailed discussion of free volume is outside the scope of this book. See Robertson's excellent review article [10] for a thorough discussion of this concept.

5. Larger-scale cooperative motions of chain segments become possible at the glass transition temperature, and the value of α manifests a discontinuous increase, along with a similar discontinuous increase in the heat capacity (discussed in Chapter 4).

Equation 3.5 was shown [11] to provide a reasonable correlation for the V(T) of an amorphous polymer or of the amorphous phase of a semicrystalline polymer in terms of V_w and T_g over a very wide temperature range below T_g:

$$V(T) \approx V_w \left[1.42 + 0.15\left(\frac{T}{T_g}\right) \right] \tag{3.5}$$

Insertion of Equation 3.5 in Equation 3.3 gives Equation 3.6, which is a good approximation for the coefficient of thermal expansion over the temperature range of 150K≤T≤T_g [11]:

$$\alpha(T) \approx \frac{1}{(T + 9.47T_g)} \tag{3.6}$$

The limits of the temperature range for the use of Equation 3.6 are imposed by the fact that it suggests that $\alpha(T)$ very slowly decreases rather than very slowly increasing [2] with increasing T. Since the $\alpha(T)$ in the solid state increases most rapidly at low temperatures, the lower limit of 150K is imposed. The value of $\alpha(T)$ for glassy polymers at T<150K can be estimated by assuming that $\alpha(0)=0$ and making use of Bueche's observation [2] that $\alpha(T)$ increases approximately proportionally to the square root of T.

The fact that Equation 3.5 is safely applicable over a much larger temperature range than Equation 3.6 is not surprising. V(T) varies much more smoothly than the differential property $\alpha(T)$ with temperature, and manifests gradual changes of slope rather than sudden increases, allowing its approximation by a straight line with reasonable accuracy.

Let α_g denote the value of $\alpha(T)$ in the "glassy" state just below the inception of the glass transition. Let α_r denote the value of $\alpha(T)$ in the "rubbery" state just above the termination of the glass transition. The Simha-Boyer relationship [4], which is given by Equation 3.7, can then be used in conjunction with Equation 3.6, to extrapolate from $T<T_g$ to $T>T_g$ and obtain rough predictions of the coefficient of thermal expansion above T_g. A much older correlation (Equation 3.8) discovered by Boyer and Spencer [3] is also useful, since it provides a rough estimate of α_r directly from a knowledge of T_g without the need to estimate α_g first:

$$(\alpha_r - \alpha_g) \cdot T_g \approx 0.113 \tag{3.7}$$

$$\alpha_r \cdot T_g \approx 0.164 \tag{3.8}$$

Equations 3.5-3.8 are very useful in estimating the molar volumes, densities (in conjunction with Equation 3.2) and thermal expansion coefficients of amorphous polymers and of amorphous phases of semicrystalline polymers. Their applicability is mainly limited by the availability of reasonable estimates for V_w and T_g. Coefficients of thermal expansion are listed in Table 3.1 for a number of polymers, both below and above T_g, from a tabulation by Seitz [11].

Table 3.1. Coefficients of thermal expansion, in units of 10^{-6}/K, for glassy polymers (α_g) and for rubbery polymers (α_r) [11].

Polymer	α_g	α_r	Polymer	α_g	α_r
Polyisobutylene	144	586	Poly(p-t-butyl styrene)	258	590
Poly(p-chloro styrene)	145	497	Poly(n-butyl acrylate)	260	600
Poly(o-methyl styrene)	159	378	Bisphenol-A polycarbonate	265	535
Poly(ethylene terephthalate)	162	442	Poly(methyl acrylate)	270	560
Poly(vinyl chloride)	175	485	Poly(ethyl acrylate)	280	610
Poly(ethylene isophthalate)	200	455	Poly(vinyl ethyl ether)	303	726
Poly(1,4-butadiene)	200	705	Poly(ethyl methacrylate)	309	540
Polyethylene	201	531	Polypropylene	343	850
Poly[oxy(2,6-dimethyl-1,4-phenylene)]	204	513	Poly(n-propyl methacrylate)	363	575
Poly(vinyl acetate)	212	583	Poly(vinyl n-hexyl ether)	375	660
Poly(methyl methacrylate)	213	490	Poly(4-methyl-1-pentene)	383	761
Poly(vinyl methyl ether)	216	645	Poly(vinyl n-butyl ether)	390	726
Poly(styrene-co-acrylonitrile) (76/24 by weight)	227	487	Poly(n-butyl methacrylate)	412	605
Poly(α-methyl styrene)	240	540	Poly(n-octyl methacrylate)	415	600
Polystyrene	250	550	Poly(n-hexyl methacrylate)	440	680

A correlation will be developed for V_w, and its predictive power will be shown, in Section 3.B. The process by which this correlation was developed, and the significance of the terms in it, will also be discussed in detail.

The prediction of T_g for a polymer of arbitrary structure is often difficult. Van Krevelen's method [1], summarized in Chapter 1, is limited by the lack of group contributions for many structural units used in new polymers. More general alternative methods for the prediction of T_g are often marred by serious accuracy and reliability problems when applied to complicated polymeric structures [12]. A new correlation will be developed for T_g in Chapter 6.

It is, however, still desirable to develop an alternative correlation, which does not require T_g to be predicted first, for the volumetric properties. A new correlation will therefore be developed for the molar volume (or equivalently, for the density) at room temperature, in terms of connectivity indices, in Section 3.C. Different authors define temperatures ranging from 293K to 300K as "room temperature". T=298K will be accepted as "room temperature" in this book. The molar volume at room temperature will be denoted by V(298K).

The combination of Equation 3.5 with this new correlation will provide an alternative expression for V(T). V(T) in this new relationship will have the same functional form as in Equation 3.5. The new equation, however, will have the important advantage of incorporating the effects of differences between the frozen-in packing volumes of different polymers at absolute zero temperature, and will therefore often provide an improvement over Equation 3.5. The final forms of the correlations we found to be most useful for describing the temperature dependences, both below and above T_g, of the volumetric properties, will be summarized in Section 3.D.

The effects of pressure on the volumetric properties will be discussed briefly in Section 3.E.

Kier and Hall [13,14] showed that the molar volumes of ordinary molecules, and other properties such as the density which are defined in terms of the molar volume, correlate well with connectivity indices. They also found [14] that the Bondi atom and group contribution values to V_w decrease linearly with increasing $(\delta+\delta^v)$ for simple atoms and their hydrides. They have used this correlation to explain the relationship between the molar volume and the connectivity indices. It is therefore not surprising that, in the next two sections, the zeroth-order and first-order connectivity indices will be found to correlate very well with V_w and V for polymers.

3.B. Correlation for the van der Waals Volume

3.B.1. Development of the Correlation

The van der Waals volume can be estimated by assuming that the impenetrable volume of a molecule is bounded by the outer surface of a number of interpenetrating spheres [1]. The radii of the spheres are assumed to be the constant (standard) atomic radii of the elements involved. The distances between the centers of the spheres are assumed to be the constant (standard) bond lengths for each type of bond involved. The contribution of each atom to V_w can then be estimated by using equations based on solid geometry. These atomic contributions can be added up to estimate the total V_w [1]. This is obviously a very tedious procedure. Several group contribution tables have hence been developed for the contributions of structural units to V_w [1].

A new correlation in terms of connectivity indices will now be developed for V_w, by fitting these indices to V_w values calculated for a large and structurally diverse set of polymers via group contributions. The advantage of this new correlation is its much greater generality, i.e., its applicability to the many polymers whose V_w cannot be calculated by group contributions because the contribution of one or more of their structural units to V_w is unknown.

V_w(group) was calculated for 110 of the polymers listed in Table 2.2, by using a combination of the information contained in different group contribution tables for V_w [1,15,16]. These V_w(group) values were fitted by linear regression equations in terms of connectivity indices. V_w(group) had a correlation coefficient of 0.9930 with $^1\chi^v$, and a correlation coefficient of 0.9876 with $^0\chi$. $^1\chi^v$ by itself was sufficient to provide a good estimate of V_w; however, the correlation was improved considerably by adding a second term proportional to $^0\chi$.

A linear regression with a weight factor of $100/V_w$(group) for each term in the sum of squares of the deviation of V_w(fit) from V_w(group) in the regression resulted in a correlation of higher quality, with a lower standard deviation and smaller relative (fractional) errors than a simple (unweighted) regression. A two-parameter regression gave a standard deviation of only 4.5 cc/mole and a correlation coefficient of 0.9957:

$$V_w \approx 3.861803 \cdot {}^0\chi + 13.748435 \cdot {}^1\chi^v \tag{3.9}$$

According to the rules of statistical analysis, the fraction of the variation of a property accounted for by a correlation is equal to the square of the correlation coefficient. The square of the correlation coefficient is hence sometimes referred to as the "coefficient of determination". A correlation coefficient of 0.9957 indicates that Equation 3.9 accounts for approximately 99.15% of the variation of the V_w values. The addition of extra terms proportional to $^0\chi^v$ and $^1\chi$ did not result in statistically significant improvements. The remaining small deviation could therefore not be corrected by using only zeroth-order and first-order connectivity indices.

A correlation coefficient of 0.9957 obtained for a very diverse set of 110 polymers by using only two adjustable parameters and remaining completely within the formalism of connectivity indices essentially guarantees that Equation 3.9 will give reasonable predictions for the V_w of all polymers, including polymers containing structural units not found in the set of test cases.

It is, nonetheless, tempting to note that most of the remaining deviation between V_w(group) and V_w(fit) is systematic, and that it can be corrected by using a "hybrid" approach in which a "correction index", consisting of a short sum of structural parameters, atomic correction terms and group correction terms, is used along with the connectivity indices. Such a linear regression resulted in a standard deviation of only 1.6 cc/mole (the standard deviation is only 1.8% of the average V_w of 87.3 cc/mole for this set of polymers) and a correlation coefficient of 0.9993:

$$V_w \approx 2.286940 \cdot {}^0\chi + 17.140570 \cdot {}^1\chi^v + 1.369231 \cdot N_{vdW} \tag{3.10}$$

$$N_{vdW} \equiv N_{menonar} + 0.5N_{mear} + N_{alamid} + N_{OH} + 2N_{cyanide} - 3N_{carbonate} - 4N_{cyc}$$
$$- 2.5N_{fused} + 2N_{C=C} + 7N_{Si} - 8N_{(-S-)} - 4N_{Br} \tag{3.11}$$

When Equation 3.10 is used to predict V_w, $^1\chi^v$ should be evaluated with all silicon atoms replaced by carbon atoms in the repeat unit. For example, the $^1\chi^v$ values listed in Table 2.3 should be used instead of the values listed in Table 2.2 for the silicon-containing polymers whose connectivity indices were listed in Chapter 2.

N_{vdW} corrects for the underestimation or overestimation of the contribution of certain structural units to V_w when $^0\chi$ and $^1\chi^v$ are correlated with V_w(group). The correction terms entering Equation 3.11 have the following meanings:

1. $N_{menonar}$ is the number of methyl groups attached to non-aromatic atoms.

2. N_{mear} is the number of methyl groups directly attached to atoms in aromatic rings.

3. N_{alamid} is the total number of linkages between amide (-CONH-) and similar (such as urea) groups and non-aromatic atoms. If an amide group is attached to non-aromatic atoms on both sides [as in poly(ϵ-caprolactam)] it contributes 2 to N_{alamid}. If it is attached to a non-aromatic atom on one side and an aromatic atom on the other side [as in poly(hexamethylene isophthalamide)] it contributes 1. If it is attached to aromatic atoms on both sides [as in poly(*p*-phenylene terephthalamide)] it does not contribute anything.

4. N_{OH} is the total number of -OH groups. These -OH groups could, for example, be in alcohol, phenol, carboxylic acid (-COOH) or sulfonic acid (-SO$_3$H) bonding environments.

5. $N_{cyanide}$ is the number of -C≡N groups.

6. $N_{carbonate}$ is the number of carbonate (-OCOO-) groups.

7. N_{cyc} is the number of non-aromatic rings (i.e., "cyclic" structures) with no double bonds along any of the edges of the ring. For example, poly(maleic anhydride), poly(N-methyl glutarimide) and poly(N-phenyl maleimide) in Figure 2.4 have N_{cyc}=1, while all other polymers shown in Figure 2.4 have N_{cyc}=0. Poly(vinyl butyral) (Figure 2.9) has N_{cyc}=1. See Figure 3.1 for an illustration of how N_{cyc} should be calculated for more complicated ring structures. Note that N_{cyc} can be fractional in structures in which several non-aromatic rings not containing any double bonds share edges, since it is necessary both to count all rings and to avoid the multiple counting of the edges of each ring which are shared with other rings.

8. N_{fused} is the number of rings in "fused" ring structures, defined in the present context as any ring structure containing at least one aromatic ring which shares at least one edge with another ring, and all of the other rings with which it shares an edge. Note that we are using the term "fused ring" in a much looser sense here than it is normally used in organic chemistry, where a fused ring structure is defined as one in which all of the edge-sharing rings are aromatic. For example, Torlon in Figure 2.4 has $N_{fused}=2$, Ultem has $N_{fused}=4$, and all other polymers shown in Figure 2.4 have $N_{fused}=0$ according to our definition. The repeat unit shown in Figure 2.10 has $N_{fused}=3$. See Figure 3.1(c) for another example.

9. $N_{C=C}$ is the number of carbon-carbon double bonds in the repeat unit, excluding any such bonds found along the edges of rings. For example, the purely formal double bonds in valence bond representations of aromatic rings, as well as carbon-carbon double bonds in any other type of ring structure, are *not* counted in $N_{C=C}$.

10. N_{Si}, $N_{(-S-)}$ and N_{Br} denote the numbers of silicon atoms, sulfur atoms in the lowest (divalent) oxidation state, and bromine atoms, respectively. The contribution of silicon atoms to V_w was estimated by using an "internal consistency" condition between the key correlations, namely by requiring that the value of V_w calculated by using equations 3.10 and 3.11, when inserted into the equations for the molar volume, should provide reasonable agreement with the observed densities [17] of silicon-containing polymers.

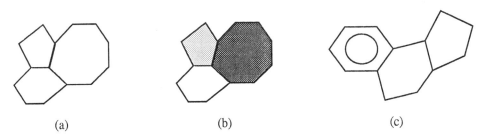

(a)   (b)   (c)

Figure 3.1. Examples of calculation of N_{cyc} and N_{fused} for some complicated ring structures.
(a) Schematic illustration. Three nonaromatic rings not containing any double bonds share edges.
(b) The largest ring (darkly shaded) is counted first, and contributes 1 to N_{cyc}. The second largest ring (unshaded) has six edges, one of which is shared with the largest ring and was already counted. This ring therefore contributes only 5/6 to N_{cyc}. The smallest ring (lightly shaded) has five edges, two of which were already counted, so that it contributes only 3/5 to N_{cyc}. The total $N_{cyc} = (1 + 5/6 + 3/5) \approx 2.4$ for the structure.
(c) An aromatic ring shares an edge with a nonaromatic ring ($N_{fused}=2$ as defined above). There is an additional nonaromatic ring not containing any double bonds and not sharing any edges with the aromatic ring ($N_{cyc}=1$).

See Section 3.B.2 for a more detailed discussion of the terms entering N_{vdW}, especially from the perspective of understanding the distinctions made between structural parameters, atomic correction terms and group correction terms in Section 2.C.

The results of the linear regression for V_w are listed in Table 3.2 and shown in Figure 3.2.

Table 3.2. Van der Waals volumes V_w calculated by group contributions, the "correction index" N_{vdW} used in the correlation for V_w, and the fitted values of V_w, for 110 polymers. V_w is in cc/mole. The list is in the order of increasing V_w(group). Polymers with equal values of V_w(group) are listed in the order of increasing V_w(fit). The values of the connectivity indices $^0\chi$ and $^1\chi^v$, which are also used in the correlation, are listed in Table 2.2.

Polymer	V_w(group)	N_{vdW}	V_w(fit)
Polyoxymethylene	15.25	0.0	13.13
Polyethylene	20.50	0.0	20.37
Poly(vinyl fluoride)	23.25	0.0	22.96
Poly(vinyl alcohol)	25.25	1.0	25.01
Polyoxyethylene	25.50	0.0	23.32
Poly(thiocarbonyl fluoride)	26.00	-8.0	23.85
Poly(vinylidene fluoride)	26.25	0.0	25.93
Polytrifluoroethylene	29.00	0.0	29.44
Poly(vinyl chloride)	29.25	0.0	30.44
Poly(ethylene sulfide)	30.50	-8.0	32.16
Polypropylene	30.75	1.0	30.48
Polytetrafluoroethylene	32.00	0.0	32.96
Poly(vinyl bromide)	32.03	-4.0	33.18
Polyacrylonitrile	32.25	2.0	32.36
Poly(vinyl methyl ether)	35.75	1.0	33.24
Polyoxytrimethylene	35.75	0.0	33.51
Poly(propylene oxide)	35.75	1.0	34.09
Poly(acrylic acid)	36.25	1.0	36.48
Poly(1,2-butadiene)	38.00	2.0	36.29
Poly(1,4-butadiene)	38.00	2.0	37.49
Polychlorotrifluoroethylene	38.00	0.0	39.44
Poly(vinylidene chloride)	38.25	0.0	38.89
Poly(β-alanine)	38.50	2.0	39.67
Poly(1-butene)	41.00	1.0	41.32

☞ ☞ ☞   TABLE 3.2 IS CONTINUED IN THE NEXT PAGE. ☞ ☞ ☞

Table 3.2. CONTINUED FROM THE PREVIOUS PAGE.

Polymer	V_W(group)	N_{vdW}	V_W(fit)
Polyisobutylene	41.25	2.0	39.33
Polymethacrylonitrile	42.75	3.0	41.87
Poly(p-phenylene)	43.50	0.0	44.61
Poly(vinylidene bromide)	43.81	-8.0	42.16
Polyfumaronitrile	44.00	4.0	44.91
Polyepichlorohydrin	44.50	0.0	45.19
Polyoxytetramethylene	46.00	0.0	43.69
Poly(methyl acrylate)	46.00	1.0	44.76
Poly(vinyl ethyl ether)	46.00	1.0	44.93
Poly(vinyl acetate)	46.00	1.0	45.42
Poly(methacrylic acid)	46.50	2.0	45.99
Polychloroprene	46.75	0.0	47.49
Polyisoprene	48.00	1.0	47.71
Poly(propylene sulfone)	51.05	1.0	51.26
Poly(methyl α-chloroacrylate)	54.76	1.0	54.05
Poly(methyl methacrylate)	56.25	2.0	54.27
Poly(ethyl acrylate)	56.25	1.0	56.44
Poly(vinyl propionate)	56.25	1.0	56.65
Poly(methyl α-cyanoacrylate)	57.75	3.0	56.80
Poly(1-butene sulfone)	61.30	1.0	62.10
Poly(4-methyl-1-pentene)	61.75	2.0	60.97
Polystyrene	63.25	0.0	64.04
Poly(p-xylylene)	64.00	0.0	64.25
Poly(ε-caprolactone)	66.25	0.0	65.60
Poly(vinyl n-butyl ether)	66.50	1.0	65.31
Poly(methyl ethacrylate)	66.50	2.0	65.50
Poly(vinyl sec-butyl ether)	66.50	2.0	65.89
Poly(ethyl methacrylate)	66.50	2.0	65.96
Poly[oxy(2,6-dimethyl-1,4-phenylene)]	69.00	1.0	68.58
Poly(ε-caprolactam)	69.25	2.0	70.23
Poly(2-hydroxyethyl methacrylate)	72.30	2.0	69.44
Poly(p-chloro styrene)	72.73	0.0	74.21
Poly(o-chloro styrene)	72.73	0.0	74.32

☞ ☞ ☞   **TABLE 3.2 IS CONTINUED IN THE NEXT PAGE.** ☞ ☞ ☞

Table 3.2. CONTINUED FROM THE PREVIOUS PAGE.

Polymer	$V_W(\text{group})$	N_{vdW}	$V_W(\text{fit})$
Poly(styrene sulfide)	73.25	-8.0	74.67
Poly(α-methyl styrene)	73.75	1.0	73.55
Poly(p-methyl styrene)	73.75	0.5	73.75
Poly(o-methyl styrene)	73.75	0.5	73.85
Poly(oxy-2,2-dichloromethyltrimethylene)	74.00	0.0	75.42
Poly(oxy-1,1-dichloromethyltrimethylene)	74.00	0.0	76.49
Poly(vinyl cyclohexane)	74.25	-4.0	74.85
Poly(p-bromo styrene)	75.85	-4.0	75.85
Poly(vinyl pivalate)	76.75	3.0	75.30
Poly(n-butyl acrylate)	76.75	1.0	76.82
Poly(vinyl benzoate)	78.25	0.0	79.64
Poly(p-methoxy styrene)	78.75	1.0	77.98
Poly(1-hexene sulfone)	81.80	1.0	82.47
Poly(t-butyl methacrylate)	87.00	4.0	84.94
Poly(n-butyl methacrylate)	87.00	2.0	86.33
Poly(sec-butyl methacrylate)	87.00	3.0	86.92
Poly(8-aminocaprylic acid)	89.75	2.0	90.61
Poly(ethylene terephthalate)	94.00	0.0	95.04
Poly(benzyl methacrylate)	99.00	1.0	98.40
Poly(cyclohexyl methacrylate)	99.75	-3.0	99.86
Poly(p-t-butyl styrene)	104.75	3.0	104.32
Poly(n-hexyl methacrylate)	107.50	2.0	106.70
Poly(2-ethylbutyl methacrylate)	107.50	3.0	107.28
Polybenzobisoxazole	112.32	-7.5	113.94
Poly(tetramethylene terephthalate)	114.50	0.0	115.42
Poly[thio bis(4-phenyl)carbonate]	116.00	-11.0	112.54
Poly(ethylene-2,6-naphthalenedicarboxylate)	122.13	-5.0	118.15
Poly(p-phenylene terephthalamide)	123.00	0.0	123.81
Polybenzobisthiazole	123.32	-23.5	120.03
Poly(n-octyl methacrylate)	128.00	2.0	127.08
Bisphenol-A polycarbonate	137.00	-1.0	135.05
Poly(hexamethylene adipamide)	138.50	4.0	140.47
Poly(hexamethylene isophthalamide)	141.00	2.0	142.33

☞ ☞ ☞   **TABLE 3.2 IS CONTINUED IN THE NEXT PAGE.** ☞ ☞ ☞

Table 3.2. CONTINUED FROM THE PREVIOUS PAGE.

Polymer	V_w(group)	N_{vdW}	V_w(fit)
Poly(1,4-cyclohexylidene dimethylene terephthalate)	147.34	-4.0	149.23
Poly(oxy-1,4-phenylene-oxy-1,4-phenylene-carbonyl-1,4-phenylene)	151.50	0.0	153.89
Poly[2,2-hexafluoropropane *bis*(4-phenyl)carbonate]	154.25	-3.0	154.61
Poly[2,2'-(*m*-phenylene)-5,5'-bibenzimidazole]	161.32	-10.0	158.28
Phenoxy resin	163.50	3.0	161.58
Poly[1,1-(1-phenylethane) *bis*(4-phenyl)carbonate]	170.25	-2.0	169.27
Poly[2,2-propane *bis*{4-(2,6-dichlorophenyl)}carbonate]	174.92	-1.0	176.17
Torlon	176.25	-5.0	183.33
Poly[2,2-propane *bis*{4-(2,6-dimethylphenyl)}carbonate]	178.00	1.0	174.32
Poly[1,1-(1-phenyltrifluoroethane) *bis*(4-phenyl)carbonate]	178.88	-3.0	179.05
Poly(hexamethylene sebacamide)	179.50	4.0	181.22
Poly[N,N'-(*p,p'*-oxydiphenylene)pyromellitimide]	186.14	-7.5	187.08
Poly[2,2-propane *bis*{4-(2,6-dibromophenyl)}carbonate]	187.40	-17.0	182.72
Resin F	199.25	3.0	198.85
Poly[diphenylmethane *bis*(4-phenyl)carbonate]	203.50	-3.0	203.48
Poly[4,4'-diphenoxy di(4-phenylene)sulfone]	204.30	0.0	207.61
Poly[1,1-biphenylethane *bis*(4-phenyl)carbonate]	213.75	-2.0	213.88
Resin G	231.73	5.0	226.68
Poly[4,4'-isopropylidene diphenoxy di(4-phenylene)sulfone]	235.30	2.0	237.49
Ultem	292.63	-8.0	307.10

The largest absolute differences between V_w(group) and V_w(fit) occur for Torlon and Ultem (see Figure 2.4), both of which contain two-ring imide groups. The V_w(fit) values of polymers containing this group are slightly more internally consistent than the V_w(group) values with the general scheme of correlations presented in this book.

The final correlations for V(T) will be expressed in Section 3.D in such a manner that the value of V_w will not be required to calculate either the molar volume or the density.

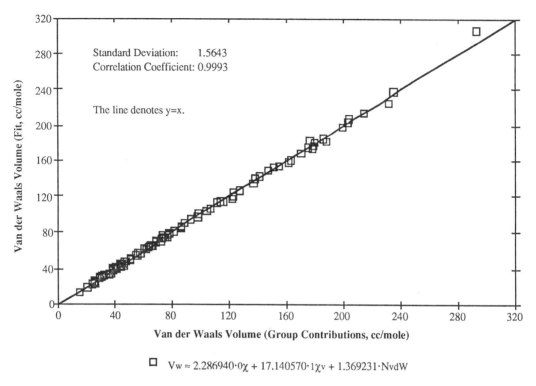

$$\square \quad V_w \approx 2.286940 \cdot {}^0\chi + 17.140570 \cdot {}^1\chi_v + 1.369231 \cdot NvdW$$

Figure 3.2. The results of a fit using connectivity indices, to the van der Waals volumes calculated by group contributions, for a set of 110 polymers.

3.B.2. Nature of the Correction Terms Used in the Correlations

The significance of the various types of correction terms incorporated into Equation 3.11 will now be discussed in greater detail. The remarks made concerning the nature of these correction terms are also valid for other correction terms which will be used throughout this manuscript.

N_{cyc} and N_{fused} are examples of *structural parameters* describing the presence of certain general types of structural features in the polymeric repeat unit, independently of the specific types of atoms or groups of atoms which constitute these structural features.

The terms proportional to N_{Si}, N_{Br}, N_{OH}, $N_{(-S-)}$ and $N_{cyanide}$ in Equation 3.11 are *atomic correction indices*. They are defined in terms of one of the following two alternatives:

1. The presence of a certain type of atom in the structure of the repeat unit. N_{Si} and N_{Br} are correction terms of this type.

2. The presence of a certain type of atom in bonding and/or electronic configurations specified completely by a given pair or a set of pairs of δ and δ^v values. N_{OH}, $N_{(-S-)}$ and $N_{cyanide}$ are correction terms of this type:

(a) N_{OH} is equal to the number of oxygen atoms with $\delta=1$ and $\delta^v=5$.

(b) $N_{(-S-)}$ is the sum of the number of sulfur atoms with $\delta=1$ and $\delta^v=5/9$, and the number of sulfur atoms with $\delta=2$ and $\delta^v=2/3$, in the structure of the polymeric repeat unit.

(c) The *only* bonding environment in which $\delta=1$ and $\delta^v=5$ for a nitrogen atom in a polymer made up of any combination of the nine elements (C, N, O, H, F, Si, S, Cl and Br) for which our correlations have been developed is the cyanide ($-C\equiv N$) group. $N_{cyanide}$ thus equals the number of nitrogen atoms with $\delta=1$ and $\delta^v=5$ in the repeat unit, despite its more convenient verbal description in terms of a group (cyanide) rather than an atom.

All of the other terms in Equation 3.11 are *group correction indices*. They *cannot* be defined completely in terms of the presence of a certain atom with a given pair of δ and δ^v values. The term proportional to $N_{carbonate}$ is very obviously a group correction. There are also cases where the differences between atomic and group correction terms are more subtle. For example, the *total* number of methyl groups in a repeat unit is equal to the number of carbon atoms with $\delta=\delta^v=1$. The total number of methyl groups is therefore an atomic correction index. On the other hand, Equation 3.11 has different correction terms for methyl groups attached to non-aromatic and aromatic atoms. It is, therefore, not sufficient to know how many carbon atoms have $\delta=\delta^v=1$ in order to calculate N_{vdW} correctly. Some additional information about the environment of each such carbon atom (namely, whether a carbon atom with $\delta=\delta^v=1$ is bonded to a non-aromatic or an aromatic atom) must also be known. The terms proportional to $N_{menonar}$ and N_{mear} in Equation 3.11 are therefore group correction terms.

It was stated in Section 2.C that the need to use atomic correction terms could be obviated by optimizing the value of δ^v for each distinct electronic configuration of each atom, for each property, and thus using a separate set of $^0\chi^v$ and $^1\chi^v$ values for each polymer in the correlation for each property. Such a procedure makes the task of developing the correlations vastly more laborious. It also creates a "bookkeeping" nightmare of trying to keep track of the different sets of atomic δ^v values and polymeric $^0\chi^v$ and $^1\chi^v$ values being used for each property. Preliminary calculations showed that the benefits (i.e., correlations of slightly improved accuracy and considerably greater elegance as a result of the omission of the atomic correction terms) were marginal compared to the costs (i.e., the extra work and bookkeeping involved). It was therefore decided to use the atomic correction terms.

An example will now be given of how atomic correction terms can be made unnecessary by optimizing δ^v. There are four polymers containing bromine atoms [$\delta=1$ and $\delta^v=(7/27)\approx0.259$], and four polymers containing sulfur atoms with $\delta=2$ and $\delta^v=(2/3)\approx0.667$, in the dataset used in developing the correlation for V_w. If the δ^v of each of these atomic configurations is varied, and equations 3.10 and 3.11 are used to predict V_w without including the terms proportional to N_{Br} and $N_{(-S-)}$ in N_{vdW}, the $^0\chi$ of each polymer remains unchanged, but the $^1\chi^v$, N_{vdW} and V_w(fit) of each polymer changes. These changes are summarized in Table 3.3. Note that there is, indeed, a

Table 3.3. V_w calculated by group contributions, $^0\chi$, and the changes in $^1\chi^v$, N_{vdW} and V_w(fit) when the δ^v of bromine and of divalent sulfur bonded to two non-hydrogen atoms are optimized and the terms proportional to N_{Br} and $N_{(-S-)}$ in N_{vdW} are omitted. The average deviation of V_w(fit) from V_w(group) is 2.70% when "standard" values of δ^v and atomic correction indices are used, but only 0.97% when the optimized values of δ^v are used.

| Polymer | Unchanged values | | Using standard δ^v | | | Using optimized δ^v | | |
	V_w(group)	$^0\chi$	$^1\chi^v$	N_{vdW}	V_w(fit)	$^1\chi^v$	N_{vdW}	V_w(fit)
Poly(thiocarbonyl fluoride)	26.00	3.2071	1.6027	-8.0	23.85	1.0680	0.0	25.64
Poly(ethylene sulfide)	30.50	2.1213	2.2320	-8.0	32.16	1.4759	0.0	30.15
Poly(vinyl bromide)	32.03	2.2845	1.9504	-4.0	33.18	1.6249	0.0	33.08
Poly(vinylidene bromide)	43.81	3.2071	2.6711	-8.0	42.16	2.1074	0.0	43.46
Poly(styrene sulfide)	73.25	6.1044	4.1807	-8.0	74.67	3.4939	0.0	73.85
Poly(p-bromo styrene)	75.85	6.2676	3.9085	-4.0	75.85	3.6267	0.0	76.50
Polybenzobisthiazole	123.32	11.6899	7.3200	-23.5	120.03	6.2506	-7.5	123.60
TBBPAPC[a]	187.40	16.9389	9.7580	-17.0	182.72	8.6306	-1.0	185.30

[a] Acronym for "tetrabromobisphenol-A polycarbonate", i.e., poly[2,2-propane *bis*{4-(2,6-dibromophenyl)}carbonate].

definite but small improvement in the quality of the fit, when δ^v is optimized instead of using the "standard" values of δ^v in combination with atomic correction terms.

The optimized values of δ^v for the calculation of V_w are 0.51 for bromine atoms with a "standard" δ^v (Table 2.1) of 0.259, and 2.1 for divalent sulfur atoms with a "standard" δ^v (Table 2.1) of 0.667. The optimized value of δ^v for bromine atoms is smaller than the value (0.778) listed for chlorine, but larger than the value (0.259) listed for bromine. Similarly, the optimized value of δ^v for sulfur atoms is smaller than the value (6) listed for oxygen atoms in the same electronic configuration, but larger than the value listed for sulfur (0.667). This observation also holds for many other properties, indicating that Equation 2.1 often overestimates the effects of going down a given column of the periodic table by adding an extra inner shell of electrons. The overestimation for silicon is often so large that the replacement of all silicon atoms by carbon atoms and then the correction for the presence of silicon atoms gives a more accurate correlation than the use of $\delta^v=1/3$ or 4/9 as appropriate and the subsequent addition of correction terms for the overestimation of the effect of the inner shell of electrons in silicon atoms.

3.B.3. Examples of the Predictive Use of the Correlation

A major advantage of the development of correlations between the physical properties and topological quantities such as connectivity indices is that such correlations can be used to predict the properties of interest for polymers containing structural units not contained in any of the polymers used in developing the correlation. This statement is especially valid if the set of polymers used in the development of the correlation is large and diverse, and the correlation coefficient of the resulting regression relationship is very high.

Unlike the tables of group contributions used as a springboard towards its derivation, the correlation given by equations 3.10 and 3.11 can thus be used to predict the V_w of any polymer, regardless of how complex its structure is. In fact, because of the cancellation of errors of opposite sign for different portions of the structure, the relative (fractional) error of V_w predicted from equations 3.10 and 3.11 is likely to become smaller with increasing size and complexity of the repeat unit. The predictive power of this correlation is shown in Table 3.4, which lists the V_w values predicted for 18 polymers whose V_w could not be calculated reliably and consistently by using the group contribution tables which were available when this work was completed.

3.C. Correlation for the Molar Volume at Room Temperature

In this section, the molar volume at room temperature will be denoted by V rather than V(298K) for the sake of brevity. Experimental values of V for amorphous polymers, as well as for the amorphous phases of a large number semicrystalline polymers, were gathered for 152 polymers. Extensive lists provided by van Krevelen [1] were used as the main sources of data, and supplemented or updated with data provided by several other sources [17-25].

The experimental values of V correlated strongly with the zeroth-order and first-order connectivity indices, and especially with the valence indices $^0\chi^v$ and $^1\chi^v$. $^1\chi^v$ had the strongest correlation with V, with a correlation coefficient of 0.9888. In the same manner as with the correlation for V_w, it was found that *the replacement of all silicon atoms by carbon atoms in the repeat unit (i.e., the use of the $^0\chi^v$ and $^1\chi^v$ values provided in Table 2.3 rather than Table 2.2 for silicon-containing polymers, along with an atomic correction index to account for the presence of silicon atoms) gave the best results.*

Weighted linear regressions were used, with weight factors of 100/V(experimental). The following simple two-parameter correlation resulted in a standard deviation of only 8.6 cc/mole and a correlation coefficient of 0.9931, indicating that the combination of $^1\chi^v$ and N_{Si} accounts for 98.62% of the variation of experimental V values:

Table 3.4. The connectivity indices and the correction index used, and the van der Waals volumes predicted in units of cc/mole, for 18 polymers whose V_w could not be calculated reliably and consistently by using any of the group contribution tables which were available when these calculations were completed. The values of $^0\chi$ are listed in Table 2.2. The values used for $^1\chi^v$ are listed in Table 2.3 for silicon-containing polymers and in Table 2.2 for all other polymers. N_{vdW} is determined by inspecting the structure of the repeat unit of the polymer for the presence of the types of structural features which are corrected for by N_{vdW}, and using Equation 3.11.

Polymer	$^0\chi$	$^1\chi^v$	N_{vdW}	V_w
Poly(maleic anhydride)	5.0165	2.0605	-4.0	41.31
Poly(dimethyl siloxane)	3.2071	1.4082	9.0	43.79
Poly(2,5-benzoxazole)	5.8449	2.8435	-5.0	55.26
Poly(p-vinyl pyridine)	5.3973	2.8656	0.0	61.46
Poly(N-vinyl pyrrolidone)	5.5605	3.1722	-4.0	61.61
Poly(vinyl trimethylsilane)	4.7845	2.6052	10.0	69.29
Poly[oxy(methyl γ-trifluoropropylsilylene)]	7.1213	2.6823	8.0	73.22
Poly(phenylmethylsilane)	5.6129	2.9107	8.0	73.68
Cellulose	8.0081	3.5170	-1.0	77.23
Poly[oxy(methylphenylsilylene)]	6.3200	3.0689	8.0	78.01
Poly(vinyl butyral)	6.9747	4.1912	-3.0	83.68
Poly(cyclohexylmethylsilane)	5.6129	3.8552	4.0	84.39
Poly[oxy(dipropylsilylene)]	6.0355	3.5296	9.0	86.63
Poly(α-vinyl naphthalene)	7.9663	4.4265	-5.0	87.25
Poly(N-phenyl maleimide)	8.9996	4.2338	-4.0	87.67
Poly(N-methyl glutarimide)	9.1463	4.2169	-1.0	91.83
Poly(N-vinyl carbazole)	9.8281	5.4266	-7.5	105.22
Poly[3,5-(4-phenyl-1,2,4-triazole)-1,4-phenylene]	11.2423	5.5501	0.0	120.84

$$V \approx 33.585960 \cdot {}^1\chi^v + 26.518075 \cdot N_{Si} \tag{3.12}$$

This result is far more remarkable than the 99.15% of the variation of V_w accounted for by Equation 3.9 which is also a two-parameter correlation. The V_w values were obtained by using group contributions, and therefore did not manifest random variations caused by experimental error. By contrast, the V values were measured by many different workers, using different instruments, in different laboratories. Even an "absolutely perfect" correlation cannot, therefore, be expected to account for 100% of the variation among the experimental V values.

A correction term N_{MV} was defined for V in the same way that N_{vdW} was defined for V_w. In addition, all four zeroth-order and first-order connectivity indices were used in the correlation:

$$V \approx 3.642770 \cdot {}^0\chi + 9.798697 \cdot {}^0\chi^v - 8.542819 \cdot {}^1\chi + 21.693912 \cdot {}^1\chi^v + 0.978655 \cdot N_{MV} \quad (3.13)$$

$$N_{MV} \equiv 24N_{Si} - 18N_{(-S-)} - 5N_{sulfone} - 7N_{Cl} - 16N_{Br} + 2N_{(backbone\ ester)} + 3N_{ether} + 5N_{carbonate}$$
$$+ 5N_{C=C} - 11N_{cyc} - 7 \cdot (N_{fused} - 1) \textit{ (last term only to be used if } N_{fused} \geq 2) \quad (3.14)$$

The use of equations 3.13 and 3.14 results in a standard deviation of only 3.2 cc/mole relative to, and a very large correlation coefficient of 0.9989 with, the experimental values of V. This correlation therefore accounts for 99.78% of the variation of the 152 experimental V values in the dataset. The standard deviation is only 2.2% of the average V of 145.4 cc/mole in this dataset. It is quite possible that most or all of the remaining 0.22% of the variation is caused by the experimental error margins. If there is any remaining systematic portion in the deviation which should be accounted for in the correlation, this systematic portion can only be uncovered after a considerable amount of additional experimental data becomes available.

Most of the eleven terms in Equation 3.14 were encountered earlier, in the correlation for V_w. The last two terms are structural parameters. The first five terms are atomic correction terms, whose use can be rendered unnecessary, at the expense of a very considerable amount of extra work, by utilizing the method described in Section 3.B.2, and optimizing the δ^v values for the five atomic configurations in question to provide the best possible fit for V. The remaining four terms are group correction terms.

Correction terms in Equation 3.14 not encountered in Section 3.B.1 are as follows:
1. $N_{sulfone}$ is the number of sulfur atoms in the highest oxidation state defined by $\delta=4$ and $\delta^v=8/3$ in the polymeric repeat unit.
2. N_{Cl} is the total number of chlorine atoms in the polymeric repeat unit.
3. $N_{(backbone\ ester)}$ is the number of ester (-COO-) groups in the *backbone* of the polymeric repeat unit, which is defined as described in Section 2.D. For example, $N_{(backbone\ ester)}$ is equal to 0 for poly(vinyl acetate) and for poly(methyl methacrylate), 1 for poly(ε-caprolactone) and for poly(glycolic acid) (see Figure 2.4), and 2 for poly(ethylene terephthalate).
4. N_{ether} is the total number of ether (-O-) linkages in the polymeric repeat unit. For example, N_{ether} is 1 for polyoxymethylene, polyepichlorohydrin (see Figure 2.4) and poly(vinyl ethyl ether), and 2 for Ultem (see Figure 2.4) and poly(vinyl butyral) (see Figure 2.9). Note that only (-O-) linkages between two carbon atoms will be counted as ether linkages. If there are silicon atoms on one or both sides of the oxygen atom, the linkage is not counted in N_{ether}. For example, $N_{ether}=0$ for poly(dimethyl siloxane).

The results of the linear regression for V are listed in Table 3.5 and shown in Figure 3.3. The observed and calculated densities of the polymers are also listed in Table 3.5. See the Appendix at the end of this book for the molecular weights per mole of repeat units of the polymers, used to calculate densities and specific volumes from molar volumes.

Table 3.5. Experimental amorphous densities ρ (g/cc) and molar volumes V (cc/mole) *at room temperature*, the "correction index" N_{MV} used in the correlation for V, and fitted values of V and ρ, for 152 polymers. Zeroth-order connectivity indices $^0\chi$ and $^0\chi^v$ and first-order connectivity indices $^1\chi$ and $^1\chi^v$, all of which are also used in the correlation, are listed in Table 2.2. The alternative set of $^0\chi^v$ and $^1\chi^v$ values listed in table 2.3 is used for the silicon-containing polymers.

Polymer	ρ(exp)	V(exp)	N_{MV}	V(fit)	ρ(fit)
Polyoxymethylene	1.250	24.0	3	23.0	1.304
Polyethylene	0.850	33.1	0	32.2	0.874
Poly(vinyl alcohol)	1.260	35.0	0	36.7	1.202
Poly(glycolic acid)	1.600	36.3	2	39.3	1.478
Polyoxyethylene	1.125	39.2	3	39.1	1.128
Poly(vinylidene fluoride)	1.600	40.0	0	39.9	1.604
Polyacrylonitrile	1.184	44.8	0	45.1	1.176
Poly(vinyl chloride)	1.385	45.1	-7	45.2	1.382
Polypropylene	0.850	49.5	0	49.0	0.859
Polytetrafluoroethylene	2.000	50.0	0	48.7	2.053
Poly(β-propiolactone)	1.360	53.0	2	55.4	1.301
Polyacrylamide	1.302	54.6	0	56.2	1.265
Poly(propylene oxide)	1.000	58.1	3	56.8	1.023
Poly(vinylidene chloride)	1.660	58.4	-14	57.4	1.689
Polychlorotrifluoroethylene	1.920	60.7	-7	57.5	2.027
Poly(1,4-butadiene)	0.892	60.7	5	59.1	0.916
Poly[oxy(methylhydrosilylene)]	0.990	60.7	24	62.1	0.968
Poly(3,3,3-trifluoropropylene)	1.580	60.8	0	60.0	1.601
Polymethacrylonitrile	1.100	61.0	0	62.3	1.077
Poly(vinyl methyl ketone)	1.120	62.6	0	66.7	1.051
Poly(vinyl methyl sulfide)	1.180	62.8	-18	70.8	1.047
Poly(1-butene)	0.860	65.2	0	65.6	0.855
Polyisobutylene	0.840	66.8	0	65.6	0.855
Poly(3-chloropropylene oxide)	1.370	67.5	-4	69.9	1.323
Poly(methyl acrylate)	1.220	70.6	0	69.3	1.243
Polychloroprene	1.243	71.2	-2	72.5	1.221
Poly(vinyl acetate)	1.190	72.4	0	70.6	1.220
Polyoxytetramethylene	0.980	73.6	3	71.2	1.013
Polyisoprene	0.906	75.2	5	76.6	0.889

☞ ☞ ☞   **TABLE 3.5 IS CONTINUED IN THE NEXT PAGE.** ☞ ☞ ☞

Table 3.5. CONTINUED FROM THE PREVIOUS PAGE.

Polymer	ρ(exp)	V(exp)	N_{MV}	V(fit)	ρ(fit)
Poly(dimethyl siloxane)	0.980	75.6	24	79.6	0.931
Poly(1,4-pentadiene)	0.890	76.5	5	75.2	0.905
Poly(vinyl ethyl ether)	0.940	76.7	3	73.4	0.982
Poly(1-pentene)	0.850	82.5	0	81.7	0.858
Poly(vinyl chloroacetate)	1.450	83.1	-7	84.2	1.431
Poly(methyl α-cyanoacrylate)	1.304	85.2	0	83.1	1.337
Poly(methyl methacrylate)	1.170	85.6	0	86.4	1.159
Poly[oxy(vinylmethylsilylene)]	0.980	88.0	29	89.4	0.965
Poly(N-vinyl pyrrolidone)	1.250	88.9	-11	91.8	1.211
Poly(ethyl acrylate)	1.120	89.4	0	87.3	1.147
Poly(α,α-dimethylpropiolactone)	1.097	91.2	2	89.5	1.118
Poly(vinyl isopropyl ether)	0.924	93.2	3	90.6	0.951
Poly(ethyl α-chloroacrylate)	1.390	96.8	-7	100.3	1.341
Poly(1-hexene)	0.860	97.9	0	97.8	0.861
Poly(vinyl propionate)	1.020	98.1	0	87.7	1.141
Polystyrene	1.050	99.1	0	97.0	1.073
Poly(4-methyl-1-pentene)	0.838	100.5	0	98.1	0.859
Poly(ethyl methacrylate)	1.119	102.0	0	104.4	1.093
Poly(ε-caprolactone)	1.095	104.2	2	103.6	1.101
Poly(styrene oxide)	1.150	104.4	3	104.7	1.147
Poly(ε-caprolactam)	1.084	104.4	0	105.0	1.078
Polyoxyhexamethylene	0.932	107.5	3	103.4	0.969
Poly(vinyl isobutyl ether)	0.930	107.7	3	105.8	0.947
Poly(vinyl n-butyl ether)	0.927	108.1	3	105.5	0.950
Poly(vinyl sec-butyl ether)	0.920	108.9	3	107.2	0.935
Poly(α-methyl styrene)	1.065	111.0	0	114.1	1.036
Poly(oxy-2,2-dichloromethyltrimethylene)	1.386	111.8	-11	115.3	1.344
Poly[oxy(2,6-dimethyl-1,4-phenylene)]	1.070	112.2	3	110.8	1.084
Poly(2-chloroethyl methacrylate)	1.320	112.6	-7	117.0	1.270
Poly(3-phenyl-1-propene)	1.046	113.0	0	112.6	1.050
Poly(p-methyl styrene)	1.040	113.7	0	114.7	1.031
Poly(n-propyl α-chloroacrylate)	1.300	114.3	-7	116.4	1.277
Poly(o-methyl styrene)	1.027	115.1	0	114.7	1.031

☞ ☞ ☞   **TABLE 3.5 IS CONTINUED IN THE NEXT PAGE.** ☞ ☞ ☞

Table 3.5. CONTINUED FROM THE PREVIOUS PAGE.

Polymer	ρ(exp)	V(exp)	N_{MV}	V(fit)	ρ(fit)
Poly(vinyl cyclohexane)	0.950	116.0	-11	113.9	0.968
Poly(isopropyl α-chloroacrylate)	1.270	117.0	-7	117.5	1.265
Poly(vinyl *n*-butyl sulfide)	0.980	118.6	-18	115.7	1.005
Poly(*n*-propyl methacrylate)	1.080	118.7	0	120.5	1.064
Poly[oxy(methyl γ-trifluoropropylsilylene)]	1.300	120.2	24	122.7	1.274
Poly(ethylene succinate)	1.175	122.6	4	110.7	1.301
Poly[oxy(methylphenylsilylene)]	1.110	122.7	24	128.1	1.063
Poly(isopropyl methacrylate)	1.033	124.1	0	121.6	1.054
Poly(vinyl *n*-pentyl ether)	0.918	124.4	3	121.6	0.939
Poly[oxy(mercaptopropylmethylsilylene)]	1.060	126.7	6	119.6	1.123
Poly(4-phenyl-1-butene)	1.041	127.0	0	128.7	1.027
Poly(*t*-butyl acrylate)	1.000	128.2	0	121.3	1.057
Poly(*m*-trifluoromethylstyrene)	1.320	130.5	0	126.6	1.361
Poly(*n*-butyl α-chloroacrylate)	1.240	131.1	-7	132.5	1.227
Poly(*sec*-butyl α-chloroacrylate)	1.240	131.1	-7	134.1	1.212
Poly(vinyl butyral)	1.083	131.3	-5	132.2	1.076
Poly(4-fluoro-2-trifluoromethylstyrene)	1.430	132.9	0	131.5	1.445
Poly(phenyl methacrylate)	1.210	134.0	0	136.7	1.186
Poly(2,2,2-trifluoro-1-methylethyl methacrylate)	1.340	134.4	0	132.6	1.358
Poly(*n*-butyl methacrylate)	1.055	134.8	0	136.6	1.041
Poly(*sec*-butyl methacrylate)	1.052	135.2	0	138.2	1.029
Poly(8-aminocaprylic acid)	1.040	135.8	0	137.1	1.030
Poly(isobutyl methacrylate)	1.045	136.1	0	136.9	1.039
Poly(vinyl *n*-hexyl ether)	0.925	138.6	3	137.7	0.931
Poly(5-phenyl-1-pentene)	1.050	139.2	0	144.8	1.009
Poly(*t*-butyl methacrylate)	1.020	139.4	0	138.3	1.028
Poly(ethylene adipate)	1.219	141.2	4	142.9	1.204
Poly(ethylene isophthalate)	1.340	143.4	4	145.9	1.317
Poly(ethylene phthalate)	1.338	143.6	4	145.9	1.317
Poly(ethylene terephthalate)	1.335	144.0	4	145.9	1.318
Poly(benzyl methacrylate)	1.179	149.4	0	151.4	1.163
Poly(cyclohexyl α-chloroacrylate)	1.250	151.0	-18	149.3	1.264
Poly(isopentyl methacrylate)	1.032	151.4	0	152.9	1.022

☞ ☞ ☞   **TABLE 3.5 IS CONTINUED IN THE NEXT PAGE.** ☞ ☞ ☞

Table 3.5. CONTINUED FROM THE PREVIOUS PAGE.

Polymer	ρ(exp)	V(exp)	N_{MV}	V(fit)	ρ(fit)
Poly(1-methylbutyl methacrylate)	1.030	151.7	0	154.3	1.013
Poly(cyclohexyl methacrylate)	1.098	153.2	-11	153.3	1.097
Poly[oxy(methyl *m*-chlorophenylsilylene)]	1.100	155.2	17	141.8	1.204
Poly[oxy(acryloxypropylmethylsilylene)]	1.110	155.2	29	158.9	1.084
Poly(neopentyl methacrylate)	0.993	157.3	0	153.1	1.020
Poly[oxy(methyl *n*-hexylsilylene)]	0.910	158.6	24	160.8	0.898
Poly(N-vinyl carbazole)	1.200	161.0	-14	158.0	1.223
Poly(2-ethylbutyl methacrylate)	1.040	163.7	0	170.0	1.001
Poly(1-methylpentyl methacrylate)	1.013	168.1	0	170.4	0.999
Poly(*n*-hexyl methacrylate)	1.010	168.5	0	168.7	1.009
Poly(1-phenylethyl methacrylate)	1.129	168.5	0	169.6	1.122
Poly(*p-t*-butyl styrene)	0.950	168.7	0	165.1	0.971
Poly(1,3-dimethylbutyl methacrylate)	1.005	169.5	0	170.6	0.999
Poly(3,3-dimethylbutyl methacrylate)	1.001	170.1	0	169.2	1.006
Poly(vinyl *n*-octyl ether)	0.914	171.0	3	169.9	0.920
Poly(1,2,2-trimethylpropyl methacrylate)	0.991	171.9	0	171.1	0.996
Poly(vinyl 2-ethylhexyl ether)	0.904	172.9	3	171.2	0.913
Poly(tetramethylene isophthalate)	1.268	173.7	4	178.1	1.237
Poly(ethylene suberate)	1.147	174.6	4	175.1	1.144
Poly[1-(*o*-chlorophenyl)ethyl methacrylate]	1.269	177.1	-7	183.2	1.227
Poly[thio *bis*(4-phenyl)carbonate]	1.355	180.3	-13	180.4	1.354
Poly(11-aminoundecanoic acid)	1.010	181.5	0	185.4	0.989
Poly(ethylene-2,6-naphthalenedicarboxylate)	1.330	182.1	-3	183.2	1.322
Poly[oxy(methyl *m*-chlorophenylethylsilylene)]	1.090	182.3	17	177.8	1.118
Poly[methane *bis*(4-phenyl)carbonate]	1.240	182.4	5	181.7	1.245
Poly[oxy(methyl *n*-octylsilylene)]	0.910	189.3	24	193.0	0.893
Poly(12-aminododecanoic acid)	0.990	199.3	0	201.4	0.980
Poly(*n*-octyl methacrylate)	0.971	204.2	0	200.9	0.987
Poly(vinyl *n*-decyl ether)	0.883	208.7	3	202.0	0.912
Poly(ethylene sebacate)	1.085	210.4	4	207.2	1.102
Poly(hexamethylene adipamide)	1.070	211.5	0	209.9	1.078
Bisphenol-A polycarbonate	1.200	211.9	5	216.5	1.175
Poly[oxy(2,6-diphenyl-1,4-phenylene)]	1.140	214.3	3	208.4	1.172

☞ ☞ ☞   TABLE 3.5 IS CONTINUED IN THE NEXT PAGE. ☞ ☞ ☞

Table 3.5. CONTINUED FROM THE PREVIOUS PAGE.

Polymer	ρ(exp)	V(exp)	N_{MV}	V(fit)	ρ(fit)
Poly(diphenylmethyl methacrylate)	1.168	216.0	0	217.5	1.160
Poly(p-cyclohexylphenyl methacrylate)	1.115	219.1	-11	220.0	1.110
Poly(1,4-cyclohexylidene dimethylene terephthalate)	1.196	229.3	-7	227.5	1.205
Poly(1,2-diphenylethyl methacrylate)	1.147	232.2	0	233.1	1.143
Poly[2,2'-(m-phenylene)-5,5'-bibenzimidazole]	1.300	237.2	-21	231.9	1.330
Poly(vinyl n-dodecyl ether)	0.892	238.1	3	234.2	0.907
Poly(2,2,2'-trimethylhexamethylene terephthalamide)	1.120	257.5	0	263.2	1.096
Poly[1,1-(1-phenylethane) bis(4-phenyl)carbonate]	1.200	263.6	5	265.0	1.194
Poly[N,N'-(p,p'-oxydiphenylene)pyromellitimide]	1.420	269.2	-11	278.2	1.374
Poly(hexamethylene sebacamide)	1.040	271.5	0	274.2	1.030
Poly(dodecyl methacrylate)	0.929	273.8	0	265.2	0.959
Poly[2,2-propane bis{4-(2,6-dichlorophenyl)}carbonate]	1.415	277.1	-23	271.2	1.446
Poly[2,2-propane bis{4-(2,6-dimethylphenyl)}carbonate]	1.083	286.6	5	287.5	1.080
Poly[oxy(methyl n-tetradecyl silylene)]	0.890	288.2	24	289.5	0.886
Poly[2,2-propane bis{4-(2,6-dibromophenyl)}carbonate]	1.953	291.8	-59	304.5	1.872
Poly(1-octadecene)	0.860	293.6	0	290.7	0.869
Poly[4,4'-diphenoxy di(4-phenylene)sulfone]	1.290	310.5	1	310.1	1.292
Poly[oxy(methyl n-hexadecylsilylene)]	0.880	323.4	24	321.6	0.885
Poly[4,4'-sulfone diphenoxy di(4-phenylene)sulfone]	1.370	339.1	-4	336.4	1.381
Poly[2,2-hexafluoropropane bis{4-(2,6-dibromophenyl)}carbonate]	1.987	341.1	-59	327.5	2.070
Poly[oxy(methyl n-octadecylsilylene)]	0.890	351.2	24	353.8	0.883
Polytriazole membrane polymer[a]	1.243	352.8	0	355.8	1.233
Poly[4,4'-isopropylidene diphenoxy di(4-phenylene)sulfone]	1.240	356.9	1	360.2	1.229
Ultem	1.270	466.6	-15	463.3	1.279

[a] Poly[3,5-(4-phenyl-1,2,4-triazole)-1,4-phenylene-3,5-(4-phenyl-1,2,4-triazole)-1,3-phenylene]. The zeroth-order and first-order connectivity indices of this polymer, which is named differently in the source [24] for the value of its density, are exactly twice the corresponding indices of poly[3,5-(4-phenyl-1,2,4-triazole)-1,4-phenylene].

The standard deviation between the observed and calculated densities listed in Table 3.5 is 0.0354 g/cc, which is only 3.1% of the average observed density of 1.1398 g/cc for the dataset. The correlation coefficient is 0.9872, which indicates that the method developed above accounts for 97.5% of the variation of the densities in the dataset. The correlation for V(298K) can therefore be combined with Equation 3.2 to predict the amorphous density of a polymer at room temperature with excellent accuracy, regardless of whether the polymer is in the glassy state (below its T_g) or in the rubbery or molten state (above its T_g) at room temperature. However, the relative (percent) standard deviation of the density calculation is slightly higher, and the correlation coefficient is slightly lower, than it is for the molar volume. See Figure 3.4 for an illustration of the quality of the density calculations.

This result is an example of a general difference between *all* corresponding pairs of extensive and intensive properties. The extensive property spans a much wider range, most of which incorporates the effects of the large differences between the sizes of polymeric repeat units. The intensive property, which reflects the true differences between the properties of the polymers independently of the sizes of their repeat units, spans a much narrower range and is therefore predicted with somewhat lower levels of accuracy and certainty than the extensive property.

The ability to predict the density at room temperature directly from a correlation for V(298K) even in the absence of any information concerning the value of T_g is very useful. The direct prediction of V(298K) also has the advantage that it generally accounts for the effects of the frozen-in packing volumes of different polymers considerably more accurately than is possible by an indirect correlation (i.e., Equation 3.5) via V_w.

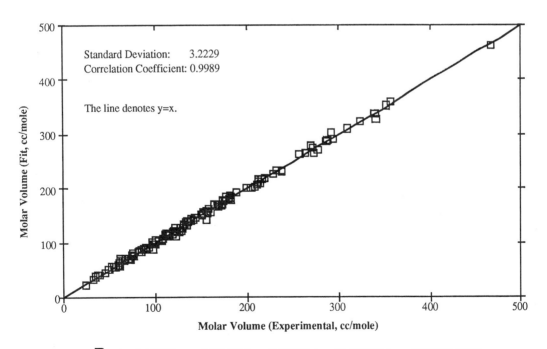

$$\square \quad V \approx 3.642770 \cdot {}^0\chi + 9.798697 \cdot {}^0\chi v - 8.542819 \cdot {}^1\chi + 21.693912 \cdot {}^1\chi v + 0.978655 \cdot NMV$$

Figure 3.3. Correlation using connectivity indices for the amorphous molar volumes of 152 polymers at room temperature.

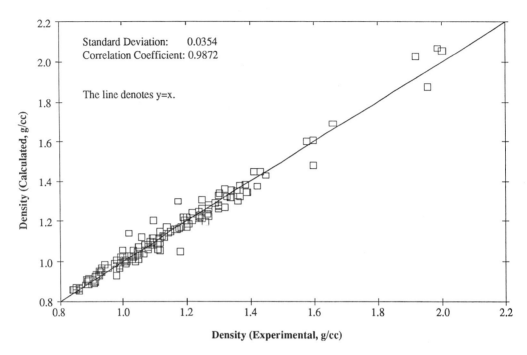

Figure 3.4. Comparison between the observed amorphous densities of 152 polymers at room temperature and the densities calculated by using the correlation developed for the molar volume.

3.D. Final Equations for Temperature Dependences of Volumetric Properties

3.D.1. Introductory Remarks

The ability to estimate V(298K) independently of V_w, and also independently of whether a polymer is glassy or rubbery at 298K, provides a very useful set of correlations to predict V(T). These equations, which will be summarized below, were derived by combining the results presented earlier in this chapter with additional data and information provided by van Krevelen [1] and by Seitz [11]. They were then used extensively in our practical work, and found to provide reasonable predictions of the temperature dependences of the molar volume and density.

These equations take the discontinuity of the thermal expansion coefficient at T_g into account. They also obviate the need to use V_w as an input parameter. The correlation derived for V_w in Section 3.B will, nonetheless, be useful in calculating other properties in later chapters of this book. The correlations presented below treat the cases of polymers with $T_g \geq 298K$ and $T_g < 298K$ separately. This is not done because there is anything inherently special about a temperature of 298K. It is done only because, as a result of the great abundance of volumetric data at 298K, this temperature was used for the reference value of the molar volume, namely V(298K), which is utilized as the starting point of the derivation of the temperature dependence.

3.D.2. Polymers With $T_g \geq 298K$

If the value of V(298K) obtained by using Equation 3.13 is substituted into Equation 3.5 and V(T) at an arbitrary $T \leq T_g$ is divided by V(298K), Equation 3.15 is obtained. The temperature dependence of V(T) as predicted by using Equation 3.15 has the same functional form as in Equation 3.5; however, the effects of the frozen-in packing volume at absolute zero temperature have essentially been incorporated into a "rescaled" effective V_w. Equation 3.15 therefore usually provides an improvement over Equation 3.5 in predicting V(T) and the density.

$$V(T) \approx V(298K)\frac{(1.42T_g + 0.15T)}{(1.42T_g + 44.7)} \qquad \text{(for $T \leq T_g$ and $T_g \geq 298K$)} \qquad (3.15)$$

When $T > T_g \geq 298K$, Equation 3.16, which incorporates the increase of the thermal expansion above T_g, is used.

$$V(T) \approx V(298K)\frac{[1.57T_g + 0.30(T - T_g)]}{(1.42T_g + 44.7)} \qquad \text{(for $T > T_g \geq 298K$)} \qquad (3.16)$$

3.D.3. Polymers With $T_g < 298K$

The coefficient of thermal expansion at 298K (which is above T_g for these polymers) is predicted by using Equation 3.17. Equation 3.18 is used when $T < T_g < 298K$, and Equation 3.19 is used when $T_g < 298K$ and $T \geq T_g$, to predict V(T).

$$\alpha_r(298K) \approx \frac{1}{(298 + 4.23T_g)} \qquad \text{(for $T_g < 298K$)} \qquad (3.17)$$

$$V(T) \approx 0.15V(298K)\frac{(T - T_g)}{(1.42T_g + 44.7)} + V(298K)\left[1 + \alpha_r(298K) \cdot (T_g - 298)\right]$$

$$\text{(for $T < T_g < 298K$)} \qquad (3.18)$$

$$V(T) \approx V(298K) \cdot [1 + \alpha_r(298K) \cdot (T - 298)] \qquad \text{(for $T_g < 298K$ and $T \geq T_g$)} \qquad (3.19)$$

3.E.   Pressure-Volume-Temperature Relationships

The volume of a material is a function of both the temperature and the pressure, as described by its *pressure-volume-temperature (PVT) relationships*. These relationships incorporate a description of the complete phase behavior of a material, such as the significant changes which take place when the material undergoes a phase transition (for example when a solid melts or sublimes, or when a liquid boils). The preceding discussion focused on the dependence of the volumetric properties on the temperature only, at atmospheric pressure, since there are no simple and general quantitative structure-property relationships to describe the full PVT behavior. This topic is of great importance, however, and will therefore now be reviewed briefly.

An *equation-of-state* is an expression of the general form given by Equation 3.20, to describe PVT behavior. For a given temperature and pressure, Equation 3.20 is solved for the volume.

$$f(p,V,T) = 0 \qquad\qquad\qquad (3.20)$$

Equations-of-state are often expressed in an alternative form (Equation 3.21), in terms of the density rather than the volume, and in terms of dimensionless *reduced variables* instead of the actual temperature, density and pressure. For a given temperature and pressure, Equation 3.21 is solved for the density (or equivalently, for the volume). In this equation, the *reduced temperature* ($\widetilde{T}$), *reduced density* ($\widetilde{\rho}$) and *reduced pressure* ($\widetilde{p}$) are defined by equations 3.22, 3.23 and 3.24, respectively, in terms of the absolute temperature T, the density ρ, the pressure p; and the *reducing parameters* T^*, ρ^* and p^* for the temperature, density and pressure, respectively.

$$f(\widetilde{T},\widetilde{\rho},\widetilde{p}) = 0 \qquad\qquad\qquad (3.21)$$

$$\widetilde{T} \equiv \frac{T}{T^*} \qquad\qquad\qquad (3.22)$$

$$\widetilde{\rho} \equiv \frac{\rho}{\rho^*} \qquad\qquad\qquad (3.23)$$

$$\widetilde{p} \equiv \frac{p}{p^*} \qquad\qquad\qquad (3.24)$$

The main advantage of using reduced variables is that they make the fundamental similarities and correspondences between different materials much more obvious than is possible otherwise. For example, many (but certainly not all) equations-of-state are *corresponding states theories*, where all materials have the same *reduced* density $\widetilde{\rho}$ at a given *reduced* temperature $\widetilde{T}$ and *reduced* pressure $\widetilde{p}$. The differences between the PVT behaviors of different materials in a corresponding states theory are described completely by the differences in the values of their *reducing* *parameters* (T^*, ρ^* and p^*). Consequently, master PVT curves can be drawn for a corresponding states theory, with the PVT curves of specific materials being obtained from these master curves by scaling according the values of their reducing parameters.

The reducing parameters are usually estimated for a given material by fitting the equation-of-state to experimental data. Empirical correlations and/or theoretical estimation methods also exist for these parameters in some equations-of-state which are used more widely, to allow the predictive use of the equation-of-state. However, such predictive uses often result in mediocre accuracy in the predicted PVT behavior because of the shortcomings of the methods used to estimate the reducing parameters in the absence of experimental data, even when the functional form of the equation-of-state is very adequate in fitting data with the correct reducing parameters.

Many important applications of equations-of-state involve the study of the PVT behavior of mixtures. For example, in the general formulation of an equation-of-state for a mixture of two components, Equation 3.21 can be rewritten in the form of Equation 3.25 for the mixed system:

$$f(\widetilde{T}_{mix}, \widetilde{P}_{mix}, \widetilde{p}_{mix}) = 0 \qquad (3.25)$$

The reducing parameters of the mixture are defined in terms of the reducing parameters and volume fractions of the pure components, and one or more *interaction parameter(s)* quantifying deviations from ideal mixing. *Ideal mixing* occurs when the reducing parameters of the mixture can be defined entirely in terms of the reducing parameters and volume fractions of the pure components, by simple rules such as arithmetic and/or geometric averaging of the parameters and/or their reciprocals weighted by volume fractions, without using any interaction parameters.

In general, mixing behavior deviates from ideal mixing, interaction parameters are needed, and they play a critical role in determining the phase diagram and PVT behavior of the mixture. Interaction parameters often depend on the temperature, pressure, and volume fractions of the components in the mixture (for example, see [26]), and are generally very difficult to predict *a priori*. Consequently, equations-of-state are more often used to correlate existing data to gain a better understanding within a theoretical framework than to make accurate predictions of mixing.

The improvement of the ability to calculate interaction parameters is an area of active research, using methods ranging from quantitative structure-property relationships to atomistic simulations. Successes in this area are certain to have far-reaching impact in many areas of polymer technology, since the mixing of components is involved in aspects of polymer technology which are as diverse as the design of polymerization processes and reactors, blending, compounding, the use of plasticizers and other additives in polymers, the extraction of residual monomers and process solvents, foaming, the use of polymers as packaging materials or separation membranes, the use of polymers in biomedical applications (such as contact lenses, surgical implants and kidney dialysis machines), thermooxidative stability, and resistance to environmental agents.

Many equations-of-state exist, to solve different types of problems [27,28]. For example, one method which is very useful for polymers is the *Sanchez-Lacombe equation-of-state theory* [29-32], whose applications to date include the complex problems of correlating and/or predicting the amount of gas sorbed by the polymer at thermodynamic equilibrium and the effects of gas sorption on the density of the polymer [33-35].

A concise review and comparison of equations-of-state for the PVT behavior of polymers was provided by Zoller [36], who compared the ability of several equations-of-state (some developed empirically and some developed based on fundamental theoretical considerations) to represent PVT data for homopolymers, copolymers and polymer blends. He concluded [36] that, among theoretically-based equations-of-state, the Simha-Somcynsky equation-of-state [37,38] gives the best representation of available data over extended ranges of temperature and pressure.

References and Notes for Chapter 3

1. D. W. van Krevelen, *Properties of Polymers*, second edition, Elsevier, Amsterdam (1976). Chapter 4 of this book deals with volumetric properties. In addition, Table V in Part VII lists the experimental densities of a large number of polymers.

2. F. Bueche, *Physical Properties of Polymers*, Interscience Publishers, New York (1962).

3. R. F. Boyer and R. S. Spencer, *J. Appl. Phys.*, *15*, 398-405 (1944).

4. R. Simha and R. F. Boyer, *J. Chem. Phys.*, *37*, 1003-1007 (1962).

5. R. Simha, J. M. Roe and V. S. Nanda, *J. Appl. Phys.*, *43*, 4312-4317 (1972).

6. M. Goldstein, *J. Chem. Phys.*, *64*, 4767-4774 (1976).

7. R. Simha, in *Molecular Basis of Transitions and Relaxations*, edited by D. J. Meier, Gordon and Breach Science Publishers, London (1978), 203-223.

8. S. M. Aharoni, *J. Appl. Polym. Sci.*, *23*, 223-228 (1979).

9. G. P. Johari, in *Lecture Notes in Physics, 277*, 90-112 (1987) (edited by T. Dorfmuller and G. Williams, Springer-Verlag, Berlin).

10. R. E. Robertson, in *Computational Modeling of Polymers*, edited by J. Bicerano, Marcel Dekker, New York (1992), Chapter 6.

11. J. T. Seitz, *J. Appl. Polym. Sci.*, *49*, 1331-1351 (1993).

12. C. J. Lee, *J. Macromol. Sci., Rev. Macromol. Chem. Phys.*, *C29*, 431-560 (1989).

13. L. B. Kier and L. H. Hall, *Molecular Connectivity in Chemistry and Drug Research*, Academic Press, New York (1976).

14. L. B. Kier and L. H. Hall, *Molecular Connectivity in Structure-Activity Analysis*, John Wiley & Sons, New York (1986).

15. D. W. van Krevelen, in *Computational Modeling of Polymers*, edited by J. Bicerano, Marcel Dekker, New York (1992), Chapter 1.

16. J. T. Seitz, unpublished calculations.

17. *Silicon Compounds: Register and Review*, Petrarch Systems, Silanes and Silicones Group, Bristol, Pennsylvania (1987).

18. J. W. Verbicky, Jr., article titled *"Polyimides"*, *Encyclopedia of Polymer Science and Engineering*, *12*, Wiley-Interscience, New York (1988).

19. J. E. Harris and R. N. Johnson, article titled *"Polysulfones"*, *Encyclopedia of Polymer Science and Engineering*, *13*, Wiley-Interscience, New York (1988).

20. J. P. Runt, article titled *"Crystallinity Determination"*, *Encyclopedia of Polymer Science and Engineering*, *4*, Wiley-Interscience, New York (1986).

21. Product literature for the Celazole™ U-60 resin, from the Hoechst Celanese Corporation (1989).

22. R. R. Light and R. W. Seymour, *Polymer Engineering and Science*, 22, 857-864 (1982).

23. N. Muruganandam, W. J. Koros and D. R. Paul, *J. Polym. Sci., Polym. Phys. Ed.*, *25*, 1999-2026 (1987).

24. B. Gebben, M. H. V. Mulder and C. A. Smolders, *J. Membrane Science*, *46*, 29-41 (1989).

25. B. R. Hahn and J. H. Wendorff, *Polymer*, *26*, 1619-1622 (1985).

26. C. Qian, S. J. Mumby and B. E. Eichinger, *Macromolecules*, *24*, 1655-1661 (1991).

27. R. C. Reid, J. M. Prausnitz and B. E. Poling, *The Properties of Gases and Liquids*, fourth edition, McGraw-Hill, New York (1987).

28. J. F. Brennecke and C. A. Eckert, *AIChE Journal*, *35*, 1409-1427 (1989).

29. I. C. Sanchez and R. H. Lacombe, *Macromolecules*, *11*, 1145-1156 (1978).

30. I. C. Sanchez and R. H. Lacombe, *J. Polym. Sci. Polym. Lett. Ed.*, *15*, 71-75 (1977).

31. I. C. Sanchez, in *Polymer Blends*, edited by D. R. Paul and S. Newman, Academic Press, New York (1978), Volume 1, Chapter 3.

32. C. Panayiotou and I. C. Sanchez, *Macromolecules*, *24*, 6231-6237 (1991).

33. M. B. Kiszka, M. A. Meilchen and M. A. McHugh, *J. Appl. Polym. Sci.*, *36*, 583-597 (1988).

34. D. S. Pope, I. C. Sanchez, W. J. Koros and G. K. Fleming, *Macromolecules*, *24*, 1779-1783 (1991).

35. J. Bicerano, *Computational Polymer Science*, *2*, 177-201 (1992).

36. P. Zoller, in *Polymer Handbook*, edited by J. Brandrup and E. H. Immergut, Wiley, New York, third edition (1989), VI/475-VI/483.

37. R. Simha and T. Somcynsky, *Macromolecules*, *2*, 342-350 (1969).

38. R. Simha, *Macromolecules*, *10*, 1025-1030 (1977).

CHAPTER 4

THERMODYNAMIC PROPERTIES

4.A. Background Information

4.A.1. Thermophysical Properties

The interaction of thermal energy (i.e., heat) with the atoms which constitute a material determines some of the most important physical properties of the material. The properties describing this interaction at the most fundamental level are often called *thermophysical properties* [1,2]. There is, unavoidably, a certain amount of arbitrariness involved in deciding which properties are most crucial in describing the effects of thermal energy on materials. The following properties, however, are most often considered to be the key thermophysical properties:

1. The *heat capacity* quantifies the amount of thermal energy absorbed by a material upon heating, or released by it upon cooling. The heat capacity can be used to calculate all of the other *thermodynamic properties*, such as the enthalpy, entropy and Gibbs free energy, as functions of the temperature and pressure. The thermodynamic properties are often called *calorimetric properties* because they are usually measured by calorimetry [1-6].

2. The *coefficient of thermal expansion* quantifies the change in the dimensions of a specimen upon the addition or removal of thermal energy, i.e., as a result of the increase or decrease of the temperature of the specimen. The coefficient of thermal expansion and the other volumetric properties of polymers were discussed in Chapter 3.

3. Two properties quantify the flow of thermal energy through a specimen at a rate dependent upon the temperature differential, the fundamental material properties, and the preparation conditions of the particular specimen:

    (a) *Thermal conductivity* describes the steady-state heat flow. The proper understanding and correlation of thermal conductivity requires information about some properties to be discussed and correlated in later chapters. Detailed treatment of thermal conductivity will therefore be postponed to Chapter 14.

    (b) *Thermal diffusivity* describes the time-dependent, non-steady-state aspects of heat flow. Thermal diffusivity can be measured directly; however, it is usually estimated indirectly, by using a simple formal definition, as the thermal conductivity divided by the product of the specific heat capacity at constant pressure and the density. A mathematical model is often used to analyze the experimental data, and to relate the measured quantitities to the actual thermal diffusivity of the test specimen. Thermal diffusivity will be discussed further in Chapter 14.

The thermodynamic properties of polymers will be discussed in the remainder of this chapter. The current state-of-the-art in the understanding and the prediction of these properties will be reviewed first. New correlations for the heat capacity, which allow its prediction by combining topological parameters (connectivity indices) with geometrical parameters ("rotational degrees of freedom" for the backbone and the side groups) will then be presented. These new correlations enable the prediction of the heat capacities of polymers, and thus the prediction of all of their thermodynamic properties, without being limited by the availability of group contributions.

4.A.2. Thermodynamic Properties

The *molar heat capacity at constant pressure* (C_p) is the increase in the total thermal energy content of one mole of a material per degree Kelvin increase in its temperature (T) at a given pressure (p). (The subscript "p" denotes constant pressure.) Ordinary atmospheric pressure is used as the constant pressure in all standard measurements and tabulations of C_p. The total thermal energy content per mole is called the *enthalpy*, and is abbreviated as H.

$$C_p(T) \equiv \left[\frac{dH(T)}{dT}\right]_p \tag{4.1}$$

H is the sum of (a) the *internal energy* U, and (b) the *reversible work* done on the system at a pressure of p as a result of the thermal expansion (change of the molar volume V) caused by the added thermal energy.

$$H \equiv U + p \cdot V \tag{4.2}$$

The *molar heat capacity at constant volume* (C_v) is the increase in the internal energy U of one mole of a material per degree Kelvin increase in its temperature at constant volume. It is defined by Equation 4.3. It is related to C_p by Equation 4.4, where α denotes the thermal expansion coefficient and κ denotes the compressibility.

$$C_v(T) \equiv \left[\frac{dH(T)}{dT}\right]_v = \frac{dU}{dT} \tag{4.3}$$

$$C_v(T) \equiv C_p(T) - \frac{TV\alpha^2}{\kappa} \tag{4.4}$$

Standard definitions from classical thermodynamics are used to calculate the enthalpy H, entropy S, and Gibbs free energy G, as functions of the temperature, from the observed $C_p(T)$:

$$H(T) - H(0) \equiv \int_0^T C_p(T')dT' \tag{4.5}$$

$$S(T) - S(0) \equiv \int_0^T \frac{C_p(T')}{T'}dT' \tag{4.6}$$

$$G(T) \equiv H(T) - T \cdot S(T) \tag{4.7}$$

C_p, C_v and S are expressed in units of energy/(mole·degrees). H and G are expressed in units of energy/mole. Units of J/(mole·K) will be used for C_p, C_v and S, and units of J/mole will be used for H and G, in this book.

H(0) and S(0) denote the residual enthalpy and residual entropy at absolute zero temperature. According to the third law of thermodynamics, the residual entropy (the temperature-independent, "frozen-in" disorder) of a perfect crystal is zero at absolute zero temperature. S(0)>0 for amorphous materials and for crystals with defects and/or imperfections. It has been shown [3] that if the structure of a polymer chain is divided up somewhat arbitrarily into "mobile beads" expected to be the smallest subunits able to perform independent motions, the contribution per mobile bead to S(0) is typically in the range of 2 to 4 J/(mole·K).

It is often more convenient to use the "specific" (per unit weight) values of thermodynamic properties, rather than the "molar" (per mole) values. Specific values can be obtained by dividing molar values by the molecular weight per mole of repeat units, which are listed in the Appendix at the end of this book. For example, the *specific heat capacity* c_p at constant pressure is the heat capacity per unit weight of material, i.e., the amount of thermal energy required to increase the temperature of a unit weight of the material by one degree Kelvin at constant pressure.

$$c_p \equiv \frac{C_p}{M} \tag{4.8}$$

Finally, local statistical fluctuations of thermodynamic and volumetric quantities [7] can also be predicted by using $C_p(T)$ and $C_v(T)$. These fluctuations are often neglected in discussions of the thermodynamic properties of polymers. They are, however, important in determining certain aspects of the performance of materials. The local fluctuations of volume, density and enthalpy in polymers and in other amorphous solids have been extensively studied experimentally and theoretically [8-21]. They have also been used to provide valid reasons for choosing between competing models for the structures of polymeric glasses and melts [22].

It can be seen from the discussion presented above that $C_p(T)$ is the most important thermodynamic property of a polymer. All other thermodynamic properties can be calculated from $C_p(T)$. As will be shown in Chapter 14, the knowledge of C_p is also useful in estimating the thermal conductivity. From a directly practical perspective, the knowledge of $C_p(T)$ is also crucial in estimating how much heat has to be transferred, i.e., how much energy has to be expended, in heating polymers to elevated temperatures in order to apply such common manufacturing techniques as melt processing (extrusion or molding) and thermal curing.

4.A.3. Heat Capacity

$C_p(T)$ is the property normally measured by calorimetry. On the other hand, $C_v(T)$ is also important, since it is directly calculated by theoretical models which express the heat capacity of a material in terms of the vibrational motions of its atoms. Vibrational motions approximated by the harmonic oscillator model are commonly accepted to be the source of the heat capacities of solids. $C_v(T)$ is estimated by integrating over the frequency spectrum of these vibrations.

More refined models of the heat capacities of polymers can be obtained by deconvoluting the "skeletal vibrations" of chain molecules from their set of discrete "atomistic group vibrations", and by further deconvoluting the "intramolecular" component of the skeletal vibrations from the "intermolecular" (i.e., interchain) component. The major portion of the heat capacity at temperatures of practical interest (i.e., temperatures which are not too low) is accounted for by the atomistic group vibrations. The remaining portion of the heat capacity arises from skeletal modes. Detailed discussion of these issues is beyond the scope of this chapter. The reader is referred to the reviews provided by references [1-3] for further details and lists of the original publications.

In the context of the heat capacity, the term "vibrational motions" should be interpreted in the most general manner possible, to encompass all possible modes of motion of atoms or groups of atoms in a macromolecule. Such motions include bond stretching, bond bending and "rocking" motions, torsional oscillations, the "flipping" of a structural unit from one equilibrium position to another, and large-scale cooperative motions. These internal motions of atoms in a material are most directly studied by vibrational (infrared and Raman) spectroscopy [23].

Each atom has three thermodynamic degrees of freedom. The maximum possible number of modes of motion of any material is therefore three per atom. The observed heat capacities of polymers usually correspond to a much smaller effective number of degrees of freedom because of the constraints imposed on possible motions. In other words, these constraints restrict the amplitudes of the vibrations within the maximum number of three modes per atom, resulting in the observed lower effective number of degrees of freedom. For example [6], the heat capacities of polymers at room temperature typically correspond to roughly one degree of freedom per atom.

Despite the great practical and theoretical importance of the heat capacity, its accurate measurement was very difficult until recently. Differences in the manner in which different authors analyzed the data caused further confusion. Consequently, it is possible to find widely varying experimental values of $C_p(T)$, even for some of the best-characterized polymers. Recent improvements in calorimetry have greatly increased the accuracy of the measurements of $C_p(T)$. The monumental work of Wunderlich et al, who analyzed all available $C_p(T)$ data, subjected these data to strict criteria of acceptability, and determined and listed extensive tables of smoothed "recommended" values of $C_p(T)$ [24-34], has also contributed greatly to progress in this area.

The heat capacities of polymers can be classified into two types, namely, the heat capacities of "solid" polymers, which will be denoted by C_p^s, and the heat capacities of "liquid" polymers, which will be denoted by C_p^l.

The heat capacity of a polymer at temperatures of practical interest (i.e., at temperatures which are not extremely low) is mainly determined by a superposition of relatively localized motions of atoms and groups of atoms. Such modes of motion are not strongly affected by the presence or absence of long range periodic crystalline order. Data for both glassy amorphous polymers and crystalline polymers can thus be considered as "solid" heat capacities. For semicrystalline polymers where the C_p^s of the crystalline and amorphous phases have both been estimated at temperatures of practical interest (T≥100K), these two phases have almost identical values of C_p^s. At most, the amorphous phase of a polymer may sometimes have a very slightly (up to one or two percent) higher C_p^s than the crystalline phase, because the usually lower density and higher unoccupied volume of the amorphous phase may facilitate certain local modes of motion in which the moving structural units sweep a relatively large volume.

Similarly, the heat capacities of both "rubbery" and "molten" polymers can be considered under the category of "liquid" heat capacities. The heat capacity increases discontinuously from its "solid" value to its "liquid" value when an amorphous phase undergoes the glass transition, or when a crystalline phase melts. The jump $\Delta C_p(T_g)$ in the heat capacity at the glass transition, which is defined by Equation 4.9, is of considerable interest:

$$\Delta C_p(T_g) \equiv C_p^l(T_g) - C_p^s(T_g) \tag{4.9}$$

Wunderlich suggested [35], on the basis of the hole theory of liquids, that the value of $\Delta C_p(T_g)$ per mobile bead in the polymeric structure should be roughly constant. According to a recent tabulation [3], an average increase of $\Delta C_p(T_g) \approx 11.5 \pm 1.7$ J/(mole·K) per bead is found at the glass transition for a large number of polymers. The $\Delta C_p(T_g)$ values of many other polymers do not correlate as well with a simple count of beads.

Finally, it should be noted that the $\Delta C_p(T_g)$ of many polymers is quite sensitive to thermal history. $\Delta C_p(T_g)$ can even be used to study the manifestations of the thermal history, such as the effects of physical aging, for such polymers [36-38]. This factor is another important source of uncertainty and potential inconsistency among the $\Delta C_p(T_g)$ values published for many polymers.

The amorphous phase of a semicrystalline polymer will follow $C_p^l(T)$, while the crystalline phase will follow $C_p^s(T)$, between the glass transition temperature T_g and the melting temperature

T_m. Measurement of $C_p(T)$ as a function of percent crystallinity can therefore enable the extrapolation of both $C_p^s(T)$ and $C_p^l(T)$, as the limits of 100% and 0% crystallinity, respectively, in this temperature range. Experimental values of both C_p^s and Cp^l can thus be determined between T_g and T_m if measurements are performed on samples of different percent crystallinity.

To a good approximation, $C_p^s(T)$ and $C_p^l(T)$ both increase linearly with temperature for most polymers. It is, therefore, possible to extrapolate $C_p^l(T)$ into the temperature range ($T<T_g$) where the polymer is a solid, and $C_p^s(T)$ into the temperature range ($T>T_g$ for amorphous polymers and $T>T_m$ for semicrystalline polymers) where the polymer is a "liquid". Such extrapolations are often performed, especially to estimate the values of C_p^s and C_p^l at room temperature, namely $C_p^s(298K)$ and $C_p^l(298K)$, for use as parameters in calculating other physical properties. The rate of increase of $C_p^l(T)$ with increasing T is smaller than the rate of increase of $C_p^s(T)$.

If neither the values of $C_p^s(298K)$ or $C_p^l(298K)$ nor the slopes of the linear approximations to $C_p^s(T)$ or $C_p^l(T)$ are known for a polymer of interest, group contributions can often be used to predict $C_p^s(298K)$ and $C_p^l(298K)$, and the following linear approximations can be used to estimate $C_p^s(T)$ and $C_p^l(T)$ for $T\geq150K$ [6]. The mean deviation of the temperature coefficient $(3.0\cdot10^{-3})$ in Equation 4.10 is only 5%, while the mean deviation of the temperature coefficient $(1.2\cdot10^{-3})$ in Equation 4.11 is much larger (30%) [6].

$$C_p^s(T) \approx C_p^s(298K)\cdot[1 + 3.0\cdot10^{-3}\cdot(T - 298)] = C_p^s(298K)\cdot(0.106 + 3.0\cdot10^{-3}\cdot T) \quad (4.10)$$

$$C_p^l(T) \approx C_p^l(298K)\cdot[1 + 1.2\cdot10^{-3}\cdot(T - 298)] = C_p^l(298K)\cdot(0.640 + 1.2\cdot10^{-3}\cdot T) \quad (4.11)$$

In Section 4.E, Equation 4.11 will be replaced by a slightly modified equation with the same type of linear dependence of $C_p^l(T)$ on T, based on our analysis of the experimental data.

4.B. Improvements in the Ability to Predict the Heat Capacities of Polymers

$C_p(T)$ is the key quantity needed to predict all of the thermodynamic properties of polymers. According to equations 4.10 and 4.11, knowledge of $C_p^s(298K)$ and $C_p^l(298K)$ allows $C_p(T)$ to be estimated. New correlations will therefore be developed for $C_p^s(298K)$ and $C_p^l(298K)$, to enable the estimation of $C_p(T)$ without being limited by the availability of group contributions. These new correlations will utilize both the topological quantities (zeroth-order and first-order connectivity indices) and geometrical quantities ("rotational degrees of freedom"). Systematic trends in the observed $C_p^s(298K)$ and $C_p^l(298K)$ can all be completely accounted for, within the limits of the amount and accuracy of available experimental data, by combining these topological and geometrical parameters.

A new pair of "rotational degrees of freedom" parameters, suitable for use in combination with the connectivity indices, will be defined for the backbones and the side groups of polymers in Section 4.C. These two parameters are determined unambiguously, by simple inspection, for a polymer of arbitrary structure. Their simplicity is a result of the fact that most of the structural variations causing the observed differences of $C_p^s(298K)$ and $C_p^l(298K)$ between different polymers are adequately described by variations of the zeroth-order and first-order connectivity indices. Once these connectivity indices have been included in the correlation equations, only a small portion of the systematic variation highly dependent on the amount of local conformational freedom remains unaccounted for. This remaining portion is then described by the rotational degrees of freedom parameters.

The new correlations for $C_p^s(298K)$ and $C_p^l(298K)$ will be developed in sections 4.D and 4.E, respectively. The same set of topological and geometrical parameters will be used to correlate $\Delta C_p(T_g)$ in Section 4.F. The tables of recommended heat capacities published by Wunderlich *et al* [3,24-34] will be used as the main sources of experimental data. Whenever data are listed for the same polymer in the same physical state in more than one of these publications, the data from the most recent publication will be used. Data for a few additional polymers will be taken from other literature sources [2,6].

C_p^s and C_p^l values linearly interpolated or linearly extrapolated to 298.15K will be utilized in sections 4.D and 4.E, when these properties were not measured at this temperature. Many more polymers of interest are in the solid state than are in the "liquid" (molten or rubbery) state at room temperature. Only a few extrapolations over a wide temperature range will thus be needed to obtain $C_p^s(298K)$ for a large and extremely diverse set of polymers. A much larger number of extrapolations over a wide temperature range will be needed to obtain a dataset for $C_p^l(298K)$ encompassing the great diversity of structural features found in polymers.

All of the data listed by Wunderlich *et al* [3,24-34] will be included in the calculations, with the following exceptions:

1. Polymers containing germanium, selenium or iodine atoms will be excluded. These elements are not among the nine elements (C, N, O, F, H, Si, S, Cl and Br) which fall within the scope of this work.

2. Polypeptides will be excluded. Biopolymers are outside the scope of this work.

3. Polyphenylsilsesquioxane will be excluded. This polymer has a ladder-type structure, two bonds connecting each repeat unit to repeat units on its left and right. Ladder-type polymers are outside the scope of this work.

4. The measurements of $C_p(T)$ for poly(oxy-2,2-dichloromethyltrimethylene) were made using a highly crystalline commercial sample whose percent crystallinity is not listed. It is, therefore, impossible to calculate $C_p^l(298K)$ and $\Delta C_p(T_g)$ from the listed experimental values of $C_p(T)$, and this polymer will therefore also be excluded.

C_p^s(298K) values measured [39] by differential scanning calorimetry (DSC) as a part of our project will be used for the following nine polymers: poly(N-vinyl pyrrolidone), poly(N-vinyl carbazole), poly(vinyl butyral), poly(α-vinyl naphthalene), Torlon, poly(hexamethylene isophthalamide), poly(1,4-cyclohexylidene dimethylene terephthalate), poly[2,2-propane *bis*{4-(2,6-dibromophenyl)}carbonate], and Ultem 1000. A Mettler DSC 30 Low Temperature Cell, connected to a Mettler TC 10A TA Processor, was used in the DSC scans on these polymers. All of the DSC scans used a nitrogen purge, and were run from liquid nitrogen temperatures up to $(T_g - 10K)$. Specimens of 11 to 19 milligrams in weight were used, in crimped aluminum pans with punctured lids. Measurements were made on several specimens of most of the polymers tested. More than one scan was made for most of the specimens. The results of all scans were averaged to determine the experimental heat capacity of each polymer tested.

4.C. Rotational Degrees of Freedom of the Backbone and the Side Groups

The separation of the repeat units of polymers into "backbone" (BB) and "side group" (SG) portions was discussed in Section 2.D. Geometrical parameters, namely the "rotational degrees of freedom" N_{BBrot} and N_{SGrot}, will now be defined as heuristic descriptors of flexibility for the backbones and side groups of polymers, respectively, by using the following simple set of rules:

1. Each single bond in the backbone contributes +1 to N_{BBrot}, provided that this single bond is not in a ring.

2. Each single bond in a side group, or connecting a side group to the backbone, contributes +1 to N_{SGrot}, provided that (a) the rotations of this bond change the coordinates of at least one atom, *and* (b) this bond is not in a ring.

3. If the coordinates of all atoms remain unchanged upon rotation of a single bond in a side group, or rotation of a single bond connecting a side group to the backbone, such a single bond does not contribute to N_{SGrot}.

4. Multiple bonds, either in the backbone or in side groups, do not contribute either to N_{BBrot} or to N_{SGrot}.

5. Bonds in "rigid" rings, either in the backbone or in side groups, do not contribute either to N_{BBrot} or to N_{SGrot}.

6. Torsional motions around bonds in "floppy" rings are generally more restricted than motions around bonds which are not in rings, but less restricted than motions around bonds in "rigid" (especially aromatic) rings:

   (a) Each single bond in a "floppy" ring in the backbone contributes +0.5 to N_{BBrot}.

   (b) Each single bond in a "floppy" ring in a side group contributes +0.5 to N_{SGrot}.

7. Torsional motions around bonds in "semi-floppy" rings, which are fused to rigid rings but are not rigid rings themselves, are taken into account by using the following two rules:

(a) Bonds of a semi-floppy ring which are directly bonded to a rigid ring do not contribute to N_{BBrot} or N_{SGrot}.

(b) Each single bond of the semi-floppy ring which is not directly bonded to any of the rigid rings fused to the semi-floppy ring contributes +0.5 to N_{BBrot} or N_{SGrot}, as appropriate.

The separate consideration of N_{BBrot} and N_{SGrot} in developing the correlations provides significant improvements in many correlations, as well as valuable physical insights. On the other hand, it is sometimes more instructive to combine N_{BBrot} and N_{SGrot} into the total number of rotational degrees of freedom parameter N_{rot}:

$$N_{rot} \equiv N_{BBrot} + N_{SGrot} \tag{4.12}$$

A few examples will now be given to illustrate the determination of N_{BBrot}, N_{SGrot} and N_{rot}.

1. Polystyrene (see Figure 1.1), poly(vinyl fluoride) (see Figure 2.3), poly(vinyl alcohol), polymethacrylonitrile, poly(acrylic acid) and poly(N-vinyl pyrrolidone) (Figure 4.1) all have vinylic backbones with $N_{BBrot}=2$, but different values of N_{SGrot} and N_{rot}:

(a) Rotation of the C-F bond in poly(vinyl fluoride) does not change the coordinates of any atom. There are no other side groups in this polymer. $N_{SGrot}=0$ and $N_{rot}=(2+0)=2$.

(b) The C-O-H bond angle is smaller than 180°, i.e., C, O and H are not collinear, in poly(vinyl alcohol). Rotation of the C-O bond changes the coordinates of the hydrogen atom. $N_{SGrot}=1$ and $N_{rot}=(2+1)=3$.

(c) Rotation of the C-C bond connecting the methyl group to the backbone changes the coordinates of the three hydrogen atoms in the methyl group of polymethacrylonitrile. Rotation of the C-C bond connecting the cyanide group to the backbone does not change the coordinates of any atom, since the carbon atom in the cyanide group has sp hybridization and the C-C-N bond angle is 180°. $N_{SGrot}=1$ and $N_{rot}=(2+1)=3$.

(d) Rotation of the bond connecting the phenyl ring to the backbone in polystyrene changes the coordinates of the eight atoms in the phenyl ring which are not collinear with this bond. $N_{SGrot}=1$ and $N_{rot}=(2+1)=3$.

(e) Rotation of the C-C bond connecting the -COOH moiety to the backbone changes the coordinates of the two oxygen atoms and the hydrogen atom, and rotation of the C-O single bond changes the coordinates of the hydrogen atom, in the -COOH group of poly(acrylic acid). $N_{SGrot}=2$ and $N_{rot}=(2+2)=4$.

(f) Rotation of the C-N bond connecting the pyrrolidone ring to the backbone in poly(N-vinyl pyrrolidone) changes the coordinates of all atoms in the ring except the nitrogen atom. The ring itself is "floppy". Rotational oscillations around the five single bonds in this ring are limited in magnitude because these bonds are "tied together" in the ring, so that none of them can carry out a full independent rotation. Each one of these five bonds thus contributes only 0.5 to N_{SGrot}. $N_{SGrot}=(1+2.5)=3.5$ and $N_{rot}=(2+3.5)=5.5$.

2. Torlon (Figure 2.4) has no floppy rings. It has six single bonds in its backbone which are not in rigid rings. $N_{BBrot}=6$. It has no side group rotations which change the coordinates of any atoms, so that $N_{SGrot}=0$ and $N_{rot}=(6+0)=6$.

3. Ultem (Figure 2.4) has no floppy rings. It has eight single bonds in its backbone which are not in rigid rings. $N_{BBrot}=8$. It has two side group rotations which change the coordinates of some atoms, i.e., rotations around the axes of the C-C bonds connecting the two methyl groups to the backbone, so that $N_{SGrot}=2$ and $N_{rot}=(8+2)=10$.

4. Poly(vinyl butyral) (Figure 2.9) has two backbone single bonds which are not in a ring, and six backbone single bonds in a "floppy" ring. $N_{BBrot}=(2+3)=5$. Rotation of each of the three C-C bonds in the *n*-propyl side group changes the coordinates of all atoms further out in this side group. $N_{SGrot}=3$ and $N_{rot}=(5+3)=8$.

5. Poly[*o*-biphenylenemethane *bis*(4-phenyl carbonate)] (Figure 2.10) has $N_{SGrot}=0$ and $N_{BBrot}=N_{rot}=6$, because each of the three single bonds in the five-membered semi-floppy ring fused to the two phenyl rings is directly bonded to at least one of the two phenyl rings, so that these single bonds do not contribute to N_{SGrot}.

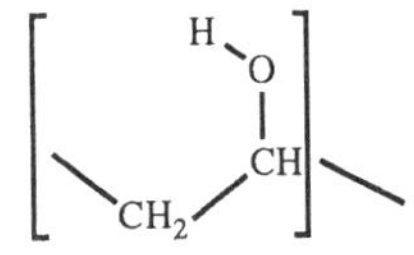

(a) Poly(vinyl alcohol).

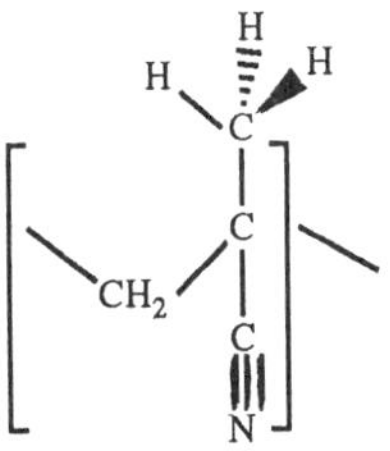

(b) Polymethacrylonitrile.

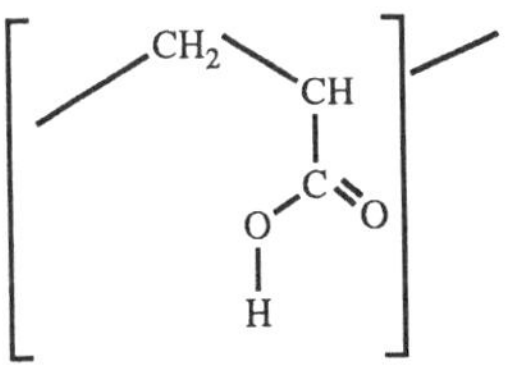

(c) Poly(acrylic acid).

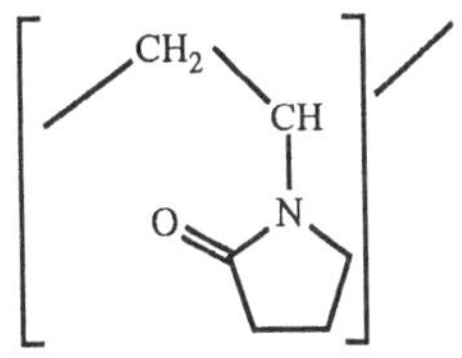

(d) Poly(N-vinyl pyrrolidone).

Figure 4.1. Repeat units of four polymers, drawn such as to facilitate immediate identification, by simple inspection, of those bond rotations which change the coordinates of some of the atoms, and therefore contribute to the count of the number of rotational degrees of freedom.

4.D. Correlation for the Heat Capacity of "Solid" Polymers at Room Temperature

The development of the correlation for $C_p{}^s$(298K) was started with the experimental $C_p{}^s$(298K) values of 97 polymers. The values of $C_p{}^s$(298K) for the different polymers were either found in the literature, or extrapolated or interpolated from the data found in the literature, or measured as a part of this project, as discussed in Section 4.B.

Several observations were made during the analysis of this dataset:

1. $C_p{}^s$(298K) correlates strongly with the zeroth-order and first-order connectivity indices. There are, however, significant differences among its correlation coefficients with these four indices.

2. The systematic and statistically significant portion of the deviation remaining between the experimental and the fitted values of $C_p{}^s$(298K) when these indices are used to correlate $C_p{}^s$(298K) can be accounted for by using N_{BBrot} and N_{SGrot} along with the connectivity indices in the linear regression equation.

3. The use of the single descriptor N_{rot} rather than separately varying the coefficients of N_{BBrot} and N_{SGrot} in the linear regression equation does not cause a statistically significant reduction in the accuracy of the correlation.

4. Two outliers were found, in terms of the percent deviation of the fitted values from the experimental values. These two outliers were removed, leaving 95 polymers in the dataset.

    (a) Poly(*p*-fluoro styrene).

    (b) Poly(imino-3,5-pyromellitolylimino-1,4-phenyleneoxy-1,4-phenylene).

5. *The replacement of silicon atoms by carbon atoms in the calculation of the first-order valence connectivity index $^1\chi^v$ (i.e., use of the alternative set of $^1\chi^v$ values listed in Table 2.3 for silicon-containing polymers) results in a much better fit than the use of the $^1\chi^v$ values listed in Table 2.2 for these polymers.*

The correlation coefficient between $^1\chi^v$ (with the replacement Si→C) and the experimental values of $C_p{}^s$(298K) is 0.9804, so that $^1\chi^v$ by itself accounts for 96.1% of the variation of the experimental $C_p{}^s$(298K) values in the final dataset of 95 polymers.

An additional connectivity index ($^0\chi$), the geometrical parameter N_{rot}, and the atomic correction term N_{Si}, were used along with $^1\chi^v$ in the final fit. Weight factors of $100/C_p{}^s$(exp) were utilized. The resulting four-parameter quantitative structure-property relationship is given by Equation 4.13. This correlation has a standard deviation of only 11.7 J(mole·K), and a correlation coefficient of 0.9938 which indicates that it accounts for 98.8% of the variation of the $C_p{}^s$(exp) values in the dataset. The average of the $C_p{}^s$(exp) values in this dataset is 235.5 J/(mole·K), so that the standard deviation is 5% of the average value.

$$C_p{}^s(298K) \approx 8.985304 \cdot {}^0\chi + 20.920972 \cdot {}^1\chi^v + 7.304602 \cdot (N_{rot} + 5 \cdot N_{Si}) \qquad (4.13)$$

The dataset used in developing this correlation contained a very large number of polymers, with a vast variety of structural features. The quality and the simplicity of the correlation for $C_p^s(298K)$ indicates that Equation 4.13 can be used for all polymers which fall within the scope of this work (as summarized in Section 4.B) with quite a high level of confidence, thus obviating the need for group contributions.

The results of these calculations are summarized in Table 4.1 and depicted in Figure 4.2.

Table 4.1. Experimental heat capacities C_p^s of solid polymers at room temperature (298K) in J/(mole·K), the geometrical parameter N_{rot} and the atomic correction index N_{Si} used in the correlation, and the fitted values of C_p^s, for 95 polymers. The connectivity indices $^0\chi$ and $^1\chi^v$, which are also used in the correlation, are listed in Table 2.2. The alternative set of $^1\chi^v$ values listed in Table 2.3 is used for the silicon-containing polymers.

Polymer	$C_p^s(298K,exp)$	N_{rot}	N_{Si}	$C_p^s(298K,fit)$
Polyoxymethylene	38.3	2.0	0	39.4
Polyethylene	43.4	2.0	0	48.2
Polyoxyethylene	55.1	3.0	0	63.5
Poly(vinyl chloride)	59.0	2.0	0	65.9
Poly(vinyl fluoride)	59.5	2.0	0	56.8
Poly(glycolic acid)	65.1	3.0	0	70.8
Poly(vinyl alcohol)	67.4	3.0	0	64.9
Polypropylene	67.8	3.0	0	71.6
Polyacrylonitrile	68.4	2.0	0	69.3
Polyoxytrimethylene	79.3	4.0	0	87.6
Poly(vinylidene chloride)	80.9	2.0	0	81.9
Poly(p-phenylene)	84.5	1.0	0	86.4
Polytrifluoroethylene	87.7	2.0	0	75.8
Poly(1,4-butadiene)	88.0	3.0	0	81.8
Polytetrafluoroethylene	89.8	2.0	0	85.8
Polyisobutylene	94.0	4.0	0	93.7
Polychlorotrifluoroethylene	99.6	2.0	0	93.7
Poly(vinyl acetate)	101.2	5.0	0	118.6
Polyisoprene	108.0	4.0	0	105.4
Poly[oxy(p-phenylene)]	108.5	2.0	0	103.4
Poly[thio(p-phenylene)]	110.2	2.0	0	120.5

☞ ☞ ☞   TABLE 4.1 IS CONTINUED IN THE NEXT PAGE. ☞ ☞ ☞

Table 4.1. CONTINUED FROM THE PREVIOUS PAGE.

Polymer	$C_p^s(298K,exp)$	N_{rot}	N_{Si}	$C_p^s(298K,fit)$
Poly(methacrylic acid)	111.9	5.0	0	118.9
Poly(methyl acrylate)	115.0	5.0	0	117.8
Poly(α-methyl acrylamide)	118.1	5.0	0	120.3
Poly(p-hydroxybenzoate)	121.8	3.0	0	134.4
Poly(propylene sulfone)	123.3	4.0	0	119.7
Polystyrene	126.5	3.0	0	133.5
Poly(ethylene oxalate)	129.1	6.0	0	142.4
Poly(p-xylylene)	134.7	3.0	0	133.8
Poly(methyl methacrylate)	137.0	6.0	0	140.7
Poly(p-chloro styrene)	141.1	3.0	0	151.3
Poly(4-methyl-1-pentene)	144.5	6.0	0	143.2
Poly(p-bromo styrene)	144.9	3.0	0	160.0
Poly(1-butene sulfone)	146.8	5.0	0	144.7
Poly[oxy(2,6-dimethyl-1,4-phenylene)]	149.0	4.0	0	151.1
Poly(α-methyl styrene)	149.8	4.0	0	156.5
Poly(p-methyl styrene)	150.0	4.0	0	157.2
Poly(N-vinyl pyrrolidone)	158.9	5.5	0	156.5
Poly(ε-caprolactone)	161.6	7.0	0	167.2
Poly(vinyl benzoate)	162.4	5.0	0	181.3
Poly(ethyl methacrylate)	166.5	7.0	0	166.7
Poly(ε-caprolactam)	168.9	7.0	0	169.6
Poly[oxy(2,6-dimethyl-5-bromo-1,4-phenylene)]	169.0	4.0	0	177.8
Poly(vinyl trimethylsilane)	179.1	6.0	1	177.8
Polyoxynaphthoate	180.6	3.0	0	186.8
Poly(α-vinyl naphthalene)	192.9	3.0	0	186.1
Poly(vinyl butyral)	204.6	8.0	0	208.8
Poly(dimethyl itaconate)	207.6	9.0	0	210.9
Poly(1-hexene sulfone)	209.5	7.0	0	192.9
Poly(n-butyl acrylate)	210.0	8.0	0	192.0
Poly(vinyl p-ethylbenzoate)	217.6	7.0	0	230.4
Poly(isobutyl methacrylate)	222.1	9.0	0	213.4
Poly(ethylene terephthalate)	223.9	7.0	0	228.9

☞ ☞ ☞   **TABLE 4.1 IS CONTINUED IN THE NEXT PAGE.** ☞ ☞ ☞

Table 4.1. CONTINUED FROM THE PREVIOUS PAGE.

Polymer	C_p^s(298K,exp)	N_{rot}	N_{Si}	C_p^s(298K,fit)
Poly(vinyl dimethylphenylsilane)	231.0	6.0	1	240.6
Poly(vinyl p-isopropylbenzoate)	235.4	8.0	0	253.5
Poly(n-butyl methacrylate)	235.9	9.0	0	214.9
Poly(N-vinyl carbazole)	245.8	3.0	0	223.8
Poly[oxy(2,6-diisopropyl-1,4-phenylene)]	246.7	8.0	0	248.1
Poly(p-methacryloxy benzoic acid)	260.2	8.0	0	253.3
Poly(tetramethylene terephthalate)	267.9	9.0	0	277.1
Poly(ethylene-2,6-naphthalenedicarboxylate)	268.5	7.0	0	281.3
Poly(m-phenylene terephthalamide)	271.2	6.0	0	272.6
Poly(vinyl p-t-butylbenzoate)	271.6	9.0	0	275.5
Poly(p-phenylene isophthalamide)	272.3	6.0	0	272.6
Poly[oxy(2,6-diphenyl-1,4-phenylene)]	272.8	4.0	0	276.5
Poly(m-phenylene isophthalamide)	280.2	6.0	0	272.6
Poly(o-phenylene terephthalamide)	286.2	6.0	0	272.7
Poly(o-phenylene isophthalamide)	291.4	6.0	0	272.7
Bisphenol-A polycarbonate	304.9	8.0	0	308.3
Poly(diethnyl diphenylsilylene)	306.9	5.0	1	293.4
Poly(1,4-cyclohexylidene dimethylene terephthalate)	312.3	11.0	0	355.6
Poly(11-aminoundecanoic acid)	318.3	12.0	0	290.2
Poly(ether ether ketone)	319.7	6.0	0	324.2
Polyheteroarylene X	323.0	4.0	0	348.3
Poly[di(n-propyl) itaconate]	328.1	13.0	0	311.1
Poly(hexamethylene adipamide)	329.2	14.0	0	339.1
Poly(12-aminododecanoic acid)	331.5	13.0	0	314.3
Poly(hexamethylene isophthalamide)	344.6	11.0	0	330.0
Torlon	351.4	6.0	0	390.8
Poly(hexamethylene azelamide)	405.7	17.0	0	411.5
Poly[N,N'-(p,p'-oxydiphenylene)pyromellitimide]	420.6	4.0	0	393.1
Poly[2,2-propane bis {4-(2,6-dibromophenyl)}carbonate]	427.7	8.0	0	414.8
Polyphenylenediamide	427.7	10.0	0	418.4
Poly(hexamethylene sebacamide)	439.0	18.0	0	435.6
Polyheteroarylene VIII	480.0	4.0	0	509.9
Nylon 6,12	491.0	20.0	0	483.8

☞ ☞ ☞   **TABLE 4.1 IS CONTINUED IN THE NEXT PAGE.** ☞ ☞ ☞

Table 4.1. CONTINUED FROM THE PREVIOUS PAGE.

Polymer	$C_p{}^s$(298K,exp)	N_{rot}	N_{Si}	$C_p{}^s$(298K,fit)
Udel	491.2	10.0	0	498.0
Polyheteroarylene XIV	516.0	6.0	0	522.2
Poly(dicyclooctyl itaconate)	538.4	17.0	0	503.2
Polyheteroarylene IX	543.7	10.0	0	577.6
Polyheteroarylene XIII	583.5	12.0	0	593.6
Polyheteroarylene XII	682.2	19.5	0	626.1
Polyheteroarylene XV	712.9	7.0	0	629.8
Ultem	720.5	10.0	0	653.5

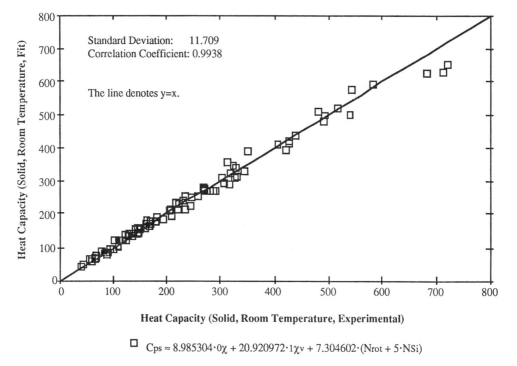

$$C_{ps} \approx 8.985304 \cdot {}^0\chi + 20.920972 \cdot {}^1\chi_v + 7.304602 \cdot (N_{rot} + 5 \cdot N_{Si})$$

Figure 4.2. Correlation using topological and geometrical parameters, for the experimental heat capacities $C_p{}^s$ of 95 "solid" (i.e., glassy or crystalline) polymers, at room temperature (298K). $C_p{}^s$(298K) is in J/(mole·K).

4.E. Correlation for the Heat Capacity of "Liquid" Polymers at Room Temperature

The development of the correlation for $C_p^l(298K)$ was started with the experimental $C_p^l(298K)$ values of 83 polymers. The values of $C_p^l(298K)$ for the different polymers were either found in the literature, or extrapolated or interpolated from the data found in the literature, as discussed in Section 4.B.

Several observations were made during the analysis of this dataset:

1. $C_p^l(298K)$ correlates strongly with the zeroth-order and first-order connectivity indices. There are, however, significant differences among its correlation coefficients with these four indices.

2. The systematic and statistically significant portion of the deviation remaining between the experimental and the fitted values of $C_p^l(298K)$ when these indices are used to correlate $C_p^s(298K)$ can be accounted for by using N_{BBrot} and N_{SGrot} along with the connectivity indices in the linear regression equation.

3. Unlike the correlation developed for $C_p^s(298K)$ in Section 4.D, there is a statistically significant improvement in the correlation for $C_p^l(298K)$ when the regression coefficients of N_{BBrot} and N_{SGrot} are varied separately rather than combining these two parameters into N_{rot} and using N_{rot} as the only geometrical parameter.

4. The regression coefficient of N_{BBrot} in the correlation equation is significantly larger than the regression coefficient of N_{SGrot}.

5. The effect of going from the solid state to the liquid state on the rotational degrees of freedom of the backbone is therefore far more pronounced than the effect on the rotational degrees of freedom of the side groups. This result is not really surprising. Rotational motions in the backbone are much more restricted than rotational motions in side groups of a solid polymer. The glass transition or melting therefore results in the "unfreezing" of a larger number of the backbone degrees of rotational freedom.

6. Two outliers [poly(α-methyl styrene) and poly(vinyl chloride)] were found in terms of the percent deviation of the fitted values from the experimental values. These two outliers, which had been obtained by extrapolating a small amount of $C_p^l(T)$ data over a wide temperature range, were removed, leaving 81 polymers in the dataset.

7. *The replacement of silicon atoms by carbon atoms in the calculation of the first-order valence connectivity index $^0\chi^v$ (i.e., use of the alternative set of $^0\chi^v$ values listed in Table 2.3 for silicon-containing polymers) results in a much better fit than the use of the $^1\chi^v$ values listed in Table 2.2 for these polymers.*

The correlation coefficient between $^0\chi^v$ (with the replacement Si→C) and the experimental values of $C_p^l(298K)$ is 0.9861, so that $^0\chi^v$ by itself accounts for 97.2% of the variation of the experimental $C_p^l(298K)$ values in the final dataset of 81 polymers.

An additional connectivity index ($^0\chi$) and the geometrical parameters N_{BBrot} and N_{SGrot} were used along with $^0\chi^v$ in the final fit. Weight factors of $100/C_p^l(exp)$ were utilized. The resulting four-parameter correlation is given by Equation 4.14. This correlation has a standard deviation of only 13.0 J(mole·K), and a correlation coefficient of 0.9955 which indicates that it accounts for 99.1% of the variation of the $C_p^l(exp)$ values in the dataset. The average of the $C_p^l(exp)$ values in this dataset is 275.7 J/(mole·K), so that the standard deviation is 4.7% of the average value.

$$C_p^l(298K) \approx 8.162061 \cdot {}^0\chi + 23.215188 \cdot {}^0\chi^v + 8.477370 \cdot N_{BBrot} + 5.350331 \cdot N_{SGrot} \qquad (4.14)$$

The dataset used in developing this correlation contained a very large number of polymers with a vast variety of structural features. The correlation for $C_p^l(298K)$ can therefore be used for all polymers which fall within the scope of this work (as summarized in Section 4.B) with a high level of confidence, and obviates the need for group contributions.

The results of these calculations are summarized in Table 4.2 and depicted in Figure 4.3.

The replacement Si→C was made in calculating $^0\chi^v$. There is no correction term proportional to N_{Si} in Equation 4.14. None of the parameters entering Equation 4.14 can distinguish between silicon and carbon atoms. Equation 4.14 therefore indicates that, within the limits of the available experimental data, the contribution of tetravalent (sp^3-hybridized) silicon atoms to $C_p^l(T)$ equals the contribution of carbon atoms in an isoelectronic configuration. This result is in contrast to the correlation for $C_p^s(298K)$, where silicon atoms make a much larger contribution than carbon atoms as a result of the term proportional to N_{Si} in Equation 4.13.

This difference between the contributions of silicon atoms to C_p^s and C_p^l may be rationalized in terms of less efficient packing in the solid around silicon atoms which have larger bond distances than carbon atoms with their neighbors. In the liquid, this packing effect becomes a less significant factor than it was in the solid, causing the contributions of isoelectronic silicon and carbon atoms to $C_p^l(298K)$ to become equal within the limits of the available experimental data.

Since the standard deviation of the temperature coefficient ($1.2 \cdot 10^{-3}$) of $C_p^l(T)$ in Equation 4.11 was very large (30%) [6], this coefficient was analyzed for the 80 polymers from our dataset for which C_p^l was available as a function of temperature. It was hoped that the coefficient would correlate with some topological and/or geometrical parameters, reducing the error involved in using Equation 4.11 to predict the $C_p^l(T)$ of an unknown polymer from its $C_p^l(298K)$. The temperature coefficients are listed in Table 4.2. No statistically significant relationship was found. The average temperature coefficient of $C_p^l(T)$ for our dataset was $1.3 \cdot 10^{-3}$, i.e., slightly larger than the value of $1.2 \cdot 10^{-3}$ found by van Krevelen [6]. Our standard deviation was 33%. Based on these results, Equation 4.11 was modified slightly, and the modified form (given by Equation 4.15) was used to estimate $C_p^l(T)$ from $C_p^l(298K)$ for unknown polymers in our work.

$$C_p^l(T) \approx C_p^l(298K) \cdot [1 + 1.3 \cdot 10^{-3} \cdot (T - 298)] = C_p^l(298K) \cdot (0.613 + 1.3 \cdot 10^{-3} \cdot T) \qquad (4.15)$$

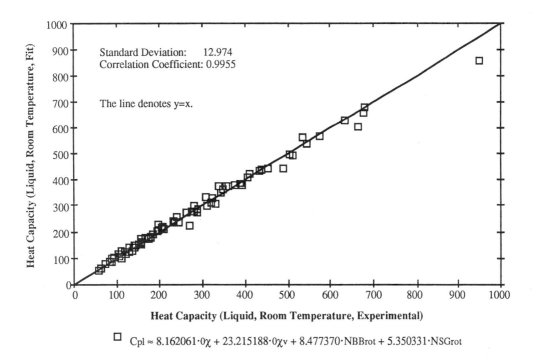

$$C_{pl} \approx 8.162061 \cdot {}^{0}\chi + 23.215188 \cdot {}^{0}\chi v + 8.477370 \cdot N_{BBrot} + 5.350331 \cdot N_{SGrot}$$

Figure 4.3. Correlation using topological and geometrical parameters, for the experimental heat capacities C_p^l of 81 "liquid" (i.e., molten or rubbery) polymers, at room temperature (298K). $C_p^l(298K)$ is in J/(mole·K).

Table 4.2. Experimental heat capacities C_p^l of "liquid" (molten or rubbery) polymers at room temperature (298K) in J/(mole·K), the geometrical parameters N_{BBrot} and N_{SGrot} used in the correlation, and the fitted values of C_p^l, for 81 polymers. The connectivity indices ${}^{0}\chi$ and ${}^{0}\chi^v$, which are also used in the correlation, are listed in Table 2.2. The alternative ${}^{0}\chi^v$ values listed in Table 2.3 are used for silicon-containing polymers. $d_{Cpl}(exp) \equiv [1000/C_p^l(298K)] \cdot (dC_p^l/dT)$ denotes the experimental value of the temperature coefficient of C_p^l.

Polymer	$C_p^l(298K,exp)$	$d_{Cpl}(exp)$	N_{BBrot}	N_{SGrot}	$C_p^l(298K,fit)$
Polyoxymethylene	57.6	0.65	2	0	54.4
Polyethylene	61.6	1.40	2	0	61.3
Poly(vinyl fluoride)	72.7	1.41	2	0	74.2
Poly(vinylidene fluoride)	82.9	1.48	2	0	88.7
Polyoxyethylene	87.8	0.76	3	0	85.1
Polypropylene	88.1	1.71	2	1	94.0

☞ ☞ ☞   **TABLE 4.2 IS CONTINUED IN THE NEXT PAGE.** ☞ ☞ ☞

Table 4.2. CONTINUED FROM THE PREVIOUS PAGE.

Polymer	C_p^l(298K,exp)	d_{Cpl}(exp)	N_{BBrot}	N_{SGrot}	C_p^l(298K,fit)
Polytrifluoroethylene	93.1	1.53	2	0	101.6
Polytetrafluoroethylene	102.6	1.50	2	0	116.1
Poly(1,4-butadiene)	105.7	1.68	3	0	108.2
Polyisobutylene	109.6	2.24	2	2	128.3
Poly(propylene oxide)	110.9	1.42	3	1	117.7
Poly(glycolic acid)	112.2	0.34	3	0	96.8
Poly(1-butene)	116.7	1.58	2	2	121.5
Poly(dimethyl siloxane)	117.8	1.02	2	2	121.3
Polyoxytrimethylene	119.2	0.85	4	0	115.7
Polyisoprene	129.8	1.85	3	1	142.0
Poly(vinylidene chloride)	130.0	------	2	0	123.8
Poly(β-propiolactone)	135.1	1.08	4	0	127.5
Poly(oxymethyleneoxyethylene)	135.9	0.81	5	0	139.4
Poly[oxy(p-phenylene)]	141.6	0.99	2	0	141.5
Polyoxytetramethylene	142.8	1.21	5	0	146.4
Poly(1-pentene)	143.8	2.04	2	3	149.1
Poly(methyl acrylate)	151.7	1.07	2	3	153.9
Poly(γ-butyrolactone)	155.7	0.66	5	0	158.2
Poly(vinyl acetate)	156.3	0.63	2	3	153.9
Poly[thio(p-phenylene)]	157.2	0.80	2	0	160.5
Poly(2-methyl-1,4-pentadiene)	157.5	1.52	4	1	172.7
Poly[oxy(diethylsilylene)]	165.0	1.38	2	4	176.4
Polystyrene	174.2	1.52	2	1	174.8
Poly(1-hexene)	178.0	2.08	2	4	176.6
Poly(ethyl acrylate)	178.5	1.21	2	4	181.4
Poly(δ-valerolactone)	183.2	0.95	6	0	188.8
Poly(methyl methacrylate)	183.7	1.30	2	4	188.2
Poly(oxymethyleneoxytetramethylene)	194.2	0.93	7	0	200.8
Poly(p-chloro styrene)	195.3	1.42	2	1	206.4
Poly(p-methyl styrene)	197.1	1.81	2	2	208.7
Poly(p-bromo styrene)	197.5	1.42	2	1	225.7
Poly(ethyl methacrylate)	203.3	2.41	2	5	215.7
Poly(ε-caprolactone)	206.5	1.41	7	0	219.5

☞ ☞ ☞   **TABLE 4.2 IS CONTINUED IN THE NEXT PAGE.** ☞ ☞ ☞

Table 4.2. CONTINUED FROM THE PREVIOUS PAGE.

Polymer	C_p^l(298K,exp)	d_{Cpl}(exp)	N_{BBrot}	N_{SGrot}	C_p^l(298K,fit)
Poly[oxy(2,6-dimethyl-1,4-phenylene)]	208.7	1.10	2	2	209.3
Poly(isobutyl acrylate)	231.5	1.46	2	6	241.6
Poly(n-butyl acrylate)	232.7	1.42	2	6	236.5
Polyoxynaphthoate	239.8	1.42	3	0	255.0
Poly(vinyl benzoate)	243.1	1.19	2	3	234.7
Poly(n-butyl methacrylate)	262.5	1.86	2	7	270.8
Poly(ε-caprolactam)	269.5	0.56	7	0	221.6
Poly(isobutyl methacrylate)	273.5	0.95	2	7	275.9
Poly(vinyl p-ethylbenzoate)	281.1	1.60	2	5	296.1
Poly(dimethyl itaconate)	281.6	0.82	2	7	275.6
Poly[ethylene-N-(β-trimethylsilyl ethyl)imine]	286.4	1.57	3	6	271.2
Poly(trimethylene succinate)	288.0	0.88	9	0	285.6
Poly(vinyl p-isopropylbenzoate)	309.4	1.55	2	6	328.8
Poly(vinyl dimethylbenzylsilane)	310.5	1.64	2	5	296.8
Poly(ethylene terephthalate)	321.2	0.87	7	0	311.5
Poly(n-hexyl methacrylate)	324.0	1.45	2	9	325.9
Poly(vinylene diphenylsilylene)	330.4	2.42	2	2	306.1
Poly[oxy(2,6-diphenyl-1,4-phenylene)]	337.2	2.05	2	2	370.9
Poly(trimethylene adipate)	343.7	1.16	11	0	347.0
Poly(vinyl p-t-butylbenzoate)	347.4	1.90	2	7	363.1
Poly(tetramethylene terephthalate)	354.5	1.28	9	0	372.8
Polyundecanolactone	360.7	1.11	12	0	372.8
Poly(11-aminoundecanoic acid)	376.7	1.92	12	0	374.9
Poly[di(n-propyl) itaconate]	388.8	0.91	2	11	385.8
Poly(ethylene-2,6-naphthalenedicarboxylate)	390.4	1.05	7	0	382.5
Poly(tetramethylene adipate)	394.0	0.83	12	0	377.6
Poly(12-aminododecanoic acid)	406.4	1.96	13	0	405.6
Bisphenol-A polycarbonate	410.8	1.39	6	2	423.1
Polytridecanolactone	433.9	0.85	14	0	434.1
Poly(ethylene sebacate)	438.3	1.27	14	0	439.0
Poly(ether ether ketone)	455.9	1.09	6	0	443.3
Poly(hexamethylene adipamide)	490.7	0.96	14	0	443.2
Polypentadecanolactone	505.1	0.77	16	0	495.5

☞ ☞ ☞   TABLE 4.2 IS CONTINUED IN THE NEXT PAGE. ☞ ☞ ☞

Table 4.2. CONTINUED FROM THE PREVIOUS PAGE.

Polymer	$C_p^l(298K,exp)$	$d_{Cpl}(exp)$	N_{BBrot}	N_{SGrot}	$C_p^l(298K,fit)$
Poly(dodecyl methacrylate)	511.5	1.41	2	15	491.1
Poly(hexamethylene sebacate)	534.3	1.41	18	0	561.6
Poly(hexamethylene azelamide)	545.8	1.69	17	0	535.2
Poly(hexamethylene sebacamide)	575.5	1.73	18	0	565.9
Nylon 6,12	634.8	1.81	20	0	627.2
Poly(dicyclooctyl itaconate)	663.6	1.21	2	15	602.5
Poly(octadecyl methacrylate)	677.8	1.67	2	21	656.3
Udel	678.1	1.33	8	2	678.4
Polyheteroarylene XV	950.4	1.47	5	2	854.3

4.F. Correlation for the Change in the Heat Capacity at the Glass Transition

The only truly consistent way to predict $\Delta C_p(T_g)$ is to compute it by inserting $C_p^s(T_g)$ and $C_p^l(T_g)$, calculated via equations 4.10 and 4.15, respectively, into Equation 4.9. This method requires the estimation of T_g (Chapter 6) prior to the extrapolation. In the computer program automating the use of our methodology, $\Delta C_p(T_g)$ is calculated by using this consistent scheme. It is, nonetheless, useful to develop an independent correlation for $\Delta C_p(T_g)$, to provide information on the effects of the glass transition on different types of motions of polymer chain segments. Such a correlation will be developed below, but it will not be used in any practical calculations.

A dataset of 89 observed $\Delta C_p(T_g)$ values was prepared from Wunderlich's tables [3,24-34], and correlated with the topological and geometrical parameters. Weight factors of $100/\Delta C_p(exp)$ were used in the linear regressions. The correlation for $\Delta C_p(T_g)$ was much weaker than the correlations for $C_p^s(298K)$ and $C_p^l(298K)$. Equation 4.16 resulted in a standard deviation of 14.5 J/(mole·K), and a correlation coefficient of 0.8698 which indicates that this equation accounts for only 75.7% of the variation of the $\Delta C_p(T_g)$ values in the dataset. The standard deviation is approximately 25% of the average $\Delta C_p(exp)$ of 57.9 J/(mole·K) for this dataset.

$$\Delta C_p(T_g) \approx 6.618544 \cdot {}^1\chi^v + 5.623265 \cdot N_{BBrot} \qquad (4.16)$$

Silicon atoms must be replaced by carbon atoms in calculating ${}^1\chi^v$, i.e., the alternative set of ${}^1\chi^v$ values listed in Table 2.3 must be used for silicon-containing polymers, in applying either Equation 4.16 or the alternative Equation 4.17 which will be derived below.

N_{BBrot} is the only geometrical parameter in Equation 4.16, which does not contain a term proportional to N_{SGrot}. The addition of a term proportional to N_{SGrot} does not result in a statistically significant improvement in the quality of the correlation. When Equation 4.16 is considered along with Equation 4.13 for $C_p^s(298K)$ and Equation 4.14 for $C_p^l(298K)$, an interesting physical picture emerges. N_{BBrot} and N_{SGrot} are weighted equally in the correlation for $C_p^s(298K)$. N_{BBrot} is weighted more heavily in the correlation for $C_p^l(298K)$. Equation 4.16 shows that the greater relative importance of the rotational motions of the backbone in the liquid state is indeed the result of the "unfreezing" of a significant number of the rotational degrees of freedom of the chain backbone by the creation of additional "free volume", and the resulting facilitation of the larger-scale and/or cooperative motions of the chain backbone. There is no significant concomitant effect on the rotational degrees of freedom of the side groups.

The results of these calculations are summarized in Table 4.3 and shown in Figure 4.4. Much of the remaining deviation between the experimental values and the fitted values of $\Delta C_p(T_g)$ appears to be completely random. There may perhaps be some systematic variation, such as the underestimation by Equation 4.16 of the $\Delta C_p(T_g)$ values of most polymers containing ester groups, by an average of approximately 10 J/(mole·K) per ester group in the repeat unit. Because of the considerable uncertainty in the experimental values of $\Delta C_p(T_g)$, and the aforementioned fact that this alternative correlation is not an internally consistent method for predicting $\Delta C_p(T_g)$, further improvements of Equation 4.16 by the addition of atomic and/or group correction terms have not been pursued.

The 33 polymers for which Wunderlich's mobile bead model is not especially successful in correlating $\Delta C_p(T_g)$ are indicated in Table 4.3, where their $\Delta C_p(exp)$ values are enclosed in parentheses. These $\Delta C_p(exp)$ values are followed by a question mark in parentheses, i.e., by the notation (?) instead of the specification of the number of beads used, in the two main tabulations [3,34] from which most of the $\Delta C_p(T_g)$ analyzed in this section were taken. Most of the polymers for which $\Delta C_p(exp)$ deviates by large amounts from the values calculated by using Equation 4.16 are polymers for which the mobile bead model is not very successful in correlating $\Delta C_p(T_g)$ either. This result is not surprising, since Equation 4.16 attempts to take exactly the same physical factors into account as the mobile bead model, albeit in a different manner.

If the 33 polymers whose $\Delta C_p(exp)$ values are enclosed in parentheses in Table 4.3 are excluded from the dataset, and the remaining 56 $\Delta C_p(exp)$ values are correlated with $^1\chi^v$ and N_{BBrot}, a much better fit, given by Equation 4.17, is found. This fit has a a standard deviation of 9.3 J/(mole·K), and a correlation coefficient of 0.9612 which indicates that it accounts for 92.4% of the variation of the $\Delta C_p(T_g)$ values in the smaller dataset of 56 polymers. The standard deviation is approximately 17% of the average $\Delta C_p(exp)$ of 53.8 J/(mole·K) for this dataset.

$$\Delta C_p(T_g) \approx 6 \cdot {^1\chi^v} + 7 \cdot N_{BBrot} \tag{4.17}$$

Table 4.3. Experimental change ΔC_p of the heat capacity at the glass transition in J/(mole·K), the geometrical parameter N_{BBrot} used in the correlation, and the fitted values of ΔC_p, for 89 polymers. The connectivity index $^1\chi^v$, which is also used in the correlation, is listed in Table 2.2. The alternative set of $^1\chi^v$ values listed in Table 2.3 is used for the silicon-containing polymers. The 33 ΔC_p(exp) values listed in parentheses are those for which there is a question mark in parentheses (?) for the appropriate number of "mobile beads" in the tabulations [3,34].

Polymer	ΔC_p(exp)	N_{BBrot}	ΔC_p(fit)
Polytetrafluoroethylene	9.4	2	19.6
Polyethylene	10.5	2	17.9
Polytrifluoroethylene	13.8	2	19.0
Poly(vinyl fluoride)	17.0	2	18.1
Poly[oxy(2,6-diisopropyl-1,4-phenylene)]	(17.6)	2	44.0
Poly[oxy(2,6-dimethyl-5-bromo-1,4-phenylene)]	(18.0)	2	37.5
Polypropylene	19.2	2	20.5
Poly(vinyl chloride)	19.4	2	21.0
Poly(vinylidene fluoride)	21.2	2	18.4
Polyisobutylene	21.3	2	22.5
Poly[oxy(p-phenylene)]	21.4	2	26.0
Poly(1-butene)	23.1	2	24.0
Poly(1-hexene)	25.1	2	30.7
Poly(α-methyl styrene)	26.3	2	33.5
Poly(1-pentene)	27.0	2	27.3
Poly(1,4-butadiene)	27.2	3	27.8
Poly(dimethyl siloxane)	27.7	2	20.6
Polyoxymethylene	28.2	2	15.1
Poly(1,4-pentadiene)	28.9	4	36.7
Poly[oxy(diethylsilylene)]	29.2	2	28.0
Poly[thio(p-phenylene)]	29.2	2	31.4
Poly(n-butyl methacrylate)	(29.7)	2	36.8
Polystyrene	30.8	2	31.2
Polyisoprene	30.9	3	30.4
Poly(p-chloro styrene)	31.1	2	34.4
Poly(ethyl methacrylate)	(31.7)	2	30.2
Poly[oxy(2,6-dimethyl-1,4-phenylene)]	31.9	2	31.5
Poly(p-bromo styrene)	31.9	2	37.1

☞ ☞ ☞   **TABLE 4.3 IS CONTINUED IN THE NEXT PAGE.** ☞ ☞ ☞

Table 4.3. CONTINUED FROM THE PREVIOUS PAGE.

Polymer	$\Delta C_p(exp)$	N_{BBrot}	$\Delta C_p(fit)$
Poly(methyl methacrylate)	(32.7)	2	26.3
Poly(propylene oxide)	33.2	3	26.9
Poly(p-fluoro styrene)	33.3	2	31.9
Polyoxynaphthoate	33.5	3	43.9
Poly(4-methyl-1-pentene)	33.7	2	29.7
Poly(p-hydroxybenzoate)	34.0	3	34.6
Poly(2-methyl-1,4-pentadiene)	(34.3)	4	39.4
Poly(p-methyl styrene)	34.6	2	33.9
Poly(isobutyl acrylate)	(36.6)	2	33.5
Poly(p-xylylene)	37.6	3	36.9
Polyoxyethylene	38.2	3	24.0
Poly(isobutyl methacrylate)	(39.0)	2	35.8
Poly(methyl acrylate)	(42.3)	2	24.0
Poly[di(n-heptyl) itaconate]	(44.0)	2	73.9
Poly(glycolic acid)	(44.4)	3	23.8
Poly(n-butyl acrylate)	(45.4)	2	34.5
Poly(ethyl acrylate)	(45.6)	2	27.9
Polyoxytrimethylene	46.6	4	32.9
Poly(vinyl acetate)	(46.7)	2	24.2
Poly[ethylene-N-(β-trimethylsilyl ethyl)imine]	46.9	3	42.6
Bisphenol-A polycarbonate	48.8	6	74.5
Poly(β-propiolactone)	(50.4)	4	32.8
Polyoxytetramethylene	52.0	5	41.9
Poly(dimethyl itaconate)	(54.2)	2	33.1
Poly(vinyl dimethylbenzylsilane)	54.2	2	42.5
Poly(ethylene oxalate)	(56.2)	6	47.9
Poly(vinyl p-ethylbenzoate)	(56.9)	2	41.6
Poly(γ-butyrolactone)	57.5	5	41.7
Poly[di(n-propyl) itaconate]	(57.8)	2	47.5
Poly(ε-caprolactone)	59.5	7	59.6
Poly(p-methacryloxybenzoic acid)	(60.0)	2	41.5
Poly(oxymethyleneoxyethylene)	60.2	5	39.1
Poly(vinyl p-t-butylbenzoate)	(60.4)	2	46.2

☞ ☞ ☞ TABLE 4.3 IS CONTINUED IN THE NEXT PAGE. ☞ ☞ ☞

Table 4.3. CONTINUED FROM THE PREVIOUS PAGE.

Polymer	$\Delta C_p(exp)$	N_{BBrot}	$\Delta C_p(fit)$
Poly(δ-valerolactone)	65.1	6	50.6
Poly(vinyl *p*-isopropylbenzoate)	(66.6)	2	44.2
Poly(dicyclooctyl itaconate)	(67.5)	2	81.3
Poly(11-aminoundecanoic acid)	(68.4)	12	104.9
Poly(vinyl benzoate)	(69.5)	2	35.2
Poly(vinylene diphenylsilylene)	70.7	2	45.9
Poly(12-aminododecanoic acid)	(74.3)	13	113.9
Poly[oxy(2,6-diphenyl-1,4-phenylene)]	76.6	2	53.5
Poly(ethylene terephthalate)	77.8	7	67.3
Poly(oxy-1,4-phenylene-oxy-1,4-phenylene-carbonyl-1,4-phenylene)	78.1	6	80.0
Poly(oxymethyleneoxytetramethylene)	81.1	7	56.9
Poly(ethylene-2,6-naphthalenedicarboxylate)	81.6	7	76.6
Poly(ε-caprolactam)	93.6	7	60.3
Polyoxyoctamethylene	95.7	9	77.6
Poly[di(*n*-octyl) itaconate]	(95.8)	2	80.6
Poly[4,4'-isopropylidene diphenoxy di(4-phenylene)sulfone]	(102.5)	8	115.9
Poly(tetramethylene terephthalate)	107.0	9	85.1
Poly(hexamethylene azelamide)	(109.5)	17	147.4
Polyundecalactone	110.6	12	104.2
Poly(hexamethylene sebacamide)	(118.0)	18	156.3
Poly(tetramethylene adipate)	(140.0)	12	101.2
Nylon 6,12	(141.4)	20	174.2
Poly(hexamethylene adipamide)	145.0	14	120.6
Polytridecanolactone	146.0	14	122.1
Poly(ethylene sebacate)	154.0	14	119.1
Polypentadecanolactone	168.0	16	140.6
Polyheteroarylene XV	177.4	5	126.6
Poly[di(*n*-nonyl) itaconate]	(177.6)	2	87.2

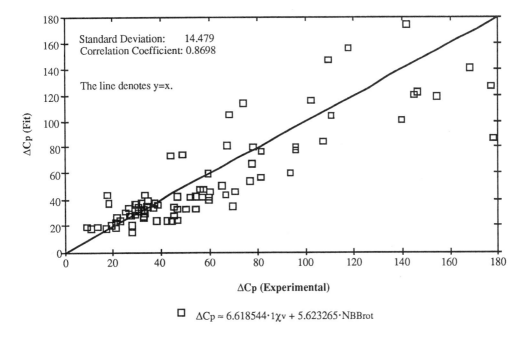

Figure 4.4. Correlation using topological and geometrical parameters, for the change ΔC_p of the heat capacity of 89 polymers at the glass transition. $\Delta C_p(T_g)$ is in J/(mole·K).

References and Notes for Chapter 4

1. Y. K. Godovsky, *Thermophysical Properties of Polymers*, Springer-Verlag, Berlin (1992).

2. E. V. Thompson, article titled *"Thermal Properties"*, in *Encyclopedia of Polymer Science and Engineering*, *16*, Wiley-Interscience, New York (1989).

3. B. Wunderlich, S. Z. D. Cheng and K. Loufakis, article titled *"Thermodynamic Properties"*, in *Encyclopedia of Polymer Science and Engineering*, *16*, Wiley-Interscience, New York (1989).

4. S. Z. D. Cheng, *J. Appl. Polym. Sci.: Appl. Polym. Symp.*, *43*, 315-371 (1989).

5. *Thermal Characterization of Polymeric Materials*, edited by E. A. Turi, Academic Press, New York (1981).

6. D. W. van Krevelen, *Properties of Polymers*, second edition, Elsevier, Amsterdam (1976). Chapter 5 of this book deals with calorimetric properties.

7. L. D. Landau and E. M. Lifshitz, *Statistical Physics, Part 1*, third edition, Pergamon Press, New York (1980).

8. V. V. Volkov, A. V. Gol'danskii, G. G. Durgar'yan, V. A. Onischuk, V. P. Shantorovich and Yu. P. Yampol'skii, *Polymer Science USSR, 29,* 217-224 (1987).

9. J. G. Victor and J. M. Torkelson, *Macromolecules, 20,* 2241-2250 (1987).

10. J. G. Victor and J. M. Torkelson, *Macromolecules, 20,* 2951-2954 (1987).

11. J. H. Wendorff and E. W. Fischer, *Kolloid-Z. u. Z. Polymere, 251,* 876-883 (1973).

12. J. Müller and J. H. Wendorff, *J. Polym. Sci., Polym. Lett. Ed., 26,* 421-427 (1988).

13. R. E. Robertson, *J. Polym. Sci., Polym. Symp., 63,* 173-183 (1978).

14. R. Simha, in *Molecular Basis of Transitions and Relaxations,* edited by D. J. Meier, Gordon and Breach Science Publishers, London (1978), 203-223.

15. C. M. Balik, A. M. Jamieson and R. Simha, *Colloid & Polymer Science, 260,* 477-486 (1982).

16. R. Simha, S. C. Jain and A. M. Jamieson, *Macromolecules, 15,* 1517-1521 (1982).

17. S. C. Jain and R. Simha, *Macromolecules, 15,* 1522-1525 (1982).

18. J. G. Curro, R. R. Lagasse and R. Simha, *Macromolecules, 15,* 1621-1626 (1982).

19. R. E. Robertson, R. Simha and J. G. Curro, *Macromolecules, 17,* 911-919 (1984).

20. J. Bicerano, *J. Polym. Sci., Polym. Phys. Ed., 29,* 1329-1343 (1991).

21. J. Bicerano, *J. Polym. Sci., Polym. Phys. Ed., 29,* 1345-1359 (1991).

22. E. W. Fischer and M. Dettenmaier, *J. Non-Crystalline Solids, 31,* 181-205 (1978).

23. E. B. Wilson, Jr., J. C. Decius and P. C. Cross, *Molecular Vibrations,* McGraw-Hill Book Company, New York (1955).

24. U. Gaur, H.-C. Shu, A. Mehta and B. Wunderlich, *J. Phys. Chem. Ref. Data, 10,* 89-117 (1981).

25. U. Gaur and B. Wunderlich, *J. Phys. Chem. Ref. Data, 10,* 119-152 (1981).

26. U. Gaur and B. Wunderlich, *J. Phys. Chem. Ref. Data, 10,* 1001-1049 (1981).

27. U. Gaur and B. Wunderlich, *J. Phys. Chem. Ref. Data, 10,* 1051-1064 (1981).

28. U. Gaur and B. Wunderlich, *J. Phys. Chem. Ref. Data, 11,* 313-325 (1982).

29. U. Gaur, S.-F. Lau, B. B. Wunderlich and B. Wunderlich, *J. Phys. Chem. Ref. Data, 11,* 1065-1089 (1982).

30. U. Gaur, B. B. Wunderlich and B. Wunderlich, *J. Phys. Chem. Ref. Data, 12,* 29-63 (1983).

31. U. Gaur, S.-F. Lau, B. B. Wunderlich and B. Wunderlich, *J. Phys. Chem. Ref. Data, 12,* 65-89 (1983).

32. U. Gaur, S.-F. Lau and B. Wunderlich, *J. Phys. Chem. Ref. Data, 12,* 91-108 (1983).

33. R. Pan, M. Y. Cao and B. Wunderlich, in *Polymer Handbook,* edited by J. Brandrup and E. H. Immergut, Wiley, New York, third edition (1989), VI/371-VI/409.

34. M. Varma-Nair and B. Wunderlich, *Heat Capacity and Other Thermodynamic Properties of Linear Macromolecules. X. Update of the ATHAS 1980 Data Bank.* This computerized

version of the data bank of heat capacities, which is now being updated continuously, is available from its authors.

35. B. Wunderlich, *J. Phys. Chem.*, *64*, 1052-1056 (1960).

36. K. Neki and P. H. Geil, *J. Macromol. Sci.-Phys.*, *B8*, 295-341 (1973).

37. G. A. Adam, A. Cross and R. N. Haward, *J. Materials Science*, *10*, 1582-1590 (1975).

38. J. R. Flick and S. E. B. Petrie, *Studies in Physical and Theoretical Chemistry*, *10*, 145-163 (1978). (Volume title: *Structure and Properties of Amorphous Polymers*. Edited by A. G. Walton. Contains the *Proceedings of the Second Symposium on Macromolecules*, held in Cleveland, Ohio, October 31-November 2, 1978).

39. Measured by K. Kuhlmann and C. A. Wedelstaedst (1990).

CHAPTER 5

COHESIVE ENERGY
AND SOLUBILITY PARAMETER

5.A. Background Information

5.A.1. Cohesive Energy

The *cohesive energy* E_{coh} of a material is the increase in the internal energy per mole of the material if all of its intermolecular forces are eliminated [1,2]. The *cohesive energy density* e_{coh}, which is defined by Equation 5.1, is the energy required to break all intermolecular physical links in a unit volume of the material. In polymers, these physical links mainly consist of interchain interactions of various types, which will be discussed below.

$$e_{coh} \equiv \frac{E_{coh}}{V} \tag{5.1}$$

These definitions imply that E_{coh} is a very fundamental property of a polymer, and that it should play a major role in the prediction of many other physical properties. This is indeed the case, as can be seen from the list below, which summarizes the properties whose prediction in this book utilizes the value of E_{coh}:

1. The most important use of E_{coh} is in calculating the *solubility parameter*, which quantifies the interactions between polymers and solvents, and which will be discussed below.
2. E_{coh} is also useful in correlating or predicting many other important properties of a polymer, such as its glass transition temperature (Chapter 6), surface tension (Chapter 7), dielectric constant (Chapter 9), mechanical properties (Chapter 11), and permeability to small molecules (Chapter 15).

For liquids of low molecular weight, the value of E_{coh} can be easily calculated from the molar heat of evaporation or from the vapor pressure as a function of the temperature. The experimental determination of E_{coh} for materials of high molecular weight (such as solid polymers) is not as straightforward, since such materials do not evaporate. The "experimental" values of E_{coh}, which are usually deduced indirectly from dissolution and/or swelling measurements at room temperature, vary over a wide range for many polymers.

Several group contribution tables have been developed to estimate the E_{coh} of polymers [2-6]. Most of these group contribution tables are based on data for small molecules. Some of them incorporate dissolution and swelling data on polymers. The best group contribution tables for the

E_{coh} of polymers are very useful for practical purposes, since they often produce reasonable trends in the prediction of properties dependent on E_{coh}.

The cohesive energy is often expressed in terms of the so-called *molar attraction constant* F and the molar volume V at room temperature [2]:

$$E_{coh} \approx \frac{F^2}{V(298K)}$$
(5.2)

F is an additive quantity both for ordinary molecules and for polymers. Group contributions for F and V(298K) can be combined to estimate E_{coh} in an indirect manner [2]. Equation 5.2 is therefore an alternative method for estimating E_{coh} via group contributions.

5.A.2. Components of the Cohesive Energy

E_{coh} encompasses all of the different types of forces of cohesion in a material:

1. Ground state oscillations of charge in a molecule can result in a random and temporary dipole moment. The polarizabilities of neighboring molecules cause *dispersion forces* to arise when this temporary random dipole interacts with the dipoles induced in all neighboring molecules because of the presence of the random dipole.

2. Permanent dipoles present in a molecule interact with neighboring dipoles and produce *polar cohesive forces*. These polar forces roughly correlate with the dipole moment, and the contribution of a given structural unit to the polar forces roughly correlates with its contribution to the dipole moment. The interactions between different structural groups resulting in the total polar component of E_{coh} are very complicated. The correlation for the polar component of E_{coh} is therefore unfortunately not a simple additive one.

3. Hydrogen bonds are considerably weaker than covalent bonds but much stronger than ordinary dipole-dipole interactions. When they are present, they add a very strong *hydrogen bonding* component to the forces of cohesion.

The total cohesive energy E_{coh} can therefore be formally divided into three parts [1,2], where E_d, E_p and E_h are the contributions of the dispersion, polar and hydrogen bonding components.

$$E_{coh} = E_d + E_p + E_h$$
(5.3)

It can also be assumed that, just like E_{coh} itself, the molar attraction constant also has three components, related to the dispersion (F_d), polar (F_p) and hydrogen bonding (F_h) forces [2]. The importance of the three components of F will be discussed further Section 5.A.4.

5.A.3.   Solubility Parameter

Solvation requires the interactions between a solvent and a solute to be strong enough to overcome the solvent-solvent and solute-solute interactions, resulting in the lowering of the Gibbs free energy of the system when the solute dissolves in the solvent. The *solubility parameter* δ quantifies the strength of the physical links between the components of a material. It is related to the cohesive energy density e_{coh} in a simple manner:

$$\delta \equiv \sqrt{e_{coh}} = \sqrt{\frac{E_{coh}}{V}} \qquad\qquad\qquad (5.4)$$

The solubility parameter is therefore a function of E_{coh} and the molar volume V. It is usually predicted by combining estimated values of E_{coh} and V [2].

The simple and straightforward calculation of δ in this manner is most likely to be reliable at or near room temperature, because the correlations on which the calculation of e_{coh} is based were developed by mainly using experimental data obtained at room temperature. For example, most of the group contribution tables for E_{coh}, including all such tables commonly used to estimate the E_{coh} of polymers, give estimates of E_{coh} at room temperature. The temperature dependence of δ has been discussed in the literature [1]. To summarize, $E_{coh}(T)$ decreases and $V(T)$ increases with increasing temperature, so that δ slowly decreases with increasing temperature.

As an example of the prediction of δ, consider polystyrene. The group contributions of van Krevelen and Hoftyzer [2] give E_{coh}=35610 J/mole. The correlation for the molar volume at room temperature (Section 3.C) gives $V \approx 97.0$ cc/mole. Substitution of these values of E_{coh} and V into Equation 5.4 gives $\delta \approx 19.2$ (J/cc)$^{0.5}$. This value is only slightly above the most reliable reported observed range [1] of 18.4 to 19.0 (J/cc)$^{0.5}$.

The solubility of a polymer in a solvent depends on how similar the δ values of the polymer and the solvent are. Crosslinking, crystallinity, and increasing molecular weight generally reduce the solubility of a polymer. In particular, the solubility parameter concept is only useful for amorphous polymers and for the amorphous phases of semicrystalline polymers. Many polymers of very high crystallinity are insoluble in solvents whose solubility parameters perfectly match their solubility parameters. Highly crystalline polymers generally only obey solubility rules based on solubility parameters for $T \geq 0.9T_m$, where T_m is the melting temperature [2].

The rate of interaction between the solvent and the polymer is also important in determining the influence of a solvent on the performance of a polymer in the time frame of a particular application. The rate of solvation increases with increasing surface/volume ratio of the polymer. For example, solvation will be faster (a) for a powder than for a pellet, (b) for a thin film than for a spherical ball, and (c) for a small spherical ball than for a large spherical ball. More generally, since the volume increases proportionally to the third power of the dimensions of a specimen while the surface area increases only proportionally to the second power of the dimensions, a small specimen of any shape will dissolve faster than a larger specimen of identical shape.

The usefulness of polymers in many technological applications is critically dependent on δ:

1. The removal of unreacted monomers, process solvents, and other synthesis or processing by-products, can both enhance the performance of the polymer and overcome health-related or environment-related objections to the use of certain types of polymers.

2. According to the Flory-Huggins solution theory, δ plays a key role in determining whether two polymers will be miscible into a blend [7]. The Flory-Huggins interaction parameter χ_{AB} (see Section 3.E for a more general discussion of interaction parameters), defined by Equation 5.5, is used as a rough quantifier of the likelihood of miscibility between Polymer A and Polymer B as a function of temperature, mole fractions of polymers, and their degrees of polymerization. V_{ref} is an appropriate "reference volume", often taken to be 100 cc/mole. R is the gas constant.

$$\chi_{AB} \equiv \frac{V_{ref}(\delta_A - \delta_B)^2}{RT} \tag{5.5}$$

The blend miscibility is assumed to decrease with increasing χ_{AB}. If strong specific interactions (such as hydrogen bonds) are present between structural units on polymers A and B, more elaborate versions of the Flory-Huggins solution theory, which attempt to account for these interactions, can be used [8]. It should be pointed out, however, that there is, at present, no simple and completely reliable method for predicting blend miscibility. Flory-Huggins interaction parameters and their more elaborate versions, and alternative methods (such as the various equation-of-state theories [9] discussed in Section 3.E), all provide correct predictions in many cases, but also unfortunately provide incorrect predictions in many other cases.

3. Environmental crazing and environmental stress cracking are often the limiting factors in determining how long the properties (and especially the strength) of the polymer will remain sufficient for a plastic part to be useful. These phenomena are dependent upon the solution and diffusion of environmental agents in the polymer [10,11], and thus upon δ.

4. In some applications, the interaction of the polymer with a specific "solvent" and/or with certain molecules carried by that solvent is not a detrimental event, but an essential aspect of the operation of the polymer. Reverse osmosis membranes and swollen hydrogels used in applications such as the desalination of water, kidney dialysis, soft contact lenses and surgical implants [12] are among such polymers.

5. Plasticization [13] is another area where the nature of the interaction of a polymer with molecules is critical to the usefulness of the polymer in many applications. Sears and Darby [13] have provided an outstanding description of the role of polymer-plasticizer compatibility in effective plasticization, and have reviewed the importance of δ and many other descriptors of compatibility in addressing this very complex problem.

5.A.4. Components of the Solubility Parameter

Hansen's three-component solubility parameter [1,2,14-16] is an attempt to separate the contributions of the different types of atomic interactions, which were summarized in Section 5.A.2, to the solubility parameter.

As discussed above, the solubility parameter is usually assumed to be equal to the square root of the cohesive energy density. It thus encompasses all of the different types of forces of cohesion in a material. Sometimes, however, the solvation of a polymer by a solvent, the behavior of plasticizers, pigments and other additives, or the blending of two polymers, requires certain specific types of interactions to exist between the polymer and the solvent, or between the two polymers that are to be blended. In such cases, the matching of the strengths of these specific types of interactions may play a crucial role in solvation or miscibility, and a more refined treatment may be necessary than can be provided by a single-valued solubility parameter. The three-component solubility parameter was developed in order to be able to predict the behavior of such systems more accurately.

The overall solubility parameter δ can be formally expressed in terms of the dispersion, polar and hydrogen bonding components δ_d, δ_p and δ_h, respectively:

$$\delta = \sqrt{\delta_d{}^2 + \delta_p{}^2 + \delta_h{}^2} \tag{5.6}$$

Equation 5.6 is a direct consequence of the definition of δ (Equation 5.4) and the expression of E_{coh} in terms of three separate components (Equation 5.3). The three components of δ have, thus far, only been accurately determined for a relatively small number of materials. In addition, the complexity of some of the interactions has prevented simple correlations using the three-component solubility parameter from providing much more than rough estimates. It has, in fact, been asserted [2] that it is impossible to derive a simple system for the accurate prediction of the polar and hydrogen-bonded components of the solubility parameter. These difficulties are even greater for polymers than they are for simple molecules.

There is a sufficient amount of data, on the other hand, to estimate the dispersion component of δ for a large number of polymers. The following relationship between δ_d and the dispersion component F_d of the molar attraction constant can be used to estimate the δ_d of many polymers. F_d is usually estimated as a sum of group contributions [2].

$$\delta_d = \frac{F_d}{V(298K)} \tag{5.7}$$

There are many common features as well as differences between physical factors determining the cohesive energy density (or the solubility parameter), and other important properties such as the surface tension (Chapter 7), dielectric constant and effective dipole moment (Chapter 9). See references [13], [17] and [18] for examples of the use of these commonalities and differences to provide both similar and complementary information concerning the behavior of polymers.

5.A.5. Improvements in the Ability to Predict the Cohesive Energies and Solubility Parameters of Polymers

The boiling temperature (T_b) of a simple liquid depends mainly on highly localized specific interactions between small subunits of the molecules in the liquid. Correlations involving only first-order connectivity indices (a sum of two terms involving $^1\chi$ and $^1\chi^v$) are therefore often sufficient to obtain excellent correlations [19] for T_b. Such correlations are especially simple and reliable for congeneric series of molecules.

The E_{coh} of a polymer depends mainly on localized intermolecular (interchain) interactions similar to the intermolecular interactions which determine the T_b of a simple liquid. E_{coh} can thus also be expected to correlate fairly well with zeroth-order and first-order connectivity indices.

On the other hand, unlike V_w (Section 3.B) which is completely determined by the *spatial arrangement* of the atoms which constitute the material, E_{coh} also depends on the *specific interactions* between the atoms. The specific interactions between atoms are not only functions of the spatial arrangement of the atoms but also very sensitively dependent upon their electronic properties. The correlation of E_{coh} with the zeroth-order and first-order connectivity indices is therefore expected to be considerably weaker than the correlation of V_w with the same indices.

The calculation of E_{coh} will be generalized in sections 5.B and 5.C. Correlations will be developed in terms of connectivity indices for E_{coh} values calculated by using the group contributions of Fedors [5,6] supplemented for a few polymers by the group contributions of Kaelble [4], and E_{coh} values calculated by using the group contributions of van Krevelen and Hoftyzer [2,3], respectively. E_{coh} values calculated by using the correlation in Section 5.B will be denoted by E_{coh1}. E_{coh} values calculated by using the correlation in Section 5.C will be denoted by E_{coh2}. The abbreviation E_{coh} will be reserved to refer to the cohesive energy in general. Unlike the group contribution tables which will be used as a springboard for their development, the new correlations will transcend the limitations of group contribution tables and will be applicable to all polymers.

Any of the available group contribution tables could be used in developing the correlation between E_{coh} and the connectivity indices. The choice of the group contribution tables of Fedors and of van Krevelen for the development of these correlations is based on the following reasons:

1. These two group contribution tables are more extensive than the others.

2. We have found both of them to be quite useful in solving practical problems.

3. These two tables seem to represent opposite extremes in the calculation of E_{coh}. The E_{coh} values calculated by using the table of Fedors are somewhat larger for most polymers. On the other hand, the values calculated by using the tables of van Krevelen and Hoftyzer are usually larger if a polymeric repeat unit contains a significant fraction of hydrogen bonding structural units or certain other extremely polar structural units.

4. Neither table always gives better predictions of properties dependent on E_{coh} for all polymers.

The molar volume V(298K) of a polymer of arbitrary repeat unit structure at room temperature can be estimated by using equations 3.13 and 3.14. The correlations which will be developed for E_{coh} in sections 5.B and 5.C will enable the estimation of E_{coh} for a polymer of arbitrary repeat unit structure, with results comparable in quality to predictions made by using two different group contribution tables. The correlations developed in this book will therefore allow two different predictions to be made for the δ (Section 5.D) of a polymer of arbitrary repeat unit structure.

F_d values calculated for a large set of polymers by using group contributions will be correlated with connectivity indices in Section 5.E. The ability to predict F_d (and therefore also the ability to predict the dispersion component δ_d of δ) will thus be extended to polymers containing structural units for which the group contributions to F_d are not available. The polar and hydrogen bonding components of the molar attraction constant F cannot, unfortunately, be predicted in a similar simple manner because of the complex specific interactions quantified by F_p and F_h. The ability to predict δ_d for all polymers therefore has very limited utility at this time.

5.B.   Correlation for the Fedors-type Cohesive Energy

E_{coh}(group) was calculated by using the group contributions of Fedors [5,6] for 118 of the polymers listed in Table 2.2. E_{coh}(group) was estimated for six additional polymers by using the group contributions of Kaelble [4], because some of the required group contributions were not available in the table of Fedors but were available in the table of Kaelble. The set of polymers to be used in developing the correlation for E_{coh1} therefore had a total of 124 members.

The strongest relationship of E_{coh}(group) with the zeroth-order and first-order connectivity indices was with $^1\chi$, and had a correlation coefficient of 0.9487. Weighted fits with weight factors of $10000/E_{coh}$(group) were used. Equation 5.8 resulted in a standard deviation of only 2292 J/mole and a correlation coefficient of 0.9974. The standard deviation is only 3.9% of the average E_{coh}(group) value of 59192 J/mole for this set of polymers.

$$E_{coh1} \approx 9882.5 \cdot {}^1\chi + 358.7 \cdot (6 \cdot N_{atomic} + 5 \cdot N_{group}) \qquad (5.8)$$

$$N_{atomic} \equiv 4N_{(-S-)} + 12N_{sulfone} - N_F + 3N_{Cl} + 5N_{Br} + 7N_{cyanide} \tag{5.9}$$

$$N_{group} \equiv 12N_{hydroxyl} + 12N_{amide} + 2N_{[non\text{-}amide\ -(NH)-\ unit]} - N_{(alkyl\ ether\ -O-)} - N_{C=C}$$
$$+ 4N_{[non\text{-}amide\ -(C=O)-\ next\ to\ nitrogen]} + 7N_{[-(C=O)-\ in\ carboxylic\ acid,\ ketone\ or\ aldehyde]}$$
$$+ 2N_{[other\ -(C=O)-]} + 4N_{(nitrogen\ atoms\ in\ six\text{-}membered\ aromatic\ rings)} \tag{5.10}$$

N_{atomic} is an atomic correction term, which is completely determined from the connectivity table of the polymeric repeat unit. N_{atomic} is expressed in terms of the total numbers of atoms of given types, with electronic configurations specified by appropriate pairs of δ and δ^v values. (Here, δ denotes the simple atomic index from graph theory as defined in Section 2.A, and not the solubility parameter.) All of the terms in the expression for N_{atomic} have been defined in earlier chapters. Recall that $N_{(-S-)}$ is the number of sulfur atoms in the lowest (divalent) oxidation state, $N_{sulfone}$ is the number of sulfur atoms in the highest oxidation state; N_F, N_{Cl} and N_{Br} are the total numbers of fluorine, chlorine and bromine atoms; and $N_{cyanide}$ is the number of nitrogen atoms with $\delta=1$ and $\delta^v=5$.

The simple first-order connectivity index $^1\chi$ cannot discriminate between elements from the same column but different rows of the periodic table in isoelectronic configurations. For example, it can be seen from Table 2.2 that poly(vinyl fluoride), poly(vinyl chloride) and poly(vinyl bromide) all have $^1\chi\approx1.3938$. Most of the terms entering N_{atomic} account for the fact that the non-hydrogen atoms in the lower rows and the right side of the periodic table, such as sulfur, chlorine and bromine, are more polarizable than atoms of the first row (C, N, O and F) or the left side (C and Si) of the relevant portion of the periodic table. Preliminary calculations showed that the use of these atomic correction terms in the correlation was more efficient than the use of additional terms proportional to the standard (unoptimized) values of the valence connectivity indices $^0\chi^v$ and $^1\chi^v$, which still left a deviation large enough to need correction. As discussed in Section 3.B.2, the use of N_{atomic} would become unnecessary if the δ^v values of the six electronic configurations specified in its definition were all optimized separately to provide the best fit for $E_{coh}(group)$, and $^0\chi^v$ and/or $^1\chi^v$ values based on the optimized δ^v values of these six electronic configurations were then used along with $^1\chi$ in developing the correlation.

N_{group} corrects for the underestimation or overestimation of the contribution of some (mostly polar) groups to E_{coh1} when $^1\chi$ is correlated with $E_{coh}(group)$. Unlike the terms in N_{atomic}, the terms in N_{group} cannot be described in terms of the electronic configuration of a single atom.

$N_{hydroxyl}$ denotes the total number of -OH moieties in alcohol or phenol environments. Unlike the term proportional to N_{OH} used in Section 3.B, the -OH groups in carboxylic acid (-COOH) or sulfonic acid ($-SO_3H$) environments are *not* counted in $N_{hydroxyl}$. This dependence on the environment of the -OH group causes $N_{hydroxyl}$ to be a group correction term, while N_{OH} is an atomic correction term.

In contrast to the amide group terms in Section 3.B, N_{amide} in Equation 5.10 denotes the *total* number of amide groups in the polymeric repeat unit, regardless of whether the amide groups are attached to aliphatic or aromatic groups. Every occurrence of adjacency between a -C=O and an -NH- unit is counted as a separate amide group. For example, a -CONH- linkage and a -CONH$_2$ (i.e., -CONH-H) unit terminating a side group both contribute 1 to N_{amide}, while -NHCONH- (as in a urea group) contributes 2 to N_{amide}. The only exception to this rule occurs for the urethane (-NHCOO-) group, which does not contribute anything to N_{amide}, but contributes 6 to N_{group} [2 from the non-amide -(NH)- unit and 4 from the non-amide -(C=O)- next to nitrogen].

$N_{[non\text{-}amide\ \text{-}(NH)\text{-}\ unit]}$ refers to the -(NH)- moieties in some rather rare polymers such as polybenzimidazoles, where there is no carbonyl group adjacent to the -(NH)- unit. For example, poly[2,2'-(m-phenylene)-5,5'-bibenzimidazole] (Figure 5.1) has two -(NH)- moieties not adjacent to carbonyl groups in its repeat unit. An -NH$_2$ (i.e., -NH-H) unit without an adjacent carbonyl unit, terminating a side group, is also counted in $N_{[non\text{-}amide\ \text{-}(NH)\text{-}\ unit]}$. As mentioned above, -(NH)- units in urethane groups are also counted as non-amide -(NH)- units.

$N_{[non\text{-}amide\ \text{-}(C=O)\text{-}\ next\ to\ nitrogen]}$ denotes the number of occurrences of carbonyl groups next to a nitrogen atom which does not have any attached hydrogen atoms. Imide and pyrrolidone units are examples of such bonding environments. For example, among the polymers shown in Figure 2.4, Torlon, poly(N-methyl glutarimide) and poly(N-phenyl maleimide) each have two such carbonyl groups, while Ultem has four of them. Poly(N-vinyl pyrrolidone) (Figure 4.1) has one such carbonyl group. As mentioned above, the carbonyl units in urethane groups are also counted as non-amide carbonyl units next to nitrogen.

$N_{[\text{-}(C=O)\text{-}\ in\ carboxylic\ acid,\ ketone\ or\ aldehyde]}$ denotes the total number of carbonyl groups in carboxylic acid (-COOH), ketone [R-(C=O)-R' with R≠H and R'≠H] and aldehyde [R-(C=O)-H with R≠H] environments. For example, poly(oxy-1,4-phenylene-oxy-1,4-phenylene-carbonyl-1,4-phenylene) (Figure 2.4) and poly(acrylic acid) (Figure 4.1) each have one such carbonyl group, in a ketone and a carboxylic acid environment, respectively.

$N_{[other\ \text{-}(C=O)\text{-}]}$ denotes the -(C=O)- moiety occurring in any other type of environment besides amide groups, other moieties in which a nitrogen atom is adjacent to the carbonyl group, carboxylic acid groups, ketone groups and aldehyde groups. Such alternative carbonyl group environments include ester and carbonate linkages, and anhydride groups. For example, poly(glycolic acid) (Figure 2.4) and bisphenol-A polycarbonate (Figure 5.1) each have one such carbonyl group, while poly(maleic anhydride) (Figure 2.4) has two such carbonyl groups.

$N_{(alkyl\ ether\ \text{-}O\text{-})}$ denotes the number of ether (R-O-R') linkages between two units R and R' *both* of which are connected to the oxygen atom via an alkyl carbon atom. For example, poly(vinyl butyral), which is shown in Figure 2.9, has two alkyl ether linkages. On the other hand, phenoxy resin (Figure 2.4) has no alkyl ether linkages, since each one of its ether linkages is attached on one side to an aromatic carbon atom. Poly(dimethyl siloxane) (Figure 5.1) has no alkyl ether linkages satisfying our definition, since its ether-type oxygen atoms are attached to silicon atoms instead of carbon atoms.

$N_{C=C}$ is the number of ordinary carbon-carbon double bonds in the repeat unit of the polymer. Purely formal double bonds, as in the common Lewis diagrams of resonance forms of phenyl rings, are *not* counted in $N_{C=C}$. For example, polybutadiene, polyisoprene and polychloroprene (Figure 5.1) each have $N_{C=C}=1$.

The nitrogen atoms in heterocyclic groups such as pyridine and quinoxaline rings are examples of nitrogen atoms in six-membered aromatic rings, corrected for by the last term in N_{group}. This correction term is used whether the six-membered ring is isolated, or a part of a fused ring moiety. Poly(p-vinyl pyridine) (Figure 5.1) is the only polymer in our dataset which has a nitrogen atom in a six-membered aromatic ring. This particular correction term was not estimated from the fit to E_{coh}(group), but by comparison with experimental data for various properties of polymers and small molecules containing such heterocyclic aromatic rings.

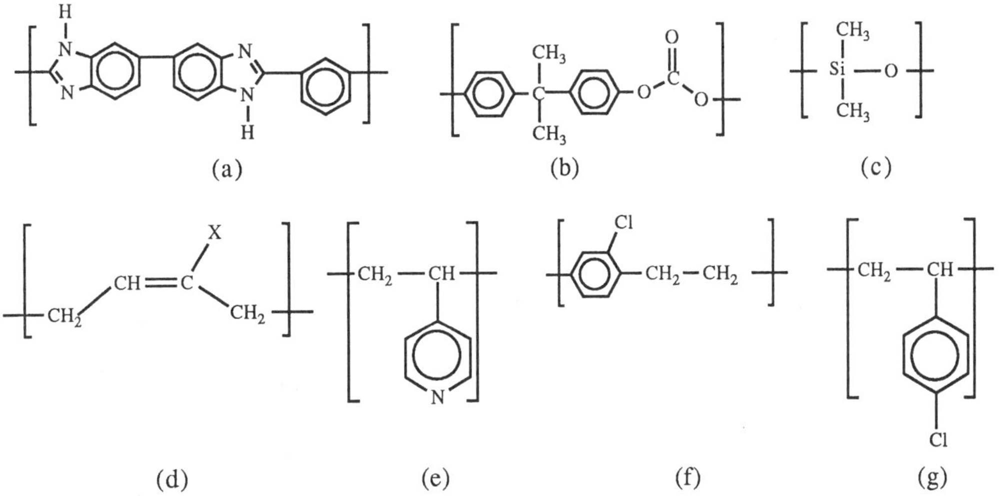

Figure 5.1. Schematic illustrations of the repeat units of several polymers discussed in the text, in the context of structural correction terms used in the correlations for the cohesive energy in sections 5.B and 5.C. (a) Poly[2,2'-(*m*-phenylene)-5,5'-bibenzimidazole]. (b) Bisphenol-A polycarbonate. (c) Poly(dimethyl siloxane). (d) Poly(1,4-butadiene) (X=H), polyisoprene (X=CH$_3$) and polychloroprene (X=Cl). Only the *cis* isomer is shown for the polydienes, which can also have *trans* isomerization. (e) Poly(p-vinyl pyridine). (f) Poly(chloro-*p*-xylylene). (g) Poly(p-chloro styrene).

The results of the linear regression are listed in Table 5.1 and shown in Figure 5.2. Table 5.1 is at the end of Section 5.C because it also contains the results of the calculations in Section 5.C.

N_{Fedors}, which is defined by Equation 5.11, is listed instead of N_{atomic} and N_{group} in Table 5.1, to prevent overcrowding of this table.

$$N_{Fedors} \equiv 6 \cdot N_{atomic} + 5 \cdot N_{group} \tag{5.11}$$

Table 5.1 also includes the predictions made by using Equation 5.8 for the cohesive energies of 16 polymers whose E_{coh} could not be calculated by using the group contributions of Fedors, even when supplemented with the group contributions of Kaelble. These additional predictions of E_{coh1} demonstrate the ability of Equation 5.8 to provide greater generality and predictive power than the group contribution tables on which it was based.

In most of the correlations to be developed in this book, the values of the physical properties of all types of polymers will be included in a single dataset for each property, and "general purpose" correlations applicable to all polymers will be obtained. If desired, it is also possible to derive "designer correlations" for each property, by selecting sets of structurally and/or compositionally related polymers, and deriving a separate correlation for each such set. Such specialized designer correlations are sometimes useful because they can provide greater accuracy than the general purpose correlations while using fewer fitting parameters. This procedure will be demonstrated in Section 17.B.3, where the E_{coh}(Fedors) values of simple hydrocarbon polymers (i.e., polymers containing only C and II, but no heteroatoms) will be correlated with $^1\chi$ and $^1\chi^v$.

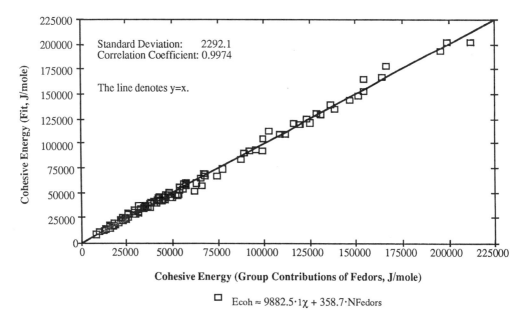

$$\square \quad E_{coh} \approx 9882.5 \cdot {}^1\chi + 358.7 \cdot N_{Fedors}$$

Figure 5.2. A fit using connectivity indices, to the cohesive energy calculated by the group contributions of Fedors supplemented by the group contributions of Kaelble, for 124 polymers.

Many of the new correlations developed in this book can be used to calculate the "group contributions" of new structural units to the properties. Although the calculation of these group contributions is not necessary, readers used to thinking about polymeric properties in terms of additive schemes can gain further insights from them. In addition, the results of calculations using correlations in terms of connectivity indices can often be combined with the available group contributions to speed up the calculations, if the calculations are being performed manually. Such calculations will be illustrated in Section 17.C, in the context of the cohesive energy.

5.C. Correlation for the van Krevelen-type Cohesive Energy

A correlation in terms of connectivity indices will now be developed for the E_{coh}(group) values of 108 polymers, obtained by combining the group contributions of van Krevelen and Hoftyzer [2] with a later set of revised group contributions by van Krevelen [3]. In a few cases, where the value of a group contribution differed significantly between these two tables, the choice of the appropriate group contribution was made based on our own experience with the relevant physical properties of polymers containing that group.

Note that some of the group contributions in Ref. [3] are not strictly additive, i.e., the group contributions listed for some of the larger structural units differ from the sum of the contributions of the smallest subunits from which they are built. This is physically reasonable, since the immediate bonding environment of a group can play a role in the strength of the nonbonded cohesive interactions of that group. The reason for pointing out this feature of this group contribution table is that, for some of the polymers in the dataset, these group contributions can be combined in different but reasonable ways to calculate somewhat different values of E_{coh}(group).

A weighted linear regression was performed with weight factors of 10000/E_{coh}(group), resulting in Equation 5.12 with the correction term N_{VKH} defined by Equation 5.13. The standard deviation is 1647 J/mole, and the correlation coefficient is 0.9988, between E_{coh}(group) and the fitted values of E_{coh2}. The standard deviation is 3.0% of the average E_{coh}(group) value of 55384 J/mole.

$$E_{coh2} \approx 10570.9 \cdot (^0\chi^v - ^0\chi) + 9072.8 \cdot (2 \cdot ^1\chi - ^1\chi^v) + 1018.2 \cdot N_{VKH} \qquad (5.12)$$

When Equation 5.12 is used to predict E_{coh2}, $^0\chi^v$ and $^1\chi^v$ should be evaluated with all silicon atoms replaced by carbon atoms in the repeat unit. For example, the $^0\chi^v$ and $^1\chi^v$ values listed in Table 2.3 should be used instead of the $^0\chi^v$ and $^1\chi^v$ values listed in Table 2.2.

$$N_{VKH} \equiv N_{Si} + 3N_{(-S-)} + 36N_{sulfone} + 4N_{Cl} + 2N_{Br} + 12N_{cyanide} + 8N_{ketone}$$

$$+ 16N_{[non-amide -(C=O)- next to nitrogen]} + 33N_{HB} - 4N_{cyc} + 19N_{anhydride}$$

$$+ 2N_{(N with \delta=2, but not adjacent to C=O, and not in a six-membered aromatic ring)}$$

$$+ 7N_{(N in six-membered aromatic rings)} + 20N_{(carboxylic acid)}$$

$$+ \Sigma(4 - N_{row})_{(substituents with \delta=1 attached to aromatic rings in backbone)} \qquad (5.13)$$

Correction terms entering N_{VKH} which have not been encountered earlier will now be discussed. In Equation 5.13, N_{HB} is the total number of alcohol-type or phenol-type hydroxyl (i.e., -OH units in carboxylic and sulfonic acid groups are not counted) and amide groups. Urea and urethane groups are not counted in N_{HB}, but instead in $N_{[non-amide -(C=O)- next to nitrogen]}$, and thus contribute only 16 instead of 33 to N_{VKH}.

N_{ketone} is the number of ketone groups in the repeat unit.

In the last term in the definition of N_{VKH}, the summation is over vertices of the hydrogen-suppressed graph with $\delta=1$ linked to aromatic rings in the backbone of the repeat unit. N_{row} is the row of the periodic table in which the atom represented by the vertex with $\delta=1$ is located. $N_{row}=1$ if the vertex represents a methyl group or a fluorine atom, 2 if it represents a chlorine atom, and 3 if it represents a bromine atom. Methyl groups and fluorine atoms therefore contribute 3 to the sum, chlorine atoms contribute 2, and bromine atoms contribute 1, if they are attached to an aromatic ring in the backbone of the repeat unit. For example, the contribution of the Cl atom in poly(chloro-p-xylylene) (Figure 5.1) to the sum is 2. Since the Cl atom also contributes 4 to N_{VKH} through the $4N_{Cl}$ term, $N_{VKH}=6$ for poly(chloro-p-xylylene). On the other hand, the Cl atom in poly(p-chloro styrene) (Figure 5.1) is not attached to a backbone aromatic ring but to an aromatic ring in a side group. It does not contribute to the summation, and $N_{VKH}=4$ because of the $4N_{Cl}$ term for poly(p-chloro styrene).

$N_{anhydride}$ is the number of anhydride groups (two carbonyl units with an oxygen between them) in the repeat unit. For example, see Figure 2.4 for the structure of poly(maleic anhdride).

The results of these calculations are listed in Table 5.1 and depicted in Figure 5.3. The group contribution tables of van Krevelen and Hoftyzer do not contain some of the group contributions required to calculate the cohesive energies of 32 of the polymers listed in Table 5.1. Predictions made for these polymers by using equations 5.12 and 5.13 are also summarized in Table 5.1.

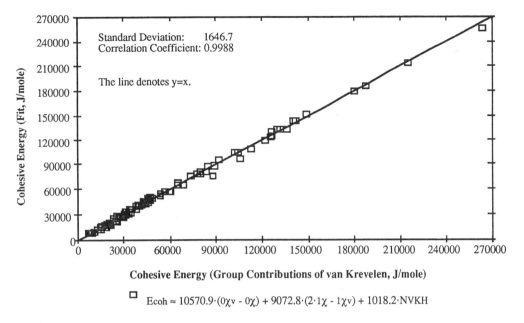

Figure 5.3. A fit using connectivity indices, to the cohesive energy calculated by combining the group contributions of van Krevelen and Hoftyzer with a later set of revised group contributions by van Krevelen, for 108 polymers.

Table 5.1. Cohesive energies calculated by group contributions, correction indices used in the linear regression procedure, and the fitted or predicted values of E_{coh}, for 140 polymers. E_{coh} is in J/mole. If no value is listed under the column labeled "Group", then the value listed under the column labeled "Fit/Pre" for that type of E_{coh} is a prediction made by utilizing the new correlation. Most of the connectivity indices used in these calculations are listed in Table 2.2. The alternative values of $^0\chi^v$ and $^1\chi^v$ for silicon-containing polymers, which were used in predicting the van Krevelen-type E_{coh}, are listed in Table 2.3.

Polymer	Fedors-type (E_{coh1})			van Krevelen-type (E_{coh2})		
	Group	N_{Fedors}	Fit/Pre	Group	N_{VKH}	Fit/Pre
Polyoxymethylene	8290	-5	8089	10490	0	9748
Polyethylene	9880	0	9883	8380	0	9073
Poly(vinyl fluoride)	12560	-6	11622	9080	0	9328
Polypropylene	13080	0	13774	14250	0	12646
Polyoxyethylene	13230	-5	13030	14680	0	14285
Poly(vinylidene fluoride)	13530	-12	12566	7550	0	7980
Polyisobutylene	15830	0	16870	17890	0	15488
Poly(dimethyl siloxane)	16160	0	16870	--------	1	16059
Polytrifluoroethylene	16210	-18	14837	8250	0	8724

☞ ☞ ☞ **TABLE 5.1 IS CONTINUED IN THE NEXT PAGE.** ☞ ☞ ☞

Table 5.1. CONTINUED FROM THE PREVIOUS PAGE.

Polymer	Fedors-type (E_{coh1})			van Krevelen-type (E_{coh2})		
	Group	N_{Fedors}	Fit/Pre	Group	N_{VKH}	Fit/Pre
Poly(propylene oxide)	16430	-5	16922	20550	0	17505
Poly(vinyl methyl ether)	16430	-5	17299	20190	0	18646
Poly(1,2-butadiene)	16990	-5	17299	--------	0	16451
Polytetrafluoroethylene	17180	-24	16097	6720	0	7668
Poly(1-butene)	18020	0	19092	18440	0	17528
Polyoxytrimethylene	18170	-5	17972	18870	0	18821
Poly(1,4-butadiene)	18500	-5	17972	18580	0	18580
Poly(vinyl chloride)	19920	18	20231	17590	4	17432
Poly(vinyl ethyl ether)	21370	-5	22240	24380	0	22388
Poly(thiocarbonyl fluoride)	22740	12	21175	12160	3	11811
Polyoxytetramethylene	23110	-5	22913	23060	0	23358
Polyisoprene	23210	-5	21863	22880	0	22644
Polychlorotrifluoroethylene	23250	0	24706	15240	4	16302
Poly(vinyl bromide)	23860	30	24536	20100	2	19821
Poly(ethylene sulfide)	24030	24	23433	17180	3	15495
Poly(vinylidene chloride)	25670	36	29784	24590	8	25250
Poly(vinyl trimethylsilane)	25890	0	25746	--------	1	24655
Poly(4-methyl-1-pentene)	26160	0	27549	23780	0	25292
Polyepichlorohydrin	28210	13	28696	28080	4	27017
Poly[oxy(methyl γ-trifluoropropylsilylene)]	29700	-18	32225	--------	1	24823
Poly(vinyl *sec*-butyl ether)	29510	-5	31073	29720	0	30146
Polychloroprene	30050	13	28320	26230	4	27525
Poly(vinyl acetate)	31080	10	31136	24690	0	22567
Poly(methyl acrylate)	31080	10	31678	24690	0	23912
Poly(vinylidene bromide)	31110	60	38392	29610	4	31194
Poly(vinyl *n*-butyl ether)	31250	-5	32122	32760	0	31460
Poly(*p*-phenylene)	31940	0	29314	25140	0	27910
Poly(vinyl cyclopentane)	32610	0	34256	28190	-4	27376
Poly(methyl methacrylate)	33830	10	35097	27590	0	26997
Polyacrylonitrile	33900	42	34157	27690	12	27185
Poly[oxy(dipropylsilylene)]	35920	0	37834	--------	1	35305
Poly(acrylic acid)	36000	35	35329	--------	20	35356
Poly(vinyl propionate)	36020	10	36453	28880	0	27242

☞ ☞ ☞   **TABLE 5.1 IS CONTINUED IN THE NEXT PAGE.** ☞ ☞ ☞

Table 5.1. CONTINUED FROM THE PREVIOUS PAGE.

Polymer	Fedors-type (E_{coh1})			van Krevelen-type (E_{coh2})		
	Group	N_{Fedors}	Fit/Pre	Group	N_{VKH}	Fit/Pre
Poly(ethyl acrylate)	36020	10	36619	28880	0	27655
Polymethacrylonitrile	36650	42	37477	31330	12	30088
Poly(vinyl cyclohexane)	37550	0	39197	32190	-4	31913
Poly(vinyl alcohol)	38170	60	35296	43190	33	43298
Poly(methacrylic acid)	38750	35	38748	--------	20	38442
Poly(ethyl methacrylate)	38770	10	40039	31780	0	30739
Poly(methyl ethacrylate)	38770	10	40637	31780	0	32082
Poly(maleic anhydride)	--------	20	39831	--------	15	39236
Polystyrene	40310	0	39197	35610	0	36933
Poly(methyl α-chloroacrylate)	40670	28	41554	30930	4	31878
Poly(p-xylylene)	41820	0	39031	33520	0	37067
Poly(vinyl pivalate)	41970	10	43107	32630	0	33205
Poly(vinyl butyral)	42510	-10	44652	--------	-4	40157
Poly(ε-caprolactone)	42700	10	42067	34360	0	32746
Poly(α-methyl styrene)	43060	0	42616	39290	0	40018
Poly(β-alanine)	43370	60	45179	53380	33	52704
Poly[oxy(methylphenylsilylene)]	43390	0	42616	--------	1	40589
Poly(2,5-benzoxazole)	--------	0	43806	--------	2	43380
Poly[oxy(2,6-dimethyl-1,4-phenylene)]	44710	0	42207	46300	6	46645
Poly(t-butyl methacrylate)	44720	10	46427	35530	0	36041
Poly(p-methyl styrene)	45020	0	43090	39390	0	40909
Poly(o-methyl styrene)	45020	0	43256	40250	0	41159
Poly(N-vinyl pyrrolidone)	--------	20	45488	--------	12	45339
Poly(ethyl α-chloroacrylate)	45610	28	46495	35120	4	35620
Poly(n-butyl acrylate)	45900	10	46502	37260	0	36738
Poly[thio(p-phenylene)]	46090	24	42698	33940	3	36361
Poly(p-vinyl pyridine)	--------	20	46371	--------	7	44048
Poly(N-methyl acrylamide)	46570	60	49613	59250	33	57234
Poly(isobutyl methacrylate)	46910	10	48496	37120	0	38504
Poly(sec-butyl methacrylate)	46910	10	48872	37120	0	38497
Poly(oxy-1,1-dichloromethyltrimethylene)	47680	31	48954	43460	8	44107
Poly(oxy-2,2-dichloromethyltrimethylene)	47680	31	48954	43460	8	44669
Poly(p-methoxy styrene)	48370	0	48405	45680	0	46490

☞ ☞ ☞   TABLE 5.1 IS CONTINUED IN THE NEXT PAGE. ☞ ☞ ☞

Table 5.1. CONTINUED FROM THE PREVIOUS PAGE.

Polymer	Fedors-type (E_{coh1})			van Krevelen-type (E_{coh2})		
	Group	N_{Fedors}	Fit/Pre	Group	N_{VKH}	Fit/Pre
Poly(*n*-butyl methacrylate)	48650	10	49921	40160	0	39812
Poly(α,α,α',α'-tetrafluoro-*p*-xylylene)	49120	-24	45353	31860	0	31840
Polyacrylamide	50230	60	44296	--------	33	49379
Poly(vinyl sulfonic acid)	--------	72	51572	--------	36	49124
Poly(*p*-chloro styrene)	51860	18	49546	42740	4	45788
Poly(*o*-chloro styrene)	51860	18	49712	43600	4	46039
Poly(α-methyl acrylamide)	52980	60	47716	--------	33	52463
Poly(chloro-*p*-xylylene)	53370	18	49546	--------	6	48211
Poly(styrene sulfide)	54460	24	52747	44410	3	43965
Poly(methyl α-cyanoacrylate)	54650	52	55703	41030	12	41594
Poly(*p*-bromo styrene)	55800	30	53851	45250	2	48761
Poly(*p*-*t*-butyl styrene)	55910	0	55060	47330	0	51548
Poly(2-ethylbutyl methacrylate)	56790	10	59130	45500	0	48266
Polyfumaronitrile	57920	84	58764	47000	24	45602
Poly(ε-caprolactam)	58190	60	60003	65950	33	66313
Poly(cyclohexyl methacrylate)	58300	10	59978	45530	-4	44620
Poly(vinyl benzoate)	58310	10	56558	46030	0	46501
Poly(*n*-hexyl methacrylate)	58530	10	59804	48540	0	48885
Poly(α-vinyl naphthalene)	--------	0	58796	--------	0	55741
Poly(propylene sulfone)[a]	62340	72	51796	--------	36	52399
Poly(3,4-dichlorostyrene)	63410	36	60062	--------	8	54896
Poly(benzyl methacrylate)	66000	10	64919	53120	0	54619
Poly(1-butene sulfone)[a]	66480	72	57113	--------	36	57282
Poly(N-methyl glutarimide)	--------	40	67617	--------	28	72539
Poly(8-aminocaprylic acid)	68070	60	69885	74330	33	75386
Poly(*n*-octyl methacrylate)	68410	10	69686	56920	0	57958
Poly(2-hydroxyethyl methacrylate)	68800	70	66502	64910	33	66578
Poly(N-vinyl carbazole)	--------	0	73620	--------	0	70322
Poly(1-hexene sulfone)[a]	74760	72	66996	--------	36	66354
Poly(N-phenyl maleimide)	--------	40	76653	--------	28	81299
Poly(ethylene terephthalate)	77820	20	74087	60340	0	57031

[a] The group contributions of Kaelble were used to supplement the group contributions of Fedors for this polymer.

☞ ☞ ☞   TABLE 5.1 IS CONTINUED IN THE NEXT PAGE. ☞ ☞ ☞

Table 5.1. CONTINUED FROM THE PREVIOUS PAGE.

Polymer	Fedors-type (E_{coh1})			van Krevelen-type (E_{coh2})		
	Group	N_{Fedors}	Fit/Pre	Group	N_{VKH}	Fit/Pre
Poly[3,5-(4-phenyl-1,2,4-triazole)-1,4-phenylene]	--------	0	83336	--------	4	83431
Polybenzobisoxazole	--------	0	87611	85140	4	86757
Poly(tetramethylene terephthalate)	87700	20	83970	68720	0	66104
Poly[1,1-ethane *bis*(4-phenyl)carbonate]	89600	10	89765	79360	0	77813
Bisphenol-A polycarbonate	92350	10	93060	83000	0	80729
Poly(ethylene-2,6-naphthalenedicarboxylate)	--------	20	93519	87700	0	75588
Poly[thio *bis*(4-phenyl)carbonate]	95610	34	94149	78100	3	77948
Poly[oxy(2,6-diphenyl-1,4-phenylene)]	99170	0	93052	89020	0	88407
Poly[2,2-hexafluoropropane *bis*(4-phenyl) carbonate]	99670	-26	104853	79380	0	80889
Cellulose	103340	170	112728	--------	95	135989
Polybenzobisthiazole	---------	48	104829	92140	10	95313
Poly[1,1-cyclohexane *bis*(4-phenyl)carbonate]	108680	10	109931	--------	-4	92145
Poly[2,2-propane *bis*{4-(2,6-dimethylphenyl)} carbonate]	111190	10	109294	112720	12	109849
Poly(hexamethylene adipamide)	116380	120	120006	131900	66	132626
Poly[1,1-(1-phenylethane) *bis*(4-phenyl) carbonate]	119580	10	118805	104400	0	105258
Poly(ether ether ketone)	119890	35	119048	105740	8	105555
Poly[1,1-(1-phenyltrifluoroethane) *bis*(4-phenyl) carbonate]	123240	-8	124703	102590	0	105340
Poly[2,2'-(*m*-phenylene)-5,5'-bibenzimidazole]	---------	20	124100	122000	8	119015
Phenoxy resin	124580	60	120878	124980	33	124274
Poly(hexamethylene isophthalamide)	128560	120	129722	141140	66	142308
Poly(*p*-phenylene terephthalamide)	130860	120	129222	140280	66	142304
Poly(hexamethylene sebacamide)	136140	120	139771	148660	66	150772
Poly[2,2-propane *bis*{4-(2,6-dichlorophenyl)} carbonate]	138550	82	135121	126120	24	125300
Poly[diphenylmethane *bis*(4-phenyl)carbonate]	146810	10	144551	125800	0	129788
Poly[1,1-biphenylethane *bis*(4-phenyl)carbonate]	151520	10	148120	129540	0	133170
Poly[4,4'-diphenoxy di(4-phenylene)sulfone][a]	154240	72	164930	---------	36	160071

[a] The group contributions of Kaelble were used to supplement the group contributions of Fedors for this polymer.

☞ ☞ ☞   TABLE 5.1 IS CONTINUED IN THE NEXT PAGE. ☞ ☞ ☞

Table 5.1. CONTINUED FROM THE PREVIOUS PAGE.

Polymer	Fedors-type (E_{coh1})			van Krevelen-type (E_{coh2})		
	Group	N_{Fedors}	Fit/Pre	Group	N_{VKH}	Fit/Pre
Poly[2,2-propane *bis*{4-(2,6-dibromophenyl)} carbonate]	154310	130	152338	136160	12	133117
Poly[N,N'-(*p,p'*-oxydiphenylene)pyromellitimide][a]	164420	80	167297	186580	64	184775
Torlon	---------	100	165305	179470	65	180030
Udel[a]	166140	72	177223	---------	36	171303
Resin F	195490	180	193834	214550	99	213217
Victrex[a]	199640	144	203050	---------	72	200153
Resin G	211560	180	201729	---------	99	222069
Ultem	---------	80	242995	263600	64	256001

[a] The group contributions of Kaelble were used to supplement the group contributions of Fedors for this polymer.

5.D. Solubility Parameter Calculations

The solubility parameters of polymers, calculated by utilizing Equation 5.4 in combination with E_{coh1} and E_{coh2} and the molar volumes computed from equations 3.13 and 3.14, are listed in Table 5.2. As mentioned earlier, it is not possible to measure the solubility parameter of a polymer directly, and this property is deduced indirectly from the results of various types of experiments, so that its value is subject to great uncertainty. The "experimental" δ values of the polymers, deduced in this manner and tabulated by van Krevelen [2], are also listed in Table 5.2.

The two different predicted values of δ differ significantly for many polymers. The reported experimental values of δ for many polymers also often span a very wide range. In fact, the range of experimental δ values is widest for polymers which have been studied most thoroughly by different workers, such as poly(propylene oxide) and poly(methyl methacrylate). Polymers for which a single experimental value of δ is listed are generally ones for which a value has only been reported once. Consequently, it is concluded that δ should be used with great caution (and preferably mainly to compare the predicted behavior of structural variants within a given family of polymers) in calculations where small differences between solubility parameters are important. Such calculations include the prediction of blend miscibility, where the effects of the uncertainties in the calculation or measurement of δ are compounded by the effects of the errors inherent in the simplications of theories which describe the complicated blending behavior of polymers in terms of just the differences between pairs of solubility parameters. It is also concluded that δ values calculated by different methods for different polymers should not be mixed in a comparison.

Table 5.2. "Experimental" values [2] of the solubility parameter (δ) deduced indirectly from various types of experiments, and predicted values (δ_1 and δ_2, respectively) computed by combining E_{coh1} and E_{coh2} with molar volumes calculated by using equations 3.13 and 3.14. The units for the solubility parameter are $(J/cc)^{0.5}$.

Polymer	$\delta("exp")$	δ_1	δ_2
Polytetrafluoroethylene	12.7	18.2	12.6
Poly(2,2,3,3,4,4,4-heptafluorobutyl acrylate)	13.7	19.9	15.4
Polychlorotrifluoroethylene	14.7 to 16.2	20.7	16.8
Poly(dimethyl siloxane)	14.9 to 15.6	14.6	14.2
Poly(propylene oxide)	15.4 to 20.3	17.3	17.6
Polyethylene	15.8 to 17.1	17.5	16.8
Polyisobutylene	16.0 to 16.6	16.0	15.4
Polyisoprene	16.2 to 20.5	16.9	17.2
Poly(1,4-butadiene)	16.6 to 17.6	17.4	17.7
Polypropylene	16.8 to 18.8	16.8	16.1
Polychloroprene	16.8 to 19.0	19.8	19.5
Poly(isobutyl methacrylate)	16.8 to 21.5	18.8	16.8
Poly(t-butyl methacrylate)	17.0	18.3	16.1
Polyoxytetramethylene	17.0 to 17.5	17.9	18.1
Polystyrene	17.4 to 19.0	20.1	19.5
Poly(n-butyl methacrylate)	17.8 to 18.4	19.1	17.1
Poly(isobutyl acrylate)	17.8 to 22.5	19.4	17.2
Poly(vinyl propionate)	18.0	20.4	17.6
Poly(n-butyl acrylate)	18.0 to 18.6	19.7	17.5
Poly(o-chloro styrene)	18.2	21.2	20.4
Poly(ethyl methacrylate)	18.2 to 18.7	19.6	17.2
Poly(ethylene sulfide)	18.4 to 19.2	21.3	17.3
Poly(2-ethoxyethyl methacrylate)	18.4 to 20.3	19.2	17.7
Poly(n-propyl acrylate)	18.5	20.0	17.6
Poly(methyl methacrylate)	18.6 to 26.2	20.2	17.7
Poly(ethyl acrylate)	18.8 to 19.2	20.5	17.8
Poly(styrene sulfide)	19.0	21.4	19.6
Poly[oxy(2,6-dimethyl-1,4-phenylene)]	19.0	19.5	20.5
Poly(vinyl acetate)	19.1 to 22.6	21.0	17.9
Polyepichlorohydrin	19.2	20.3	19.7
Poly(vinyl chloride)	19.2 to 22.1	21.2	19.6

☞ ☞ ☞ TABLE 5.2 IS CONTINUED IN THE NEXT PAGE. ☞ ☞ ☞

Table 5.2. CONTINUED FROM THE PREVIOUS PAGE.

Polymer	δ("exp")	δ$_1$	δ$_2$
Poly(vinyl bromide)	19.4	21.1	19.0
Poly[oxy(2,6-diphenyl-1,4-phenylene)]	19.6	21.1	20.6
Poly(methyl acrylate)	19.9 to 21.3	21.4	18.6
Poly(ethylene terephthalate)	19.9 to 21.9	22.5	19.8
Poly(benzyl methacrylate)	20.1 to 20.5	20.7	19.0
Bisphenol-A polycarbonate	20.3	20.7	19.3
Poly[4,4'-isopropylidene diphenoxy di(4-phenylene)sulfone]	20.3	22.2	21.8
Poly(vinylidene chloride)	20.3 to 25.0	22.8	21.0
Polyoxymethylene	20.9 to 22.5	18.8	20.6
Polymethacrylonitrile	21.9	24.5	22.0
Poly(ε-caprolactam)	22.5	23.9	25.1
Polyacrylonitrile	25.6 to 31.5	27.5	24.6
Poly(vinyl alcohol)	25.8 to 29.1	31.0	34.4
Poly(8-aminocaprylic acid)	26.0	22.6	23.4
Poly(hexamethylene adipamide)	27.8	23.9	25.1
Poly(methyl α-cyanoacrylate)	28.7 to 29.7	25.9	22.4
Poly(dimethyl silazane)	----	15.7	14.9
Poly[oxy(methyl γ-trifluoropropylsilylene)]	----	16.2	14.2
Poly(4-methyl-1-pentene)	----	16.8	16.1
Poly(vinyl *sec*-butyl ether)	----	17.0	16.8
Poly(1-butene)	----	17.1	16.3
Poly(1,2-butadiene)	----	17.2	16.8
Poly(vinyl ethyl ether)	----	17.4	17.5
Poly(vinyl *n*-butyl ether)	----	17.4	17.3
Poly(vinylidene fluoride)	----	17.7	14.1
Polyoxytrimethylene	----	18.1	18.5
Poly[oxy(methylphenylsilylene)]	----	18.2	17.8
Poly(*p-t*-butyl styrene)	----	18.3	17.7
Poly(vinyl butyral)	----	18.4	17.4
Poly(vinyl cyclohexane)	----	18.6	16.7
Poly(*n*-octyl methacrylate)	----	18.6	17.0
Poly(*sec*-butyl methacrylate)	----	18.8	16.7
Poly(*n*-hexyl methacrylate)	----	18.8	17.0
Poly(α-methyl styrene)	----	19.3	18.7

☞ ☞ ☞   **TABLE 5.2 IS CONTINUED IN THE NEXT PAGE.** ☞ ☞ ☞

Table 5.2. CONTINUED FROM THE PREVIOUS PAGE.

Polymer	$\delta("exp")$	δ_1	δ_2
Poly(p-methyl styrene)	----	19.4	18.9
Poly[2,2-propane bis{4-(2,6-dimethylphenyl)}carbonate]	----	19.5	19.5
Poly(cyclohexyl methacrylate)	----	19.8	17.1
Poly(p-fluoro styrene)	----	20.0	19.1
Poly(p-xylylene)	----	20.0	19.5
Poly(ε-caprolactone)	----	20.2	17.8
Cellulose tributyrate	----	20.4	17.9
Poly(oxy-2,2-dichloromethyltrimethylene)	----	20.6	19.7
Poly[2,2-propane bis{4-(2,6-difluorophenyl)}carbonate]	----	20.7	20.0
Poly(α-vinyl naphthalene)	----	20.9	20.4
Poly[1,1-cyclohexane bis(4-phenyl)carbonate]	----	21.0	19.2
Poly(chloro-p-xylylene)	----	21.1	20.8
Poly[1,1-(1-phenylethane) bis(4-phenyl)carbonate]	----	21.2	19.9
Poly(p-chloro styrene)	----	21.2	20.3
Poly(p-bromo styrene)	----	21.3	20.2
Poly(ethyl α-chloroacrylate)	----	21.5	18.8
Poly[diphenylmethane bis(4-phenyl)carbonate]	----	21.5	20.4
Poly(N-vinyl carbazole)	----	21.6	21.1
Poly[3,5-(4-phenyl-1,2,4-triazole)-1,4-phenylene-3,5-(4-phenyl-1,2,4-triazole)-1,3-phenylene]	----	21.6	21.7
Cellulose triacetate	----	21.6	18.5
Poly(dimethyl itaconate)	----	21.8	18.7
Phenoxy resin	----	21.9	22.2
Poly[o-biphenylenemethane bis(4-phenyl)carbonate]	----	22.0	20.9
Poly(N-vinyl pyrrolidone)	----	22.3	22.2
Poly[2,2-propane bis{4-(2,6-dichlorophenyl)}carbonate]	----	22.3	21.5[a]
Poly[1,1-dichloroethylene bis(4-phenyl)carbonate]	----	22.3	21.0
Poly[2,2-propane bis{4-(2,6-dibromophenyl)}carbonate]	----	22.4	20.9[a]

[a] When the experimental rather than the predicted molar volumes (see Table 3.5) of poly[2,2-propane bis{4-(2,6-dichlorophenyl)}carbonate] and poly[2,2-propane bis{4-(2,6-dibromophenyl)}carbonate] are used in calculating δ_2 for these polymers, their δ_2's become 21.3 and 21.4 $(J/cc)^{0.5}$, respectively, reversing their order and giving a prediction in better agreement with physical intuition. This example shows how small errors in input parameters can sometimes result in qualitatively incorrect conclusions. It is, therefore, in general desirable to refine the calculations whenever possible by incorporating any available reliable experimental data into them.

☞ ☞ ☞   **TABLE 5.2 IS CONTINUED IN THE NEXT PAGE.** ☞ ☞ ☞

Table 5.2. CONTINUED FROM THE PREVIOUS PAGE.

Polymer	δ("exp")	δ_1	δ_2
Poly(p-vinyl pyridine)	----	22.4	21.8
Poly(ethylene-2,6-naphthalenedicarboxylate)	----	22.6	20.3
Poly(hexamethylene sebacamide)	----	22.6	23.4
Poly(ether ether ketone)	----	22.6	21.2
Poly[thio bis(4-phenyl)carbonate]	----	22.8	20.8
Poly(vinylidene bromide)	----	22.8	20.5
Ultem	----	22.9	23.5
Poly[2,2'-(m-phenylene)-5,5'-bibenzimidazole]	----	23.1	22.7
Poly[4,4'-diphenoxy di(4-phenylene)sulfone]	----	23.1	22.7
Poly(thiocarbonyl fluoride)	----	23.4	17.5
Poly[N,N'-(p,p'-oxydiphenylene)pyromellitimide]	----	24.5	25.8
Torlon	----	24.5	25.6
Poly[4,4'-sulfone diphenoxy di(4-phenylene)sulfone]	----	24.6	24.4
Poly(hexamethylene isophthalamide)	----	24.7	25.9
Poly(acrylic acid)	----	25.7	25.7
Polyacrylamide	----	28.1	29.6
Cellulose	----	31.7	34.8

5.E. Correlation for Dispersion Component of the Molar Attraction Constant

Group contributions [2] were used to estimate the dispersion component F_d (in units of $J^{0.5} \cdot cm^{1.5}$/mole) of the molar attraction constant F for 107 polymers. These 107 F_d(group) values have the exceptionally large correlation coefficient of 0.9990 with the standard zeroth-order and first-order connectivity indices in the combination of [- $^0\chi$ + 2·($^0\chi^v$ + $^1\chi$ + $^1\chi^v$)], without requiring any correction terms at all. When a very simple correction term N_{Fd} is added, and a weighted linear regression with weight factors of 1000/F_d(group) is performed, a fit with a standard deviation of 30.355 and correlation coefficient of 0.9997 is found:

$$F_d \approx 97.95 \cdot [- \, ^0\chi + 2 \cdot (^0\chi^v + \, ^1\chi + \, ^1\chi^v)] + 134.61 \cdot N_{Fd} \qquad (5.14)$$

$$N_{Fd} \equiv N_{Si} - N_{Br} - N_{cyc} \qquad (5.15)$$

When Equation 5.14 is used to predict F_d, $^0\chi^v$ and $^1\chi^v$ should be evaluated with all silicon atoms replaced by carbon atoms in the repeat unit. For example, the $^0\chi^v$ and $^1\chi^v$ values listed in Table 2.3 should be used instead of the $^0\chi^v$ and $^1\chi^v$ values listed in Table 2.2.

The results of these calculations are listed in Table 5.3 and illustrated in Figure 5.4. Predictions made by using equations 5.14 and 5.15 for an additional 31 polymers whose F_d values could not be calculated via group contributions because of the lack of some of the required group contribution values are also listed in Table 5.3, to demonstrate the additional predictive power beyond group contributions provided by Equation 5.14.

Table 5.3. The dispersion component F_d of the "molar attraction constant" calculated by group contributions, correction index N_{Fd} used in the correlation, and fitted or predicted (fit/pre) values of F_d, for 138 polymers. For 31 of the polymers, F_d was not calculated by group contributions because of missing values, but was calculated by the new correlation. Most of the connectivity indices used in these calculations are listed in Table 2.2. The alternative values of $^0\chi^v$ and $^1\chi^v$ used for silicon-containing polymers are listed in Table 2.3. F_d is in units of $J^{0.5} \cdot cm^{1.5}/mole$.

Polymer	F_d(group)	N_{Fd}	F_d(fit/pre)
Polyoxymethylene	370	0	389
Polyethylene	540	0	530
Poly(vinyl alcohol)	560	0	599
Poly(vinyl fluoride)	570	0	578
Poly(vinylidene fluoride)	640	0	617
Polyoxyethylene	640	0	654
Polytrifluoroethylene	670	0	686
Polytetrafluoroethylene	740	0	738
Polypropylene	770	0	770
Polyacrylonitrile	780	0	783
Poly(vinyl chloride)	800	0	811
Poly(thiocarbonyl fluoride)	810	0	820
Poly(vinyl methyl ether)	870	0	899
Poly(propylene oxide)	870	0	901
Poly(acrylic acid)	880	0	891
Poly(vinyl bromide)	900	-1	933
Polyoxytrimethylene	910	0	919
Polyacrylamide	920	0	929
Poly(1,4-butadiene)	940	0	941
Poly(1,2-butadiene)	950	0	894

☞ ☞ ☞   **TABLE 5.3 IS CONTINUED IN THE NEXT PAGE.** ☞ ☞ ☞

Table 5.3. CONTINUED FROM THE PREVIOUS PAGE.

Polymer	F_d(group)	N_{Fd}	F_d(fit/pre)
Polychlorotrifluoroethylene	970	0	960
Poly(ethylene sulfide)	980	0	1040
Poly(β-alanine)	990	0	985
Poly(dimethyl siloxane)	-----	1	1000
Polyfumaronitrile	1020	0	1049
Poly(maleic anhydride)	1030	-1	1087
Polyisobutylene	1040	0	983
Poly(1-butene)	1040	0	1050
Poly(vinyl sulfonic acid)	-----	0	1042
Polymethacrylonitrile	1050	0	1008
Poly(vinylidene chloride)	1100	0	1062
Poly(vinyl ethyl ether)	1140	0	1181
Poly(methacrylic acid)	1150	0	1118
Poly(vinyl acetate)	1160	0	1188
Poly(methyl acrylate)	1160	0	1191
Polyepichlorohydrin	1170	0	1226
Polyoxytetramethylene	1180	0	1184
Poly(α-methyl acrylamide)	1190	0	1156
Poly(propylene sulfone)	-----	0	1219
Poly(N-methyl acrylamide)	1220	0	1236
Polyisoprene	1230	0	1192
Polychloroprene	1260	0	1232
Poly(p-phenylene)	1270	0	1245
Poly(vinylidene bromide)	1300	-2	1280
Poly(methyl methacrylate)	1430	0	1418
Poly(vinyl propionate)	1430	0	1473
Poly(ethyl acrylate)	1430	0	1474
Poly(methyl α-cyanoacrylate)	1440	0	1443
Poly(methyl α-chloroacrylate)	1460	0	1458
Poly(1-butene sulfone)	-----	0	1499
Poly(4-methyl-1-pentene)	1540	0	1540
Poly(vinyl trimethylsilane)	-----	1	1624
Poly(vinyl *sec*-butyl ether)	1640	0	1694
Poly(N-vinyl pyrrolidone)	1660	-1	1634

☞ ☞ ☞   TABLE 5.3 IS CONTINUED IN THE NEXT PAGE. ☞ ☞ ☞

Table 5.3. CONTINUED FROM THE PREVIOUS PAGE.

Polymer	F_d(group)	N_{Fd}	F_d(fit/pre)
Poly(vinyl n-butyl ether)	1680	0	1712
Poly(p-vinyl pyridine)	-----	0	1699
Poly(vinyl cyclopentane)	1700	-1	1683
Poly(ethyl methacrylate)	1700	0	1701
Poly(methyl ethacrylate)	1700	0	1707
Poly[thio(p-phenylene)]	1710	0	1701
Poly(ethyl α-chloroacrylate)	1730	0	1740
Poly(ε-caprolactone)	1740	0	1741
Poly(2,5-benzoxazole)	-----	0	1752
Poly(2-hydroxyethyl methacrylate)	1760	0	1781
Polystyrene	1780	0	1754
Poly(ε-caprolactam)	1800	0	1780
Poly(p-xylylene)	1810	0	1764
Poly[oxy(2,6-dimethyl-1,4-phenylene)]	-----	0	1897
Cellulose	-----	-1	1923
Poly(vinyl pivalate)	1930	0	1915
Poly(vinyl cyclohexane)	1970	-1	1948
Poly(n-butyl acrylate)	1970	0	2004
Poly[oxy(methylphenylsilylene)]	-----	1	1999
Poly(oxy-2,2-dichloromethyltrimethylene)	2010	0	2039
Poly(oxy-1,1-dichloromethyltrimethylene)	2010	0	2051
Poly($\alpha,\alpha,\alpha',\alpha'$-tetrafluoro-$p$-xylylene)	2010	0	2060
Poly(1-hexene sulfone)	-----	0	2029
Poly(p-methyl styrene)	2040	0	2007
Poly(o-methyl styrene)	2040	0	2012
Poly(α-methyl styrene)	2050	0	1981
Poly(chloro-p-xylylene)	-----	0	2061
Poly(p-chloro styrene)	2070	0	2047
Poly(o-chloro styrene)	2070	0	2051
Poly[oxy(dipropylsilylene)]	-----	1	2109
Poly(vinyl butyral)	2130	-1	2209
Poly(p-methoxy styrene)	2140	0	2145
Poly(p-bromo styrene)	2170	-1	2156
Poly(vinyl benzoate)	2170	0	2180

☞ ☞ ☞ **TABLE 5.3 IS CONTINUED IN THE NEXT PAGE.** ☞ ☞ ☞

Table 5.3. CONTINUED FROM THE PREVIOUS PAGE.

Polymer	F_d(group)	N_{Fd}	F_d(fit/pre)
Poly(*t*-butyl methacrylate)	2200	0	2142
Poly(isobutyl methacrylate)	2200	0	2190
Poly(*sec*-butyl methacrylate)	2200	0	2213
Poly(styrene sulfide)	2220	0	2251
Poly(*n*-butyl methacrylate)	2240	0	2231
Poly(8-aminocaprylic acid)	2340	0	2311
Poly(3,4-dichlorostyrene)	-----	0	2344
Poly(N-phenyl maleimide)	2380	-1	2381
Poly(N-methyl glutarimide)	2450	-1	2356
Poly(ethylene terephthalate)	2590	0	2617
Poly(α-vinyl naphthalene)	-----	0	2590
Poly(cyclohexyl methacrylate)	2630	-1	2600
Poly(benzyl methacrylate)	2710	0	2662
Poly(2-ethylbutyl methacrylate)	2740	0	2751
Poly(*n*-hexyl methacrylate)	2780	0	2761
Poly(*p*-*t*-butyl styrene)	2810	0	2734
Poly(tetramethylene terephthalate)	3130	0	3147
Poly(N-vinyl carbazole)	-----	0	3196
Poly(*n*-octyl methacrylate)	3320	0	3292
Poly[3,5-(4-phenyl-1,2,4-triazole)-1,4-phenylene]	-----	0	3409
Poly(*p*-phenylene terephthalamide)	3440	0	3416
Poly(ethylene-2,6-naphthalenedicarboxylate)	-----	0	3448
Polybenzobisoxazole	-----	0	3503
Poly[1,1-ethane *bis*(4-phenyl)carbonate]	3530	0	3564
Poly(hexamethylene adipamide)	3600	0	3561
Poly(hexamethylene isophthalamide)	3790	0	3757
Bisphenol-A polycarbonate	3800	0	3787
Poly[oxy(2,6-phenyl-1,4-phenylene)]	-----	0	3880
Poly[2,2-hexafluoropropane *bis*(4-phenyl)carbonate]	4140	0	4160
Polybenzobisthiazole	-----	0	4143
Poly(oxy-1,4-phenylene-oxy-1,4-phenylene-carbonyl-1,4-phenylene)	4300	0	4299
Phenoxy resin	4340	0	4348
Poly[1,1-cyclohexane *bis*(4-phenyl)carbonate]	4500	-1	4472
Poly(hexamethylene sebacamide)	4680	0	4622

☞ ☞ ☞   **TABLE 5.3 IS CONTINUED IN THE NEXT PAGE.** ☞ ☞ ☞

Table 5.3. CONTINUED FROM THE PREVIOUS PAGE.

Polymer	F_d(group)	N_{Fd}	F_d(fit/pre)
Poly[1,1-(1-phenylethane) *bis*(4-phenyl)carbonate]	4810	0	4785
Poly[2,2-propane *bis*{4-(2,6-dimethylphenyl)}carbonate]	-----	0	4817
Poly[2,2'-(*m*-phenylene)-5,5'-bibenzimidazole]	-----	0	4820
Poly[2,2-propane *bis*{4-(2,6-dichlorophenyl)}carbonate]	-----	0	4975
Poly[1,1-(1-phenyltrifluoroethane) *bis*(4-phenyl)carbonate]	4980	0	4972
Torlon	-----	0	5226
Resin F	5280	0	5324
Poly[2,2-propane *bis*{4-(2,6-dibromophenyl)}carbonate]	-----	-4	5412
Poly[N,N'-(*p,p*'-oxydiphenylene)pyromellitimide]	-----	0	5420
Poly[4,4'-diphenoxy di(4-phenylene)sulfone]	-----	0	5701
Poly[diphenylmethane *bis*(4-phenyl)carbonate]	5820	0	5784
Resin G	5940	0	5998
Poly[1,1-biphenylethane *bis*(4-phenyl)carbonate]	6080	0	6030
Poly[4,4'-sulfone diphenoxy di(4-phenylene)sulfone]	-----	0	6148
Poly[4,4'-isopropylidene diphenoxy di(4-phenylene)sulfone]	-----	0	6434
Ultem	-----	0	8780

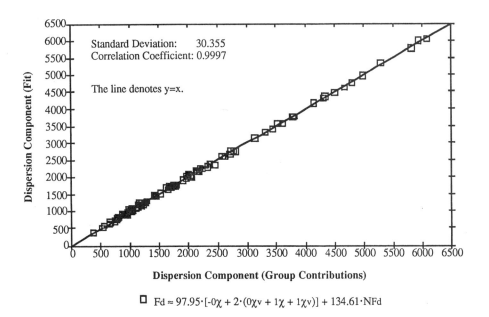

Standard Deviation: 30.355
Correlation Coefficient: 0.9997

The line denotes y=x.

$\square$ $F_d \approx 97.95 \cdot [-0\chi + 2 \cdot (0\chi v + 1\chi + 1\chi v)] + 134.61 \cdot N_{Fd}$

Figure 5.4. A fit using connectivity indices, to dispersion component F_d of molar attraction constant F calculated by group contributions, for 107 polymers. F_d is in $J^{0.5} \cdot cm^{1.5}$/mole.

References and Notes for Chapter 5

1. A. F. M. Barton, *CRC Handbook of Solubility Parameters and Other Cohesion Parameters*, CRC Press, Boca Raton, Florida (1983).

2. D. W. van Krevelen, *Properties of Polymers*, second edition, Elsevier, Amsterdam (1976). Chapter 7 of this book deals with the cohesive energy and the solubility parameter, and includes lists of the group contribution tables of several authors.

3. D. W. van Krevelen, in *Computational Modeling of Polymers*, edited by J. Bicerano, Marcel Dekker, New York (1992), Chapter 1.

4. D. H. Kaelble, *Computer Aided Design of Polymers and Composites*, Marcel Dekker, New York (1985).

5. R. F. Fedors, *Polymer Engineering and Science*, *14*, 147-154 (1974).

6. R. F. Fedors, *Polymer Engineering and Science*, *14*, 472 (1974).

7. S. Krause, in *Polymer Blends*, edited by D. R. Paul and S. Newman, Academic Press, New York (1978), Volume 1, Chapter 2.

8. M. M. Coleman, C. J. Serman, D. E. Bhagwagar and P. C. Painter, *Polymer*, *31*, 1187-1203 (1990), and references therein.

9. I. C. Sanchez, in *Polymer Blends*, edited by D. R. Paul and S. Newman, Academic Press, New York (1978), Volume 1, Chapter 3.

10. E. H. Andrews and L. Bevan, *Polymer*, *13*, 337-346 (1972).

11. E. H. Andrews, G. M. Levy and J. Willis, *J. Materials Science*, *8*, 1000-1008 (1973).

12. S.-T. Hwang and K. Kammermeyer, *Membranes in Separations*, Robert E. Krieger Publishing Company, Malabar, Florida (1984).

13. J. K. Sears and J. R. Darby, *The Technology of Plasticizers*, Wiley, New York (1982).

14. C. M. Hansen, *J. Paint Technology*, *39*, 104-117 (1967).

15. C. M. Hansen, *J. Paint Technology*, *39*, 505-510 (1967).

16. C. M. Hansen and K. Skaarup, *J. Paint Technology*, *39*, 511-514 (1967).

17. J. D. Crowley, G. S. Teague, Jr. and J. W. Lowe, Jr., *J. Paint Technology*, *38*, 269-280 (1966).

18. J. L. Gardon, *J. Colloid Interface Science*, *59*, 582-596 (1977).

19. L. B. Kier and L. H. Hall, *Molecular Connectivity in Chemistry and Drug Research*, Academic Press, New York (1976).

CHAPTER 6

TRANSITION AND RELAXATION TEMPERATURES

6.A. Background Information

The glass transition is by far the most important one among the many transitions and relaxations [1] observed in amorphous polymers. When an amorphous polymer undergoes the glass transition, almost all of its properties that relate to its processing and/or performance change dramatically. The following are some examples:

1. The coefficient of thermal expansion (Section 3.A) increases from its value for the "glassy" polymer to its much larger value for the "rubbery" polymer when the temperature increases to above the glass transition temperature (T_g), so that the rate of decrease of the density with increasing temperature becomes much faster.

2. The heat capacity (Chapter 4) of an amorphous polymer jumps from its value for the "solid" polymer to its significantly larger value for the "liquid" (molten or rubbery) polymer at T_g.

3. Mechanical properties (Chapter 11) undergo catastrophic deterioration. Structural rigidity is lost, the moduli decrease by several orders of magnitude, and the yield stress decreases to zero, slightly above T_g.

4. The "heat distortion temperature", which is used more frequently than T_g in the product literature of commercial polymers as an indicator of the softening temperature, is closely related to (and usually slightly lower than) T_g.

5. The rate of change of the refractive index with increasing temperature (Chapter 8), the dielectric constant (Chapter 9), and many other optical and/or electrical properties, often (but not always) change considerably.

6. The viscosity (Chapter 13) above T_g is lower than the viscosity below T_g by many orders of magnitude. Melt processing, such as extrusion, injection molding and compression molding, requires temperatures significantly above T_g. If T_g is in degrees Kelvin, the optimum melt processing temperature is normally at least $1.2T_g$.

The glass transition also plays a role in determining the physical properties of semicrystalline polymers, whose amorphous portions "melt" or "soften" at T_g while the crystalline portions remain "solid" up to the melting temperature T_m. A semicrystalline polymer can be treated as a solid below T_g, as a composite consisting of solid and rubbery phases of the same chemical composition above T_g but below T_m, and as a fluid above T_m. The effect of the glass transition on the physical properties of semicrystalline polymers decreases with increasing crystallinity.

The observed value of T_g is a function of the rate of measurement. For example, when the glass transition is approached from below, heating a specimen very fast results in a higher apparent T_g than heating it very slowly. Conversely, when the glass transition is approached from above, cooling a specimen very fast results in a lower apparent T_g than cooling it very slowly. There is, therefore, an important rate-dependent (kinetic) aspect of the glass transition. Nonetheless, the glass transition undoubtedly has an underlying fundamental thermodynamic basis. Theories of the glass transition [2-15] invariably treat the observed value of T_g as a kinetic (rate-dependent) manifestation of an underlying thermodynamic phenomenon; however, they differ significantly in their description of the nature of this phenomenon at a fundamental level. Differences of opinion also exist concerning the issue of whether or not the discontinuities observed at T_g in the second derivatives of the Gibbs free energy (i.e., the coefficient of thermal expansion and the heat capacity) justify referring to the glass transition as a "second-order phase transition". The detailed treatment of these fundamental issues is outside the scope of this book.

At a relatively simple-minded practical and operational (and thus theoretically nonrigorous) level of treatment, we can define T_g as *the temperature at which the forces holding the distinct components of an amorphous solid together are overcome, so that these components become able to undergo large-scale viscous flow, limited mainly by the inherent resistance of each component to such flow.*

Despite its apparent simplicity, this operational definition actually comprehends both of the key aspects of the physics of the glass transition. It states that, when a solid is heated up to T_g, it acquires enough thermal energy to be able to overcome two types of resistance to the large-scale motions of its components:

1. The *cohesive forces* holding its different components together. The relevant components for the glass transition in amorphous polymers are chain segments. The cohesive forces can be quantified in terms of properties such as the cohesive energy density or solubility parameter.
2. Attributes of the individual components (chain segments in polymers) which resist viscous flow. Resistance to the viscous flow of polymer chain segments is related to the topological and geometrical arrangement of their atoms, especially as expressed by the somewhat nebulous concept of *chain stiffness*. The glass transition occurs when there is enough freedom of motion for chain segments of up to several "statistical chain segments" (Kuhn segments) in length to be able to execute cooperative motions [16,17]. As a general rule, the length of the Kuhn segment increases with increasing chain stiffness. See the two classic textbooks by Flory [18,19] for background information on statistical chain segments and on other configurational properties of polymer chains.

The "intrachain" effect of the stiffness of individual chain segments is generally somewhat more important than the "interchain" effect of the cohesive (attractive) forces between different chains in determining the value of T_g.

The effects of chain stiffness and of cohesive forces on the value of T_g are different from each other. For example, a pendant hydroxyl group has relatively little effect on the chain stiffness, but can increase the cohesive forces very significantly by hydrogen bonding. On the other hand, the chain stiffness and the interchain cohesive forces are not completely independent of each other. Many interrelationships exist between the structural features increasing chain stiffness and the structural features increasing interchain cohesion. For example, phenyl rings are much more rigid than cyclohexyl rings, and therefore result in greater chain stiffness. Phenyl rings are also much more dense than cyclohexyl rings, and result in much higher cohesive energy densities.

Based on the considerations summarized above, it is not surprising that most theories of the glass transition describe this phenomenon in terms of key physical ingredients whose values strongly depend on the chain stiffness and/or the cohesive forces. The same statement can also be made for all empirical correlations for T_g, which either explicitly or implicitly attempt to account for the effects of chain stiffness and cohesive forces. One such empirical correlation is the relationship of van Krevelen [20], which was briefly reviewed in Section 1.B. Many other empirical correlations, which usually express T_g as a function of quantities calculated via group contributions, have also been used with limited success.

An excellent review article by Lee [21] provides detailed quantitative critical assessments, and extensive lists of the original references, for some of the best-known empirical correlations for T_g. Some of the many other interesting attempts to estimate T_g, which were not reviewed by Lee [21], include the method of Askadskii and Slonimskii [22,23], a more elaborate version of this method developed by Wiff et al [24], and the combination of molecular modeling and group contributions in the method of Hopfinger et al [25].

The factors determining the measured values of T_g for polymers are summarized below, and additional references are given for factors not discussed in the references [1-25] provided above.

1. The *rate of measurement*, which was discussed above.

2. *Structural and compositional factors*, the most fundamental of which are *chain stiffness* and interchain *cohesive forces*. These factors will be discussed in detail in Section 6.B.

3. *Conformational factors*. The most important conformational factor is the *tacticity* of vinyl-type polymers. The same polymer, for example poly(methyl methacrylate) (PMMA), can manifest significantly different values of T_g depending on whether it is isotactic, syndiotactic or atactic.

4. *Crosslinking* is another structural factor which can affect T_g. T_g increases with increasing crosslink density. For a few of the many examples of this effect, see references [15,26-32].

5. *Average molecular weight* of polymer chains. T_g increases asymptotically, with increasing number-average molecular weight M_n, to its limiting value for the "high polymer" as $M_n \rightarrow \infty$.

6. The presence of *additives, fillers, unreacted residual monomers and/or impurities*, whether deliberately included in the formulation of a resin, or left over as undesirable by-products of synthesis. For example, plasticizers of low molecular weight generally decrease T_g [20,33].

7. *Thermal history*. The annealing (or "physical aging") of test specimens at elevated temperatures below T_g usually results in an increase of T_g. For example, this effect has been studied [34] in detail for bisphenol-A polycarbonate.

8. *Thermal, thermooxidative and/or photochemical degradation.* The onset of rapid degradation sometimes occurs in the temperature range of the glass transition, obscuring the distinction between the glass transition and degradation. For example, T_g values of 700K or above, reported in the literature for some polymers with very stiff chains, are often not true T_g values, since degradation and softening take place simultaneously and inextricably.

9. Morphological effects, and especially *crystallinity*. The presence of the rigid crystallites, and of the interphase regions ("tie molecules") between amorphous and crystalline regions, often increases the observed value of T_g. In addition, the decrease of the amorphous fraction of the polymer naturally leads to a decrease in the strength (intensity) of its amorphous relaxations, with the decrease in the strength of the glass transition at a given percent crystallinity normally being larger than the decrease in the strength of the secondary (sub-T_g) relaxations [35].

10. The *pressure* also affects the value of T_g. Most measurements of T_g are performed under normal atmospheric pressure, so that the effect of pressure is very seldom considered in any detail in correlations for T_g.

This list of factors affecting T_g demonstrates that many "extraneous" factors, not related either to the composition or to the structure of a polymer, can all significantly affect the measured value of T_g. Some internal inconsistency, and the need to exercise some judgment and to make choices, are therefore inherent in the preparation of any dataset collected from a variety of different sources and used to develop an empirical correlation for T_g. For example, one major source of data [20] for our work lists the following very different experimental values collected from the literature for some common polymers: 188K and 243K for polyoxymethylene; 143K, 195K and 253K for polyethylene; 238K and 299K for polypropylene; 247K and 354K for poly(vinyl chloride); and 266K and 399K for PMMA. Under the most ideal set of conditions, the developer of a correlation for T_g would synthesize all of the polymers which will be used in the dataset, characterize them very carefully, and then measure their T_g's under identical test conditions. For practical reasons, however, the use of data from many sources is unavoidable. This is one of the major causes of the considerable standard deviation found between the measured and fitted values of T_g whenever a general correlation, applicable to all polymers, is developed for T_g.

A new correlation will be developed in Section 6.B, to enable the prediction of reasonable values of T_g without requiring and thus being limited by the availability of group contributions. This correlation will account for the effect of the composition and the structure of a polymer, as reflected in the chain stiffness and the cohesive forces between different chains, on T_g. A new correlation will be presented for the effects of the average molecular weight on T_g in Section 6.C. The effects of plasticization on T_g will be discussed in Section 6.D. A new correlation will be presented for the effects of crosslinking on T_g in Section 6.E. The effects of tacticity on T_g will be discussed in Section 6.F. Secondary relaxations will be discussed in Section 6.G. Finally, the crystalline melting temperature will be discussed in Section 6.H.

6.B.   Correlation for the Glass Transition Temperature

6.B.1.   Outline

Measured values of T_g from many different sources [20,21,36-43] were combined into a dataset of 320 polymers containing a vast variety of compositions and structural features. Only uncrosslinked polymers were included in the dataset, since the detailed treatment of the effects of crosslinking on the properties is outside the scope of this book. The T_g's of the most common atactic forms were used, whenever available, for polymers manifesting different tacticities.

The dataset was developed with the characterization of the "incremental differences" between the T_g's of structurally related polymers used as a foremost consideration. Addition of successive backbone methylene units in the linear polyoxides, and addition of successive methylene units in side groups as in the poly(vinyl ethers), are examples of incremental structural changes. Such incremental changes usually cause incremental (gradual) differences between successive structural variants over a wide range of systematic structural variation. Since the practical design of new polymers often involves the prediction of the properties of sets of structural variants of a certain basic type of structure, it is especially important to be able to characterize such incremental differences properly.

Most of the polymers in this dataset will be identified by their names in this chapter. No attempt will be made to use a consistent system of nomenclature. Many of the polymers in the dataset have very cumbersome and/or unfamiliar names. Several other polymers are used as examples to illustrate the rules for calculating some of the parameters used in developing the correlation for T_g. A number of such polymers are shown in figures 6.1 to 6.8.

The correlation for T_g was developed by analyzing the dataset for relationships between the polymeric structure and the two important physical factors summarized in Section 6.A, namely chain stiffness and cohesive forces. Chain stiffness is, admittedly, a somewhat nebulous concept, which has been quantified in different ways by different authors. It is hoped that the reader will agree that the manner in which this key physical factor will be incorporated into our correlation for T_g makes sense at an intuitive level.

The parameters to be used in the correlation were defined in such a manner that they can all be determined without needing group contributions. Consequently, the correlation which will be developed below is applicable to all polymers constructed from the nine elements (C, N, O, H, F, Si, S, Cl and Br) which are of interest in this book.

Thirteen *structural parameters* (x_1 to x_{13}) will now be defined and used in the correlation for T_g. These parameters were developed by detailed analysis of the structure-property relationships determining the 320 experimental T_g values in the dataset. The definitions of these structural parameters make extensive use of concepts such as the backbone and the side groups of the polymeric repeat unit (Section 2.D) and the shortest path across the chain backbone (Section 2.E).

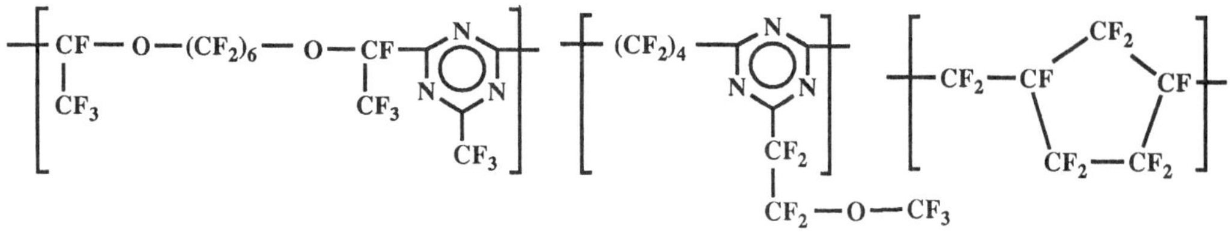

(a) Perfluoropolymer 1.   (b) Perfluoropolymer 2.   (c) Perfluoropolymer 3.

Figure 6.1. Schematic illustration of some perfluoropolymers.

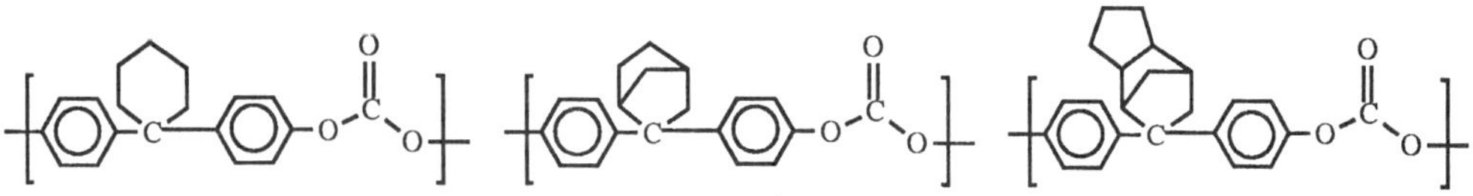

(a)   (b)   (c)

Figure 6.2. Schematic illustration of some polycarbonates: (a) Poly[1,1-cyclohexane *bis*(4-phenyl)carbonate]. (b) Polycarbonate 2. (c) Polycarbonate 3.

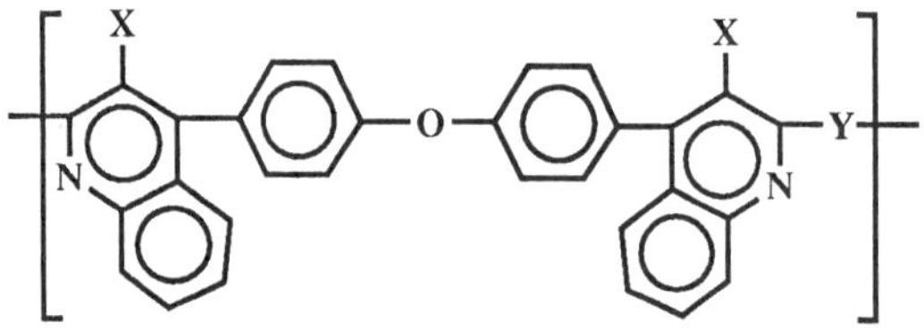

Polyquinoline 1: X = H and Y = *m*-phenylene.
Polyquinoline 2: X = phenyl and Y = *m*-phenylene.
Polyquinoline 3: X = H and Y = *p,p*'-oxydiphenylene.
Polyquinoline 4: X = phenyl and Y = *p,p*'-oxydiphenylene.

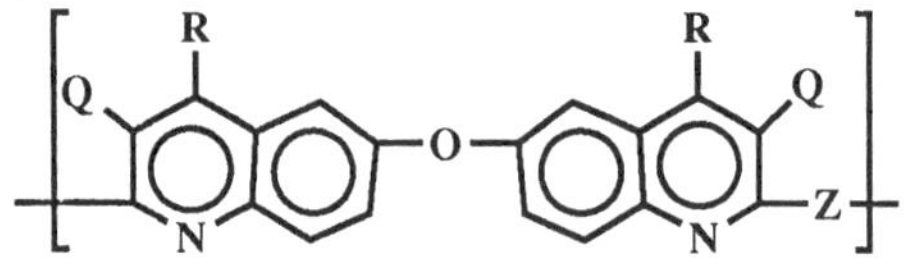

Polyquinoline 5: Q = H, R = phenyl and Z = *p,p*'-oxydiphenylene.
Polyquinoline 6: Q = R = phenyl and Z = *p,p*'-oxydiphenylene.
Polyquinoline 7: Q = H, R = phenyl and Z = *p,p*'-biphenylene.
Polyquinoline 8: Q = R = phenyl and Z = *p,p*'-biphenylene.
Polyquinoline 9: Q = H, R = phenyl and Z = *p*-phenylene.
Polyquinoline 10: Q = R = phenyl and Z = *p*-phenylene.

Figure 6.3. Schematic illustration of polyquinolines 1 to 4 (top) and 5 to 10 (bottom).

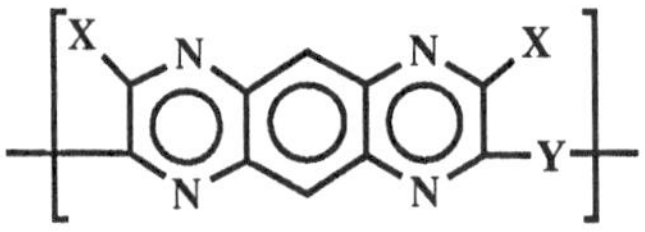

Polytricyclic 1: X = H and Y = *p,p*'-oxydiphenylene.
Polytricyclic 2: X = phenyl and Y = *p*-phenylene.
Polytricyclic 3: X = phenyl and Y = *m*-phenylene.

Figure 6.4. Schematic illustration of some polymers containing tricyclic quinoxaline moieties.

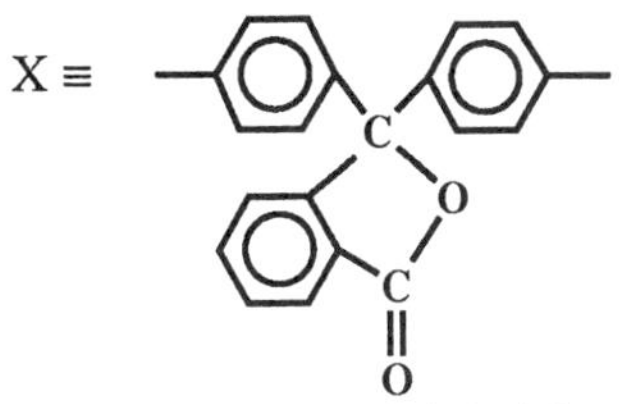

(a) X is the phenolphthalein unit. 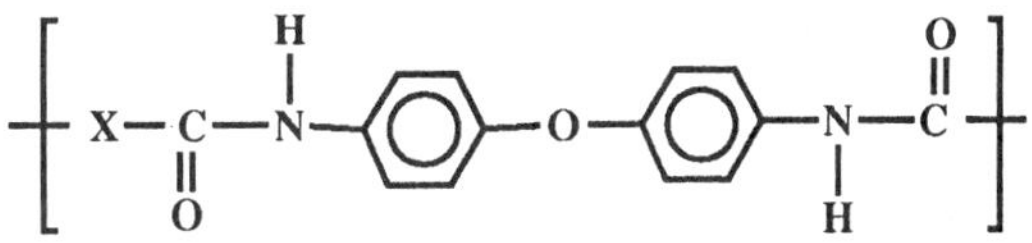 (b) Polyphenolphthalein 1.

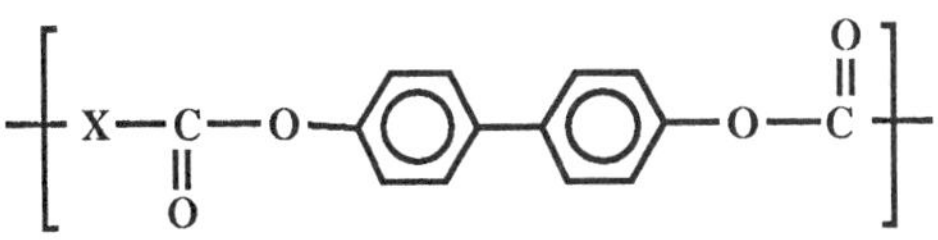

(c) Polyphenolphthalein 2. 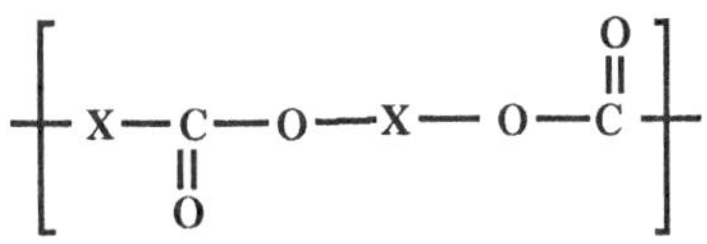 (d) Polyphenolphthalein 3.

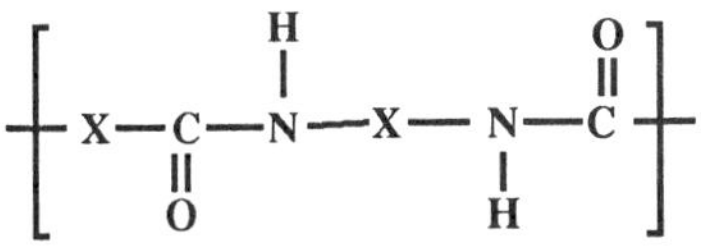

(e) Polyphenolphthalein 4. 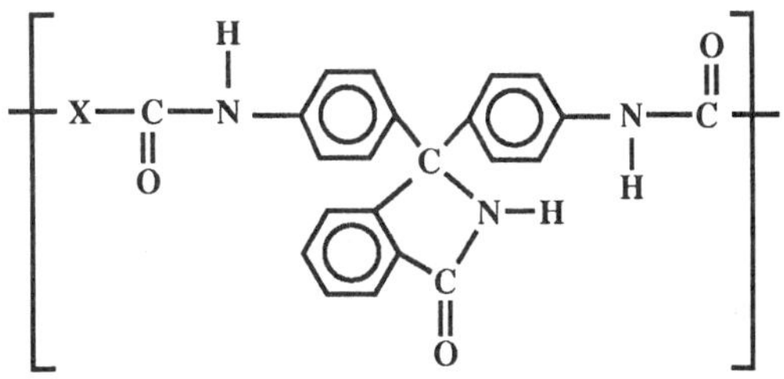 (e) Polyphenolphthalein 5 (-O- replaced by -NH- in an X).

Figure 6.5. Schematic illustration of some polyphenolphthaleins.

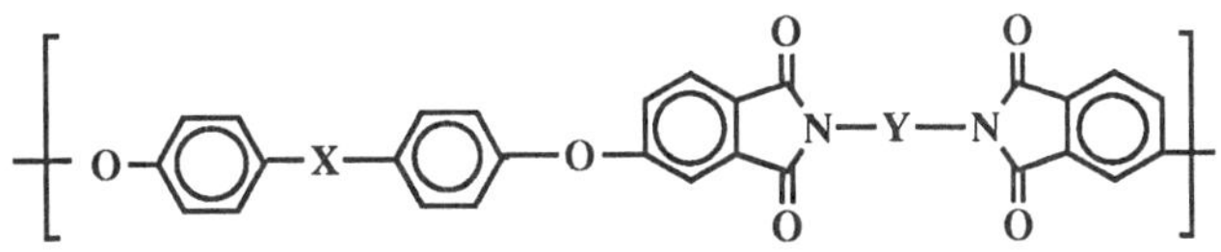

Polyetherimide 1: X = O and Y = $(CH_2)_6$.
Polyetherimide 2: X = S and Y = *m*-phenylene.
Polyetherimide 3: X = O and Y = *m*-phenylene.
Polyetherimide 4: X = carbonyl and Y = *m*-phenylene.
Polyetherimide 5: No X (rings connected directly), Y = *m*-phenylene.

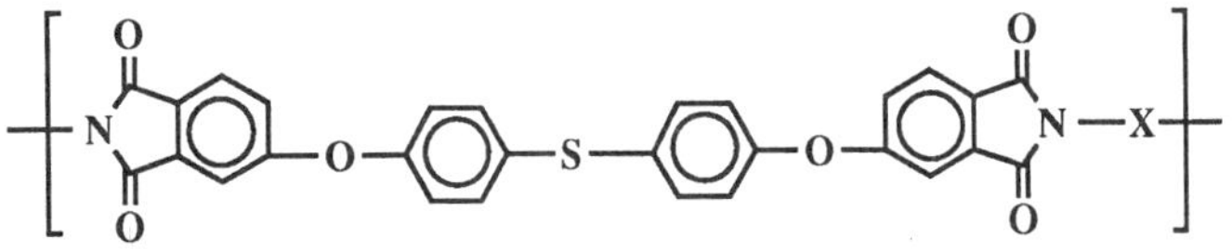

Polyetherimide 6: X = *m,m'*-carbonyldiphenylene.
Polyetherimide 7: X = *o,p'*-carbonyldiphenylene.
Polyetherimide 8: X = *m,p'*-carbonyldiphenylene.
Polyetherimide 9: X = *p,p'*-carbonyldiphenylene.

Figure 6.6. Schematic illustration of polyetherimides 1 to 5 (top) and 6 to 9 (bottom).

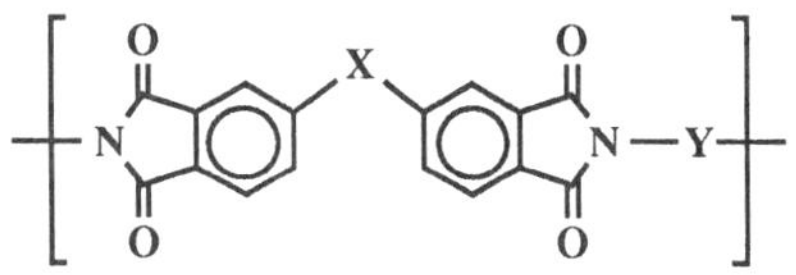

Polyimide 10: X = carbonyl and Y = *m,m'*-carbonyldiphenylene.
Polyimide 11: X = carbonyl and Y = *p*-phenylene.
Polyetherimide 12: X = O and Y = *p*-phenylene.

Figure 6.7. Schematic illustration of some polyimides.

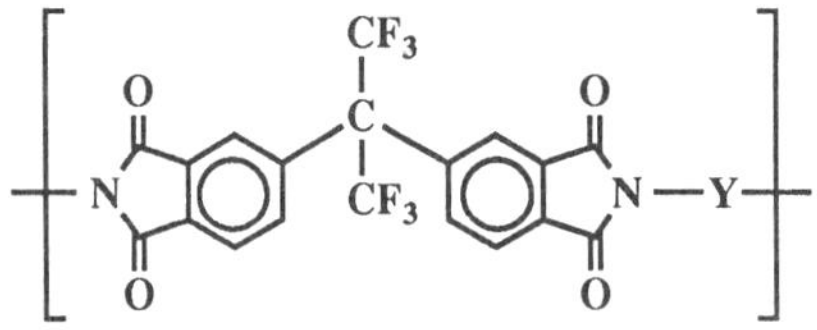

Polyimide 1F: Y = *m,m'*-carbonyldiphenylene.
Polyimide 2F: Y = *o,p'*-carbonyldiphenylene.
Polyimide 3F: Y = *m,p'*-carbonyldiphenylene.
Polyimide 4F: Y = *p,p'*-carbonyldiphenylene.

Figure 6.8. Schematic illustration of some fluorine-containing polyimides.

The rules for determining the values of these structural parameters may, at first, appear somewhat cumbersome. The reader will find, however, that after a little practice, these rules become very easy to apply. While the use of thirteen structural parameters may seem excessive, it should be remembered that these parameters are general, that they incorporate the very complex structural factors determining T_g, and that they obviate the need for group contributions. Furthermore, even with a vastly larger number of group contributions, the T_g of many new polymers cannot be estimated because of the lack of some of the required group contributions.

Before developing the correlation for T_g, we will describe its general form:

$$T_g \approx a + b \cdot \delta + c \cdot (\text{weighted sum of structural parameters}) \qquad (6.1)$$

In Equation 6.1, a, b and c are correlation coefficients, while δ is the solubility parameter. The weight factors for the different structural parameters are determined by the requirement that the correlation should give the best possible fit between the experimental and the calculated values of T_g. The parameters used in Equation 6.1 have both complementary and combination aspects, as discussed in Section 6.B.2. For example, δ mainly quantifies the cohesive forces, while the structural parameters mainly quantify chain stiffness. On the other hand, there is some overlap between the structural factors determining the cohesive forces and determining the chain stiffness.

6.B.2. The Structural Parameters

Structural parameters x_1 to x_4 mainly account for the effect on T_g of the presence and local environment of rigid rings and hydrogen-bonding moieties on the chain backbone. The rules for determining these parameters for most polymers are summarized below. See Section 17.F for generalizations of these rules to cover some especially complicated types of structural features, which were incorporated into the software package implementing our method.

The structural parameter x_1 is determined by the following rules, illustrated in Figure 6.9:

1. Rigid rings which are in side groups do not contribute to x_1.

2. Rigid backbone rings which are not located on the shortest path across the backbone do not contribute to x_1.

3. A distinct rigid ring (i.e., a rigid ring not fused to any other ring) along the shortest path across the chain backbone contributes +1 to x_1 if it is in a *para* bonding configuration, and does not contribute anything to x_1 if it is in a *meta* or *ortho* bonding configuration.

4. If there is a rigid fused ring unit of $n \geq 2$ rings, containing exactly two of its rigid rings along the shortest path across the chain backbone, the contribution of this rigid fused ring unit to x_1 is determined as follows:

   (a) It contributes +2 if the two bonds connecting it to the rest of the polymer chain are close to being collinear.

   (b) It contributes only +1 if the two bonds connecting it to the rest of the chain are far from being collinear, causing large chain displacements to occur for even a small rotation of the unit around either bond.

5. If a rigid fused ring unit contains $n \geq 3$ rings along the shortest path across the chain backbone, it contributes $(n+1)$ to x_1, regardless of the relative orientation of the two bonds connecting it to the rest of the backbone.

6. A hydrogen-bonding moiety containing non-hydrogen atoms on the chain backbone contributes +1 to x_1. For example, the carbon and nitrogen atoms of amide, urea and urethane groups along the chain backbone are on the backbone, so that these groups contribute +1 to x_1. On the other hand, the non-hydrogen atoms of pendant amide or hydroxyl groups are not on the chain backbone, so that these moieties do not contribute to x_1.

The structural parameter x_2 is determined by the following rules, illustrated in Figure 6.9:

1. Each distinct rigid ring, and each fused ring unit containing two rigid rings along the shortest path across the chain backbone, located in a *meta* or equivalent bonding configuration, contributes +1 to x_2. "Equivalence" to a *meta* bonding configuration is defined in terms of the relative orientation of the two bonds connecting the ring structure of interest to the rest of the chain backbone. It is a useful concept since it allows the calculation of x_2 for rigid ring units which are not simple six-membered rings.

2. Each *ortho* isolated rigid ring along the shortest path across the chain backbone contributes +5.

3. Each two-ring fused rigid ring unit along the shortest path across the chain backbone in an *ortho* or equivalent bonding configuration contributes +8 to x_2.

4. Rigid fused ring units with $n \geq 3$ rigid rings along the shortest path do not contribute to x_2.

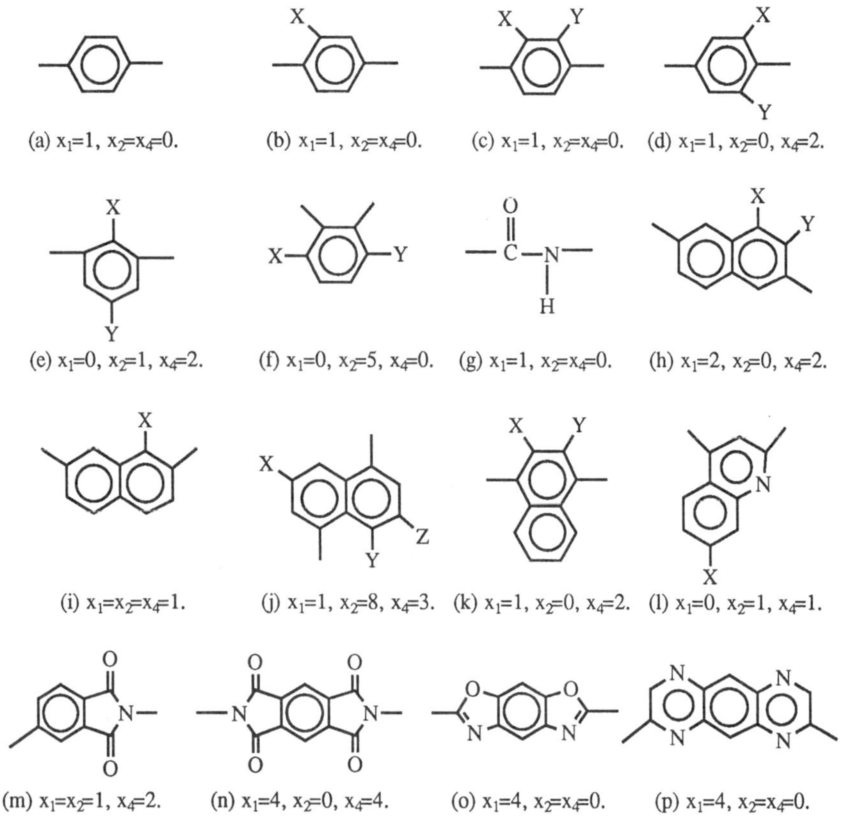

(a) $x_1=1$, $x_2=x_4=0$. (b) $x_1=1$, $x_2=x_4=0$. (c) $x_1=1$, $x_2=x_4=0$. (d) $x_1=1$, $x_2=0$, $x_4=2$.

(e) $x_1=0$, $x_2=1$, $x_4=2$. (f) $x_1=0$, $x_2=5$, $x_4=0$. (g) $x_1=1$, $x_2=x_4=0$. (h) $x_1=2$, $x_2=0$, $x_4=2$.

(i) $x_1=x_2=x_4=1$. (j) $x_1=1$, $x_2=8$, $x_4=3$. (k) $x_1=1$, $x_2=0$, $x_4=2$. (l) $x_1=0$, $x_2=1$, $x_4=1$.

(m) $x_1=x_2=1$, $x_4=2$. (n) $x_1=4$, $x_2=0$, $x_4=4$. (o) $x_1=4$, $x_2=x_4=0$. (p) $x_1=4$, $x_2=x_4=0$.

Figure 6.9. Examples of the application of the rules for calculating the structural parameters x_1, x_2 and x_4, for the relevant types of structural units (rigid rings and hydrogen-bonding groups) along the chain backbone. X, Y and Z are substituents on the rigid rings. The rules listed in the text for calculating x_1, x_2 and x_4 are completely general, and can be applied to any other structural unit of interest. See Figure 6.10 for examples of the calculation of the structural parameter x_3, which accounts for the environments of the rigid rings along the chain backbone.

The structural parameter x_3 accounts for the differences between the various possible bonding environments of the rigid rings located on the shortest path across the chain backbone. If such a rigid ring is surrounded by other rigid structural units on both sides along the chain backbone, a synergistic enhancement occurs in the chain stiffness. On the other hand, if such a rigid ring is surrounded by relatively flexible structural units on both sides along the chain backbone, the effect of the rigid ring on the overall chain stiffness is diminished. The value of x_3 is thus determined by the following two rules, illustrated in Figure 6.10:

1. Each rigid ring along the shortest path across the chain backbone, which is surrounded by rigid rings and/or by hydrogen-bonding backbone groups *on both sides*, contributes +1 to x_3. The same contribution of +1 is made whether neighboring rigid rings are distinct from the ring of interest, or members of the same fused ring unit.

2. Each rigid ring along the shortest path across the chain backbone, which is surrounded by any combination of ether (-O-), thioether (-S-), and/or methylene (-CH$_2$-) moieties *on both sides*, contributes -1 to x_3.

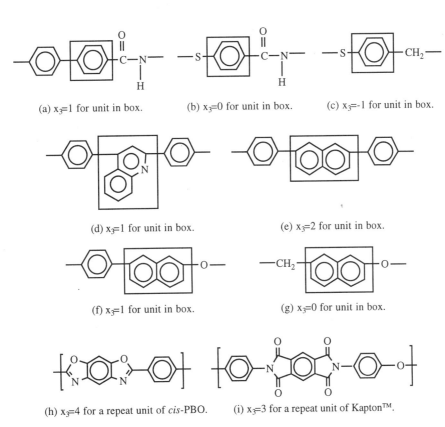

(a) x_3=1 for unit in box.   (b) x_3=0 for unit in box.   (c) x_3=-1 for unit in box.

(d) x_3=1 for unit in box.   (e) x_3=2 for unit in box.

(f) x_3=1 for unit in box.   (g) x_3=0 for unit in box.

(h) x_3=4 for a repeat unit of *cis*-PBO.   (i) x_3=3 for a repeat unit of Kapton™.

Figure 6.10. Examples of the application of the rules for calculating the structural parameter x_3. In (a)-(g), the quantity of interest is the contribution of the rigid backbone ring unit enclosed in the box to x_3, as a result of its bonding environment. In (h) and (i), x_3 is calculated for complete polymeric repeat units. The rules listed in the text for calculating x_3 are completely general, and can be applied to any other structural unit of interest.

The presence of atoms larger than hydrogen, or of groups containing more than one atom, "sticking out" from a ring located on the chain backbone, usually enhances the chain stiffness. This effect is more pronounced if the ring is in a fused rigid ring moiety. The structural parameter x_4 accounts for the presence, layout and number of such substituents. Details of the sizes and shapes of these substituents will be taken into account by structural parameters which will be introduced later. As illustrated in Figure 6.9, x_4 is determined by the following rules:

1. If a rigid ring is "distinct", i.e., if it is not a part of fused ring unit, then the number of non-hydrogen atoms or groups sticking out from it is counted in determining x_4, but only if (a) two or more atoms and/or groups are sticking out from this ring, *and* (b) atoms and/or groups are sticking out along both of the "paths" traversing this ring across the backbone.

2. If a distinct rigid ring has only one non-hydrogen substituent, or if it has more than one substituent all attached to ring atoms located on the same chain backbone path traversing the ring, no contribution is made to x_4.

3. Each non-hydrogen substituent sticking out of a rigid ring located in a fused ring unit along the chain backbone contributes +1 to x_4, regardless of the number and location of such substituents on the ring. The contribution of the rigid rings in each fused ring unit along the chain backbone to x_4 is therefore equal to the total number of non-hydrogen substituents sticking out from them.

4. If a non-rigid ring is fused to a rigid ring, each non-hydrogen substituent sticking out of the two atoms of the non-rigid ring which connect this ring to the rigid ring also contributes +1 to x_4.

The number of rotational degrees of freedom along the chain backbone affects chain stiffness. The presence of rigid rings along the chain backbone reduces the number of backbone rotational degrees of freedom. This effect has therefore already been at least partially taken into account by the structural parameters x_1, x_2, x_3 and x_4. The structural parameter x_5 considers the number of local rotational degrees of freedom of the chain backbone, and the effect of such degrees of freedom on the observed T_g, more explicitly. It is, therefore, not linearly independent from x_1, x_2, x_3 and x_4, but nonetheless useful. The value of x_5 is determined by the following two rules:

1. If there are no "floppy" backbone rings along the chain backbone, then $x_5 = N_{BBrot}$, where N_{BBrot} was defined in Section 4.C. For example, $x_5 = 2$ for poly(vinyl alcohol) (Figure 4.1) and $x_5 = 8$ for Ultem (Figure 2.4).

2. Localized torsional motions of relatively limited amplitude, such as those around single bonds in floppy rings, can contribute significantly to the heat capacity, and thus were partially counted among the degrees of freedom for N_{BBrot}. On the other hand, the glass transition requires larger-scale cooperative chain motions. Atoms "tied down" in a ring are not free to participate in such larger-scale motions even if the ring is floppy. Consequently, if there are "floppy" backbone rings along the chain backbone, bonds located in such rings do not contribute to x_5.

For example, $N_{BBrot}=5$ but $x_5=2$ for poly(vinyl butyral) (Figure 2.9), since the six-membered backbone ring contributes 3 to N_{BBrot}, but does not contribute anything to x_5.

The number of rotational degrees of freedom in the side groups, and the relative ease of rotational motions of the side groups, also affect the chain stiffness. If two side groups have equal volumes, but one of them is much less flexible than the other one, the use of the less flexible side group will normally result in a greater enhancement of chain stiffness. The structural parameter x_6 accounts for these factors, and is determined by the following rules:

1. If there are no floppy rings in any side groups, and no silicon atoms in the chain backbone, then $x_6=N_{SGrot}$, where N_{SGrot} was defined in Section 4.C. For example, $x_6=0$ for poly(vinyl fluoride) (Figure 2.3), $x_6=1$ for polymethacrylonitrile (Figure 4.1), and $x_6=3$ for poly(vinyl butyral) (Figure 2.9).

2. If there are floppy rings in the side groups, these rings do not contribute to x_6 although they contribute to N_{SGrot}. For example, $x_6=1$ for poly(N-vinyl pyrrolidone) (Figure 4.1), which has $N_{SGrot}=3.5$, since the five-membered pendant ring contributes 2.5 to N_{SGrot}, but does not contribute anything to x_6. This rule can be rationalized by an argument similar to the justification given above for the different treatment of the effect of floppy rings along the chain backbone in counting degrees of freedom for N_{BBrot} and x_5.

3. There are many qualitatively significant differences between bonds involving silicon atoms and bonds involving carbon atoms [44-48]. For example, silicon atoms manifest much longer bonds than carbon atoms to their neighbors, and also generally result in much lower force constants for bond stretching, bending and torsional motions. Rotations of side groups attached to backbone silicon atoms flanked by "small" backbone atoms or groups are therefore especially unhindered, so that the contribution of these side groups to the chain stiffness is relatively small. This factor is accounted for in an unambiguous manner in determining x_6:

    (a) If a backbone silicon atom has an ether (-O-), thioether (-S-), methylene (-CH_2-), or non-hydrogen-bonding (i.e., non-amide, non-urea and non-urethane) -NH- moiety on at least one side along the chain backbone, each non-hydrogen atom in the side groups attached to that silicon atom contributes +1 to x_6, regardless of the actual number of rotational degrees of freedom in those side groups. For example, $x_6=N_{SGrot}=2$ for poly(dimethyl siloxane) (Figure 5.1), where N_{SGrot} equals the number of non-hydrogen atoms in the two side (methyl) groups. On the other hand, for poly[oxy(methylphenylsilylene)] (Figure 6.11), whose pendant phenyl ring only slightly enhances the chain stiffness, $N_{SGrot}=2$ but $x_6=7$.

    (b) If a backbone silicon atom does not have any of the types of neighboring moieties enumerated in (a) adjacent to it on either side along the chain backbone, then the side groups attached to it are treated like side groups attached to any other backbone atom, and only their degrees of freedom are counted in determining x_6.

Symmetry of disubstitution of larger atoms or groups for H atoms on backbone C or Si atoms generally lowers T_g. The parameter x_7 accounts for this effect, making exceptions for bonding environments where this symmetric substitution effect does not hold. Each symmetrically disubstituted backbone C or Si contributes +1 to x_7, with the exceptions of such atoms which are (a) located in rings, or (b) flanked on both sides by rigid backbone rings and/or backbone hydrogen bonding groups. For example, the Si atom in the backbone of poly(dimethyl siloxane) (Figure 5.1) is disubstituted with methyl (-CH$_3$) groups, so that x_7=1. Perfluoropolymer 1 [Figure 6.1(a)] has six -CF$_2$- groups, so that x_7=6. On the other hand, the polyimides shown in Figure 6.8 have x_7=0 since the backbone C atom symmetrically disubstituted with the trifluoromethyl (-CF$_3$) groups is flanked by the rigid phthalimide rings on both sides.

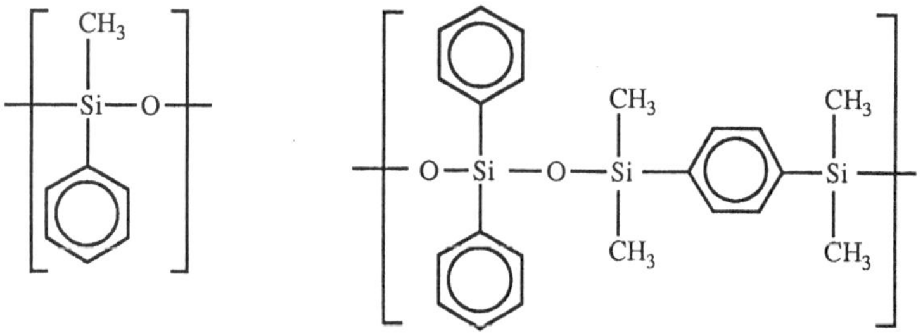

Figure 6.11. Schematic illustration of two silicon-containing polymers. (a) Poly[oxy(methyl-phenylsilylene)]. (b) Poly(oxydiphenylsilyleneoxydimethylsilylene-1,4-phenylenedimethylsilylene).

Methylene (-CH$_2$-) units in side groups generally cause "internal plasticization" and lower T_g. The only common exceptions to this rule are hydrocarbon polymers with very long side chains, where side chain crystallization and/or other nonbonded interactions between side groups may counter this effect, and cause a slight increase of T_g with increasing number of methylene units after a certain side chain length is exceeded.

Tertiary carbon atoms in side groups generally cause "crowding" of the atoms of the side groups, hinder the motions (i.e., the effectiveness of the nominal rotational degrees of freedom) of the side group, increase the chain stiffness, and cause T_g to increase. For example, *n*-butyl, *sec*-butyl and *t*-butyl side groups, each of which contributes +4 to x_6, affect T_g quite differently.

The effects of tertiary carbon atoms, and of methylene groups in the most common situation of the reduction of T_g with increasing side chain length for side chains which are not too long, are taken into account by the structural parameter x_8, whose value is determined by the following set of rules, as illustrated in Figure 6.12:

1. Each methylene unit (i.e., carbon atom with $\delta=\delta^v=2$) in a side group contributes -1 to x_8 unless it is located in a ring. (Terminal $=CH_2$ groups, where carbon has $\delta^v=2$ but $\delta=1$, are not counted as methylene units.)

2. Methylene units (carbon atoms with $\delta=\delta^v=2$) located in rings in the side groups do not contribute to x_8.

3. Each tertiary carbon atom ($\delta=\delta^v=4$) in a side group contributes +1 to x_8, whether it is in a ring or not in a ring.

4. Similarly, each tertiary silicon atom in a side group contributes +1 to x_8.

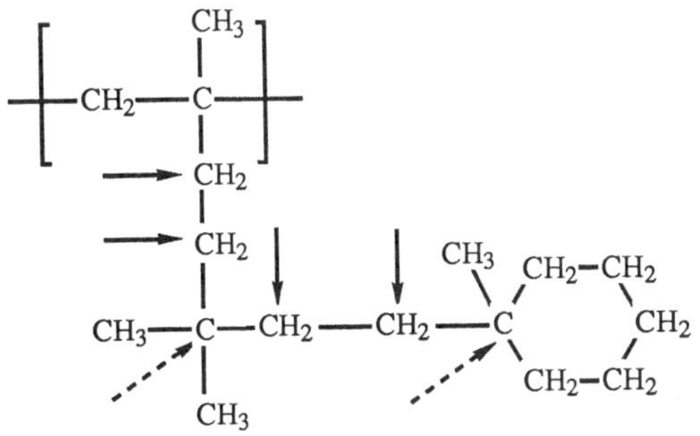

Figure 6.12. Schematic illustration of the repeat unit of a hypothetical polymer, illustrating the application of the rules for determining the value of the structural parameter x_8. There are four methylene units, pointed out by solid arrows, which are in the side group but not in the ring. There are also two tertiary carbon atoms, pointed out by dashed arrows, in the side group. Consequently, $x_8 = (-4+2) = -2$ for this polymer.

Asymmetry of disubstitution on a backbone carbon atom generally increases T_g. This effect is taken into account by the structural parameter x_9, which equals the number of asymmetrically disubstituted backbone carbon atoms. For example, there are two asymmetrically disubstituted (disubstituted by an -F and a $-CF_3$) backbone carbon atoms in Perfluoropolymer 1 [Figure 6.1(a)], which therefore has $x_9=2$. Asymmetrically disubstituted backbone silicon atoms are *not* counted in x_9. For example, poly[oxy(methylphenylsilylene)] [Figure 6.11(a)] has $x_9=0$.

In determining x_9, a backbone carbon atom attached by two bonds to a ring all of whose other atoms are in a side group, is also considered to be asymmetrically disubstituted. For example, the backbone carbon atom attached to the pendant ring portion of the phenolphthalein unit [Figure 6.5(a)] by two bonds contributes +1 to x_9. Since Polyphenolphthalein 3 [Figure 6.5(d)] has two phenolphthalein units, it has $x_9=2$.

When a backbone carbon atom is attached by two bonds to a multiple-ring unit, all of whose other atoms are in a side group, overcrowding of the atoms of the side group occurs in the immediate vicinity of the chain backbone, resulting in increased chain stiffness and increased T_g.

The structural parameter x_{10} takes this effect into account. The contribution of each such multiple-ring moiety to x_{10} is equal to the number of non-hydrogen side group atoms in it. For example, the phenolphthalein unit [Figure 6.5(a)] has nine non-hydrogen side group atoms in its pendant ring, which is attached to the same backbone carbon atom by two bonds. It thus contributes 9 to x_{10}, so that $x_{10}=9$ for Polyphenolphthalein 2 [Figure 6.5(c)], while $x_{10}=(2\cdot9)=18$ for Polyphenolphthalein 3 which contains two phenolphthalein units. Polycarbonate 2 [Figure 6.2(b)] has $x_{10}=6$, Polycarbonate 3 [Figure 6.2(c)] has $x_{10}=9$, and poly[o-biphenylenemethane bis(4-phenyl)carbonate] (Figure 2.10) has $x_{10}=12$. On the other hand, poly[1,1-cyclohexane bis(4-phenyl)carbonate] [Figure 6.2(a)] has $x_{10}=0$, since the ring attached with two bonds to the same backbone carbon atom is a simple cyclohexyl ring, and not a multiple-ring moiety.

A double bond between two carbon atoms which are both located on the chain backbone, with the double bond itself *not* located inside a ring, causes inefficiency of packing, which generally results in some lowering of T_g. This effect is taken into account by the structural parameter x_{11}, which is the number of such C=C bonds located on the chain backbone. For example, see Figure 6.13. The C=C bond in poly(1,4-butadiene) is on the chain backbone, so that $x_{11}=1$. On the other hand, the C=C bonds in poly(1,2-butadiene) and in poly[1,1-dichloroethylene bis(4-phenyl) carbonate] are not on the chain backbone, so that these polymers have $x_{11}=0$.

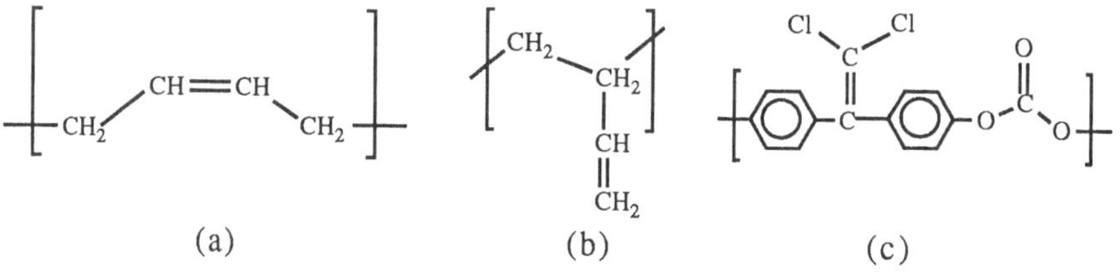

Figure 6.13. Schematic illustration of the repeat units of some polymers containing double bonds between carbon atoms. (a) Poly(1,4-butadiene). (b) Poly(1,2-butadiene). (c) Poly[1,1-dichloroethylene bis(4-phenyl)carbonate].

The structural parameter x_{12} is defined as the number of fluorine atoms attached to non-aromatic atoms in side groups, where the backbone atom to which the side group itself is attached is *not* a silicon atom. For example, Perfluoropolymer 1 [Figure 6.1(a)] has $x_{12}=9$, since there are nine fluorine atoms attached to non-aromatic side group atoms in the three -CF$_3$ side groups. Perfluoropolymer 2 [Figure 6.1(b)] has $x_{12}=7$. Perfluoropolymer 3 [Figure 6.1(c)] has $x_{12}=0$, since all of its fluorine atoms are attached to backbone atoms. Poly(2,3,4,5,6-pentafluoro styrene) also has $x_{12}=0$, because all five of its fluorine atoms are attached to the aromatic carbon atoms of the phenyl ring. Finally, poly[oxy(methyl γ-trifluoropropylsilylene)] [see Figure 9.3(b) in Chapter 9] has $x_{12}=0$, because the side group containing the fluorine atoms is attached to a backbone silicon atom.

The values of the structural parameters x_1 to x_{12} are listed in Table 6.1 for some of the polymers in the dataset used for developing the correlation for T_g. With the exception of x_5 and x_6, most of the structural parameters are equal to zero for most of the polymers. This fact underscores the simplicity of these structural parameters, as well as the fact that the few nonzero structural parameters for any given polymer can all play a very significant role in determining the value of T_g, so that the counting rules presented above have to be applied very carefully.

The twelve structural parameters defined above are all extensive variables. In order to convert them into intensive variables for use in the correlation for T_g, which is an intensive property, they will all be scaled by the number N of vertices in the hydrogen-suppressed graph of the repeat unit, as described by Equation 2.8 in Section 2.C. In other words, x_1/N, x_2/N, ..., x_{12}/N, will be used as linear regression variables in the correlation for T_g.

The final structural parameter x_{13} accounts for the extreme torsional freedom around backbone bonds involving siloxane-type silicon atoms, which are flanked by ether (-O-) linkages on both sides in the path along the chain backbone. The value of x_{13} equals the total number of non-hydrogen side group atoms attached to the backbone silicon atoms flanked by ether (-O-) linkages on both sides along the chain backbone, divided by the number N_{BB} of atoms on the chain backbone. The structural parameter x_{13} calculated by using this definition is automatically an intensive quantity, with N_{BB} rather than N as the scaling factor. Here are two examples:

1. Poly[oxy(methylphenylsilylene)] [Figure 6.11(a)] has $N_{BB}=2$, and also has a siloxane-type silicon atom with two side groups containing a total of seven non-hydrogen atoms attached to it, so that $x_{13}=(7/2)=3.5$.

2. Poly(oxydiphenylsilyleneoxydimethylsilylene-1,4-phenylenedimethylsilylene) [Figure 6.11(b)] has $N_{BB}=11$, contains only one siloxane-type silicon atom surrounded on both sides by ether linkages, and has twelve non-hydrogen atoms in the two side groups attached to this silicon atom, resulting in $x_{13}=(12/11)\approx1.1$.

6.B.3. The Correlation

The parameter N_{Tg} defined by Equation 6.2, by itself, provides a fairly reasonable correlation for T_g. The correlation coefficient between the intensive variable N_{Tg}/N and the 320 experimental values of T_g in the dataset is 0.9650. This parameter will be used, along with two other parameters as discussed below, to develop the final correlation equation for T_g.

$$N_{Tg} \equiv 15x_1 - 4x_2 + 23x_3 + 12x_4 - 8x_5 - 4x_6 - 8x_7 + 5x_8 + 11x_9 + 8x_{10} - 11x_{11} - 4x_{12} \qquad (6.2)$$

Table 6.1. The values of the structural parameters x_1 to x_{12} for some polymers.

Polymer	x_1	x_2	x_3	x_4	x_5	x_6	x_7	x_8	x_9	x_{10}	x_{11}	x_{12}
Poly(dimethyl siloxane)	0	0	0	0	2	2	1	0	0	0	0	0
Poly[oxy(methyl γ-trifluoropropylsilylene)]	0	0	0	0	2	7	0	-1	0	0	0	0
Poly(cyclohexylmethylsilane)	0	0	0	0	1	2	0	0	0	0	0	0
Poly(1-butene)	0	0	0	0	2	2	0	-1	0	0	0	0
Polyisobutylene	0	0	0	0	2	2	1	0	0	0	0	0
Poly(4-methyl-1-pentene)	0	0	0	0	2	4	0	-1	0	0	0	0
Polystyrene	0	0	0	0	2	1	0	0	0	0	0	0
Poly(o-ethyl styrene)	0	0	0	0	2	3	0	-1	0	0	0	0
Poly(p-sec-butyl styrene)	0	0	0	0	2	5	0	-1	0	0	0	0
Poly(p-t-butyl styrene)	0	0	0	0	2	5	0	1	0	0	0	0
Poly(α-vinyl naphthalene)	0	0	0	0	2	1	0	0	0	0	0	0
Poly(1,2-butadiene) *(See Fig. 6.13)*	0	0	0	0	2	1	0	0	0	0	0	0
Poly(1,4-butadiene) *(See Fig. 6.13)*	0	0	0	0	3	0	0	0	0	0	1	0
Polyisoprene	0	0	0	0	3	1	0	0	0	0	1	0
Polychloroprene	0	0	0	0	3	0	0	0	0	0	1	0
Poly(vinylidene chloride)	0	0	0	0	2	0	1	0	0	0	0	0
Polyoxyethylene	0	0	0	0	3	0	0	0	0	0	0	0
Poly(vinyl alcohol)	0	0	0	0	2	1	0	0	0	0	0	0
Poly(p-xylylene)	1	0	-1	0	3	0	0	0	0	0	0	0
Poly(ethyl-p-xylylene)	1	0	-1	0	3	2	0	-1	0	0	0	0
Poly[oxy(m-phenylene)]	0	1	-1	0	2	0	0	0	0	0	0	0
Poly[oxy(p-phenylene)]	1	0	-1	0	2	0	0	0	0	0	0	0
Poly[oxy(2,6-diphenyl-1,4-phenylene)]	1	0	-1	2	2	2	0	0	0	0	0	0
Poly(3,3,3-trifluoropropylene)	0	0	0	0	2	1	0	1	0	0	0	3
Poly(2,5-difluoro styrene)	0	0	0	0	2	1	0	0	0	0	0	0
Perfluoropolymer 1 *(See Fig. 6.1)*	0	1	0	0	11	3	6	3	2	0	0	9
Poly(vinyl butyral)	0	0	0	0	2	3	0	-2	0	0	0	0
Poly(cyclohexyl methacrylate)	0	0	0	0	2	4	0	0	1	0	0	0
Poly(ethylene-1,4-naphthalenedicarboxylate)	1	0	0	0	7	0	0	0	0	0	0	0
Poly(ethylene-1,5-naphthalenedicarboxylate)	1	8	0	0	7	0	0	0	0	0	0	0
Poly(ethylene-2,6-naphthalenedicarboxylate)	2	0	0	0	7	0	0	-1	0	0	0	0
Poly(ε-caprolactone)	0	0	0	0	7	0	0	0	0	0	0	0
Poly(ε-caprolactam)	1	0	0	0	7	0	0	0	0	0	0	0

☞ ☞ ☞ **TABLE 6.1 IS CONTINUED IN THE NEXT PAGE.** ☞ ☞ ☞

Table 6.1. CONTINUED FROM THE PREVIOUS PAGE.

Polymer	x_1	x_2	x_3	x_4	x_5	x_6	x_7	x_8	x_9	x_{10}	x_{11}	x_{12}
Poly(methyl methacrylate)	0	0	0	0	2	4	0	0	1	0	0	0
Poly[methane bis(4-phenyl) carbonate]	2	0	0	0	6	0	0	0	0	0	0	0
Bisphenol-A polycarbonate	2	0	0	0	6	2	0	0	0	0	0	0
Poly[1,1-dichloroethylene bis(4-phenyl)carbonate] *(See Fig. 6.13)*	2	0	0	0	6	0	0	0	0	0	0	0
Poly[2,2-hexafluoropropane bis(4-phenyl)carbonate]	2	0	0	0	6	2	0	2	0	0	0	6
Poly[2,2-propane bis{4-(2,6-dichlorophenyl)}carbonate]	2	0	0	4	6	2	0	0	0	0	0	0
Poly[2,2-butane bis(4-phenyl) carbonate]	2	0	0	0	6	3	0	-1	1	0	0	0
Poly[1,1-cyclohexane bis(4-phenyl) carbonate] *(See Fig. 6.2)*	2	0	0	0	6	0	0	0	1	0	0	0
Poly[o-biphenylenemethane bis(4-phenyl)carbonate]	2	0	0	0	6	0	0	0	1	12	0	0
Polycarbonate 3 *(See Fig. 6.2)*	2	0	0	0	6	0	0	0	1	9	0	0
Ultem	4	3	3	4	8	2	0	0	0	0	0	0
Torlon	4	1	2	2	6	0	0	0	0	0	0	0
Polyquinoline 1 *(See Fig. 6.3)*	2	3	3	0	6	0	0	0	0	0	0	0
Polyquinoline 5 *(See Fig. 6.3)*	6	0	2	2	6	2	0	0	0	0	0	0
Polyquinoline 10 *(See Fig. 6.3)*	5	0	3	4	4	4	0	0	0	0	0	0
Polytricyclic 1 *(See Fig. 6.4)*	6	0	3	0	4	0	0	0	0	0	0	0
Polyphenolphthalein 3 *(See Fig. 6.5)*	4	0	0	0	10	0	0	0	2	18	0	0
Poly(3-phenylquinoxaline-2,7-diyl-3-phenylquinoxaline-7,2-diyl-1,4-phenylene)	3	2	5	2	3	2	0	0	0	0	0	0
Poly[N,N'-(p,p'-oxydiphenylene)pyromellitimide]	6	0	3	4	4	0	0	0	0	0	0	0

The correlation is improved quite significantly by using the solubility parameter δ and the structural parameter x_{13} in addition to N_{Tg}/N in the linear regression procedure. The solubility parameter δ, by itself, correlates rather weakly with the experimental values of T_g, with a correlation coefficient of only 0.6285. On the other hand, it complements the structural parameters. It allows physically significant distinctions to be made between many polymers which have identical values for the structural parameters x_1/N, x_2/N ..., x_{12}/N, x_{13}, but which manifest different cohesive energy densities and solubility parameters, and different glass transition temperatures, mainly as a result of differences in composition. For example, polypropylene and poly(vinyl alcohol) have identical values of all thirteen structural parameters; however, the replacement of the pendant methyl group with the pendant hydroxyl group imparts poly(vinyl alcohol) with a much higher solubility parameter, and thus also with a much higher T_g.

The use of δ results in a refinement, albeit an important one, of the basic correlation between T_g and the thirteen structural parameters, affecting the trends between a relatively small fraction of polymers in a significant manner. The utilization of a combination of cohesive energies and molar volumes obtained via slightly different methods to speed up the calculations thus has a negligible effect on the final correlation for T_g. In calculating δ, van Krevelen's group contributions were used whenever available, and the correlation developed for them in terms of connectivity indices in Section 5.C (i.e., E_{coh2} values) were used otherwise, for the cohesive energy.

Experimental values or van Krevelen's group contributions were used when available, and correlations developed in terms of connectivity indices in Chapter 3 were used otherwise, for the molar volume at room temperature. It is important to note, however, that it would not have been possible to calculate the solubility parameters of a very large number of the polymers in our dataset in the absence of the new correlations developed in chapters 3 and 5.

The final correlation for T_g, which was obtained by using a linear regression procedure with weight factors of $1000/T_g(exp)$, is given by Equation 6.3. This correlation has a standard deviation of 24.65K, which equals 6.7% of the average $T_g(exp)$ value of 365.8K for the 320 polymers in the dataset. The correlation coefficient of 0.9749 indicates that Equation 6.3 accounts for 95.0% of the variation of the $T_g(exp)$ values in the dataset. The results of these calculations are summarized in Table 6.2 and shown in Figure 6.14.

$$T_g \approx 351.00 + 5.63 \cdot \delta + 31.68 \cdot \frac{N_{Tg}}{N} - 23.94 \cdot x_{13} \tag{6.3}$$

It is seen from Table 6.2 that Equation 6.3 overestimates the T_g's of all polymers containing phthalimide groups and completely built from C, N, O, S and H atoms, such as Torlon and Ultem (Figure 2.4) and the polymers shown in figures 6.6 and 6.7. In Section 17.B.1, a specialized "designer correlation" will be developed for the T_g's of this important class of polymers. The polyesters are another class of polymers for which the development of a designer correlation, which is described in Section 17.B.2, has been found to improve the agreement with experiment.

Table 6.2. Experimental glass transition temperatures $T_g(exp)$ in degrees Kelvin, the quantities (δ, N, N_{Tg} and x_{13}) used in the correlation equation for T_g and defined in the text, and the fitted values of T_g, for 320 polymers.

Polymer	$T_g(exp)$	δ	N	N_{Tg}	x_{13}	$T_g(fit)$
Poly[oxy(diethylsilylene)]	130	15.0	6	-50	2.0	124
Poly(dimethyl siloxane)	152	14.6	4	-32	1.0	156
Poly(1,4-butadiene) *(cis, see Fig. 6.13)*	171	17.5	4	-35	0.0	172

☞ ☞ ☞   **TABLE 6.2 IS CONTINUED IN THE NEXT PAGE.** ☞ ☞ ☞

Table 6.2. CONTINUED FROM THE PREVIOUS PAGE.

Polymer	$T_g(exp)$	δ	N	N_{Tg}	x_{13}	$T_g(fit)$
Poly(dimethylsilylenemethylene)	173	12.9	4	-32	0.0	170
Poly[oxy(methylphenylsilylene)] *(See Fig. 6.11)*	187	18.2	9	-44	3.5	215
Poly(3-hexoxypropylene oxide)	188	17.8	11	-86	0.0	204
Polyoxytetramethylene	190	17.7	5	-40	0.0	197
Poly(1,1-dimethylsilazane)	191	14.8	4	-32	0.0	181
Poly(2-*n*-butyl-1,4-butadiene)	192	16.9	8	-66	0.0	185
Poly(vinyl *n*-octyl ether)	194	17.0	11	-87	0.0	196
Poly(3-butoxypropylene oxide)	194	18.3	9	-68	0.0	214
Polyethylene	195	15.9	2	-16	0.0	187
Polyoxytrimethylene	195	18.5	4	-32	0.0	201
Poly(2-*n*-propyl-1,4-butadiene)	196	17.0	7	-57	0.0	189
Poly(vinyl n-decyl ether)	197	16.7	13	-105	0.0	189
Poly(2-ethyl-1,4-butadiene)	197	17.2	6	-48	0.0	194
Poly[oxy(methyl γ-trifluoropropylsilylene)]	199	14.4	9	-49	3.5	176
Polyisobutylene	199	16.4	4	-32	0.0	190
Poly(dimethylsilylenetrimethylene)	203	13.9	6	-48	0.0	176
Polyoxyoctamethylene	203	16.9	9	-72	0.0	193
Polyisoprene	203	17.4	5	-39	0.0	202
Polyoxyhexamethylene	204	17.1	7	-56	0.0	194
Poly(tetramethylene adipate)	205	18.6	14	-96	0.0	239
Polyoxyethylene	206	19.4	3	-24	0.0	207
Poly(propylene oxide)	206	18.8	4	-28	0.0	235
Poly(vinyl *n*-pentyl ether)	207	17.2	8	-60	0.0	210
Poly(vinyl 2-ethylhexyl ether)	207	17.2	11	-77	0.0	226
Poly(*n*-octyl acrylate)	208	17.0	13	-91	0.0	225
Poly(vinyl *n*-hexyl ether)	209	17.2	9	-69	0.0	205
Poly(3-methoxypropylene oxide)	211	19.4	6	-41	0.0	244
Polypentadiene	213	17.9	5	-39	0.0	205
Poly(*n*-heptyl acrylate)	213	17.1	12	-82	0.0	231
Poly(ε-caprolactone)	213	18.1	8	-56	0.0	231
Poly(*n*-nonyl acrylate)	215	16.9	14	-100	0.0	220
Poly(*n*-hexyl acrylate)	216	17.2	11	-73	0.0	238
Poly(decamethylene adipate)	217	17.8	20	-144	0.0	223

☞ ☞ ☞   **TABLE 6.2 IS CONTINUED IN THE NEXT PAGE.** ☞ ☞ ☞

Table 6.2. CONTINUED FROM THE PREVIOUS PAGE.

Polymer	T_g(exp)	δ	N	N_{Tg}	x_{13}	T_g(fit)
Polyoxymethylene	218	20.9	2	-16	0.0	215
Poly(dodecyl methacrylate)	218	16.4	18	-120	0.0	232
Poly(*n*-butyl acrylate)	219	17.5	9	-55	0.0	256
Poly(1-heptene)	220	16.5	7	-56	0.0	190
Poly(oxycarbonyl-3-methylpentamethylene)	220	18.2	9	-60	0.0	242
Poly(vinyl *n*-butyl ether)	221	17.4	7	-51	0.0	218
Poly(2-isopropyl-1,4-butadiene)	221	17.5	7	-47	0.0	237
Poly(1-hexene)	223	16.6	6	-47	0.0	196
Poly(1-pentene)	223	16.6	5	-38	0.0	203
Polychloroprene	225	19.2	5	-35	0.0	237
Poly(propylene sulfide)	226	18.5	4	-28	0.0	233
Poly(1-butene)	228	16.8	4	-29	0.0	216
Poly(ethylene azelate)	228	18.4	15	-104	0.0	235
Poly(2-octyl acrylate)	228	16.5	13	-81	0.0	247
Poly(*n*-propyl acrylate)	229	17.7	8	-46	0.0	268
Polypropylene	233	17.0	3	-20	0.0	235
Poly(vinylidene fluoride)	233	14.3	4	-24	0.0	242
Poly(ethylene adipate)	233	19.2	12	-80	0.0	248
Poly(2-heptyl acrylate)	235	16.6	12	-72	0.0	254
Poly(6-methyl-1-heptene)	239	15.6	8	-55	0.0	221
Poly(oxycarbonyl-1,5-dimethylpentamethylene)	240	18.3	10	-64	0.0	251
Poly(2-bromo-1,4-butadiene)	241	18.8	5	-35	0.0	235
Poly(ethylene sebacate)	243	18.1	16	-112	0.0	231
Poly[(methyl)phenylsilylenetrimethylene]	243	16.2	11	-60	0.0	269
Poly(isobutyl acrylate)	249	16.8	9	-45	0.0	287
Poly(vinyl isobutyl ether)	251	16.6	7	-41	0.0	259
Poly(ethyl acrylate)	251	18.0	7	-37	0.0	285
Poly(*n*-octyl methacrylate)	253	16.7	14	-84	0.0	255
Poly(vinyl *sec*-butyl ether)	253	16.5	7	-41	0.0	258
Poly(*sec*-butyl acrylate)	253	16.8	9	-45	0.0	287
Poly(vinyl ethyl ether)	254	17.8	5	-33	0.0	242
Perfluoropolymer 2 *(See Fig. 6.1)*	255	17.9	29	-105	0.0	337
Poly(vinylidene chloride)	256	20.5	4	-24	0.0	276

☞ ☞ ☞   **TABLE 6.2 IS CONTINUED IN THE NEXT PAGE.** ☞ ☞ ☞

Table 6.2. CONTINUED FROM THE PREVIOUS PAGE.

Polymer	T_g(exp)	δ	N	N_{Tg}	x_{13}	T_g(fit)
Poly(3-pentyl acrylate)	257	16.7	10	-54	0.0	274
Poly(5-methyl-1-hexene)	259	15.5	7	-46	0.0	230
Perfluoropolymer 1 *(See Fig. 6.1)*	260	16.7	42	-151	0.0	331
Poly(oxy-2,2-dichloromethyltrimethylene)	265	19.7	8	-58	0.0	232
Poly[(4-dimethylaminophenyl)methylsilylenetrimethylene]	267	16.9	14	-72	0.0	283
Poly(n-hexyl methacrylate)	268	17.0	12	-66	0.0	272
Poly(1,2-butadiene) *(See Fig. 6.13)*	269	16.6	4	-20	0.0	286
Poly(vinyl isopropyl ether)	270	16.6	6	-32	0.0	275
Poly(ethylene succinate)	272	18.8	10	-64	0.0	254
Poly(vinyl methyl sulfide)	272	19.2	4	-24	0.0	269
Poly(oxydiphenylsilyleneoxydimethylsilylene- 1,4-phenylenedimethylsilylene) *(See Fig. 6.11)*	273	17.5	27	-121	1.1	281
Poly(vinyl butyrate)	278	17.0	8	-46	0.0	265
Poly(p-n-hexoxymethyl styrene)	278	18.0	16	-82	0.0	290
Poly(p-n-butyl styrene)	279	17.9	12	-51	0.0	317
Poly(methyl acrylate)	281	18.7	6	-28	0.0	308
Poly(vinyl propionate)	283	17.2	7	-37	0.0	280
Poly(2-ethylbutyl methacrylate)	284	16.7	12	-56	0.0	297
Poly(o-n-octoxy styrene)	286	17.9	17	-91	0.0	282
Poly(2-t-butyl-1,4-butadiene)	293	17.2	8	-46	0.0	266
Poly(n-butyl methacrylate)	293	17.3	10	-48	0.0	296
Poly(2-methoxyethyl methacrylate)	293	18.3	10	-43	0.0	318
Poly(p-n-propoxymethyl styrene)	295	18.5	13	-55	0.0	321
Poly(ethyl-p-xylylene)	298	18.4	10	-45	0.0	312
Poly(3,3,3-trifluoropropylene)	300	14.1	6	-27	0.0	288
Poly(vinyl acetate)	301	18.5	6	-28	0.0	307
Poly(4-methyl-1-pentene)	302	15.4	6	-37	0.0	242
Poly(vinyl formate)	304	19.1	5	-24	0.0	306
Poly(vinyl chloroacetate)	304	19.7	7	-33	0.0	313
Poly(neopentyl methacrylate)	306	15.9	11	-37	0.0	334
Poly(n-propyl methacrylate)	308	17.4	9	-39	0.0	312
Poly(12-aminododecanoic acid)	310	21.4	14	-89	0.0	270
Poly[di(p-tolyl)silylenetrimethylene]	311	17.2	18	-96	0.0	279

☞ ☞ ☞ TABLE 6.2 IS CONTINUED IN THE NEXT PAGE. ☞ ☞ ☞

Table 6.2. CONTINUED FROM THE PREVIOUS PAGE.

Polymer	$T_g(exp)$	δ	N	N_{Tg}	x_{13}	$T_g(fit)$
Poly(hexamethylene sebacamide)	313	23.4	20	-114	0.0	302
Poly(4-cyclohexyl-1-butene)	313	16.7	10	-38	0.0	324
Poly[(pentafluoroethyl)ethylene]	314	13.4	9	-34	0.0	307
Poly(11-aminoundecanoic acid)	315	21.9	13	-81	0.0	277
Poly(*t*-butyl acrylate)	315	16.0	9	-35	0.0	318
Poly(3-phenoxypropylene oxide)	315	20.0	11	-41	0.0	345
Poly(2,3,3,3-tetrafluoropropylene)	315	12.7	7	-16	0.0	350
Poly(10-aminodecanoic acid)	316	22.2	12	-73	0.0	283
Poly[oxy(*m*-phenylene)]	318	20.6	7	-43	0.0	272
Poly(3,3-dimethylbutyl methacrylate)	318	16.1	12	-46	0.0	320
Poly(decamethylene sebacamide)	319	22.2	24	-146	0.0	283
Poly(N-butyl acrylamide)	319	24.0	9	-55	0.0	292
Poly(vinyl trifluoroacetate)	319	16.5	9	-35	0.0	321
Poly(*p-n*-butoxy styrene)	320	18.5	13	-55	0.0	321
Poly(isobutyl methacrylate)	321	16.5	10	-38	0.0	324
Poly(3-methyl-1-butene)	323	15.2	5	-28	0.0	259
Poly(9-aminononanoic acid)	324	22.7	11	-65	0.0	292
Poly(8-aminocaprylic acid)	324	23.4	10	-57	0.0	302
Poly(vinyl butyral)	324	17.5	10	-38	0.0	329
Poly(ethylene isophthalate)	324	20.7	14	-60	0.0	332
Poly(ethyl methacrylate)	324	17.7	8	-30	0.0	332
Poly[(4-dimethylaminophenyl)phenylsilylenetrimethylene)]	325	17.6	19	-92	0.0	297
Poly(isopropyl methacrylate)	327	16.3	9	-29	0.0	341
Poly(methyl-*p*-xylylene)	328	18.7	9	-36	0.0	330
Poly(vinyl isobutyral)	329	16.8	10	-38	0.0	325
Poly(*n*-butyl α-chloroacrylate)	330	18.2	10	-44	0.0	314
Poly(7-aminoheptanoic acid)	330	24.1	9	-49	0.0	314
Poly(*sec*-butyl methacrylate)	330	16.6	10	-38	0.0	324
Poly(hexamethylene adipamide)	330	25.0	16	-82	0.0	329
Poly(*p*-isopentoxy styrene)	330	17.8	14	-54	0.0	329
Poly[(heptafluoropropyl)ethylene]	331	13.0	12	-41	0.0	316
Poly(oxydiphenylsilylene-1,3-phenylene)	331	19.6	20	-84	0.0	328
Poly(*p*-xylylene)	333	18.6	8	-32	0.0	329

☞ ☞ ☞   **TABLE 6.2 IS CONTINUED IN THE NEXT PAGE.** ☞ ☞ ☞

Table 6.2. CONTINUED FROM THE PREVIOUS PAGE.

Polymer	$T_g(exp)$	δ	N	N_{Tg}	x_{13}	$T_g(fit)$
Poly(3-cyclopentyl-1-propene)	333	16.8	8	-29	0.0	331
Poly(3-phenyl-1-propene)	333	18.8	9	-29	0.0	355
Poly(ε-caprolactam)	335	25.1	8	-41	0.0	330
Poly(ethylene-1,4-naphthalenedicarboxylate)	337	21.9	18	-41	0.0	402
Poly(p-n-propoxy styrene)	343	18.7	12	-46	0.0	335
Poly(n-propyl α-chloroacrylate)	344	18.5	9	-35	0.0	332
Poly(ethylene-1,5-naphthalenedicarboxylate)	344	21.9	18	-73	0.0	346
Poly(vinyl propional)	345	17.7	9	-29	0.0	348
Poly(ethylene terephthalate)	345	20.5	14	-41	0.0	373
Poly(sec-butyl α-chloroacrylate)	347	17.6	10	-34	0.0	342
Poly(vinyl chloride)	348	19.7	3	-16	0.0	293
Poly(3-cyclohexyl-1-propene)	348	16.7	9	-29	0.0	343
Poly(vinyl cyclopentane)	348	17.0	7	-20	0.0	356
Poly(2-hydroxypropyl methacrylate)	349	23.7	10	-48	0.0	332
Poly(p-methoxymethyl styrene)	350	19.0	11	-37	0.0	352
Resin G	351	25.3	29	-104	0.0	380
Poly(chloro-p-xylylene)	353	19.8	9	-32	0.0	350
Poly(bromo-p-xylylene)	353	20.0	9	-32	0.0	351
Poly(vinyl acetal)	355	17.9	8	-20	0.0	373
Poly(ethylene oxybenzoate)	355	19.8	11	-25	0.0	390
Poly(vinyl alcohol)	358	35.1	3	-20	0.0	338
Poly[oxy(p-phenylene)]	358	20.8	7	-24	0.0	359
Poly(p-sec-butyl styrene)	359	18.0	12	-41	0.0	344
Poly(p-ethoxy styrene)	359	19.0	11	-37	0.0	352
Poly(2-hydroxyethyl methacrylate)	359	24.6	9	-39	0.0	352
Poly[thio(p-phenylene)]	360	20.3	7	-24	0.0	357
Poly(p-isopropyl styrene)	360	17.5	11	-32	0.0	357
Poly(2-methyl-5-t-butyl styrene)	360	17.7	13	-35	0.0	365
Poly(p-methoxy styrene)	362	19.4	10	-28	0.0	371
Poly(isopropyl α-chloroacrylate)	363	17.6	9	-25	0.0	362
Poly(4-methoxy-2-methyl styrene)	363	19.1	11	-32	0.0	366
Poly(vinyl cyclohexane)	363	16.8	8	-20	0.0	366
Poly(cyano-p-xylylene)	363	21.6	10	-32	0.0	371

☞ ☞ ☞ **TABLE 6.2 IS CONTINUED IN THE NEXT PAGE.** ☞ ☞ ☞

Table 6.2. CONTINUED FROM THE PREVIOUS PAGE.

Polymer	T_g(exp)	δ	N	N_{T_g}	x_{13}	T_g(fit)
Poly($\alpha,\alpha,\alpha',\alpha'$-tetrafluoro-$p$-xylylene)	363	16.5	12	-25	0.0	378
Poly(m-xylylene adipamide)	363	25.4	18	-62	0.0	385
Poly(m-chloro styrene)	363	19.6	9	-20	0.0	391
Poly(2-chloroethyl methacrylate)	365	18.7	9	-35	0.0	333
Poly(ethyl α-chloroacrylate)	366	19.0	8	-26	0.0	355
Poly(cyclohexylmethylsilane)	366	15.2	8	-16	0.0	373
Poly(1,4-cyclohexylidene dimethylene terephthalate)	368	19.1	20	-49	0.0	381
Poly(m-methyl styrene)	370	18.7	9	-24	0.0	372
Poly(2,5-dimethyl-p-xylylene)	373	18.9	10	-28	0.0	368
Polychlorotrifluoroethylene	373	15.8	6	-13	0.0	372
Polystyrene	373	19.0	8	-20	0.0	379
Phenoxy resin	373	22.3	21	-46	0.0	407
Poly(p-methyl styrene)	374	18.6	9	-24	0.0	371
Poly(2,5-difluoro styrene)	374	17.5	10	-20	0.0	386
Poly(o-ethyl styrene)	376	18.4	10	-33	0.0	350
Poly(3,5-dimethyl styrene)	377	18.1	10	-28	0.0	364
Poly(cyclohexyl methacrylate)	377	17.2	12	-21	0.0	393
Poly(o-vinyl pyridine)	377	21.8	8	-20	0.0	395
Poly(methyl methacrylate)	378	18.0	7	-21	0.0	357
Polyacrylonitrile	378	24.9	4	-16	0.0	364
Poly(vinyl formal)	378	18.2	7	-16	0.0	381
Poly(o-fluoro styrene)	378	18.4	9	-20	0.0	384
Poly(2,3,4,5,6-pentafluoro styrene)	378	18.0	13	-20	0.0	404
Poly(acrylic acid)	379	25.7	5	-24	0.0	344
Poly(p-fluoro styrene)	379	18.2	9	-20	0.0	383
Poly(t-butyl methacrylate)	380	16.0	10	-28	0.0	352
Poly(3,4-dimethyl styrene)	384	18.1	10	-28	0.0	364
Poly(2-fluoro-5-methyl styrene)	384	17.7	10	-24	0.0	375
Resin F	384	26.7	27	-94	0.0	391
Poly(2,4-dimethyl styrene)	385	18.1	10	-28	0.0	364
Poly(p-methoxycarbonyl styrene)	386	19.3	12	-32	0.0	375
Poly(3-methyl-4-chloro styrene)	387	19.0	10	-24	0.0	382
Poly(cyclohexyl α-chloroacrylate)	387	18.0	12	-17	0.0	407

☞ ☞ ☞   **TABLE 6.2 IS CONTINUED IN THE NEXT PAGE.** ☞ ☞ ☞

Table 6.2. CONTINUED FROM THE PREVIOUS PAGE.

Polymer	$T_g(exp)$	δ	N	N_{Tg}	x_{13}	$T_g(fit)$
Poly(p-xylylene sebacamide)	388	23.6	22	-75	0.0	376
Poly[thio bis(4-phenyl)carbonate]	388	20.8	17	-18	0.0	435
Poly(p-chloro styrene)	389	19.6	9	-20	0.0	391
Perfluoropolymer 3 (See Fig. 6.1)	390	13.8	16	-24	0.0	381
Poly(phenylmethylsilane)	390	17.3	8	-16	0.0	385
Poly(o-chloro styrene)	392	19.6	9	-20	0.0	391
Poly[2,2-butane bis{4-(2-methylphenyl)}carbonate]	392	19.6	22	-32	0.0	415
Poly(2,5-dichloro styrene)	393	19.9	10	-20	0.0	400
Poly(phenyl methacrylate)	393	19.1	12	-21	0.0	403
Polymethacrylonitrile	393	22.7	5	-9	0.0	422
Poly(α-p-dimethyl styrene)	394	18.2	10	-17	0.0	400
Poly(3-fluoro-4-chloro styrene)	395	18.8	10	-20	0.0	394
Poly[1,1-butane bis(4-phenyl)carbonate]	396	19.6	20	-29	0.0	415
Poly(ethylene-2,6-naphthalenedicarboxylate)	397	21.9	18	-26	0.0	429
Poly(m-hydroxymethyl styrene)	398	24.7	10	-33	0.0	385
Poly(3,4-dichlorostyrene)	401	19.9	10	-20	0.0	400
Polyetherimide 1 (See Fig. 6.6)	401	23.5	43	-50	0.0	447
Poly(p-t-butyl styrene)	402	17.7	12	-31	0.0	369
Poly(hexamethylene isophthalamide)	403	25.3	18	-39	0.0	425
Poly[1,1-ethane bis(4-phenyl)carbonate]	403	20.1	18	-22	0.0	426
Poly(2,4-dichloro styrene)	406	19.9	10	-20	0.0	400
Poly[2,2-butane bis(4-phenyl)carbonate]	407	19.6	20	-24	0.0	423
Poly(o-methyl styrene)	409	18.7	9	-24	0.0	372
Poly(α-methyl styrene)	409	18.8	9	-13	0.0	411
Poly[2,2-pentane bis(4-phenyl)carbonate]	410	19.3	21	-33	0.0	410
Poly(m-phenylene isophthalate)	411	21.0	18	-56	0.0	371
Poly(p-phenyl styrene)	411	19.2	14	-24	0.0	405
Poly(oxycarbonyloxy-2-chloro-1,4-phenyleneisopropylidene-2-methyl-1,4-phenylene)	411	20.2	21	-30	0.0	420
Poly(p-hydroxymethyl styrene)	413	25.0	10	-33	0.0	387
Poly(p-vinyl pyridine)	415	21.6	8	-20	0.0	395
Poly[2,2-butane bis{4-(2-chlorophenyl)}carbonate]	415	20.4	22	-24	0.0	431
Poly(2,5-dimethyl styrene)	416	18.1	10	-28	0.0	364

☞ ☞ ☞   TABLE 6.2 IS CONTINUED IN THE NEXT PAGE. ☞ ☞ ☞

Table 6.2. CONTINUED FROM THE PREVIOUS PAGE.

Polymer	T_g(exp)	δ	N	N_{Tg}	x_{13}	T_g(fit)
Poly(p-bromo styrene)	417	19.5	9	-20	0.0	390
Poly(2-methyl-4-chloro styrene)	418	19.0	10	-24	0.0	382
Poly(N-vinyl pyrrolidone)	418	22.6	8	-20	0.0	399
Trichlorostyrene *(mixed isomer)*	418	20.4	11	-20	0.0	408
Poly(oxycarbonyloxy-2-chloro-1,4-phenyleneisopropylidene-1,4-phenylene)	419	20.3	20	-26	0.0	424
Poly[2,2-propane *bis*{4-(2-chlorophenyl)}carbonate]	419	20.7	21	-26	0.0	428
Poly(oxy-1,4-phenylene-oxy-1,4-phenylene-carbonyl-1,4-phenylene)	419	21.2	22	-26	0.0	433
Poly[methane *bis*(4-phenyl)carbonate]	420	20.1	17	-18	0.0	430
Poly(p-hydroxybenzoate)	420	21.0	9	-9	0.0	437
Poly[4,4-heptane *bis*(4-phenyl)carbonate]	421	19.0	23	-62	0.0	373
Poly[1,1-(2-methyl propane) *bis*(4-phenyl)carbonate]	422	19.3	20	-30	0.0	412
Bisphenol-A polycarbonate	423	19.8	19	-26	0.0	419
Poly(N-vinyl carbazole)	423	20.9	15	-20	0.0	426
Poly(β-vinyl naphthalene)	424	20.3	12	-20	0.0	413
Polyhexafluoropropylene	425	11.6	9	-24	0.0	332
Poly[1,1-dichloroethylene *bis*(4-phenyl)carbonate] *(See Fig. 6.13)*	430	21.2	20	-18	0.0	442
Poly(α-vinyl naphthalene)	432	20.4	12	-20	0.0	413
Poly(o-hydroxymethyl styrene)	433	24.8	10	-33	0.0	386
Poly(methyl α-cyanoacrylate)	433	21.9	8	-17	0.0	407
Poly[1,1-cyclopentane *bis*(4-phenyl)carbonate]	440	19.5	21	-7	0.0	450
Poly(oxyterephthaloyloxy-2-methyl-1,4-phenyleneisopropylidene-3-methyl-1,4-phenylene)	444	20.1	29	-35	0.0	426
Poly[1,1-cyclohexane *bis*(4-phenyl)carbonate] *(See Fig. 6.2)*	444	19.3	22	-7	0.0	450
Poly[2,2-hexafluoropropane *bis*(4-phenyl)carbonate]	449	18.4	25	-40	0.0	404
Poly[1,1-(1-phenylethane) *bis*(4-phenyl)carbonate]	449	19.9	24	-15	0.0	443
Poly[2,2-(1,3-dichloro-1,1,3,3-tetrafluoro)propane *bis*(4-phenyl)carbonate]	457	19.5	25	-32	0.0	420
Poly[4,4'-isopropylidene diphenoxy di(4-phenylene)sulfone]	458	21.9	32	-12	0.0	462
Poly(oxycarbonyloxy-2,6-dichloro-1,4-phenyleneisopropylidene-1,4-phenylene)	459	20.7	21	-2	0.0	464
Poly(perfluorostyrene)	467	16.5	16	-17	0.0	410
Poly[2,2-propane *bis*{4-(2,6-dimethylphenyl)}carbonate]	473	19.8	23	6	0.0	471

☞ ☞ ☞   **TABLE 6.2 IS CONTINUED IN THE NEXT PAGE.** ☞ ☞ ☞

Table 6.2. CONTINUED FROM THE PREVIOUS PAGE.

Polymer	T_g(exp)	δ	N	N_{T_g}	x_{13}	T_g(fit)
Polyetherimide 6 *(See Fig. 6.6)*	473	24.3	51	12	0.0	495
Poly(α,β,β-trifluoro styrene)	475	17.2	11	-17	0.0	399
Poly(bisphenol-A terephthalate)	478	20.0	27	-27	0.0	432
Poly[oxy(2,6-dimethyl-1,4-phenylene)]	482	20.3	9	-8	0.0	437
Polyetherimide 2 *(See Fig. 6.6)*	482	24.3	43	55	0.0	528
Polyetherimide 7 *(See Fig. 6.6)*	485	24.4	51	3	0.0	490
Polyetherimide 8 *(See Fig. 6.6)*	486	24.4	51	31	0.0	507
Poly[oxy(2,6-diphenyl-1,4-phenylene)]	493	20.4	19	-8	0.0	452
Poly[4,4'-diphenoxy di(4-phenylene)sulfone]	493	22.7	29	4	0.0	483
Poly[4,4'-sulfone diphenoxy di(4-phenylene)sulfone]	493	24.3	32	-4	0.0	484
Ultem	493	23.7	45	93	0.0	550
Polyetherimide 9 *(See Fig. 6.6)*	494	24.4	51	50	0.0	519
Poly[N,N'-(*m,m'*-oxydiphenylene-oxy-*m*-phenylene)pyromellitimide]	494	25.1	36	94	0.0	575
Poly(oxyterephthaloyloxy-2,6-dimethyl-1,4-phenylene-isopropylidene-3,5-dimethyl-1,4-phenylene)	498	20.3	31	5	0.0	470
Polyetherimide 3 *(See Fig. 6.6)*	500	24.3	43	55	0.0	529
Poly[2,2-propane *bis*{4-(2,6-dichlorophenyl)}carbonate]	503	21.3	23	22	0.0	501
Polycarbonate 2 *(See Fig. 6.2)*	505	19.2	23	41	0.0	515
Polyetherimide 4 *(See Fig. 6.6)*	512	24.7	44	101	0.0	563
Polyimide 10 *(See Fig. 6.7)*	513	26.1	38	60	0.0	548
Polycarbonate 3 *(See Fig. 6.2)*	520	18.5	26	65	0.0	534
Polyetherimide 5 *(See Fig. 6.6)*	520	24.2	42	109	0.0	570
Poly[2,2-propane *bis*{4-(2,6-dibromophenyl)}carbonate]	523	21.6	23	22	0.0	503
Polyimide 1F *(See Fig. 6.8)*	533	23.9	45	38	0.0	512
Polyquinoline 5 *(See Fig. 6.3)*	539	21.8	46	104	0.0	546
Polyquinoline 1 *(See Fig. 6.3)*	541	21.9	39	39	0.0	506
Poly[3,5-(4-phenyl-1,2,4-triazole)-1,4-phenylene-3,5-(4-phenyl-1,2,4-triazole)-1,3-phenylene]	543	21.8	34	55	0.0	525
Poly(*m*-phenylene isophthalamide)	545	27.3	18	20	0.0	540
Polyquinoline 2 *(See Fig. 6.3)*	546	21.5	51	55	0.0	506
Poly[*o*-biphenylenemethane *bis*(4-phenyl)carbonate]	548	20.5	29	89	0.0	564
Torlon	550	25.7	27	78	0.0	587
Polyimide 3F *(See Fig. 6.8)*	561	23.9	45	57	0.0	526

☞ ☞ ☞   **TABLE 6.2 IS CONTINUED IN THE NEXT PAGE.** ☞ ☞ ☞

Table 6.2. CONTINUED FROM THE PREVIOUS PAGE.

Polymer	T_g(exp)	δ	N	N_{Tg}	x_{13}	T_g(fit)
Polyimide 2F *(See Fig. 6.8)*	562	23.9	45	29	0.0	506
Polyquinoline 9 *(See Fig. 6.3)*	573	22.0	39	128	0.0	579
Polyquinoline 6 *(See Fig. 6.3)*	578	21.5	58	120	0.0	537
Poly(quinoxaline-2,7-diylquinoxaline-7,2-diyl-*p*-terphenyl-4,4'-ylene)	578	21.9	38	188	0.0	631
Poly(quinoxaline-2,7-diyloxyquinoxaline-7,2-diyl-1,4-phenylene)	579	22.9	27	74	0.0	567
Polyphenolphthalein 2 *(See Fig. 6.5)*	580	20.5	40	71	0.0	523
Polyquinoline 7 *(See Fig. 6.3)*	581	21.7	45	158	0.0	585
Polyphenolphthalein 3 *(See Fig. 6.5)*	583	20.6	50	146	0.0	560
Polyimide 4F *(See Fig. 6.8)*	584	23.9	45	76	0.0	539
Poly(quinoxaline-2,7-diylcarbonylquinoxaline-7,2-diyl-1,4-phenylene)	591	23.3	28	74	0.0	566
Polyphenolphthalein 1 *(See Fig. 6.5)*	593	23.9	41	139	0.0	593
Polyquinoline 3 *(See Fig. 6.3)*	599	21.7	46	34	0.0	497
Poly(*p*-phenylene terephthalamide)	600	27.5	18	58	0.0	608
Polyimide 11 *(See Fig. 6.7)*	606	27.1	30	122	0.0	633
Poly(quinoxaline-2,7-diylsulfonylquinoxaline-7,2-diyl-1,4-phenylene)	615	24.5	29	74	0.0	570
Polyetherimide 12 *(See Fig. 6.7)*	615	26.8	29	122	0.0	635
Polyquinoline 4 *(See Fig. 6.3)*	618	21.4	58	50	0.0	499
Polyquinoline 10 *(See Fig. 6.3)*	618	21.7	51	144	0.0	562
Polyquinoline 8 *(See Fig. 6.3)*	624	21.5	57	174	0.0	569
Polytricyclic 1 *(See Fig. 6.4)*	626	23.9	27	127	0.0	634
Poly(3-phenylquinoxaline-2,7-diyl-3-phenylquinoxaline-7,2-diyl-1,4-phenylene)	645	22.0	38	144	0.0	595
Poly(quinoxaline-2,7-diylquinoxaline-7,2-diyl-1,4-phenylene)	649	22.7	26	128	0.0	635
Polyphenolphthalein 4 *(See Fig. 6.5)*	658	23.4	50	268	0.0	652
Polytricyclic 3 *(See Fig. 6.4)*	668	23.0	32	148	0.0	627
Polytricyclic 2 *(See Fig. 6.4)*	668	23.1	32	167	0.0	646
Poly[N,N'-(*p,p*'-oxydiphenylene)pyromellitimide]	672	26.3	29	175	0.0	690
Polyphenolphthalein 5 *(See Fig. 6.5)*	673	24.7	50	268	0.0	660
Poly[N,N'-(*p,p*'-carbonyldiphenylene)pyromellitimide]	685	26.7	30	175	0.0	686

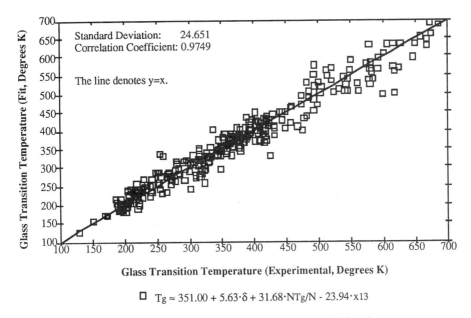

The line denotes y=x.

Standard Deviation: 24.651
Correlation Coefficient: 0.9749

$\square$ $T_g \approx 351.00 + 5.63 \cdot \delta + 31.68 \cdot NTg/N - 23.94 \cdot x13$

Figure 6.14. Correlation for the glass transition temperatures of 320 polymers.

6.C.   Effects of Number-Average Molecular Weight

A vast amount of work has been done to estimate the dependence of T_g on M_n by both empirical and fundamental theoretical methods. Three methods, which appear to be of greatest potential practical utility, will be discussed below. The first two methods are empirical, with some theoretical justification for the forms of the equations. The third method was developed from rigorous theoretical considerations, but is more difficult to use.

The most widely utilized empirical expression (Equation 6.4) was developed by Fox and Flory [49] from considerations of free volume. In this equation, $T_{g\infty}$ is the limiting value of T_g for $M_n \rightarrow \infty$. $T_{g\infty}$ is often larger than the commonly accepted values of the T_g of a polymer measured at ordinary molecular weights. For example, $T_{g\infty} \approx 382K$ for polystyrene and $T_{g\infty} \approx 434K$ for bisphenol-A polycarbonate, in comparison with the commonly accepted values of $T_g \approx 373K$ and $T_g \approx 423K$, respectively. K_g is an additional empirical parameter which is also usually obtained by fitting Equation 6.3 to experimental data for a given polymer.

$$T_g \approx T_{g\infty} - \frac{K_g}{M_n} \qquad (6.4)$$

There have been several attempts to relate K_g to the critical molecular weight (M_{cr}, which is discussed in Chapter 13) based on considerations of the effects of polymer chain entanglements on the large-scale cooperative motions of chain segments involved in the glass transition; for example, see references [50,51,52]. It was also suggested [53,54], from considerations of chain stiffness and the statistics of chain conformations, that K_g should be proportional to a power of $T_{g\infty}$, most likely between $T_{g\infty}^2$ and $T_{g\infty}^4$. These correlations were not, however, expressed as quantitative relationships in the previous literature.

A dataset containing 35 polymers was, therefore, assembled from many sources [53-71] and analyzed statistically. A weighted nonlinear regression procedure was used, with weight factors of $10000/K_g(exp)$. The best relationship (Equation 6.5) was found between K_g and $T_{g\infty}^3$. The details are provided in Table 6.3, Figure 6.15 and Figure 6.16. The structures of some of the polymers are shown in Figure 6.17. K_g values calculated by using Equation 6.5 have a standard deviation of 20.6% relative to those obtained from the experimental data, with a correlation coefficient of 0.8710 which indicates that Equation 6.5 accounts for roughly 76% of the variation of the K_g values in the dataset. The substitution of Equation 6.5 into Equation 6.4 results in a rough predictive version of the equation of Fox and Flory, given by Equation 6.6.

$$K_g \approx 0.002715 \cdot T_{g\infty}^3 \tag{6.5}$$

$$T_g \approx T_{g\infty} - 0.002715 \cdot \frac{T_{g\infty}^3}{M_n} \tag{6.6}$$

The equation of Fox and Flory is not flexible enough to describe the behavior of T_g over the entire range of possible M_n values with great accuracy. Deviations occur between the measured and fitted values of T_g, especially at very low values of M_n. The empirical relationship of Fox and Loshaek [26], given by Equation 6.7, has the additional flexibility needed to provide an almost perfect fit to data on the dependence of T_g on M_n over the entire range of M_n. In Equation 6.7, K_g' and K_g'' are empirical parameters obtained by fitting this equation to experimental data. It is not at all surprising that Equation 6.7 provides greater flexibility than Equation 6.4, since it includes Equation 6.4 as a special case where K_g'' is taken to be zero. The positive values usually found for K_g'' [64,65] can be ascribed to the effects of chain ends which become a major portion of a polymer chain as $M_n \to 0$ and which can thus have a significant effect on T_g in this limit [72].

$$T_g \approx T_{g\infty} - \frac{K_g'}{(K_g'' + M_n)} \tag{6.7}$$

Our analysis of the data for the dependence of T_g on M_n failed to reveal any quantitative structure-property relationships of sufficient accuracy for K_g' and K_g'' to allow the predictive use of Equation 6.7. There is considerable mathematical interaction between the effects of K_g' and K_g''. Small variations in the locations of the data points on the (M_n, T_g)-plane can cause large changes in the magnitudes of K_g' and K_g'', especially if T_g was only measured at a small number of values of M_n. Consequently, Equation 6.7 cannot be used to predict T_g as a function of M_n if experimental data are unavailable. However, if data are available, Equation 6.7 is usually preferable to Equation 6.4 in interpolating between data points to estimate T_g at values of M_n for which it was not measured.

Table 6.3. Data [53-71] gathered to develop a relationship for the parameter K_g of the equation of Fox and Flory, the polymer property ($T_{g\infty}$) used in fitting the values of K_g, and the fitted values of K_g, for 35 polymers. $T_{g\infty}$ values were rounded off to the nearest integer and K_g values were rounded off to the nearest thousand for simplicity in this table, but not in the original dataset analyzed in order to develop the relationship. $T_{g\infty}$ is in degrees Kelvin, while K_g is K·g/mole.

Polymer	$10^{-4} \cdot K_g$	$T_{g\infty}$	$10^{-4} \cdot K_g(fit)$
Poly(dimethyl siloxane)	0.6	150	0.9
n-Alkanes (oligomers of linear polyethylene)	1.2	176	1.5
Polyisoprene	1.2	207	2.4
Fomblin™ Z (Sample B): CF_3-[(O-CF_2-CF_2)$_m$-(O-CF_2)$_n$]-OCF_3	1.2	145	0.8
Polybutadiene	1.2	174	1.4
Poly(ethylene adipate)	1.3	228	3.2
Fomblin™ Z (Series II): CF_3-[(O-CF_2-CF_2)$_m$-(O-CF_2)$_n$]-OCF_3	1.5	142	0.8
Fomblin™ Z (Sample C): CF_3-[(O-CF_2-CF_2)$_m$-(O-CF_2)$_n$]-OCF_3	2.2	153	1.0
Fomblin™ Z (Sample A): CF_3-[(O-CF_2-CF_2)$_m$-(O-CF_2)$_n$]-OCF_3	2.4	142	0.8
Poly(propylene oxide)	2.5	198	2.1
Demnum™: CF_3-CF_2-CF_2(-O-CF_2-CF_2-CF_2)$_n$-O-CF_2-CF_3	3.3	172	1.4
Polypropylene (average of two listed K_g values)	3.9	266	5.1
Poly(3,3-dimethylthietane): -[S-CH_2-C(CH_3)$_2$-CH_2-]$_n$	4.4	225	3.1
Poly(tetramethylene terephthalate)	4.6	295	6.9
Poly(ethylene terephthalate)	5.1	342	10.9
Fomblin™ Y: CF_3-{[O-CF_2-CF(CF_3)]$_m$-(O-CF_2)$_n$}-OCF_3	5.4	217	2.8
Polyisobutylene	6.4	243	3.9
Poly(vinyl acetate) (average of two listed K_g values)	8.9	305	7.7
PMET *(See Fig. 6.17)*	10.1	399	17.2
Isotactic poly(methyl methacrylate)	11.0	318	8.7
PMiBT *(See Fig. 6.17)*	11.3	408	18.5
Poly(glycidyl methacrylate)	11.3	350	11.6

☞ ☞ ☞ **TABLE 6.3 IS CONTINUED IN THE NEXT PAGE.** ☞ ☞ ☞

Table 6.3. CONTINUED FROM THE PREVIOUS PAGE.

Polymer	$10^{-4} \cdot K_g$	$T_{g\infty}$	$10^{-4} \cdot K_g$(fit)
Poly(vinyl chloride) (average of two listed K_g values)	12.3	351	11.7
Polyacrylonitrile	14.0	371	13.9
Bisphenol-A polycarbonate	18.7	434	22.2
Polystyrene (combination of two datasets)	20.0	382	15.1
Atactic poly(methyl methacrylate)	21.0	388	15.9
Poly(N-vinyl carbazole)	22.8	500	33.9
Syndiotactic poly(methyl methacrylate)	25.6	405	18.0
Poly(p-methyl styrene)	26.5	384	15.4
Syndiotactic poly(α-methyl styrene)	31.0	453	25.2
Atactic poly(α-methyl styrene)	36.0	446	24.1
Poly(p-tert-butyl styrene) (from data at 10K/minute)	38.5	430	21.6
PMPhT (See Fig. 6.17)	41.8	477	29.5
PMMT (See Fig. 6.17)	44.7	435	22.3

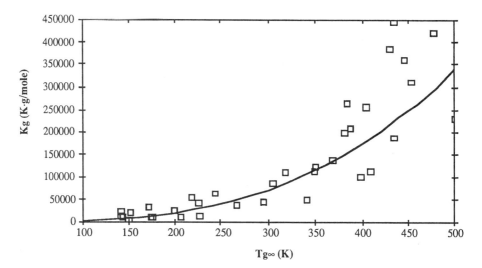

Figure 6.15. Experimental values of K_g as a function of $T_{g\infty}$. The squares indicate the data points. The curve shows the equation giving the best fit: $K_g \approx 0.002715 \cdot T_{g\infty}^3$.

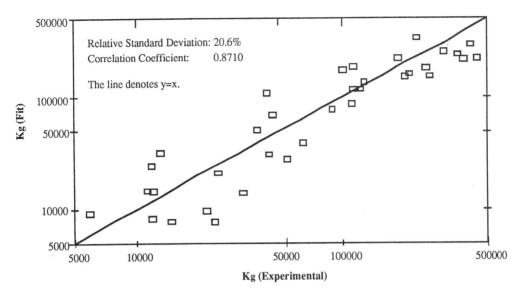

Figure 6.16. Comparison of experimental values of K_g with values calculated from the equation relating this parameter to $T_{g\infty}$. Note that a logarithmic scale is used for both axes.

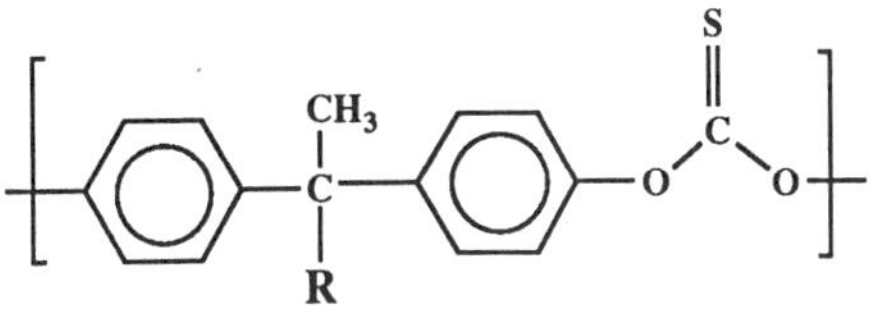

R	Polymer
Methyl	PMMT
Ethyl	PMET
Isobutyl	PMiBT
Phenyl	PMPhT

Figure 6.17. The structures of the repeat units of some of the polymers listed in Table 6.3.

Another very useful method for relating T_g to M_n was developed by Lu and Jiang [53] from rigorous considerations of polymer chain statistics and its relationship to chain stiffness and hence to T_g. This method is applicable to polymers with simple chain backbones (polymers whose backbones do not contain either rings or other rigid moieties such as double bonds). The application of this method to a polymer chain of arbitrary backbone structure requires calculations by using rotational isomeric state theory ([18] and [19]) to obtain some of the input parameters. However, for polymers with a vinyl-type backbone, the authors have worked out the details [53], and the key relationship is Equation 6.8 (obtained by combining two of the equations given in [53] and rearranging the combined equation algebraically). Here, n is the number-average degree of polymerization. F(y) is defined by Equation 6.9, and y is given by Equation 6.10 where θ is the angle between two successive backbone bonds. It is usually a very good approximation to use the value of the tetrahedral angle (109.47122°) for the θ of a vinyl-type chain backbone.

$$T_g = \frac{69.5 \cdot n \cdot T_{g\infty}}{\left[69.5 \cdot n + T_{g\infty} \cdot F(y)\right]} \tag{6.8}$$

$$F(y) = 1 - 3y + 6y^2 - 6y^3 \cdot \left[1 - \exp\left(-\frac{1}{y}\right)\right] \tag{6.9}$$

$$y = \frac{T_{g\infty}}{139 \cdot n \cdot \left[\sin\left(\frac{\theta}{2}\right)\right]^2} \tag{6.10}$$

The method of Lu and Jiang gives excellent agreement with experimental data for the dependence of T_g on M_n for polymers with a vinyl-type chain backbone [53], and is therefore preferable to the use of Equation 6.6 for such polymers.

In summary, the recommended procedure to estimate T_g as a function of M_n conveniently is to use Equation 6.7 to interpolate between the existing data points if a sufficient number of data points is already available, to use equations 6.8, 6.9 and 6.10 if experimental data are unavailable and the polymer has a vinyl-type chain backbone, and to use Equation 6.6 if experimental data are unavailable and the polymer has any type of chain backbone other than a vinyl-type backbone.

6.D. Effects of Plasticization

The methods used to estimate the effects of plasticization on T_g fall into two general classes; namely, semi-empirical equations based on considerations of "free volume" and the much more sophisticated theoretical treatments based on statistical thermodynamics. Neither type of approach is able to provide accurate predictions on a consistent basis. The equations based on free volume [20,33,73] have generally been favored over the statistical mechanical methods (see [74] for an example) for use in practical applications, because of their greater simplicity. For example, the T_g of a plasticized polymer can be estimated *very roughly* by using the following two equations:

$$T_g \approx \frac{\left\{T_{gp} + \left(X \cdot T_{gs} - T_{gp}\right) \cdot \left(1 - \Phi_p\right)\right\}}{\left\{1 + (X - 1) \cdot \left(1 - \Phi_p\right)\right\}} \tag{6.11}$$

$$X = \frac{(\alpha_{ls} - \alpha_{gs})}{(\alpha_{lp} - \alpha_{gp})} \tag{6.12}$$

In these equations, T_g is the glass transition temperature of the plasticized polymer, T_{gp} and T_{gs} are the glass transition temperatures of the pure polymer and the solvent, and α is the coefficient of volumetric thermal expansion (subscripts: l denotes above the glass transition temperature, g denotes below the glass transition temperature, p denotes the polymer, and s denotes the solvent). A rough general estimate for (α_{lp} - α_{gp}) is given by Equation 3.7, although it is preferable to use experimental values in general if they are available.

If the difference between the coefficients of thermal expansion of the solvent above and below T_{gs} is unknown, Equation 3.7 can also be used to provide a first estimate for the solvent by substituting the quantities referring to the solvent instead of those referring to the polymer into this equation. This procedure amounts to making the assumption that the free volume arguments underlying Equation 3.7 are just as valid for simple molecular liquids as they are for amorphous polymers. If this assumption is made, Equation 6.12 is simplified into Equation 6.13 which should only be used if the necessary thermal expansion data are unavailable for the solvent.

$$X \approx \frac{T_{gp}}{T_{gs}} \qquad (6.13)$$

T_{gs} has only been measured for a very small fraction of the common solvents. If T_{gs} has not been measured for the solvent of interest, the following empirical relationship [20] can be used to obtain a reasonable estimate for it in terms of the melting temperature (T_m):

$$T_g \approx \frac{2 \cdot T_m}{3} \qquad (6.14)$$

6.E. Effects of Crosslinking

Many commercial polymers are crosslinked, ranging from lightly-crosslinked elastomers to very densely crosslinked thermosets. The effects of crosslinking on the properties of polymers can be roughly classified as follows [26,27]:

1. *Topological effect* caused by topological constraints introduced by crosslinks on the properties. This effect is referred to simply as the "crosslinking effect" by many authors.
2. *Copolymerization effect* (also referred to as the "copolymer effect") related to the change of the fractions of two or more types of repeat units with increasing crosslinking. Depending on the types of monomers involved, this effect may either strengthen or weaken the trends expected on the basis of the topological effect, and may even reverse them in some cases.

We will focus here mainly on the more general topological effect of crosslinking on T_g. The copolymerization effect can be easily superimposed on the topological effect when necessary. The topological effect of crosslinking on T_g is to increase its value. The rate of increase of T_g

accelerates as the average distance between crosslinks (quantified by the average molecular weight M_c, or the average number n of "repeat units" between them as defined by Equation 6.15) decreases, as shown in Figure 6.18 from data [75] on copolymers of styrene and divinylbenzene.

$$n = \frac{M_c}{M} \qquad\qquad (6.15)$$

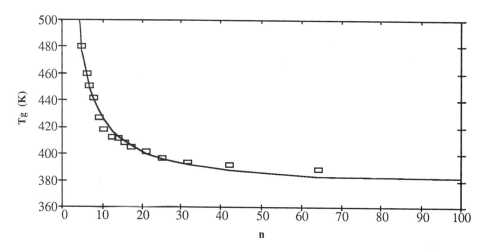

Figure 6.18. Glass transition temperature (T_g) of styrene-divinylbenzene copolymers as a function of the average number (n) of repeat units between crosslinks.

Ten datasets were assembled. A wide variety of polymers were included in the analysis: crosslinked natural rubber [15], styrene-divinylbenzene copolymers [75a], poly(ethyl acrylate) [29], poly(methyl methacrylate) [30], epoxies [31], poly(divinylsiloxane *bis*-benzocyclobutene) (DVS-BCB$_2$, see Figure 6.19) [75a], poly(diketone - *bis* benzocyclobutene) (DK-BCB$_2$, see Figure 6.19) [75a], poly(bisphenol-A carbonate - *bis* BCB) (BPAC-BCB$_2$) [75b], polyurethanes synthesized by reacting 4,4'-methylenediisocyanate (MDI) with various glycols [75a], and polyurethanes synthesized by the reaction of VORANOL™ polyol with BCB isocyanates [75a].

These data (see Table 6.4) were analyzed, showing that the effect of adding a crosslink to the polymeric structure becomes larger with increasing stiffness of the polymer chains prior to the addition of this crosslink, as quantified by the parameter N_{rot} defined in Chapter 4, with $T_g(\infty)$ representing the glass transition temperature of the uncrosslinked limit (n→∞):

$$T_g(n) \approx T_g(\infty) \cdot \left[1 + \frac{c}{\left(n \cdot N_{rot}\right)}\right] \qquad\qquad (6.16)$$

Equation 6.16 is consistent in general form with previous treatments [15,26,27,29] of the effects of crosslinks on T_g, and with expectations based on statistical thermodynamic theories of the glass transition [4-8]. The new concept here is the use of N_{rot} as a normalization factor, to multiply the value of n. The product $n \cdot N_{rot}$ is the average number of rotational degrees of freedom between crosslinks. Equation 6.16 states that, if we start from the uncrosslinked limit and add crosslinks, T_g increases, and its rate of increase is larger if the repeat unit is rigid (has a small N_{rot}) than if the repeat unit is flexible (has a large N_{rot}). The glass transition involves the cooperative motions of sizeable chain segments. The more stiff the chains are, the more difficult such large-scale motions will be. For a repeat unit of given length, chain stiffness increases, and the number of degrees of freedom available for chain motions decreases, with decreasing N_{rot}. A large N_{rot} implies that there are many degrees of freedom for chain motion, so that the effect of tying down a degree of freedom by a crosslink is less drastic than when N_{rot} is small. The effect of introducing a crosslink on the value of T_g is thus roughly *inversely* proportional to N_{rot}.

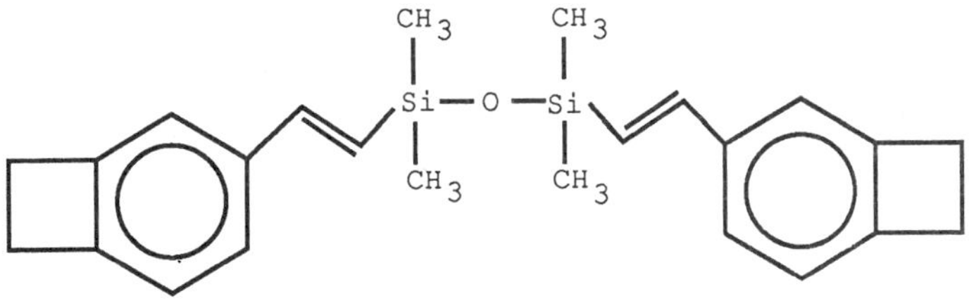

(a) Monomer of DVS-BCB$_2$.

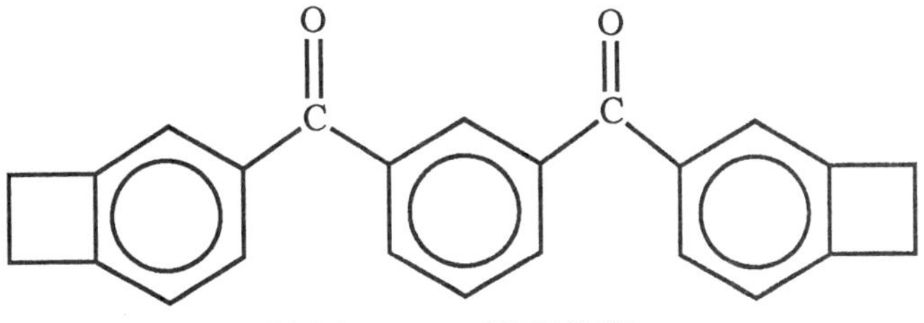

(b) Monomer of DK-BCB$_2$.

Figure 6.19. Some monomers which can be cured thermally to produce thermoset resins.

Table 6.4. T_g as a function of the average number n of repeat units between crosslinks. The rotational degrees of freedom parameter (N_{rot}) used in the calculations is also listed.

(a) Natural Rubber: $N_{rot}=4$							
n	$T_g(K)$	n	$T_g(K)$	n	$T_g(K)$	n	$T_g(K)$
66.67	223	13.16	237	5.56	273	3.03	328
40.00	228	10.00	251	4.17	280	2.04	385
18.18	232	6.41	253	3.70	296		

(b) Styrene-divinylbenzene copolymers:

Contribution of 3 from each styrene and 5 from each divinylbenzene to N_{rot}

n	$T_g(K)$	n	$T_g(K)$	n	$T_g(K)$	n	$T_g(K)$
∞	373.1	25.272	396.8	14.058	411.6	7.709	442.2
64.168	388.5	21.210	401.4	12.631	412.8	6.915	450.6
42.045	391.4	17.144	405.5	10.389	417.9	6.168	459.3
31.787	394.0	15.820	408.6	8.976	426.8	4.895	480.0

(c) Poly(ethyl methacrylate): Contribution of 6 from each ethyl acrylate and 12 from each ethylene glycol dimethacrylate (EGDMA, used as hardener) to N_{rot}

n	$T_g(K)$	n	$T_g(K)$	n	$T_g(K)$	n	$T_g(K)$
∞	270.5	25.00	277.5	8.33	293	5.00	307.5
50.00	274	12.50	286	6.25	300.5		

(d) Poly(methyl methacrylate):

Contribution of 6 from each methyl methacrylate and 12 from each EGDMA to N_{rot}

n	$T_g(K)$	n	$T_g(K)$	n	$T_g(K)$
∞	383	39.0	399	12.0	428
138.9	388	24.0	403		

(e) Epoxy resins: $N_{rot}=11$ (corresponding to phenoxy resin)

n	$T_g(K)$	n	$T_g(K)$	n	$T_g(K)$	n	$T_g(K)$
15.30	373	4.08	393	5.63	378	1.07	493

(f) Poly(divinylsiloxane - *bis* BCB): $N_{rot}=12.5$

n	$T_g(K)$	n	$T_g(K)$	n	$T_g(K)$	n	$T_g(K)$
75.060	317	7.000	368	3.051	403	1.885	414
20.432	321	3.366	417	2.189	436	1.159	458

☞ ☞ ☞   TABLE 6.4 IS CONTINUED IN THE NEXT PAGE. ☞ ☞ ☞

Table 6.4. CONTINUED FROM THE PREVIOUS PAGE.

(g) Poly(diketone - *bis* BCB): $N_{rot}=6.5$							
n	T_g(K)	n	T_g(K)	n	T_g(K)	n	T_g(K)
2052.0	413	62.0	428	16.0	439	4.0	463
283.0	418	21.0	432	7.0	456	1.9	592

(h) BPAC-BCB$_2$: $N_{rot}=8$ (corresponding to bisphenol-A polycarbonate)							
n	T_g(K)	n	T_g(K)	n	T_g(K)	n	T_g(K)
∞	423	26.65	443	5.47	453	2.95	480
69.63	433	12.22	449	4.24	467		
41.00	438	8.31	454	3.06	471		

(i) Polyurethanes from 4,4'-MDI and glycols: $N_{rot}=(20+12/n)$					
n	T_g(K)	n	T_g(K)	n	T_g(K)
∞	364	1.00	385	0.50	411
2.00	378	0.67	401		

(j) Polyurethanes from VORANOL™ polyol and BCB isocyanate: $N_{rot}=4$					
n	T_g(K)	n	T_g(K)	n	T_g(K)
34.44	220	17.22	236	8.61	258

The "uncrosslinked limit" of many crosslinked polymers can be defined quite easily. For example, polystyrene is the uncrosslinked limit of styrene-divinylbenzene copolymers. On the other hand, the uncrosslinked limit of many other crosslinked polymers cannot be defined unambiguously, and is an idealization. In such cases, there is no unique uncrosslinked limit for a densely crosslinked polymer synthesized from a monomer or monomers with functionalities able to react by several mechanisms. DVS-BCB$_2$ is an example of such a complicated network polymer. However, even for such cases, reasonable choices of uncrosslinked limit can be made. Furthermore, the general form for the dependence of T_g on the crosslink density can be combined with a very small amount of data (such as one or two datapoints) to give the proper scale for the T_g values and to provide a good estimate for T_g at other crosslink densities.

Several experimental factors can have a significant effect on the measured values of M_c and T_g, and thus also on the quality of an empirical correlation for T_g, such as the purity of monomers used during synthesis, how the crosslinked polymers were formed (especially the extent of curing), and the methods used to measure M_c (indirectly) and also to measure T_g.

When crosslinking was accomplished by reacting monomers which incorporated repeat units of different N_{rot} into the polymer, the variation of N_{rot} with composition (a manifestation of the copolymerization effect) was taken into account. On the other hand, a single $T_g(\infty)$ value was used for each polymer, so that the copolymerization effect on $T_g(\infty)$ was not considered.

The data were analyzed to obtain the best general value of c to use in applying Equation 6.16 to predict the T_g of an unknown polymer as a function of crosslinking. Details of the analysis are summarized in Table 6.5. The number of data points (n_{data}) available for different polymers used in developing the correlation ranged from 3 to 16. It was, consequently, necessary to take into account systematically the fact that the c values for all ten polymers may not be of equal statistical significance. The optimum value of c was hence determined by weighted averaging of the c values for the best fits for the ten polymers, with weight factors equal to the number of data points from which each c value was obtained. It was calculated that c≈5 is the optimum value by using this procedure. The correlation coefficient between $T_g(n)/T_g(\infty)$ and $1/(n \cdot N_{rot})$ is 0.9354, so that this relationship accounts for roughly 87.5% of the variation of the $T_g(n)/T_g(\infty)$ values in the ten datasets taken all together. The standard deviation is 2, so that c≈5±2 has a large variance. The resulting final relationship is given by Equation 6.17 and depicted in Figure 6.20.

$$T_g(n) \approx T_g(\infty) \cdot \left[1 + \frac{5}{(n \cdot N_{rot})} \right] \qquad\qquad (6.17)$$

Table 6.5. Best fits, by using Equation 6.16, to the T_g's of ten polymers with variable amounts of crosslinking. $T_g(\infty)$ and c are fitting parameters (see text for details). The number of data points, the standard deviation between the observed and fitted values of T_g, and the correlation coefficient, are denoted by n_{data}, σ, and R, respectively. $T_g(\infty)$ and σ are listed in degrees K.

Polymer	n_{data}	$T_g(\infty)$	c	σ	R
Natural rubber	11	212.7	6.28	7.9	0.9886
Styrene-divinylbenzene copolymers	16	374.7	4.49	1.7	0.9924
Poly(ethyl acrylate)	7	270.0	4.56	0.4	0.9997
Poly(methyl methacrylate)	5	383.2	8.91	2.1	0.9947
Epoxy resins	4	358.2	4.41	5.6	0.9967
Poly(divinylsiloxane - *bis* benzocyclobutene)	8	335.7	6.23	23.7	0.9067
Poly(diketone - *bis* benzocyclobutene)	8	414.2	4.90	14.0	0.9748
Poly(bisphenol-A carbonate - *bis* benzocyclobutene)	10	433.0	2.45	5.5	0.9566
Polyurethanes from 4,4'-MDI and glycols	5	360.8	2.76	5.4	0.9680
Polyurethanes from VORANOL™ polyols and BCB isocyanate	3	209.0	8.19	2.7	0.9951

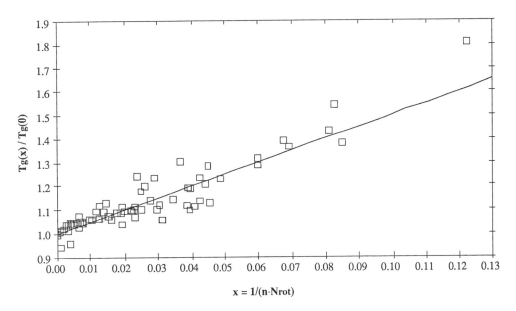

Figure 6.20. Effects of crosslinking on the glass transition temperature, expressed in terms of the ratio $[T_g(x)/T_g(\infty)]$ as a function of $x \equiv [1/(n \cdot N_{rot})]$. The squares denote the data points, while the line denotes the model equation: $[T_g(x)/T_g(\infty)] \approx (1 + 5 \cdot x)$.

The dependence of T_g on the rotational degrees of freedom of the backbone and the side groups (N_{BBrot} and N_{SGrot}, respectively) may actually be somewhat different, with the effect of N_{BBrot} being stronger because N_{BBrot} is more directly related to the intrinsic rigidity of the backbone. However, it is impossible to draw any statistically significant conclusions concerning this point from the available experimental data, which correlates just as well with N_{rot}. Additional data, on the effects of crosslinking polymers whose uncrosslinked limits contain very large and flexible side groups (with $N_{SGrot} \gg N_{BBrot}$), may be needed to shed light on this issue.

Attempts have been made to develop group contribution techniques to correlate and/or predict T_g for thermosets, mainly by using data for epoxy resins of different structures [28,76,77]. The study by Kreibich and Batzer [28] is a valuable reference, since it lists the T_g's of a large number of thermosets as a function of their molecular structures. The study by Won *et al* [77] attempts to distinguish carefully between the topological effect and the copolymerization effect.

The effects of using different curing agents (hardeners), different curing rates, and/or different thermal histories (such as physical aging), on the physical properties (including T_g) of epoxy resins, have been investigated [78-81]. Such studies, if extended to a wider variety of types of polymers, may allow the development of refined correlations in the future.

Finally, there are ongoing efforts to provide molecular interpretations, and predictive capabilities derived from such interpretations, for the glass transition temperature and other important physical properties of crosslinked polymers. See the work of Chow [82-84] for an example of such efforts which have, so far, shown only very limited success.

The work summarized above has been presented elsewhere in greater detail [75].

6.F. Effects of Tacticity

The extent and type of stereoregularity of the polymer chains can have a significant impact on many properties. Stereoregularity is usually characterized in terms of the *tacticity* [85], which is depicted in Figure 6.21 for vinyl-type chain backbones where it has been studied in the greatest detail. The ideal limiting cases of perfectly syndiotactic and perfectly isotactic polymers are rare. Polymers usually contain statistical mixtures of different types of diads and triads, with a stereoregular polymer containing a very large fraction of one specific type.

One example of the effects of tacticity on polymer properties is that stereoregular polymers can have high crystallinity, while atactic polymers are amorphous. This tendency is the result of the greater ease of packing stereoregular chains than of packing irregular chains into crystallites.

Tacticity can also have a large effect on T_g, and hence on all properties which depend on T_g. Literature data [53,86-94] on the effects of tacticity on T_g are summarized in Table 6.6.

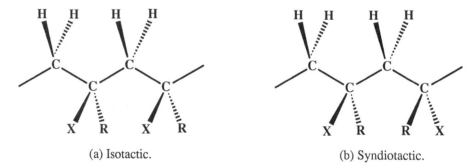

(a) Isotactic.   (b) Syndiotactic.

Figure 6.21. Illustration of the two limiting cases of perfect stereoregularity, for a vinyl-type chain backbone. One diad (pair of -CXR- groups) is shown. X and R denote atoms or groups attached to backbone carbon atoms, with X≠R. The thick lines represent bonds coming out of the plane of the paper, while the dashed lines represent bonds going below the plane of the paper. X's and the R's point above and below this plane, respectively, for the *isotactic* chain. They alternate regularly in direction (handedness) for the *syndiotactic* chain. The limit of complete lack of stereoregularity (*atactic* chain) occurs when the isotactic and syndiotactic diads are equal in number on a polymer chain and are distributed completely randomly over the length of the chain.

Table 6.6. Glass transition temperatures of syndiotactic, isotactic and atactic polymers, in degrees Kelvin. Many of the values listed for the syndiotactic and isotactic polymers are extrapolations from measurements on series of polymers with differing tacticities. Some of the listed T_g values are averages of two or more published values. Some of the T_g values listed for atactic polymers differ from those listed in Table 6.2 because they are from different sources.

Polymer	T_g(syndiotactic)	T_g(isotactic)	T_g(atactic)
Poly(methyl methacrylate)	433	316	378
Poly(ethyl methacrylate)	393	281	324
Poly(isopropyl methacrylate)	412	300	327
Poly(n-butyl methacrylate)	361	249	293
Poly(isobutyl methacrylate)	393	281	321
Poly(cyclohexyl methacrylate)	436	324	377
Poly(2-hydroxyethyl methacrylate)	377	308	359
Poly(methyl acrylate)	-----	283	281
Poly(ethyl acrylate)	-----	248	249
Poly(isopropyl acrylate)	270	262	267
Poly(sec-butyl acrylate)	-----	250	251
Poly(cyclohexyl acrylate)	-----	285	292
Poly(methyl α-chloroacrylate)	450	358	416
Poly(ethyl α-chloroacrylate)	393	310	366
Poly(isopropyl α-chloroacrylate)	409	341	363
Polystyrene	378	360	373
Poly(α-methyl styrene)	453	-----	446
Polypropylene	269	255	267
Poly(N-vinyl carbazole)	549	399	423
Poly(vinyl chloride)	T_g increased with syndiotactic triad content (28% to 43%), and decreased with isotactic triad content (13% to 21%), with lowest and highest values of 352K and 370K, respectively, observed for the complete set of 19 samples [92].		

The following observations can be made from these data:

1. The effects of tacticity on T_g have been studied most extensively for acrylic polymers, whose structural variants make up 15 of the 20 polymers listed in Table 6.6.

2. Substitution of a larger group (such as methyl) or a larger atom (such as chlorine) for the hydrogen atom on the α-carbon results in a large tacticity effect on the T_g's of acrylic polymers, while polyacrylates with unsubstituted α-carbon atoms show little tacticity effect.

3. Even in the absence of substitution on the α-carbon, extremely large side groups attached directly to the polymer chain, as in poly(N-vinyl carbazole), can cause a large tacticity effect.

4. The T_g's of styrenic polymers do not vary much with tacticity. The effects of tacticity on the properties of these polymers are mainly due to its effects on the crystallinity. For example, the high crystallinity and the detailed structure of the crystalline phase of syndiotactic polystyrene [94-97] are both important in the differences between this polymer and ordinary polystyrene.

5. In every case where tacticity results in a significant variation of T_g (i.e., more than a few degrees, and ranging up to 150K in the most extreme cases), the following order is observed: T_g(syndiotactic)$>>$$T_g$(atactic)$>>$$T_g$(isotactic).

The effects of tacticity on T_g are associated intimately with its effects on chain stiffness, which was shown earlier to be a key factor in determining T_g. These effects are difficult to quantify purely on the basis of empirical quantitative structure-property relationships, and require instead the examination of chain stiffness from a computationally more sophisticated perspective.

For example, Lu and Jiang showed [53], based on a statistical thermodynamic derivation in which some of the parameters of the model were calibrated by using experimental data, that the following approximation can be made for polymers with a vinyl-type chain backbone:

$$T_g \approx 34.75 \cdot C_\infty(T_g) \tag{6.18}$$

$C_\infty(T_g)$ is the value of the *characteristic ratio* at T_g. C_∞, which will be discussed in Chapter 12, depends both on polymer chain stereoregularity and on the temperature. Direct applications of the method which will be presented in Chapter 12 for its prediction as a function of polymeric structure is, unfortunately, mainly limited to atactic polymers in dilute solutions under "theta conditions", and cannot be used to determine the effects of tacticity or of temperature variation.

The use of more sophisticated calculations based on rotational isomeric state theory [18,19] allows the calculation of $C_\infty(T)$, but this capability still does not enable us to predict T_g accurately since $C_\infty(T_g)$ cannot be calculated without having some idea of what T_g should be before doing the calculation. Despite this limitation, Equation 6.18 and the general theoretical framework which led to its development hold out the hope that further extensions of this approach or similar methods may be able to allow accurate predictions of the effects of tacticity on T_g in the future.

6.G. Secondary Relaxations

In addition to the glass transition temperature (amorphous phases) and melting temperature (crystalline phases), polymers also manifest secondary relaxations at temperatures below those of major relaxations (T_g or T_m, which will collectively be referred to as T_α). The main secondary relaxation temperature will be generically designated as T_β although it may be labeled differently in the literature on specific polymers. For example, it is commonly labeled as T_γ for bisphenol-A polycarbonate where T_γ is a relaxation of higher intensity than T_β, and occurring at a lower temperature, which is the main secondary relaxation temperature of this particular polymer.

The secondary relaxations are very important because they are often associated with the onset of ductile behavior in polymers with increasing temperature under mechanical deformation. This semiquantitative relationship between the "ductile-brittle transition temperature" and T_β has stimulated much experimental and theoretical work on secondary relaxations. Secondary relaxations are usually measured either by mechanical methods such as dynamic mechanical spectroscopy or (somewhat less often) by electrical methods such as dielectric relaxation spectroscopy [98]. The existence of T_β is generally ascribed to the onset of a significant amount of some kind of motion of the polymer chains and/or the side groups attached to them, on a much smaller and more localized scale than the large-scale cooperative motions of chain segments associated with T_α. These motions are usually inferred from the results of measurements using methods such as nuclear magnetic resonance spectroscopy. See references [99,100] for a phenomenological model (one of several in existence) which attempts to relate the local motions of small groups of atoms to the observed secondary relaxations from fundamental considerations, as well as providing a large number of references to the previous literature.

A shortcoming of fundamental models of secondary relaxations is that they do not, generally, provide a simple and reliable way to estimate T_β. The complications include the significant dependence of T_β on the measurement frequency. The development of a simple correlation for T_β was addressed by Wu [101], who provided the following empirical relationships to correlate data (Table 6.7) for T_β and T_α measured by mechanical and electrical tests at a frequency of 1 Hz:

$$T_\beta \approx (0.135 + 0.082 \cdot C_\infty) \cdot T_\alpha \qquad\qquad \textit{(if } C_\infty \leq 10.5 \textit{)} \qquad\qquad (6.19)$$

$$T_\beta \approx T_\alpha \qquad\qquad \textit{(no distinct } T_\beta, \textit{ if } C_\infty > 10.5 \textit{)} \qquad\qquad (6.20)$$

Again, C_∞ is the characteristic ratio of the polymer. The values of C_∞ calculated by the method of Chapter 12 can be used in equations 6.19 and 6.20 for polymers which are not stereoregular, to obtain an estimate of T_β if T_α (T_g for an amorphous polymer) can be predicted.

Some of the T_α values listed in Table 6.7 differ significantly from T_g values listed earlier in this chapter. This is partially due to the use of data from mechanical and electrical measurements in Table 6.7, instead of the calorimetric data commonly preferred in reporting T_g. Some of the other differences (such as those for polyethylene) are also caused by the different conventions used by different authors in labeling the complex relaxation spectra of semicrystalline polymers.

Some important questions still remain unanswered, such as the prediction of the intensity and the width of a secondary relaxation and the dependence of its temperature on the test frequency. Finally, there are several secondary relaxations in many polymers, and these equations can only be used to predict the temperature of one of these relaxations (albeit the most important one).

Table 6.7. Temperatures of the main relaxation (T_α) and the largest secondary relaxation (T_β), in degrees Kelvin, as measured by mechanical and electrical tests at a frequency of 1 Hz [101].

Polymer	T_α	T_β
Polyoxymethylene	260	203
Polyethylene	275	165
Poly(n-butyl methacrylate)	307	293
Poly(n-propyl methacrylate)	332	300
Poly(hexamethylene adipamide)	341	208
Poly(ethylene terephthalate)	355	200
Poly(methyl methacrylate) plasticized with 25% dibutyl phthalate	355	308
Poly(ethyl methacrylate)	362	300
Poly(methyl methacrylate) plasticized with 10% dibutyl phthalate	363	308
Isotactic poly(methyl methacrylate)	373	342
Poly(methyl methacrylate) plasticized with 5% dibutyl phthalate	373	308
Polystyrene	373	363
Atactic poly(methyl methacrylate)	395	302
Poly(t-butyl methacrylate)	395	395
Syndiotactic poly(methyl methacrylate)	403	313
Poly(cyclohexyl methacrylate)	406	406
Poly(phenyl methacrylate)	407	407
Bisphenol-A polycarbonate	428	163
Bisphenol-A isophthalate	457	187
Poly[4,4'-isopropylidene diphenoxy di(4-phenylene)sulfone]	459	169
Bisphenol-A terephthalate-co-isophthalate (3/7)	466	187
Bisphenol-A terephthalate	471	180
Bisphenol-A terephthalate-co-isophthalate (1/1)	472	185
Ultem	485	171
Poly[oxy(2,6-dimethyl-1,4-phenylene)]	493	158

6.H. Crystalline Melting Temperature

The crystalline melting temperature (T_m) is very important in predicting the processing behavior and the properties of semicrystalline polymers. Unlike the glass transition, melting is a first-order thermodynamic transition where the first derivatives of the free energy (the enthalpy and the entropy) change discontinuously. T_m, the enthalpy of fusion (ΔH_m), and the entropy of fusion (ΔS_m) are related by Equation 6.21, which states that the Gibbs free energy change during a first-order phase transition (the free energy of fusion ΔG_m in the case of melting) is zero:

$$\Delta G_m = (\Delta H_m - T_m \cdot \Delta S_m) = 0 \tag{6.21}$$

ΔH_m and ΔS_m were tabulated for polymers, and group contributions were provided for their calculation [20]. It was also indicated that it is very improbable that a method can be derived for the accurate prediction of ΔH_m by a simple addition of group contributions, while ΔS_m appears to show a much more regular relation to structure than ΔH_m does.

Melting is a first-order phase transition free of the kinetic (measurement rate dependence) effects which occur at the glass transition. Furthermore, T_m is a property of a well-ordered crystalline material while T_g is a property of an amorphous material. It may, therefore, be imagined that the prediction of T_m should be easier than the prediction of T_g. In fact, just the reverse is true. While the two key factors determining T_g (chain stiffness and interchain cohesive forces) are also important in determining T_m, the details of the interchain packing (which did not have to be considered in calculating T_g) are also of paramount importance in determining T_m.

The details of the interchain packing determine whether there will be any crystalline phase at all, what the maximum possible crystallinity is, and what the rate of crystallization is under a given set of conditions [20]. It is, for example, entirely possible and very common for a polymer to be predicted to have a very high T_m but to be completely amorphous because it has irregular chains which do not pack well in a crystalline lattice. An example of this phenomenon occurs in atactic polymers with a vinyl-type chain backbone and large side groups, where the absence of stereoregularity makes crystalline packing impossible. Furthermore, even when a polymer has a crystalline phase, its T_m will often depend on the morphology, both at the microscopic scale (such as whether the crystalline domains are lamellar, spherulitic or extended chain crystallites) and at the crystalline unit cell scale (such as which polymorph has been formed).

Van Krevelen [20] defined the empirically-developed quantity Y_m, which he named the *molar melt transition function*, and showed how T_m can be estimated by using Equation 6.22.

$$T_m \approx \frac{Y_m}{M} \tag{6.22}$$

Equation 6.22 is isomorphous to Equation 1.1, which van Krevelen [20] had developed to estimate T_g in terms of the molar glass transition function. The calculation of Y_m is done in terms of a combination of group contributions and structural correction terms, the latter taking into account the fact that the contributions of many groups to T_m depend on their local environments.

It had been observed previously by many workers that the values of T_g and T_m are very approximately proportional to each other. Van Krevelen [20] suggested that this relationship (Equation 6.23) can be used to obtain rough estimates of T_m in practical calculations.

$$T_m \approx \text{constant} \cdot T_g \tag{6.23}$$

The main difficulty with Equation 6.23 is that the "constant" factor varies over an enormous range (from 1.06 to 4). The following general empirical rules have been formulated [20]:
1. Polymers with $T_m > 2 \cdot T_g$ are highly symmetrical and have short repeat units consisting of one or two chain backbone atoms, each carrying either no substituents at all or single-atom substituents. These polymers can manifest very high crystallinity. Examples include polyethylene, polyoxymethylene and polytetrafluoroethylene.
2. Polymers with $T_m < 1.32 \cdot T_g$ are unsymmetrical (contain a chain backbone atom which does *not* have two identical substituents). They generally have a much more complex repeat unit structure than polymers with $T_m > 2 \cdot T_g$. They can, however, still be highly crystalline if they have long sequences of methylene groups or if they are highly stereoregular.
3. For the majority of the semicrystalline polymers (both symmetrical and unsymmetrical), $1.32 \cdot T_g < T_m < 1.79 \cdot T_g$, with a most probable value centered around $T_m \approx 1.5 \cdot T_g$ when Equation 6.23 becomes identical to Equation 6.14.
4. In random copolymers, T_m is depressed because of the difficulty of crystallization caused by the irregularity of the structure, while T_g usually has a normal value between those of the pure homopolymers. The result is usually a lowered crystallinity and a lowered T_m/T_g ratio.
5. In block copolymers [102], long sequences of identical repeat units may crystallize in the same manner as in the homopolymers. This phenomenon can be utilized to develop block polymers with important technological applications where the T_m/T_g ratio is much higher than in semicrystalline homopolymers, by using an amorphous "soft segment" block which has a very low T_g and a highly crystalline "hard segment" block which has a very high T_m.

Equation 6.23, when combined with these rules, can be useful if Equation 6.22 cannot be utilized for a novel polymer for which some of the group contributions to Y_m are unknown. However, Equation 6.22 is strongly preferred if all of the group contributions to Y_m are known.

The melting temperatures of a large number of polymers, taken from the extensive tabulations of van Krevelen [20], are summarized in Table 6.8. Readers interested in performing calculations of T_m are referred to this book for further information.

Table 6.8. Melting temperatures (T_m, second and third editions of Ref. [20]), in degrees Kelvin (K). Wide ranges of T_m values are reported for many polymers because of factors such as differences in types, sizes and quality of crystalline regions, difficulties in discerning T_m in some polymers of very low maximum percent crystallinity, and differences in tacticity. In fact, many of the polymers for which only one value of T_m is listed are polymers for which a measured value of T_m has only been reported once in the literature. The list is ordered by polymer types (polyolefins, polydienes, polystyrenes, polyxylylenes, polyethers, polyacrylates, polymethacrylates, various polyvinyls, polysulfides, polyesters, polyamides, polycarbonates, and others).

Polymer	T_m	Polymer	T_m
Polyethylene *(limit of perfect lamellae)*	414	Polypropylene	385 to 481
Polyisobutylene	275 to 317	Poly(1-butene)	379 to 415
Poly(3-methyl-1-butene)	573 to 583	Poly(1-pentene)	384 to 403
Poly(4-methyl-1-pentene)	501 to 523	Poly(1-hexene)	321
Poly(5-methyl-1-hexene)	383 to 403	Poly(1-octene)	235
Poly(1-octadecene)	314 to 383	Poly(1,4-butadiene) *(cis)*	275 to 285
Poly(1,4-butadiene) *(trans)*	415 to 421	Polyisoprene *(cis)*	287 to 309
Polyisoprene *(trans)*	347	Poly(2-*t*-butyl-1,4-butadiene) *(cis)*	379
Polychloroprene *(trans)*	353 to 388	Polystyrene	498 to 523
Poly(*o*-methyl styrene)	633	Poly(*p*-methoxy styrene)	511
Poly(fluoro styrene)	523 to 543	Poly(3-phenyl-1-propene)	503 to 513
Poly(*p*-xylylene)	648 to 713	Poly(chloro-*p*-xylylene)	552 to 572
Polyoxymethylene	333 to 473	Polyoxyethylene	335 to 349
Polyoxytrimethylene	308	Polyoxytetramethylene	308 to 333
Poly(propylene oxide)	333 to 348	Poly(hexene oxide)	345
Poly(octene oxide)	347 to 360	Poly(3-methoxypropylene oxide)	330
Poly(3-butoxypropylene oxide)	300	Poly(3-hexoxypropylene oxide)	317
Poly(3-phenoxypropylene oxide)	485	Poly(3-chloropropylene oxide)	390 to 408
Poly[oxy(2,6-dimethyl-1,4-phenylene)]	534 to 580	Poly[oxy(2,6-diphenyl-1,4-phenylene)]	730 to 770
Poly(vinyl methyl ether)	417 to 423	Poly(vinyl ethyl ether)	359
Poly(vinyl heptadecyl ether)	333	Poly(*n*-propyl acrylate)	388 to 435
Poly(*n*-butyl acrylate)	320	Poly(isobutyl acrylate)	354
Poly(*sec*-butyl acrylate)	403	Poly(*t*-butyl acrylate)	466 to 473
Poly(methyl methacrylate)	433 to 473	Poly(*t*-butyl methacrylate)	377 to 438
Poly(dodecyl methacrylate)	239	Poly(octadecyl methacrylate)	309
Poly(ethylene sulfide)	418 to 483	Poly(decamethylene sulfide)	351 to 365
Poly(propylene sulfide)	313 to 326	Poly[thio(*p*-phenylene)]	527 to 630
Poly(vinyl dodecanoate)	274 to 302	Poly(*o*-vinyl pyridine)	488
Poly(vinyl cyclopentane)	565	Poly(vinyl cyclohexane)	575 to 656

☞ ☞ ☞ **TABLE 6.8 IS CONTINUED IN THE NEXT PAGE.** ☞ ☞ ☞

Table 6.8. CONTINUED FROM THE PREVIOUS PAGE.

Polymer	T_m	Polymer	T_m
Poly(α-vinyl naphthalene)	633	Poly(vinyl alcohol)	505 to 538
Poly(vinyl fluoride)	503	Poly(vinylidene fluoride)	410 to 511
Polytetrafluoroethylene	292 to 672	Poly(vinyl chloride)	485 to 583
Poly(vinylidene chloride)	463 to 483	Polychlorotrifluoroethylene	483 to 533
Poly(ethylene adipate)	320 to 338	Poly(decamethylene sebacate)	344 to 358
Poly(ethylene terephthalate)	538 to 557	Poly(ethylene isophthalate)	416 to 513
Poly(tetramethylene terephthalate)	505	Poly(tetramethylene isophthalate)	426
Poly(hexamethylene terephthalate)	427 to 434	Poly(decamethylene terephthalate)	396 to 411
Poly(ethylene-1,5-naphthalenedicarboxylate)	503	Poly(ethylene-2,6-naphthalenedicarboxylate)	533 to 541
Poly(1,4-cyclohexylidene dimethylene terephthalate) *(cis)*	530	Poly(1,4-cyclohexylidene dimethylene terephthalate) *(trans)*	590
Poly(α,α-dimethylpropiolactone)	513	Poly(ethylene oxybenzoate)	475 to 500
Poly(4-aminobutyric acid)	523 to 538	Poly(ε-caprolactam)	533
Poly(hexamethylene adipamide)	523 to 545	Poly(decamethylene sebacamide)	467 to 489
Poly(11-aminoundecanoic acid)	455 to 493	Poly(nonamethylene azelamide)	438 to 462
Poly(ethylene terephthalamide)	728	Poly(hexamethylene terephthalamide)	623 to 644
Poly(octadecamethylene terephthalamide)	528	Poly(*p*-phenylene terephthalamide)	770 to 870
Poly(*m*-phenylene isophthalamide)	660 to 700	Poly(*m*-xylylene adipamide)	518
Poly(*p*-xylylene sebacamide)	541 to 573	Poly[thio *bis*(4-phenyl)carbonate]	513
Poly[methane *bis*(4-phenyl)carbonate]	513 to 573	Poly[1,1-ethane *bis*(4-phenyl)carbonate)]	468
Bisphenol-A polycarbonate	498 to 608	Poly[1,1-butane *bis*(4-phenyl)carbonate]	443
Poly[2,2-butane *bis*(4-phenyl)carbonate]	495	Poly[2,2-pentane *bis*(4-phenyl)carbonate]	493
Poly[2,2-propane *bis*{4-(2-methylphenyl)}carbonate]	443	Poly[2,2-propane *bis*{4-(2-chlorophenyl)}carbonate]	483
Poly[2,2-propane *bis*{4-(2,6-dichlorophenyl)}carbonate]	533	Poly[2,2-propane *bis*{4-(2,6-dibromophenyl)}carbonate]	533
Poly[diphenylmethane *bis*(4-phenyl)carbonate]	503	Poly[1,1-(1-phenylethane) *bis*(4-phenyl)carbonate]	503
Poly[1,1-cyclopentane *bis*(4-phenyl)carbonate]	523	Poly[1,1-cyclohexane *bis*(4-phenyl)carbonate]	533
Poly(tetramethylene carbonate)	332	Poly(decamethylene carbonate)	328 to 378
Poly(dimethyl siloxane)	234 to 244	Poly[4,4'-isopropylidene diphenoxy di(4-phenylene)sulfone]	570
Poly[N,N'-(*p*,*p*'-oxydiphenylene)pyromellitimide]	770		

References and Notes for Chapter 6

1. *Molecular Basis of Transitions and Relaxations*, edited by D. J. Meier, Gordon and Breach Science Publishers, London (1978).
2. R. Zallen, *The Physics of Amorphous Solids*, John Wiley & Sons, New York (1983).
3. J. Bicerano and D. Adler, *Pure & Applied Chemistry*, *59*, 101-144 (1987).
4. J. H. Gibbs and E. A. DiMarzio, *J. Chem. Phys.*, *28*, 373-383 (1958).
5. E. A. DiMarzio and J. H. Gibbs, *J. Chem. Phys.*, *28*, 807-813 (1958).
6. G. Adam and J. H. Gibbs, *J. Chem. Phys.*, *43*, 139-146 (1965).
7. R. P. Kusy and A. R. Greenberg, *Polymer*, *23*, 36-38 (1982).
8. A. R. Greenberg and R. P. Kusy, *Polymer*, *24*, 513-518 (1983).
9. P. R. Couchman, *Polymer Engineering and Science*, *24*, 135-143 (1984).
10. M. Goldstein, *Annals of the New York Academy of Sciences*, *279*, 68-77 (1976).
11. M. Goldstein, *J. Chem. Phys.*, *64*, 4767-4774 (1976).
12. G. P. Johari, in *Plastic Deformation of Amorphous and Semicrystalline Materials*, Les Houches Lectures, edited by B. Escaig and C. G.'Sell, *Les Editions de Physique* (1982), 109-141.
13. M. H. Cohen and G. S. Grest, *Phys. Rev. B*, *20*, 1077-1098 (1979).
14. G. S. Grest and M. H. Cohen, *Advances in Chemical Physics*, *48*, 455-525 (1981).
15. H. Stutz, K.-H. Illers and J. Mertes, *J. Polym. Sci., Polym. Phys. Ed.*, *28*, 1483-1498 (1990).
16. V. A. Bershtein and V. M. Yegorov, *Polymer Science USSR*, *27*, 2743-2757 (1985).
17. V. A. Bershtein, V. M. Yegorov and Yu. A. Yemel'yanov, *Polymer Science USSR*, *27*, 2757-2764 (1985).
18. P. J. Flory, *The Principles of Polymer Chemistry*, Cornell University Press, Ithaca, New York (1953).
19. P. J. Flory, *Statistical Mechanics of Chain Molecules*, Interscience Publishers, New York (1969).
20. D. W. van Krevelen, *Properties of Polymers*, second edition, Elsevier, Amsterdam (1976). The tables of group contributions for various additive functions used in calculating transition temperatures were expanded significantly in the third edition of this book, published in 1990.
21. C. J. Lee, *J. Macromol. Sci. - Rev. Macromol. Chem. Phys.*, *C29*, 431-560 (1989).
22. A. A. Askadskii, *Polymer Science USSR*, *9*, 471-487 (1967).
23. A. A. Askadskii and G. L. Slonimskii, *Polymer Science USSR*, *13*, 2158-2160 (1971).
24. D. R. Wiff, M. S. Altieri and I. J. Goldfarb, *J. Polym. Sci., Polym. Phys. Ed.*, *23*, 1165-1176 (1985).
25. A. J. Hopfinger, M. G. Koehler, R. A. Pearlstein and S. K. Tripathy, *J. Polym. Sci., Polym. Phys. Ed.*, *26*, 2007-2028 (1988).

26. T. G. Fox and S. Loshaek, *J. Polym. Sci.*, *15*, 371-390 (1955).

27. S. Loshaek, *J. Polym. Sci.*, *15*, 391-404 (1955).

28. U. T. Kreibich and H. Batzer, *Die Angewandte Makromolekulare Chemie*, *83*, 57-112 (1979).

29. C. G. Reid and A. R. Greenberg, *ACS Division of Polymeric Materials Science and Engineering Preprints*, *56*, 764-768 (1987).

30. G. C. Martin and M. Shen, *ACS Polymer Preprints*, *20*, 786-789 (1979).

31. R. A. Pearson and A. F. Yee, *J. Materials Science*, *24*, 2571-2580 (1989).

32. V. Bellenger, B. Mortaigne and J. Verdu, *J. Appl. Polym. Sci.*, *44*, 653-661 (1992).

33. J. K. Sears and J. R. Darby, *The Technology of Plasticizers*, Wiley, New York (1982).

34. K. Neki and P. H. Geil, *J. Macromol. Sci.-Phys.*, *B8*, 295-341 (1973).

35. T. Alfrey, Jr. and R. F. Boyer, Ref. [1], pp. 193-202.

36. Articles on special types of polymers in various volumes of the *Encyclopedia of Polymer Science and Engineering*, tabulations in the *Polymer Handbook*, and product brochures and catalogs on commercial polymers, were among the sources of experimental data used for the glass transition temperature.

37. D. R. Wiff, database of experimental T_g values used in the work of Wiff *et al* [24].

38. D. W. Brown and L. A. Wall, *J. Polym. Sci., A-2*, *7*, 601-608 (1969).

39. A. F. Yee and S. A. Smith, *Macromolecules*, *14*, 54-64 (1981).

40. R. R. Light and R. W. Seymour, *Polymer Engineering and Science*, *22*, 857-864 (1982).

41. N. Muruganandam, W. J. Koros and D. R. Paul, *J. Polym. Sci., Polym. Phys. Ed.*, *25*, 1999-2026 (1987).

42. B. Gebben, M. H. V. Mulder and C. A. Smolders, *J. Membrane Science*, *46*, 29-41 (1989).

43. Measured by T. Sangster and C. A. Wedelstaedt (1990).

44. J. E. Huheey, *Inorganic Chemistry*, Harper and Row, New York (1972).

45. R. Walsh, *Acc. Chem. Res.*, *14*, 246-252 (1981).

46. J. L. Margrave and P. W. Wilson, *Acc. Chem. Res.*, *4*, 145-151 (1971).

47. J. Bunnell and T. A. Ford, *J. Mol. Spectrosc.*, *100*, 215-233 (1983).

48. R. J. Abraham and G. H. Grant, *J. Comput. Chem.*, *9*, 709-718 (1988).

49. T. G. Fox and P. J. Flory, *J. Appl. Phys.*, *21*, 581-591 (1950).

50. D. T. Turner, *Polymer*, *19*, 789-796 (1978).

51. Y.-H. Lin, *Macromolecules*, *23*, 5292-5294 (1990).

52. G. Marchionni, G. Ajroldi, M. C. Righetti and G. Pezzin, *Polymer Communications*, *32*, 71-73 (1991).

53. X. Lu and B. Jiang, *Polymer*, *32*, 471-478 (1991).

54. R. F. Boyer, *Macromolecules*, *7*, 142-143 (1974).

55. J. M. G. Cowie, *European Polymer Journal*, *11*, 297-300 (1975).

56. S. L. Malhotra, P. Lessard and L. P. Blanchard, *J. Macromol. Sci.-Chem.*, *A15*, 121-141 (1981).

57. G. Marchionni, G. Ajroldi, P. Cinquina, E. Tampellini and G. Pezzin, *Polymer Engineering and Science*, *30*, 829-834 (1990).

58. G. Marchionni, G. Ajroldi and G. Pezzin, *European Polymer Journal*, *24*, 1211-1216 (1988).

59. B. Boutevin, Y. Pietrasanta, L. Sarraf and H. Snoussi, *European Polymer Journal*, *24*, 539-545 (1988).

60. S. Montserrat and P. Colomer, *Polymer Bullletin*, *12*, 173-180 (1984).

61. L. Gargallo, E. Soto, L. H. Tagle and D. Radic, *Thermochimica Acta*, *130*, 289-297 (1988).

62. Z. Dobrowski, *European Polymer Journal*, *18*, 563-567 (1982).

63. J. A. Bergfjord, R. C. Penwell and M. Stolka, *J. Polym. Sci., Polym. Phys. Ed.*, *17*, 711-713 (1979).

64. R. F. Fedors, *Polymer*, *20*, 518-519 (1979).

65. R. F. Fedors, *Polymer*, *20*, 1055-1056 (1979).

66. R. H. Colby, L. J. Fetters and W. W. Graessley, *Macromolecules*, *20*, 2226-2237 (1987).

67. K. Onder, R. H. Peters and L. C. Spark, *Polymer*, *13*, 133-139 (1972).

68. S. Lazcano, C. Marco, J. G. Fatou and A. Bello, *European Polymer Journal*, *24*, 991-997 (1988).

69. R. B. Beevers and E. F. T. White, *Transactions of the Faraday Society*, *56*, 1529-1534 (1960).

70. S. L. Malhotra, P. Lessard, L. Minh and L. P. Blanchard, *J. Macromol. Sci.-Chem.*, *A14*, 517-540 (1980).

71. J. M. G. Cowie and P. M. Toporowski, *European Polymer Journal*, *4*, 621-625 (1968).

72. F. Danusso, M. Levi, G. Gianotti and S. Turri, *Polymer*, *34*, 3687-3693 (1993).

73. F. Bueche, *Physical Properties of Polymers*, Interscience Publishers, New York (1962).

74. T. S. Chow, *Macromolecules*, *13*, 362-364 (1980).

75. (a) J. Bicerano, R. L. Sammler, C. J. Carriere and J. T. Seitz, to be published. The data measured at Dow on the effects of crosslinking on T_g were obtained by M. J. Marks, W. Lidy, R. A. Kirchhoff, G. H. Stopyak III, J. J. Curphy, C. J. Carriere, R. L. Sammler, A. Erskine, J. J. Curphy, C. F. Broomall, J. K. Sekinger and M. D. Joseph. (b) M. J. Marks and J. K. Sekinger, *Macromolecules*, 1994, *27*, 4106-4113.

76. T. I. Ponomareva, V. I. Irzhak and B. A. Rozenberg, *Polymer Science USSR*, *20*, 673-679 (1979).

77. Y.-G. Won, J. Galy, J.-P. Pascault and J. Verdu, *J. Polym. Sci., Polym. Phys. Ed.*, *29*, 981-987 (1991).

78. A. Kanno and K. Kurashiki, *J. Soc. Mater. Sci. Japan*, *33*, 102-108 (1984).

79. J. S. Tira, *SAMPE Journal*, 18-22 (July/August 1987).

80. P. Förster, K. R. Hauschildt and D. Wilhelm, *Makromol. Chem., Macromol. Symp.*, *41*, 141-151 (1991).

81. B. Fuller, J. T. Gotro and G. C. Martin, *Advances in Chemistry*, *227* (1990), Chapter 12.

82. T. S. Chow, *Polym. Mater. Sci. Eng.*, *56*, 248-252 (1987).

83. T. S. Chow, *Polymer*, *29*, 1447-1451 (1988).

84. T. S. Chow, *ACS Symposium Series*, *367*, *Crosslinked Polymers* (1988), Chapter 10.

85. H.-G. Elias, *Macromolecules, Volume 1: Structure and Properties*, second edition, Plenum Press, New York (1984).

86. F. E. Karasz and W. J. MacKnight, *Macromolecules*, *1*, 537-540 (1968).

87. J. M. G. Cowie, *European Polymer Journal*, *9*, 1041-1049 (1973).

88. G. R. Dever, F. E. Karasz, W. J. MacKnight and R. W. Lenz, *J. Polym. Sci., Polym. Chem. Ed.*, *13*, 2151-2179 (1975).

89. G. A. Russell, P. A. Hiltner, D. E. Gregonis, A. C. deVisser and J. D. Andrade, *J. Polym. Sci., Polym. Phys. Ed.*, *18*, 1271-1283 (1980).

90. D. R. Terrell, F. Evers, H. Smoorenburg and H. M. van den Bogaert, *J. Polym. Sci., Polym. Phys. Ed.*, *20*, 1933-1945 (1982).

91. D. R. Burfield and Y. Doi, *Macromolecules*, *16*, 702-704 (1983).

92. C. Mijangos, G. Martinez and J.-L. Millan, *Makromol. Chem.*, *189*, 567-572 (1988).

93. M. Naumann and R. Duran, *ACS Polymer Preprints*, *32*, 96-97 (1991).

94. Y. Chatani, Y. Shimane, Y. Inoue, T. Inagaki, T. Ishioka, T. Ijitsu and T. Yukinari, *Polymer*, *33*, 488-492 (1992).

95. Y. Chatani, Y. Shimane, T. Inagaki, T. Ijitsu, T. Yukinari and H. Shikuma, *Polymer*, *34*, 1620-1624 (1993).

96. Y. Chatani, Y. Shimane, T. Ijitsu and T. Yukinari, *Polymer*, *34*, 1625-1629 (1993).

97. R. Napolitano and B. Pirozzi, *Macromolecules*, *26*, 7225-7228 (1993).

98. N. G. McCrum, B. E. Read and G. Williams, *Anelastic and Dielectric Effects in Polymeric Solids*, Wiley, New York (1967).

99. J. Bicerano, *J. Polym. Sci., Polym. Phys. Ed.*, *29*, 1329-1343 (1991).

100. J. Bicerano, *J. Polym. Sci., Polym. Phys. Ed.*, *29*, 1345-1359 (1991).

101. S. Wu, *J. Appl. Polym. Sci.*, *46*, 619-624 (1992).

102. *Thermoplastic Elastomers: A Comprehensive Review*, edited by N. R. Legge, G. Holden and H. E. Schroeder, Hanser Publishers, Munich (1987).

CHAPTER 7

SURFACE TENSION AND INTERFACIAL TENSION

7.A. Background Information

The *surface tension* γ is the reversible work (i.e., the reversible increase in the Gibbs free energy G) as a result of the creation of a unit surface area, at constant temperature, pressure and composition [1-4]. The increase in G upon the creation of new surface area is caused by the imbalance of the molecular forces acting on the molecules at the surface, compared with the forces acting on molecules in the interior (bulk) of a liquid or a solid. The surface tension is the key physical property of the surface of a material. It is expressed in units of energy/area, such as J/m^2 which is identical to 1000 dyn/cm. Units of dyn/cm are quoted more commonly than J/m^2.

If the new surface is in contact with another material instead of being exposed to air, the area of contact is referred to as an *interface*, and the reversible increase in G is called the *interfacial tension*. The interfacial tension is the key physical property of an interface. The interfacial tension between two materials can be expressed in terms of the surface tensions of the individual materials. Consequently, our work has focused mainly on the development of new correlations to calculate the surface tension.

The surface tension and the interfacial tension play a crucial role in the design, processing, performance and durability of polymeric systems intended for a variety of applications. The following are some examples:

1. All *multicomponent systems* contain one or more interfaces. The proper performance of such a system requires the components of the system to bond strongly to each other, and failure at the interface to be avoided during use [1,5]. Such multicomponent systems span a vast range polymeric materials, such as polymer blends, thermoplastics and thermosets toughened by the addition of rubber particles, rubbers reinforced by the addition of rigid fillers, other types of composites which contain at least one polymeric component, thermoplastic elastomers, laminates, coatings, and adhesive joints. (Adhesion is, however, a very complex behavior, in which "engineering" factors such as the sizes, shapes and layouts of the various components of a multicomponent system are also very important. Consideration of all aspects of the problem [1] is therefore often required for an adequate understanding of adhesive behavior.)

2. The ability of a liquid to wet a solid depends on the surface tensions of the solid and liquid.

3. The failure of plastic specimens by crazing, both in air [6] and in the presence of any type of environmental agent [7,8], requires the creation of new surface area, and is therefore resisted by the surface tension of the polymer.

4. During melt processing, it is necessary to prevent the adhesion of the material being processed to the mold or to the barrel of the extruder or injection molding machine.

There are many independent techniques for measuring the surface tensions of liquids, including both ordinary molecular liquids and polymer melts. Melts of oligomers of relatively low average molecular weight are commonly used instead of polymers of high molecular weight in measuring the surface tension of a polymer melt.

On the other hand, the surface tension of a solid polymer cannot be measured directly. It has to be extrapolated from directly observed quantities, often with a fairly large uncertainty, by using any one or a combination of the following four methods [1,3]:

1. Extrapolation of the surface tension data for polymer melts to room temperature. This is often considered to be the most reliable method for estimating the γ of a solid polymer. See Table 7.1 for a tabulation of the surface tensions of polymers [2], measured by using this technique.

2. Extrapolation of the surface tension data for liquid homologs of the polymer as a function of molecular weight. It is assumed that the surface tension increases asymptotically to a limiting value as the molecular weight increases. An empirical relationship is used in the extrapolation.

3. Measurement of the contact angle between the solid and different liquids, and application of equations which express the surface tension of the solid as a function of the contact angle and the surface tension of the liquid.

4. Measurement of the "critical surface tension of wetting" γ_{cr}. A liquid with a surface tension of less than γ_{cr} will spread on the surface of the solid. It is assumed that $\gamma \approx \gamma_{cr}$ to estimate the surface tension of the solid.

There are many theories [1] and empirical correlations [1,3] to calculate γ. The intermolecular interactions which play a role in determining γ and the cohesive energy density e_{coh} are quite similar. The surface tension at room temperature can therefore be estimated by Equation 7.1 [3]:

$$\gamma(298K) \approx 0.75 \cdot e_{coh}{}^{(2/3)} \tag{7.1}$$

The cohesive energy density e_{coh} is expressed in units of J/cc, and γ is expressed in units of dyn/cm, in Equation 7.1. The use of different units would not change the form of the functional dependence of γ on e_{coh}, but would alter the proportionality constant. The power of 2/3 is appropriate because e_{coh} is an average per unit volume (i.e., in three dimensions), while γ is an average per unit area (i.e., in two dimensions).

As an example of the use of Equation 7.1, consider poly(methyl methacrylate). The correlation in terms of connectivity indices for the cohesive energy according to the group contributions of Fedors (Section 5.B) gives $E_{coh1} \approx 35097$ J/mole. The correlation for the molar volume at room temperature (Section 3.C) gives $V \approx 86.4$ cc/mole. The cohesive energy density is therefore estimated to be $e_{coh1} \approx (35097/86.4) \approx 406.2$ J/cc. The surface tension predicted by using Equation 7.1 is $\gamma(298K) \approx 41.1$ dyn/cm, in agreement with the experimental value of $\gamma(298K) \approx 41.1$ dyn/cm [2] extrapolated to room temperature from the melt.

Table 7.1. Surface tensions (γ) of polymers at three different temperatures, in dyn/cm, from measurements on polymer melts and liquids and extrapolations from them unless indicated otherwise [1,2]. The available information on the polymer molecular weight is also listed, as M_n (number-average), M_w (weight-average), M_v (viscosity-average), or M (an average molecular weight not specified in greater detail), all in g/mole; or viscosity (η), in centistokes (cS). If the temperature differs from the one listed in the column heading, it is indicated in parentheses. The temperature derivatives of γ provided in Ref. [2] were used to convert the γ values listed in that reference at T=150°C to T=140°C, and the γ values listed at T=200°C to T=180°C.

Polymer	Molecular Weight	$\gamma(20°C)$	$\gamma(140°C)$	$\gamma(180°C)$
Linear polyethylene	M_w=67000	35.7	28.8	26.5
Ideal *n*-alkane limit of polyethylene	M≡∞	36.8	30.0	27.7
Branched polyethylene	M_n=7000	35.3	27.3	24.6
Branched polyethylene	M_n=2000	33.7	26.5	24.1
Atactic polypropylene	M_n=3000	28.3	23.5	21.9
Polyisobutylene	M_n=2700	33.6	25.9	23.4
Polystyrene	M_v=44000	40.7	32.1	29.2
Polystyrene	M_n=9290	39.4	31.6	29.0
Polystyrene	M_n=1680	39.3	30.0	26.9
Poly(α-methyl styrene)	M_n=3000	38.7	31.7	29.4
Poly(*m*-methyl styrene)	M_n=3000	38.7	31.7	29.4
Polychloroprene	M_v=30000	43.6	33.2	29.8
Polychlorotrifluoroethylene	M_n=1280	30.9	22.9	20.2
Polytetrafluoroethylene	M≡∞	23.9	16.9	14.6
Polytetrafluoroethylene	M=1088	21.5	13.7	11.1
Poly(vinyl acetate)	M_w=11000	36.5	28.6	25.9
Poly(methyl methacrylate)	M_v=3000	41.1	32.0	28.9
Poly(ethyl methacrylate)	M_v=5200	35.9	27.5	24.7
Poly(*n*-propyl methacrylate)	M_v=8500	33.2	25.4	22.8
Poly(*n*-butyl methacrylate)	M_v=37000	31.2	24.1	21.7
Poly(isobutyl methacrylate)	M_v=35000	30.9	23.7	21.3
Poly(*t*-butyl methacrylate)	M_v=6000	30.4	23.3	21.0
Poly(*n*-hexyl methacrylate)	M_v=52000	30.0	22.6	20.1
Poly(2-ethylhexyl methacrylate)	M_v=64000	28.8	21.4	18.9
Poly(methyl acrylate)	M_n=25000	41.0	31.8	28.7
Poly(ethyl acrylate)	M_n=28000	37.0	27.8	24.7
Poly(*n*-butyl acrylate)	M_n=32000	33.7	25.3	22.5
Poly(2-ethylhexyl acrylate)	M_n=34000	30.2	21.8	19.0

☞ ☞ ☞   **TABLE 7.1 IS CONTINUED IN THE NEXT PAGE.** ☞ ☞ ☞

Table 7.1. CONTINUED FROM THE PREVIOUS PAGE.

Polymer	Molecular Weight	γ(20°C)	γ(140°C)	γ(180°C)
Poly(ethylene oxide)	M=6000	42.9	33.8	30.7
Poly(propylene oxide)	400≤M≤4100	30.4	20.8	17.6
Polyoxytetramethylene	M_n=43000	31.9	24.6	22.2
Polyoxydecamethylene	Unspecified	36.1	27.9	25.2
Polyepichlorohydrin	M=1500	43.2 (25°C)	----	----
Phenoxy resin	M=30000	43.0[a]	----	----
Bisphenol-A polycarbonate	Unspecified	42.9	35.7	33.3
Poly(ethylene terephthalate)	M_n=25000	44.6	36.7	34.2
Poly(propylene isophthalate)	Unspecified	49.3	39.3	36.0
Poly(tetramethylene isophthalate)	Unspecified	47.8	38.2	35.0
Poly(hexamethylene isophthalate)	Unspecified	45.6	36.4	33.2
Poly(decamethylene isophthalate)	Unspecified	42.7	34.0	31.1
Poly(dodecamethylene isophthalate)	Unspecified	40.0	31.6	28.8
Poly(vinyl propionate)	Unspecified	34.0	25.4	22.5
Poly(vinyl butyrate)	Unspecified	31.1	22.7	19.9
Poly(vinyl hexanoate)	Unspecified	29.4	21.8	19.4
Poly(vinyl octanoate)	Unspecified	28.7	21.3	18.9
Poly(vinyl decanoate)	Unspecified	28.9	21.5	19.0
Poly(vinyl dodecanoate)	Unspecified	29.1	21.5	19.1
Poly(vinyl hexadecanoate)	Unspecified	30.9	23.2	20.3
Poly(hexamethylene adipamide)	M_n=19000	46.5	38.8	36.1
Poly(ε-caprolactam)	Unspecified	----	----	36.1 (265°C)
Poly(hexamethylene sebacamide)	Unspecified	----	----	37.0 (265°C)
Poly(11-aminoundecanoic acid)	Unspecified	----	----	22.6 (225°C)
Poly(dimethyl siloxane)	M_n=75000	20.9	14.3	12.3
Poly(dimethyl siloxane)	η=60000	19.8	14.0	12.1
Poly(dimethyl siloxane)	M_n=1274	19.9	----	----
Poly(dimethyl siloxane)	M_n=607	18.8	----	----
Poly(dimethyl siloxane)	M_n=310	17.6	----	----
Poly(dimethyl siloxane)	M_n=162	15.7	----	----
Poly[oxy(diethylsilylene)]	η=158	25.7	16.9	14.0
Poly[oxy(methylphenylsilylene)]	η=102	26.1	12.9	8.5
Poly[4,4'-isopropylidene diphenoxy di(4-phenylene)sulfone]	Unspecified	46.6[a]	----	----

[a] From contact angle measurements.

Equation 7.1 usually provides reasonable estimates of γ at room temperature, although normally not the perfect agreement fortuitously found with the experimental value for poly(methyl methacrylate) in our example. On the other hand, it cannot be used to estimate the temperature dependence of $\gamma(T)$. Furthermore, the effect of hydrogen bonding on γ is generally smaller than its effect on e_{coh}. Values of γ calculated by using Equation 7.1 are therefore often significantly larger than the experimental values of γ for polymers containing structural units such as amide, urea, and/or hydroxyl groups, which form strong hydrogen bonds [3].

An additive and temperature-independent quantity, named the *molar parachor* and represented by P_S, is often used [1,3,4] along with the amorphous molar volume $V(T)$, to estimate $\gamma(T)$:

$$\gamma(T) \approx \left[\frac{P_S}{V(T)}\right]^4 \tag{7.2}$$

The units to be used in Equation 7.2 are cc/mole for $V(T)$, (cc/mole)·(dyn/cm)$^{1/4}$ for P_S, and dyn/cm for $\gamma(T)$. The fourth power dependence of $\gamma(T)$ on the ratio $P_S/V(T)$ causes any error made in estimating this quotient to introduce an error at least four times as large in the predicted value of $\gamma(T)$. For example, an error of 3% in $P_S/V(T)$ causes an error of approximately 12.6% in the prediction of $\gamma(T)$. Equation 7.2 must therefore be used with caution.

When T increases, $V(T)$ increases because of thermal expansion, and $\gamma(T)$ decreases as a result of its dependence on the fourth power of $1/V(T)$. Equation 7.3 shows that the rate of decrease of γ with increasing temperature mainly comes from a factor proportional to the coefficient of volumetric thermal expansion α:

$$\frac{d\gamma(T)}{dT} \approx -4 \cdot \alpha(T) \cdot \gamma(T) \tag{7.3}$$

As discussed in Chapter 3, α changes slowly with changing temperature, and only manifests a discontinuous increase when a material undergoes the glass transition. $\gamma(T)$ can therefore be approximated over a wide range of temperatures as being a linear function of temperature, i.e., as having a small and almost constant negative slope.

Several atomic and structural contribution tables, derived from the observed surface tensions of organic liquids, are available for the calculation of P_S [9-11]. Van Krevelen [3,4] found that when these tables are used to estimate the surface tensions of solid polymers instead of liquids and melts, the original table of group contributions developed by Sugden [9] provides better agreement than the two revised and improved tables [10,11] with the "experimental" values of γ extrapolated for solid polymers from various types of measurements. He also provided an extensive set of group contributions [3,4], primarily based upon Sugden's values, for calculating the P_S of polymers.

The general form of the temperature dependence of the surface tension (its dependence on the inverse of the fourth power of the molar volume as given by Equation 7.2) can be used to obtain $\gamma(T)$ starting from a value of γ calculated at any given reference temperature by any method. For example, the combination of Equation 7.1 with the inverse fourth power dependence of $\gamma(T)$ on $V(T)$ gives an alternative estimate for $\gamma(T)$:

$$\gamma(T) \approx 0.75 \cdot e_{coh}^{(2/3)} \cdot \left[\frac{V(298K)}{V(T)}\right]^4 \tag{7.4}$$

7.B. Improvements in the Ability to Predict the Surface Tensions of Polymers

The key contributions of this book to the calculation of the surface tensions of polymers can be summarized as follows:

1. The ability to estimate both the molar volume (Section 3.C) and the cohesive energy (Chapter 5) of a polymer of arbitrary structure at room temperature without having to rely on the availability of group contributions implies that Equation 7.1 can now be used to estimate the surface tension of any polymer at room temperature. Such estimates will, of course, still be subject to the intrinsic limitations of Equation 7.1, such as its tendency to overestimate the contribution of hydrogen bonds to γ.

2. Equation 7.2 can now be used to predict the $\gamma(T)$ of a polymer of arbitrary structure:

    (a) The ability to predict the molar volume as a function of temperature (Section 3.D) by combining the predicted values of $V(298K)$ (Section 3.C) and the glass transition temperature (Chapter 6) facilitates the use of Equation 7.2.

    (b) A new correlation was developed for P_S, to enable the estimation of P_S even when a polymer contains structural units for which the group contributions to P_S are not available.

A rough but instructive approximation for the temperature dependence of $\gamma(T)$ below T_g will be presented in Section 7.C. The new correlation for P_S will be developed, and its predictive use demonstrated, in Section 7.D. The interfacial tension will be discussed briefly in Section 7.E.

7.C. Approximate "Master Curve" as a Function of Reduced Temperature

$\gamma(T)$ can be very roughly estimated from an empirically derived "master curve" for $T \leq T_g$. Let x (defined by Equation 7.5) denote a "reduced temperature" parameter. The relationships obtained for $\gamma(x)$ by combining equations 7.2, 7.5 and 3.5 are summarized by equations 7.6-7.8.

$$x \equiv \frac{T}{T_g} \tag{7.5}$$

$$\gamma(0) \approx 0.24595 \left(\frac{P_S}{V_w}\right)^4 \tag{7.6}$$

$$\frac{\gamma(x)}{\gamma(0)} \approx \frac{1}{(1 + 0.10563x)^4} \tag{7.7}$$

$$\gamma(x) \approx 0.24595 \left(\frac{P_S}{V_w}\right)^4 \frac{1}{(1 + 0.10563x)^4} \tag{7.8}$$

Equation 7.6 provides an estimate for γ in the limit of absolute zero temperature in terms of the ratio P_S/V_w. Equation 7.7 states that, *as a first approximation*, a "principle of corresponding states" holds for the quotient $\gamma(x)/\gamma(0)$, so that this ratio is constant for all polymers at a given $x \leq 1$. $\gamma(x)/\gamma(0)$ in the temperature range of $0 \leq T \leq T_g$ can thus be approximated by a master curve, which is shown in Figure 7.1. Equation 7.8 combines equations 7.6 and 7.7, and suggests that much of the difference between the $\gamma(T)$ values of different polymers at a given temperature below T_g can be accounted for in terms of a superposition of two factors:
1. Differences between T_g cause polymers to have different values of x at the same temperature.
2. Differences between P_S/V_w cause polymers to have different values of $\gamma(x)$ at the same x.

In general, Equation 7.8 is not nearly as accurate as Equation 7.2 with the best possible predicted value of V(T) inserted in it. Equation 7.8 should, thus, only be considered as a rough but rather instructive generalization for the temperature dependence of γ below T_g. It is not used in the computer program automating the application of our quantitative structure-property relationships. Equation 7.2 is used to predict $\gamma(T)$ both above and below T_g in this software package, with the equations from Section 3.D being used to estimate V(T).

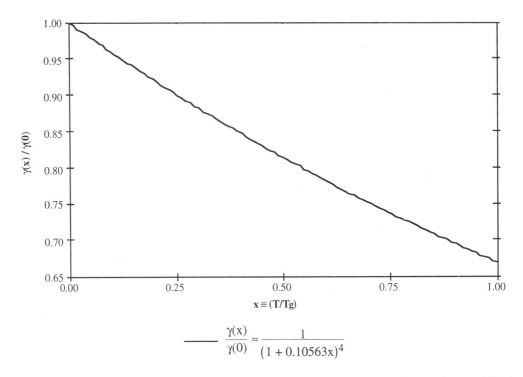

$$\frac{\gamma(x)}{\gamma(0)} \approx \frac{1}{\left(1 + 0.10563x\right)^4}$$

Figure 7.1. Approximate "master curve" for $\gamma(x)/\gamma(0)$, where γ is the surface tension, $x \equiv (T/T_g)$ is the "reduced temperature", and $\gamma(0)$ is the value of γ extrapolated to absolute zero temperature.

7.D. Correlation for the Molar Parachor

A new correlation will now be developed for P_S. In addition to providing increased predictive power, this quantitative structure-property relationship will also show how the systematic fitting of physical properties with a few connectivity indices can reveal interconnections between physical properties which are not easily recognized from long tables of group contributions.

Connectivity indices were used at first, to fit the P_S values of 115 of the polymers listed in Table 2.2, for which all of the necessary atomic or group contribution values were available. Van Krevelen's group contributions were used as the main source, and supplemented with values calculated by using the atomic and structural contribution table of Sugden [3a,4,9].

P_S correlated very strongly with all four of the zeroth-order and first-order connectivity indices. The best two-parameter fit for P_S depended on the same two connectivity indices, namely $^0\chi$ and $^1\chi^v$, as in Equation 3.9 which gives the best two-parameter fit to V_w without using correction terms. Furthermore, the ratios of the regression coefficient multiplying $^1\chi^v$ to the regression coefficient multiplying $^0\chi$ for these two properties were very similar. These observations led immediately to the conclusion that P_S must be closely correlated with V_w.

This conclusion was verified by correlating the values of V_w from Section 3.B with P_S. The V_w values obtained by using group contributions (whenever available and apparently reliable) were combined with V_w values calculated via equations 3.10 and 3.11 in terms of connectivity indices, to obtain V_w values for all polymers in the dataset. It was found that the correlation between P_S and V_w was, in fact, much stronger than the correlation of either property with a linear combination of $^0\chi$ and $^1\chi^v$.

Weighted fits with weight factors of $100/P_S(group)$ were used to develop a regression relationship. Equation 7.9 has a standard deviation of 6.0 and a correlation coefficient of 0.9995:

$$P_S \approx 3.945868 \cdot V_w \tag{7.9}$$

The small deviation of P_S as estimated by Equation 7.9 from $P_S(group)$ is very systematic. It depends mainly on the atoms present and the number of bonds emanating from each atom, with relatively little effect of the details of the bonding environment. This result is not really surprising, since the group contribution table used for P_S is essentially based on an atomic contribution table supplemented by correction terms to adjust for the presence of a few special types of structural features such as multiple bonds and rings. The following correction index was defined to correct for almost half of the standard deviation in the correlation between P_S and V_w:

$$N_{PS} \equiv -3N_{(\text{carbon atoms with } \delta=\delta^v)} + 6N_{(\text{carbonyl groups})} + 16N_{(-S-)} + 6N_{Br} - 6N_F \tag{7.10}$$

$N_{(\text{carbon atoms with } \delta=\delta^v)}$ equals the number of carbon atoms which are singly bonded to all of their neighbors. It can be seen from Table 2.1 that sp^3 hybridization, which is necessary if a carbon atom is to be able to saturate its maximum valence of four without any multiple bonds, results in the condition $\delta=\delta^v$ in graph theoretical terms.

$N_{(\text{carbonyl groups})}$ is the total number of carbonyl (-C=O) moieties in *any* type of bonding environment, such as keto, aldehyde, amide, imide, ester, carbonate, urea and carboxylic acid.

$N_{(-S-)}$, N_{Br} and N_F are the numbers of sulfur atoms in the lowest (divalent) oxidation state, bromine atoms, and fluorine atoms, respectively, in the repeat unit.

The final correlation is given by Equation 7.11. It has a standard deviation of only 3.2 and a correlation coefficient of 0.9998. The standard deviation is less than 1% of the average $P_S(group)$ value of 333.0 for the 115 polymers in the dataset used for developing the correlation. More than two thirds of $\gamma(T)$ values predicted by substituting $P_S(fit)$ into Equation 7.2 will thus differ by less than 4% from $\gamma(T)$ values predicted by using $P_S(group)$. A deviation of less than 4% in the computation of a property whose experimental determination is subject to considerable uncertainty implies that, for most practical purposes, Equation 7.11 reproduces the predictions of $\gamma(T)$ made by using group contributions for P_S.

$$P_S \approx 3.989792 \cdot V_w + 0.502094 \cdot N_{PS} \tag{7.11}$$

The results of these calculations are summarized in Table 7.2 and shown in Figure 7.2.

Table 7.2. The molar parachor P_S calculated by the group contributions of van Krevelen as supplemented with the atomic and structural contributions of Sugden, the quantities (V_w and N_{PS}) used in the correlation for P_S, and the fitted values of P_S, for 115 polymers.

Polymer	P_S(group)	V_w	N_{PS}	P_S(fit)
Polyoxymethylene	59.0	15.25	-3	59.3
Polyethylene	78.0	20.50	-6	78.8
Poly(vinyl fluoride)	86.6	23.25	-12	86.7
Poly(vinylidene fluoride)	95.2	26.25	-18	95.7
Poly(vinyl alcohol)	98.0	25.25	-6	97.7
Polyoxyethylene	98.0	25.50	-6	98.7
Polytrifluoroethylene	103.8	29.00	-24	103.7
Poly(thiocarbonyl fluoride)	104.4	26.00	1	104.2
Polytetrafluoroethylene	112.4	32.00	-30	112.6
Poly(vinyl chloride)	115.2	29.25	-6	113.7
Polypropylene	117.0	30.75	-9	118.2
Polyacrylonitrile	124.6	32.25	-6	125.7
Poly(ethylene sulfide)	126.2	30.50	10	126.7
Poly(vinyl bromide)	128.9	32.03	0	127.8
Polyoxytrimethylene	137.0	35.75	-9	138.1
Poly(vinyl methyl ether)	137.0	35.75	-9	138.1
Poly(propylene oxide)	137.0	35.75	-9	138.1
Polychlorotrifluoroethylene	141.0	38.00	-24	139.6
Poly(acrylic acid)	142.8	36.25	0	144.6
Poly(1,4-butadiene)	145.0	38.00	-6	148.6
Poly(1,2-butadiene)	145.0	38.00	-6	148.6
Poly(vinylidene chloride)	152.4	38.25	-6	149.6
Poly(β-alanine)	156.0	38.50	0	153.6

☞ ☞ ☞   **TABLE 7.2 IS CONTINUED IN THE NEXT PAGE.** ☞ ☞ ☞

Table 7.2. CONTINUED FROM THE PREVIOUS PAGE.

Polymer	P_S(group)	V_w	N_{PS}	P_S(fit)
Poly(1-butene)	156.0	41.00	-12	157.6
Polyisobutylene	156.0	41.25	-12	158.6
Polymethacrylonitrile	163.6	42.75	-9	166.0
Poly(maleic anhydride)	165.3	41.31	6	167.8
Polyfumaronitrile	171.2	44.00	-6	172.5
Poly(*p*-phenylene)	172.9	43.50	0	173.6
Polyepichlorohydrin	174.2	44.50	-9	173.0
Polyoxytetramethylene	176.0	46.00	-12	177.5
Poly(vinyl ethyl ether)	176.0	46.00	-12	177.5
Poly(vinylidene bromide)	179.8	43.81	6	177.8
Poly(methacrylic acid)	181.8	46.50	-3	184.0
Poly(methyl acrylate)	182.0	46.00	-3	182.0
Poly(vinyl acetate)	182.0	46.00	-3	182.0
Polychloroprene	182.2	46.75	-6	183.5
Polyisoprene	184.0	48.00	-9	187.0
Poly(methyl α-chloroacrylate)	219.0	54.76	-3	217.0
Poly(methyl methacrylate)	220.8	56.25	-6	221.4
Poly(ethyl acrylate)	221.0	56.25	-6	221.4
Poly(vinyl propionate)	221.0	56.25	-6	221.4
Poly(2,5-benzoxazole)	224.8	55.26	0	220.5
Poly(methyl α-cyanoacrylate)	228.4	57.75	-3	228.9
Poly(4-methyl-1-pentene)	234.0	61.75	-18	237.3
Poly(*p*-vinyl pyridine)	241.5	61.46	-6	242.2
Poly(N-vinyl pyrrolidone)	246.9	61.61	-9	241.3
Polystyrene	250.9	63.25	-6	249.3
Poly(*p*-xylylene)	250.9	64.00	-6	252.3
Poly(vinyl *n*-butyl ether)	254.0	66.50	-18	256.3
Poly(vinyl *sec*-butyl ether)	254.0	66.50	-18	256.3
Poly(ε-caprolactone)	259.8	66.25	-9	259.8
Poly(ethyl methacrylate)	259.8	66.50	-9	260.8
Poly(methyl ethacrylate)	259.8	66.50	-9	260.8
Poly[oxy(2,6-dimethyl-1,4-phenylene)]	271.0	69.00	-6	272.3
Poly(ε-caprolactam)	273.0	69.25	-9	271.8

☞ ☞ ☞ **TABLE 7.2 IS CONTINUED IN THE NEXT PAGE.** ☞ ☞ ☞

Table 7.2. CONTINUED FROM THE PREVIOUS PAGE.

Polymer	P_S(group)	V_w	N_{PS}	P_S(fit)
Poly(2-hydroxyethyl methacrylate)	279.8	72.30	-9	283.9
Poly(vinyl cyclohexane)	283.9	74.25	-24	284.2
Poly(o-chloro styrene)	288.1	72.73	-6	287.2
Poly(p-chloro styrene)	288.1	72.73	-6	287.2
Poly(oxy-2,2-dichloromethyltrimethylene)	289.4	74.00	-15	287.7
Poly(oxy-1,1-dichloromethyltrimethylene)	289.4	74.00	-15	287.7
Poly(o-methyl styrene)	289.9	73.75	-9	289.7
Poly(p-methyl styrene)	289.9	73.75	-9	289.7
Poly(α-methyl styrene)	289.9	73.75	-9	289.7
Poly(vinyl pivalate)	298.8	76.75	-12	300.2
Poly(n-butyl acrylate)	299.0	76.75	-12	300.2
Poly(styrene sulfide)	299.1	73.25	10	297.3
Poly(p-bromo styrene)	301.8	75.85	0	302.6
Cellulose	305.9	77.23	-18	299.1
Poly(p-methoxy styrene)	309.9	78.75	-9	309.7
Poly(vinyl benzoate)	315.9	78.25	0	312.2
Poly(vinyl butyral)	323.9	83.68	-24	321.8
Poly(n-butyl methacrylate)	337.8	87.00	-15	339.6
Poly(sec-butyl methacrylate)	337.8	87.00	-15	339.6
Poly(t-butyl methacrylate)	337.8	87.00	-15	339.6
Poly(N-phenyl maleimide)	347.8	87.67	6	352.8
Poly(8-aminocaprylic acid)	351.0	89.75	-15	350.6
Poly(α-vinyl naphthalene)	360.0	87.25	-6	345.1
Poly(N-methyl glutarimide)	370.5	91.83	-9	361.9
Poly(ethylene terephthalate)	380.5	94.00	6	378.1
Poly(cyclohexyl methacrylate)	387.7	99.75	-21	387.4
Poly(benzyl methacrylate)	393.7	99.00	-6	392.0
Poly(p-t-butyl styrene)	406.9	104.75	-18	408.9
Poly(n-hexyl methacrylate)	415.8	107.50	-21	418.4
Poly(2-ethylbutyl methacrylate)	415.8	107.50	-21	418.4
Poly(N-vinyl carbazole)	427.7	105.22	-6	416.8
Polybenzobisoxazole	449.6	112.32	0	448.1
Poly(tetramethylene terephthalate)	458.5	114.50	0	456.8

☞ ☞ ☞   **TABLE 7.2 IS CONTINUED IN THE NEXT PAGE.** ☞ ☞ ☞

Table 7.2. CONTINUED FROM THE PREVIOUS PAGE.

Polymer	P_S(group)	V_w	N_{PS}	P_S(fit)
Poly[3,5-(4-phenyl-1,2,4-triazole)-1,4-phenylene]	464.9	120.84	0	482.1
Poly[thio *bis*(4-phenyl)carbonate]	479.0	116.00	22	473.9
Poly(ethylene-2,6-naphthalenedicarboxylate)	489.6	122.13	6	490.3
Poly(*n*-octyl methacrylate)	493.8	128.00	-27	497.1
Poly(*p*-phenylene terephthalamide)	501.8	123.00	12	496.8
Polybenzobisthiazole	506.0	123.32	32	508.1
Poly(hexamethylene adipamide)	546.0	138.50	-18	543.5
Bisphenol-A polycarbonate	548.0	137.00	-3	545.1
Poly(hexamethylene isophthalamide)	562.9	141.00	-6	559.5
Poly(1,4-cyclohexylidene dimethylene terephthalate)	586.4	147.34	-12	581.8
Poly[2,2-hexafluoropropane *bis*(4-phenyl)carbonate]	599.6	154.25	-39	595.8
Poly(oxy-1,4-phenylene-oxy-1,4-phenylene-carbonyl-1,4-phenylene)	606.9	151.50	6	607.5
Phenoxy resin	640.0	163.50	-18	643.3
Poly[1,1-(1-phenylethane) *bis*(4-phenyl)carbonate]	681.9	170.25	0	679.3
Poly[2,2-propane *bis*{4-(2,6-dichlorophenyl)}carbonate]	697.0	174.92	-3	696.4
Poly(hexamethylene sebacamide)	702.0	179.50	-30	701.1
Poly[2,2-propane *bis*{4-(2,6-dimethylphenyl)}carbonate]	704.2	178.00	-15	702.7
Poly[1,1-(1-phenyltrifluoroethane *bis*(4-phenyl)carbonate]	707.7	178.88	-18	704.7
Torlon	735.8	183.33	15	739.0
Poly[N,N'-(*p,p'*-oxydiphenylene)pyromellitimide]	738.5	186.14	24	754.7
Poly[2,2-propane *bis*{4-(2,6-dibromophenyl)}carbonate]	751.8	187.40	21	758.2
Resin F	795.8	199.25	-6	792.0
Poly[diphenylmethane *bis*(4-phenyl)carbonate]	815.8	203.50	3	813.4
Poly[1,1-biphenylethane *bis*(4-phenyl)carbonate]	854.8	213.75	0	852.8
Resin G	904.6	231.73	-33	908.0
Ultem	1221.5	307.10	15	1232.8

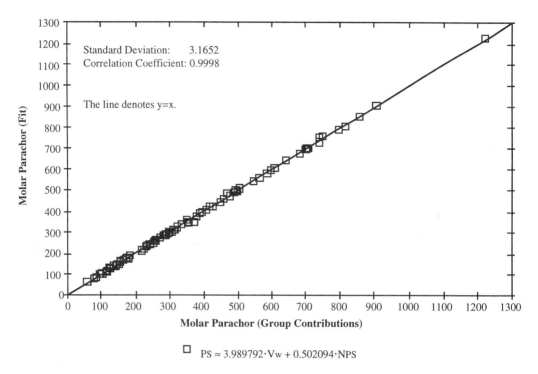

$$PS \approx 3.989792 \cdot Vw + 0.502094 \cdot NPS$$

Figure 7.2. The results of a fit using the van der Waals volume V_w and a correction index N_{PS}, to the molar parachor P_S calculated by the group contributions of van Krevelen supplemented by the atomic and structural contributions of Sugden, for a set of 115 polymers.

As with almost all of the new correlations developed in this monograph, the main purpose of developing Equation 7.11 was to transcend the intrinsic limitations of the predictive capabilities of group contribution techniques, which were discussed in Chapter 1. Examples of the predictive use of Equation 7.11 are given in Table 7.3, where this equation is combined with equations 3.10 and 3.11 for the van der Waals volume, and with equations 3.13 and 3.14 for the molar volume at room temperature, to provide a mutually consistent set of predictions of the γ of ten polymers at room temperature. No values were available for the contributions of silicon atoms and sulfone groups to P_S in the group contribution tables [3a,4] which provide the best values of P_S for polymers. The ten P_S values calculated by using Equation 7.11 and listed in Table 7.3 could therefore not have been estimated from the group contribution tables for the P_S of polymers.

The predicted γ values follow the same order as the experimental values of γ extrapolated from the melt [2] for the four polymers for which such experimental data are available. In addition, the absolute magnitudes of three of the predicted values are very close to the experimental values. Poly[oxy(methylphenylsilylene)] is the only polymer for which there is a significant difference between the predicted and measured values of γ. The surface tension increases asymptotically to a limiting value with increasing number-average molecular weight M_n. The relatively low experimental value for poly[oxy(methylphenylsilylene)] may therefore be at least partially due to the possibly very low M_n of the sample, as suggested by the very low viscosity of this polymer.

The predicted surface tensions of the remaining six polymers listed in Table 7.3 cannot be compared with experimental data due to the lack of such data. They do, however, follow trends which may be expected from basic physical considerations. They are predicted to increase with increasing fractions of (a) units of high cohesive energy density and (b) aromatic moieties in the hydrocarbon portions of the polymeric repeat units, and to decrease with increasing fraction of saturated aliphatic moieties.

Table 7.3. The quantities (V_w and N_{PS}) used to estimate the molar parachor P_S, the predicted value of P_S, and the predicted molar volume V and surface tension γ(pred) at room temperature, for ten polymers. Available experimental values [2] γ(exp) for these polymers at 20°C are also listed. The units are cc/mole for V_w and V, (cc/mole)·(dyn/cm)$^{1/4}$ for P_S, and dyn/cm for γ.

Polymer	V_w	N_{PS}	P_S	V	γ(pred)	γ(exp)
Poly(dimethyl siloxane)	43.79	-6	171.7	79.6	21.6	20.9[a]
Poly[oxy(methyl γ-trifluoropropylsilylene)]	73.22	-30	277.1	122.7	26.0	------
Poly[oxy(diethylsilylene)]	66.25	-12	258.3	113.4	26.9	25.7[b]
Poly[oxy(methylphenylsilylene)]	78.01	-3	309.7	128.1	34.2	26.1[c]
Poly(1-hexene sulfone)	82.47	-18	320.0	124.5	43.6	------
Poly[4,4'-isopropylidene diphenoxy di(4-phenylene)sulfone]	237.49	-9	943.0	360.2	47.0	46.6
Poly(1-butene sulfone)	62.10	-12	241.7	92.3	47.1	------
Poly(propylene sulfone)	51.26	-9	200.0	75.8	48.5	------
Poly[4,4'-diphenoxy di(4-phenylene)sulfone]	207.61	0	828.3	310.1	50.9	------
Poly[4,4'-sulfone diphenoxy di(4-phenylene)sulfone]	228.11	0	910.1	336.4	53.6	------

[a] Average of three very similar values listed [2] for the polymer of highest molecular weight (75000).

[b] The molecular weight is not listed. The viscosity is 158 cS [2].

[c] The molecular weight is not listed. The viscosity is 102 cS [2], suggesting a very low average molecular weight.

7.E. Interfacial Tension and Adhesion

The main contribution of this book to the calculation of the interfacial properties of polymers is its generalization of two different methods to calculate the surface tension (from the cohesive energy density and from the molar parachor), to allow its prediction without being limited by the lack of group contributions. It is, nonetheless, desirable to make this chapter more self-contained by reviewing the most useful methods for estimating the interfacial tension from surface tensions.

Theories of the interfacial tension were reviewed by Wu [1]. Wu also provided an extensive tabulation of experimental values for amorphous polymer-polymer interfaces [2], including both homopolymers and copolymers, at several temperatures. For example, the interfacial tensions between a number of pairs of homopolymers [2] are listed in Table 7.4. Note that they are always much smaller than the surface tensions listed in Table 7.1 for polymers. Van Krevelen [3] also examined the methods used to estimate the interfacial tension in practical calculations.

A simple equation was developed by Girifalco and Good [12-14] to relate the interfacial tension of a solid-liquid interface (where both surfaces are molecular) to the surface tensions of the components. This equation contained a factor of the order of unity related to the molecular volumes of the components. It was generalized later by assuming that its general form is applicable to all types of interfaces and that the factor of order unity can be dropped out. In Equation 7.12, γ_{12} is the interfacial tension, and γ_1 and γ_2 are the surface tensions of the surfaces.

$$\gamma_{12} \approx (\gamma_1{}^{0.5} - \gamma_2{}^{0.5})^2 \tag{7.12}$$

Equation 7.12 has the advantage of great simplicity. For example, if γ_1=40 dyn/cm and γ_2=30 dyn/cm, it can be easily estimated that $\gamma_{12} \approx 0.72$ dyn/cm. It was, however, found to suffer from the limitations that it does not work well for materials with strong polar interactions and fails completely when extremely polar interactions such as hydrogen bonds are present [15].

It was, then, suggested [15] that the total free energy at a surface should be expressed as the sum of the contributions from the different types of intermolecular forces at the surface:

$$\gamma \approx \gamma_d + \gamma_x \tag{7.13}$$

Here, γ_d is the dispersion force component of γ, while γ_x combines the polar (dipole and induced dipole) and hydrogen bonding components. Note the similarity to the separation of the cohesive energy and the solubility parameter into dispersion, polar and hydrogen bonding components in Chapter 5. The main difference is that polar and hydrogen bonding force components have been lumped together into a single term (γ_x) in Equation 7.13.

When this refinement is made, γ_{12} can be written in the following more general form [16], where each surface can be either a solid or a liquid, and either molecular or polymeric:

Table 7.4. Homopolymer-homopolymer interfacial tensions (γ_{12}), in dyn/cm [2].

Polymer 1	Polymer 2	$\gamma_{12}(20^\circ C)$	$\gamma_{12}(150^\circ C)$	$\gamma_{12}(200^\circ C)$
Polybutadiene (M_n=2350)	Poly(dimethyl siloxane) (M_n=5200)	3.98	2.85	2.42
Polybutadiene (M_n=960)	Poly[oxy(diethylsilylene)] (M_n=5200)	2.58	2.25	2.13
Polyethylene (linear)	Polystyrene	8.3	5.7	4.7
Polyethylene (linear)	Poly(methyl acrylate)	10.6	8.2	7.3
Polyethylene (linear)	Poly(ethyl acrylate)	7.5	5.4	4.6
Polyethylene (linear)	Poly(n-butyl acrylate)	5.0	3.3	2.7
Polyethylene (linear)	Poly(2-ethylhexyl acrylate)	3.1	1.8	1.3
Polyethylene (linear)	Poly(methyl methacrylate)	11.8	9.5	8.6
Polyethylene (linear)	Poly(n-butyl methacrylate)	7.1	5.1	4.4
Polyethylene (linear)	Poly(vinyl acetate)	14.6	11.0	9.7
Polyethylene (linear)	Poly(ethylene terephthalate)	9.4	----	----
Polyethylene (linear)	Poly(hexamethylene adipamide)	14.9	----	----
Polyisobutylene	Poly(vinyl acetate)	9.9	7.3	6.3
Polyisobutylene	Poly(dimethyl siloxane)	4.9	4.1	3.8
Polypropylene	Poly(dimethyl siloxane)	3.2	2.9	2.8
Polychloroprene	Poly(n-butyl methacrylate)	2.2	1.5	1.3
Polychloroprene	Poly(dimethyl siloxane)	7.1	6.4	6.2
Polystyrene	Polychloroprene	0.7	0.5	0.4
Polystyrene	Poly(vinyl acetate)	4.2	3.7	3.4
Polystyrene	Poly(methyl methacrylate)	3.2	1.5	0.8
Polystyrene	Poly(dimethyl siloxane)	6.1	6.1	6.1
Poly(methyl acrylate)	Poly(ethyl acrylate)	2.4	1.3	0.9
Poly(methyl acrylate)	Poly(n-butyl acrylate)	4.0	3.0	2.6
Poly(methyl acrylate)	Poly(2-ethylhexyl acrylate)	6.6	5.7	5.4
Poly(ethyl acrylate)	Poly(2-ethylhexyl acrylate)	3.9	3.2	3.0
Poly(n-butyl acrylate)	Poly(2-ethylhexyl acrylate)	1.8	1.2	1.0
Poly(methyl methacrylate)	Poly(n-butyl methacrylate)	3.4	1.9	1.3
Poly(methyl methacrylate)	Poly(t-butyl methacrylate)	3.0	2.2	1.9
Polyoxyethylene	Poly(2-ethylhexyl acrylate)	6.6	4.8	4.1
Polyoxyethylene	Poly(n-butyl acrylate)	2.8	2.0	1.7
Polyoxyethylene	Polyoxytetramethylene	4.5	3.8	3.6
Poly(dimethyl siloxane)	Poly(n-butyl methacrylate)	4.2	3.8	3.6
Poly(dimethyl siloxane)	Poly(t-butyl methacrylate)	3.6	3.3	3.2
Poly(dimethyl siloxane)	Polyoxyethylene	10.8	9.8	9.4
Poly(dimethyl siloxane)	Polyoxytetramethylene	6.4	6.3	6.2

$$\gamma_{12} \approx (\gamma_{d1}^{0.5} - \gamma_{d2}^{0.5})^2 + (\gamma_{x1}^{0.5} - \gamma_{x2}^{0.5})^2 \tag{7.14}$$

Equation 7.14 works much better than Equation 7.12 for interfaces of materials containing polar (and especially hydrogen bonding) interactions. Once γ_d and γ_x are determined for a substance by using a limited amount of interfacial tension data, they can be used in calculating the interfacial tensions of other interfaces of this substance. The main difficulty with this procedure is that, at this time, γ_d and γ_x for each material must be initially estimated from experimental data, preventing the use of Equation 7.14 for completely novel types of polymers.

This difficulty is similar to the one mentioned in Chapter 5 concerning the calculation of the separate polar and hydrogen bonding components of the cohesive energy and the solubility parameter. It is clear, by inspecting the experimentally-determined γ_d and γ_x values, that while these values qualitatively follow trends of the type that one would expect intuitively, such as γ_x in general being much larger for substances which can form strong hydrogen bonds, no general and robust quantitative structure-property relationships can be developed from these values.

For example, one could envision developing a quantitative structure-property relationship for γ_x, and then calculating γ_d by using Equation 7.13 with either experimentally determined total values of γ such as those listed in Table 7.1 or with values of γ predicted in terms of either the cohesive energy density or the molar parachor as described earlier in this chapter. (This procedure would be much safer than trying to develop a correlation for γ_d and then calculating γ_x from γ_d, since typically $\gamma_d \gg \gamma_x$ and therefore this alternative procedure would be very likely to lead to large errors in the values of γ_x calculated by subtracting one large number from another.) However, as can be seen from the values of γ_x listed in Table 7.5, it is impossible to develop a reliable correlation for γ_x in terms of the polymeric structure with the currently available values of γ_x. The best that can be done is to use Equation 7.14 semi-empirically to estimate γ_x for additional polymers, with the hope of developing such a relationship in the future.

There is, clearly, room for significant further improvements in our ability to predict the interfacial tension from the surface tensions of the components in a simple manner. The simplest expression (Equation 7.12) fails for many important types of materials, while the more general expression (Equation 7.14) requires some experimental data on each material to estimate the values of its input parameters and thus cannot be used for completely novel types of materials.

The *reversible work of adhesion* (W_{adh}) is an important quantity in determining whether an interface will be stable (require the expenditure of energy for its separation) or will separate spontaneously (with the free energy of the system being reduced by the separation). W_{adh} is the free energy difference between the separated surfaces and the interface:

$$W_{adh} = \gamma_1 + \gamma_2 - \gamma_{12} \tag{7.15}$$

Table 7.5. Combined polar and hydrogen bonding components (γ_x) of the surface tensions of polymers, in dyn/cm [1,3b,16]. Values in parentheses may be less reliable than other values.

Polymer	γ_x	Polymer	γ_x
Polyethylene	0.0, 1.1	Poly(methyl acrylate)	10.3
Polypropylene	0.0	Poly(ethyl acrylate)	6.3
Polyisobutylene	0.0	Poly(methyl methacrylate)	4.3, 11.5
Polystyrene	0.6, 6.1	Poly(ethyl methacrylate)	9.0
Poly(α-methyl styrene)	4.0	Poly(n-butyl methacrylate)	5.0
Poly(vinyl fluoride)	5.5	Poly(isobutyl methacrylate)	4.3
Poly(vinylidene fluoride)	7.1	Poly(t-butyl methacrylate)	3.7
Polytrifluoroethylene	4.1	Poly(n-hexyl methacrylate)	3.0
Polytetrafluoroethylene	0.5, 1.5, 1.6	Polyoxyethylene	12.0
Poly(vinyl chloride)	2.0	Polyoxytetramethylene	4.5
Poly(vinylidene chloride)	4.5	Poly(ethylene terephthalate)	3.5, 4.1, 9.0
Polychlorotrifluoroethylene	8.6	Poly(hexamethylene adipamide)	6.2, 9.1, 14.0
Poly(vinyl acetate)	1.2	Poly(dimethyl siloxane)	0.8, 1.1

If W_{adh} is positive, it quantifies the work needed to separate the surfaces. If it is negative, it quantifies the lowering of the free energy by the spontaneous separation of the two surfaces.

It must be emphasized, however, that adhesion can be only partially understood in terms of the interfacial tension and the work of adhesion. The importance of "engineering" aspects of the problem, such as the sizes, shapes and layouts of the components of a multicomponent system, was pointed out in Section 7.A. Furthermore, other key material properties besides the interfacial tension may also play a key role in determining the strength of an adhesive joint. For example, according to equations 7.2, 7.3 and 7.4, the surface tension (and thus also the interfacial tension) changes gradually with the temperature, without a discontinuity at T_g where the slope of γ changes discontinuously but not γ itself. On the other hand, if the test temperature is made higher than the T_g of an adhesive (either by increasing the temperature or by using an additive which lowers the T_g of the adhesive), the strength of the adhesive interface can be reduced as a result of easy failure within the adhesive [17]. Failure at an interface of an adhesive joint is called *adhesive failure*, while failure inside any of the materials being joined is called *cohesive failure*.

References and Notes for Chapter 7

1. S. Wu, *Polymer Interface and Adhesion*, Marcel Dekker, New York (1982).

2. S. Wu, in *Polymer Handbook*, edited by J. Brandrup and E. H. Immergut, Wiley, New York, third edition (1989), VI/411-VI/449.

3. (a) D. W. van Krevelen, *Properties of Polymers*, second edition, Elsevier, Amsterdam (1976). Chapter 8 of this book deals with the properties of surfaces and interfaces. In addition, Table VII in Part VII lists the group contributions to the parachor. (b) D. W. van Krevelen, *Properties of Polymers*, third edition, Elsevier, Amsterdam (1990).

4. D. W. van Krevelen, in *Computational Modeling of Polymers*, edited by J. Bicerano, Marcel Dekker, New York (1992), Chapter 1.

5. A. J. Kinloch and R. J. Young, *Fracture Behaviour of Polymers*, Elsevier Applied Science Publishers, London (1983).

6. E. J. Kramer and L. L. Berger, *Advances in Polymer Science*, *91/92*, 1-68 (1990).

7. E. H. Andrews and L. Bevan, *Polymer*, *13*, 337-346 (1972).

8. E. H. Andrews, G. M. Levy and J. Willis, *J. Materials Science*, *8*, 1000-1008 (1973).

9. S. Sugden, *J. Chem. Soc.*, *125*, 1177-1189 (1924).

10. S. A. Mumford and J. W. C. Phillips, *J. Chem. Soc.*, *130*, 2112-2133 (1929).

11. O. R. Quayle, *Chemical Reviews*, *53*, 439-589 (1953).

12. L. A. Girifalco and R. J. Good, *J. Phys. Chem.*, *61*, 904-909 (1957).

13. L. A. Girifalco and R. J. Good, *J. Phys. Chem.*, *62*, 1418-1421 (1958).

14. L. A. Girifalco and R. J. Good, *J. Phys. Chem.*, *64*, 561-565 (1960).

15. F. M. Fowkes, *Ind. Eng. Chem.*, *56*, 40-52 (1964).

16. D. K. Owens and R. C. Wendt, *J. Appl. Polym. Sci.*, *13*, 1741-1747 (1969).

17. P. E. Cassidy, J. M. Johnson and C. E. Locke, *J. Adhesion*, *4*, 183-191 (1972).

CHAPTER 8

OPTICAL PROPERTIES

8.A. Background Information

8.A.1. Types of Optical Properties

The interactions of materials with electromagnetic fields are described by their optical, electrical and magnetic properties. These three types of properties are interrelated. For example, the complex dielectric constant ε^* (see also the much more extensive discussion in Chapter 9), the complex refractive index n^*, the refractive index n, the absorption index K, and the real (storage) and imaginary (loss) components ε' and ε'' of ε^* are interrelated by the following equations:

$$\varepsilon^* = (n^*)^2 \tag{8.1}$$

$$n^* = n \cdot (1 - i \cdot K) \tag{8.2}$$

$$\varepsilon' = n^2 \cdot (1 - K^2) \tag{8.3}$$

$$\varepsilon'' = 2n^2 \cdot K \tag{8.4}$$

The optical properties, which describe the interaction of materials with light, will be treated in this chapter. The electrical properties, which describe the interaction of materials with electric fields, will be discussed in Chapter 9, along with some of the additional information which can be gained by combining the optical and electrical properties. The magnetic properties, which describe the interaction of materials with magnetic fields, will be discussed in Chapter 10.

Polymers have many important optical properties, such as the refractive index, reflection, scattering, absorption, clarity, gloss, haze, birefringence, stress-optic coefficient, the yellowing induced by photochemical degradation, and the specific refractive index increment in dilute solutions. The optical properties of a polymer have to be considered in evaluating its potential usefulness in a wide variety of applications. Compact disk coatings, lenses used in eyeglasses, and certain food and beverage packaging materials, are just a few examples of such applications. The stress-optic coefficient is important in determining whether a polymer will be useful as a compact disk coating. The optical clarity determines whether a polymer can be used in lenses for eyeglasses. (Toughness is also important in eyeglass lenses, which for reasons of safety must be able to withstand an impact of considerable strength without shattering.) The optical properties of a packaging material determine whether the package will look attractive to a customer and

213

therefore help in the sale of its contents. The total optical loss of a polymer has to be reduced to the minimum amount possible for the polymer to be potentially useful as an optical fiber material.

8.A.2.   Refractive Index and Molar Refraction

This chapter will focus mainly on the *refractive index* n, which is the most fundamental optical property of a polymer. It is also used as a parameter in estimating many other optical and electrical properties. *The refractive index of a material is the ratio of the velocity of light in vacuum to the velocity of light in the material.* It is always larger than 1.0 because light slows down a result of its interaction with the atoms from which the material is built, while the vacuum by definition contains no atoms and thus contains nothing that could slow down a beam of light.

The refractive index is generally estimated in terms the *molar refraction* R, which quantifies the intrinsic refractive power of the structural units constituting the material. Several definitions have been proposed for R. Two of these definitions incorporate both of the key physical factors determining the refractive index, and are thus especially useful. Equation 8.5 expresses n in terms of the molar refraction R_{LL} according to Lorentz and Lorenz [1,2], and Equation 8.6 expresses n in terms of the molar refraction R_{GD} according to Gladstone and Dale [3].

$$n = \sqrt{\frac{(V + 2R_{LL})}{(V - R_{LL})}}$$
(8.5)

$$n = 1 + \frac{R_{GD}}{V}$$
(8.6)

Since n is dimensionless, R_{LL} and R_{GD} have the same units as the molar volume V, namely cc/mole. According to equations 8.5 and 8.6, the value of n is a manifestation of the combined effect of two key physical factors:

1. The value of n increases with increasing intrinsic refractive power of a material, as quantified by its molar refraction R. For example, if two polymers have identical V, then the polymer with the larger intrinsic refractive power (larger R) will have a larger n. The values of the molar refraction are determined by the interactions of electromagnetic waves with atoms, and therefore cannot be rationalized in terms of very simple and general arguments.

2. The value of n increases with increasing amount of material per unit volume, as quantified by decreasing molar volume. The atoms act as "obstacles" to a beam of light traversing a material, and slow it down. For example, if two polymers have identical R, then the polymer with the smaller V will present a larger number of such obstacles per unit length of the beam of light, and will therefore have a larger n. The importance of the density as a key factor in the determination of n can thus be rationalized in terms of a very simple and general argument.

The common assumption of a simple correlation between high density and high refractive index has very little quantitative merit in predicting the effects of structural and compositional variations on the refractive indices of a structurally diverse set of polymers. For example, fluorinated polymers often have very high densities, but very low intrinsic refractivities and therefore very low refractive indices. By contrast, the densities of many polymers containing aromatic rings are not especially high, but these polymers usually have large intrinsic refractivities and therefore large refractive indices. It suffices to compare polytetrafluoroethylene (amorphous density of 2.00 grams/cc but a refractive index of only 1.350) with polystyrene (amorphous density of only 1.05 grams/cc but a refractive index of 1.592) as an example. It is, therefore, necessary to consider *both* the *intrinsic refractivity* of the structural units *and* the *density* as key physical factors determining the refractive index. The intrinsic refractivities and densities of polymers are not necessarily determined by the same fundamental causes.

R_{LL} and R_{GD} have relatively little dependence on temperature and percent crystallinity, while V changes significantly with changing temperature and/or crystallinity. V(T) normally increases with increasing temperature and decreases with increasing crystallinity. Consequently, n normally decreases with increasing temperature and increases with increasing crystallinity. These effects can all be estimated by using either Equation 8.5 or Equation 8.6. For example, it can be seen that n will decrease with increasing molar volume, asymptotically approaching its limiting absolute minimum value of 1.0 for the vacuum as the molar volume approaches infinity.

The temperature dependence of n is, however, quite similar for all polymers, and therefore usually not investigated in great detail. A very simple and useful empirical "rule of thumb" [4] is that n decreases by about $1 \cdot 10^{-4}$ to $2 \cdot 10^{-4}$ per degree Kelvin increase of the temperature for glassy polymers, and by about $3 \cdot 10^{-4}$ to $5 \cdot 10^{-4}$ per degree Kelvin increase of the temperature for isotropic (unoriented) amorphous polymers above T_g, at normal temperatures. Since n can never become smaller than 1.0, however, the rate of decrease of n with increasing temperature above T_g must eventually level off and asymptotically approach zero as suggested by equations 8.5 and 8.6, in order for n to remain greater than 1.0. Most polymers undergo thermal or thermooxidative degradation long before reaching such extremely high temperatures, so that the empirical rule of thumb can generally be used for temperature ranges of practical interest without much need for concern for this ultimate asymptotic behavior.

Equations 8.7 and 8.8, which are obtained by inverting equations 8.5 and 8.6, respectively, and expressing R_{LL} and R_{GD} in terms of V and n, are also very useful. These equations enable the estimation of the R_{LL} and R_{GD} of a polymer if its n and V are both known, or can both be predicted, at a given temperature and percent crystallinity. Values of R_{LL} and R_{GD} calculated in this manner can then be substituted into equations 8.5 and 8.6 to estimate the effects of crystallinity or of temperature variation.

$$R_{LL} = V \frac{\left(n^2 - 1\right)}{\left(n^2 + 2\right)} \tag{8.7}$$

$$R_{GD} = V \cdot (n - 1) \tag{8.8}$$

The refractive indices of polymers are usually predicted as follows:

1. Use group contributions to estimate V and either R_{LL} or R_{GD}.
2. Use Equation 8.5 or Equation 8.6 to estimate n in terms of V and either R_{LL} or R_{GD}.

Available group contributions were all tabulated, compared with each other, and extensively discussed, by van Krevelen [5,6]. Most of the group contributions for R were developed from data on liquid organic compounds rather than polymers [7]. As shown by van Krevelen [5], these group contributions can, nonetheless, be used to calculate the refractive indices of polymers with reasonable accuracy. The reason for this transferability of R values from liquid organic compounds to high polymers is that the intrinsic refractive power of a given structural unit is only affected very slightly by whether this structural unit is located in a small molecule or in a polymer chain. Changes in the molar volume account for most of the differences between the refractive indices of liquid organic compounds and high polymers containing the same structural units.

The value of n also depends on the wavelength of measurement, decreasing very slowly with increasing wavelength. This dependence is usually characterized by the Abbé number A, which is defined by Equation 8.9. In this equation, the refractive indices measured at three standard wavelengths are denoted by n_C (measured at 6563Å), n_F (measured at 4861Å), and n_D (measured at 5890Å). While the Abbé number is important in certain specialized applications such as the design of achromatic lenses which focus a range of wavelengths to the same spot, the variation of n with the wavelength of visible light is normally quite small, and will not be discussed further.

$$A = \frac{\left(n_D - 1\right)}{\left(n_F - n_C\right)} \tag{8.9}$$

8.A.3. Optical Losses

Although the detailed discussion of the prediction of optical losses is beyond the scope of this chapter, it should also be pointed out that some very significant work has been done in this area in recent years [8-14]. The calculations start by subdividing the total optical loss (α_{total}) into its intrinsic and extrinsic components ($\alpha_{intrinsic}$ and $\alpha_{extrinsic}$, respectively):

$$\alpha_{total} = \alpha_{intrinsic} + \alpha_{extrinsic} \tag{8.10}$$

Extrinsic losses are caused by both impurities in the polymer and imperfections introduced during specimen fabrication. The total loss is always greater than $\alpha_{intrinsic}$, approaching $\alpha_{intrinsic}$ for specimens fabricated of exceptional quality manufactured from resins of utmost purity. The calculated value of $\alpha_{intrinsic}$ is, hence, used to screen materials, providing a performance target for a material when the factors leading to extrinsic losses are reduced to the minimum possible.

The intrinsic loss is expressed as the sum of the vibrational absorption loss (α_{vibr}), electronic transition absorption loss (α_{elec}) and Rayleigh scattering loss (α_{scat}):

$$\alpha_{intrinsic} = \alpha_{vibr} + \alpha_{elec} + \alpha_{scat} \tag{8.11}$$

The most important contribution to $\alpha_{intrinsic}$ is usually α_{vibr} in the near-infrared wavelength range for polymers containing hydrogen atoms in their repeat units. The importance of α_{vibr} decreases rapidly with decreasing wavelength. It becomes negligible in the ultraviolet wavelength range, where α_{elec} and α_{scat} typically become much more important and can get very large for polymers whose repeat units contain such structural features as aromatic rings. The method [8-14] involves the use of a set of semi-empirical relationships relating each component of $\alpha_{intrinsic}$ both to the structure of the polymeric repeat unit and to the wavelength of measurement.

The value of α_{vibr} is estimated by considering the overtones of the stretching vibrations of bonds involving hydrogen atoms, as well as the less intense combined stretch-bend vibration for each overtone. These contributions are estimated by taking advantage of the fact that the overtone wavelengths for a given type of hydrogen stretching vibration generally vary over a relatively small range with variation of the local environment of the bond [15], so that generic values can be used for the frequencies and the absorption strengths of the overtones of each type of vibration for purposes of estimating a rough spectrum for a semiquantitative evaluation. The theoretical framework used to estimate these frequencies and strengths is based on a quantum mechanical model, with some of the parameters of the model obtained from empirical considerations. The details have been worked out for C-H, N-H and O-H bonds. The small vibrational absorption losses arising from bonds not involving hydrogen atoms are neglected in these calculations.

Electronic transition absorption losses of organic materials in the visible wavelength region are caused by multiple bonds (such as double bonds, triple bonds, and aromatic rings) whose electronic transitions involve the electrons in the π-bonds. The accurate estimation of $\alpha_{electronic}$ therefore requires very detailed quantum mechanical calculations which can use up large amounts of computer time and memory. Instead, a simple empirical correlation is used for α_{elec}.

The scattering component of $\alpha_{intrinsic}$ is caused by the heterogeneities in a material, such as local fluctuations in its density or its refractive index. It can be subdivided into isotropic and anisotropic scattering components, each of which can then be estimated by using semiempirical expressions based on theoretical considerations but containing empirical parameters.

8.B. Improvements in the Ability to Predict the Refractive Indices of Polymers

The ability to predict n as summarized in Section 8.A is limited by the need for the simultaneous availability of reliable group contributions to the molar volume and to the molar refraction for all structural units of a polymer. These group contributions are not available for many structural units incorporated into novel types of polymers. The refractive indices of these polymers therefore cannot be predicted by using this standard procedure.

A new correlation will consequently be developed for n(298K), i.e., for the refractive index at room temperature, in Section 8.C. The use of this correlation for the prediction of n(298K) does not require the prior estimation of the molar refraction. In addition, this correlation for n(298K) has the same attractive features as the correlation developed for the amorphous volume V(298K) in Section 3.C. It does not require group contributions, so that it is applicable to all polymers constructed from the nine elements (C, N, O, H, F, Si, S, Cl and Br) which fall within the scope of our work. Furthermore, it is independent of whether a polymer is glassy (below T_g) or rubbery (above T_g) at room temperature.

This correlation will be tested in Section 8.D, where it will be used in the calculation of the specific refractive increments of a large number of polymer-solvent combinations. Specific refractive index increments calculated by using n(298K) values obtained from the new correlation will be shown to agree well with the experimental values. The specific refractive index increment is proportional to the relatively small difference between two much larger numbers, namely the refractive indices of the polymer and the solvent. It is, consequently, sensitive even to small errors in the estimation of the refractive index of the polymer. Its calculation is therefore a stringent test of the accuracy of the new correlation for n(298K) and of the usefulness of this correlation in practical applications.

Finally, in Section 8.E, R_{LL} values calculated by substituting the correlations developed in this book for the n(298K) and V(298K) of polymers into Equation 8.3 will be compared with R_{LL} values calculated by using the standard group contributions developed for liquid organic compounds. This is not an independent test of the quality of the correlations for n(298K) and V(298K). It is mainly an illustration of how the new correlations for n(298K) and V(298K) can be used to calculate R_{LL} or R_{GD}, for subsequent use in estimating the effects of temperature variation or crystallinity. Since the new correlations for n(298K) and V(298K) are valid for all polymers, R_{LL} and R_{GD} can be calculated in this manner for all polymers without using any group contributions.

The ability to calculate R_{LL} (or R_{GD}) for all polymers by inserting V(298K) and n(298K) values predicted via independent correlations into Equation 8.7 (or 8.8) is very useful. If n(T) needs to be predicted at an alternative temperature, R_{LL} (or R_{GD}) values calculated in this manner can be inserted into Equation 8.5 (or 8.6), along with V(T) values calculated by using the equations listed in Section 3.D, to predict n(T) for T≠298K.

8.C. Correlation for the Refractive Index at Room Temperature

A dataset containing the experimental n(298K) values of 183 polymers, all measured at "standard" wavelengths of light, was prepared. "Room temperature" is defined differently by different authors. This dataset therefore contains n values measured at various temperatures in the narrow range of 293K to 303K. The largest number of n(298K) values was taken from the *Polymer Handbook* [4]. The second major source of data was van Krevelen's book [5]. All of the experimental n(298K) values for silicon-containing polymers were taken from a catalog by Petrarch Systems [16], which contains comprehensive tables of the properties of such molecules and polymers. Most of the polymers in the dataset are nearly or completely amorphous. Either a single experimental n(298K) value, or several very similar n(298K) values, have been reported for most amorphous polymers.

Ranges of values of n(298K) have been reported in the literature for some semicrystalline polymers which can be prepared with different percent crystallinities. For example, the n(298K) of polyethylene is 1.476 in the limit of 0% crystallinity (complete amorphicity), and 1.565 in the limit of 100% crystallinity [4]. For such polymers, the n(298K) value corresponding to the limit of 0% crystallinity was used in the calculations. This procedure, in general, resulted in the use of the lowest n(298K) values reported for semicrystalline polymers which can attain high crystallinities. Once the refractive index of the amorphous limit of a semicrystalline polymer is predicted with reasonable accuracy, the effects of crystallinity can be estimated from Equation 8.5 or Equation 8.6, by using R_{LL} or R_{GD} values calculated as will be described in Section 8.E, combined with an estimate of the change in volume (usually a considerable reduction of the molar volume) caused by crystallinity. For a few polymers, such as poly(p-xylylene) and poly(vinylidene fluoride), only a single value of n for a highly crystalline sample was reported, so that this value had to be used in the development of the new correlation for n(298K).

The experimental values of n(298K) correlated well with a combination of the "intensive" versions (see Section 2.C) of the following variables:

1. Three connectivity indices ($^0\chi$, $^0\chi^v$ and $^1\chi^v$).

2. The total number of "rotational degrees of freedom" parameter N_{rot} introduced in Chapter 4.

3. The correction index N_{ref} defined by Equation 8.12.

$$N_{ref} \equiv -11N_F - 3N_{(Cl\ bonded\ to\ aromatic\ ring\ atoms)} + 18N_S + 9N_{fused} + 12N_{HB}$$
$$+ 32N_{(Si-Si\ bonds)} \qquad (8.12)$$

The final correlation, given by Equation 8.13, has a standard deviation of 0.0157, and a correlation coefficient of 0.9770 which indicates that it accounts for 95.5% of the variation of n(298K). The average experimental n(298K) value of the dataset is 1.5218, so that the standard deviation is approximately 1% of the average value.

$$n(298K) \approx 1.885312 + 0.024558 \cdot (17 \cdot {}^{0}\chi^{v} - 20 \cdot {}^{0}\chi - 12 \cdot {}^{1}\chi^{v} - 9 \cdot N_{rot} + N_{ref}) / N \qquad (8.13)$$

There were fundamental physical reasons (Chapter 4) for the use of N_{rot} in the correlations for the heat capacity. The use of N_{rot} in Equation 8.13 does not have any such obvious physical justification. It is mainly a convenient way of accounting systematically for the effect of rigid structural units which simultaneously decrease N_{rot} and increase n(298K) more than expected on the basis of the variations of the zeroth-order and first-order connectivity indices.

Most of the terms in the definition of N_{ref} were previously encountered. N_F is the number of fluorine atoms, and N_S is the total number of sulfur atoms (regardless of oxidation state or bonding environment), in the repeat unit. N_{fused} was defined in Section 3.B.1. N_{HB} is the total number of strongly hydrogen-bonding moieties, which, for purposes of calculating the refractive index, include structural units such as amide, urea, urethane, hydroxyl, carboxylic acid and sulfonic acid moieties. $N_{(Si-Si\ bonds)}$ is the only term in N_{ref} not encountered earlier. It accounts for the delocalization and conjugation of the silicon-silicon sigma-bonding orbitals [17-19], which cause a significant increase in the intrinsic refractive power of molecules and polymers containing such orbitals in polysilanes where the entire chain backbone consists of silicon atoms.

The results of these calculations are summarized in Table 8.1 and illustrated in Figure 8.1.

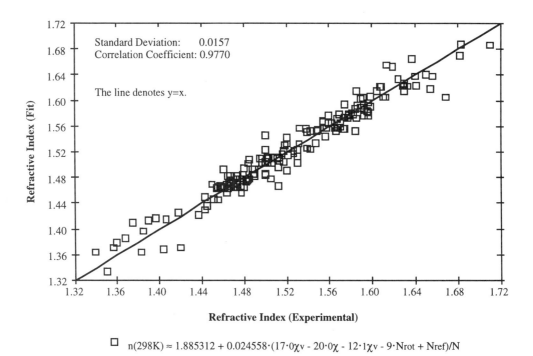

$$\square \quad n(298K) \approx 1.885312 + 0.024558 \cdot (17 \cdot {}^{0}\chi v - 20 \cdot {}^{0}\chi - 12 \cdot {}^{1}\chi v - 9 \cdot Nrot + Nref)/N$$

Figure 8.1. Correlation for the refractive indices of 183 polymers at room temperature.

Table 8.1. Experimental refractive index n at room temperature, number of rotational degrees of freedom N_{rot} and correction index N_{ref} used in the correlation for n, and the fitted value of n, for 183 polymers. The number N of vertices in the hydrogen-suppressed graph of the polymeric repeat unit, and the connectivity indices $^0\chi$, $^0\chi^v$ and $^1\chi^v$, all of which are also used in the correlation equation for n, are listed in Table 2.2.

Polymer	n(exp)	N_{rot}	N_{ref}	n(fit)
Poly(pentadecafluorooctyl acrylate)	1.3390	12.0	-165	1.3641
Polytetrafluoroethylene	1.3500	2.0	-44	1.3353
Poly(undecafluorohexyl acrylate)	1.3560	10.0	-121	1.3719
Poly(nonafluoropentyl acrylate)	1.3600	9.0	-99	1.3777
Poly(heptafluorobutyl acrylate)	1.3670	8.0	-77	1.3856
Poly(trifluorovinyl acetate)	1.3750	5.0	-33	1.4100
Poly[oxy(methyl γ-trifluoropropylsilylene)]	1.3830	6.0	-33	1.3651
Poly(pentafluoropropyl acrylate)	1.3850	7.0	-55	1.3972
Polychlorotrifluoroethylene	1.3900	2.0	-33	1.4144
Poly[oxy(methylhydrosilylene)]	1.3970	3.0	0	1.4182
Poly(dimethyl siloxane)	1.4035	4.0	0	1.3671
Poly(trifluoroethyl acrylate)	1.4070	6.0	-33	1.4158
Poly(2,2,2-trifluoro-1-methylethyl methacrylate)	1.4185	8.0	-33	1.4263
Poly(vinylidene fluoride)	1.4200	2.0	-22	1.3709
Poly(trifluoroethyl methacrylate)	1.4370	7.0	-33	1.4228
Poly[oxy(methyl *n*-hexylsilylene)]	1.4430	9.0	0	1.4290
Poly[oxy(methyl *n*-octadecylsilylene)]	1.4430	21.0	0	1.4495
Poly[oxy(methyl *n*-octylsilylene)]	1.4450	11.0	0	1.4355
Poly(vinyl isobutyl ether)	1.4507	7.0	0	1.4687
Poly[oxy(methyl *n*-hexadecylsilylene)]	1.4510	19.0	0	1.4479
Poly(vinyl ethyl ether)	1.4540	5.0	0	1.4641
Poly[oxy(methyl *n*-tetradecyl silylene)]	1.4550	17.0	0	1.4459
Poly(vinyl *n*-butyl ether)	1.4563	7.0	0	1.4643
Polyoxyethylene	1.4563	3.0	0	1.4648
Poly(propylene oxide)	1.4570	4.0	0	1.4667
Poly(3-butoxypropylene oxide)	1.4580	9.0	0	1.4644
Poly(vinyl *n*-pentyl ether)	1.4590	8.0	0	1.4644
Poly(3-hexoxypropylene oxide)	1.4590	11.0	0	1.4645
Poly(vinyl *n*-hexyl ether)	1.4591	9.0	0	1.4644
Poly(4-fluoro-2-trifluoromethylstyrene)	1.4600	4.0	-44	1.4925

☞ ☞ ☞   **TABLE 8.1 IS CONTINUED IN THE NEXT PAGE.** ☞ ☞ ☞

Table 8.1. CONTINUED FROM THE PREVIOUS PAGE.

Polymer	n(exp)	N_{rot}	N_{ref}	n(fit)
Poly(vinyl *n*-octyl ether)	1.4613	11.0	0	1.4645
Poly(vinyl 2-ethylhexyl ether)	1.4626	11.0	0	1.4653
Poly(vinyl *n*-decyl ether)	1.4628	13.0	0	1.4646
Poly[oxy(acryloxypropylmethylsilylene)]	1.4630	9.0	0	1.4570
Poly(3-methoxypropylene oxide)	1.4630	6.0	0	1.4685
Poly(*t*-butyl methacrylate)	1.4638	9.0	0	1.4819
Poly(vinyl *n*-dodecyl ether)	1.4640	15.0	0	1.4646
Poly(4-methyl-1-pentene)	1.4650	6.0	0	1.4713
Poly(*n*-butyl acrylate)	1.4660	8.0	0	1.4743
Poly(vinyl propionate)	1.4665	6.0	0	1.4765
Poly(vinyl methyl ether)	1.4670	4.0	0	1.4704
Poly(vinyl acetate)	1.4670	5.0	0	1.4814
Poly(ethyl acrylate)	1.4685	6.0	0	1.4770
Polyoxyoctamethylene	1.4690	9.0	0	1.4648
Poly(1-octadecene)	1.4710	18.0	0	1.4653
Poly(isopropyl methacrylate)	1.4728	8.0	0	1.4795
Poly(1-decene)	1.4730	10.0	0	1.4657
Polypropylene *(atactic)*	1.4735	3.0	0	1.4713
Poly(vinyl *sec*-butyl ether) *(isotactic)*	1.4740	7.0	0	1.4655
Poly(dodecyl methacrylate)	1.4740	17.0	0	1.4711
Poly(ethylene succinate)	1.4744	8.0	0	1.4818
Poly(vinyl formate)	1.4757	4.0	0	1.4790
Polyethylene	1.4760	2.0	0	1.4648
Poly(2-fluoroethyl methacrylate)	1.4768	7.0	-11	1.4574
Poly(isobutyl methacrylate)	1.4770	9.0	0	1.4792
Poly(octylmethylsilane)	1.4780	10.0	32	1.4787
Poly(methyl acrylate)	1.4790	5.0	0	1.4833
Ethyl cellulose	1.4790	15.0	0	1.4890
Polyoxymethylene	1.4800	2.0	0	1.4647
Poly[oxy(dicyanopropylsilylene)]	1.4800	8.0	0	1.4954
Poly(*n*-hexyl methacrylate)	1.4813	11.0	0	1.4742
Poly(*n*-butyl methacrylate)	1.4830	9.0	0	1.4761
Poly(*n*-propyl methacrylate)	1.4840	8.0	0	1.4774

☞ ☞ ☞   TABLE 8.1 IS CONTINUED IN THE NEXT PAGE. ☞ ☞ ☞

Table 8.1. CONTINUED FROM THE PREVIOUS PAGE.

Polymer	n(exp)	N_{rot}	N_{ref}	n(fit)
Poly(ethylene maleate)	1.4840	7.0	0	1.5018
Poly(ethyl methacrylate)	1.4850	7.0	0	1.4789
Poly(3,3,5-trimethylcyclohexyl methacrylate)	1.4850	12.0	0	1.5080
Poly(vinyl butyral)	1.4850	8.0	0	1.5086
Poly(2-nitro-2-methylpropyl methacrylate)	1.4868	10.0	0	1.4917
Poly(methyl methacrylate)	1.4893	6.0	0	1.4846
Poly(2-decyl-1,4-butadiene)	1.4899	13.0	0	1.4818
Poly[oxy(mercaptopropylmethylsilylene)]	1.4900	7.0	18	1.4921
Poly(3-methylcyclohexyl methacrylate)	1.4947	10.0	0	1.5101
Poly(cyclohexyl α-ethoxyacrylate)	1.4969	11.0	0	1.5029
Methyl cellulose	1.4970	12.0	0	1.4997
Poly(4-methylcyclohexyl methacrylate)	1.4975	10.0	0	1.5101
Poly(vinyl methyl ketone)	1.5000	4.0	0	1.4843
Poly(2-heptyl-1,4-butadiene)	1.5000	10.0	0	1.4864
Poly(sec-butyl α-chloroacrylate)	1.5000	8.0	0	1.5026
Poly(1,2-butadiene)	1.5000	3.0	0	1.5055
Poly(vinyl alcohol)	1.5000	3.0	12	1.5239
Poly(2-bromo-4-trifluoromethyl styrene)	1.5000	4.0	-33	1.5462
Poly(2-isopropyl-1,4-butadiene)	1.5020	6.0	0	1.5019
Poly(ethyl α-chloroacrylate)	1.5020	6.0	0	1.5111
Poly(2-methylcyclohexyl methacrylate)	1.5028	10.0	0	1.5097
Polyisobutylene	1.5050	4.0	0	1.4795
Poly(2-t-butyl-1,4-butadiene)	1.5060	7.0	0	1.5024
Poly(cyclohexyl methacrylate)	1.5066	9.0	0	1.5123
Poly(1-methylcyclohexyl methacrylate)	1.5111	10.0	0	1.5109
Poly(2-hydroxyethyl methacrylate)	1.5119	8.0	12	1.4973
Poly(1-butene) (isotactic)	1.5125	4.0	0	1.4669
Poly(vinyl chloroacetate)	1.5130	5.0	0	1.5120
Poly(N-butyl methacrylamide)	1.5135	9.0	12	1.5061
Poly(1,4-butadiene)	1.5160	3.0	0	1.5188
Poly(methyl α-chloroacrylate)	1.5170	5.0	0	1.5214
Poly(2-chloroethyl methacrylate)	1.5170	7.0	0	1.5050
Poly(2-chlorocyclohexyl methacrylate)	1.5179	9.0	0	1.5293

☞ ☞ ☞   TABLE 8.1 IS CONTINUED IN THE NEXT PAGE. ☞ ☞ ☞

Table 8.1. CONTINUED FROM THE PREVIOUS PAGE.

Polymer	n(exp)	N_{rot}	N_{ref}	n(fit)
Poly(methyl isopropenyl ketone)	1.5200	5.0	0	1.4897
Polymethacrylonitrile	1.5200	3.0	0	1.5325
Polyacrylonitrile	1.5200	2.0	0	1.5425
Polyisoprene	1.5210	4.0	0	1.5158
Poly(acrylic acid)	1.5270	4.0	12	1.5182
Poly(1,3-dichloropropyl methacrylate)	1.5270	8.0	0	1.5215
Poly(hexamethylene sebacamide)	1.5300	18.0	24	1.5034
Poly(ε-caprolactam)	1.5300	7.0	12	1.5130
Poly(hexamethylene adipamide)	1.5300	14.0	24	1.5130
Poly(N-vinyl pyrrolidone)	1.5300	5.5	0	1.5236
Poly(cyclohexyl α-chloroacrylate)	1.5320	8.0	0	1.5337
Poly[oxy(methylphenylsilylene)]	1.5330	4.0	0	1.5578
Poly(vinyl chloride)	1.5390	2.0	0	1.5560
Poly(N-methyl methacrylamide)	1.5398	6.0	12	1.5264
Cellulose	1.5400	9.0	36	1.5514
Poly(*sec*-butyl α-bromoacrylate)	1.5420	8.0	0	1.5250
Poly(cyclohexyl α-bromoacrylate)	1.5420	8.0	0	1.5524
Poly(2-bromoethyl methacrylate)	1.5426	7.0	0	1.5243
Poly(ethylmercaptyl methacrylate)	1.5470	8.0	18	1.5344
Poly(1-phenylethyl methacrylate)	1.5487	8.0	0	1.5522
Poly[oxy(methyl *m*-chlorophenylethylsilylene)]	1.5500	6.0	-3	1.5489
Poly(*p*-isopropyl styrene)	1.5540	6.0	0	1.5696
Poly(cyclohexylmethylsilane)	1.5570	6.0	32	1.5445
Poly(*p*-cyclohexylphenyl methacrylate)	1.5575	10.0	0	1.5584
Polychloroprene	1.5580	3.0	0	1.5673
Poly[oxy(methyl *m*-chlorophenylsilylene)]	1.5600	4.0	-3	1.5705
Poly[4,4-heptane *bis*(4-phenyl)carbonate]	1.5602	12.0	0	1.5642
Poly[1-(*o*-chlorophenyl)ethyl methacrylate]	1.5624	8.0	-3	1.5610
Poly(2,2,2'-trimethylhexamethylene terephthalamide)	1.5660	14.0	24	1.5578
Poly(methyl α-bromoacrylate)	1.5672	5.0	0	1.5534
Poly(benzyl methacrylate)	1.5679	7.0	0	1.5590
Poly[2-(phenylsulfonyl)ethyl methacrylate]	1.5682	9.0	18	1.5715
Poly(*m*-cresyl methacrylate)	1.5683	7.0	0	1.5603

☞ ☞ ☞   **TABLE 8.1 IS CONTINUED IN THE NEXT PAGE.** ☞ ☞ ☞

Table 8.1. CONTINUED FROM THE PREVIOUS PAGE.

Polymer	n(exp)	N_{rot}	N_{ref}	n(fit)
Poly[1,1-(2-methyl propane) bis(4-phenyl)carbonate]	1.5702	9.0	0	1.5801
Poly(phenyl methacrylate)	1.5706	6.0	0	1.5653
Poly(o-cresyl methacrylate)	1.5707	7.0	0	1.5602
Poly(2,3-dibromopropyl methacrylate)	1.5739	8.0	0	1.5570
Poly[2,2-pentane bis(4-phenyl)carbonate]	1.5745	10.0	0	1.5745
Poly(ethylene terephthalate)	1.5750	7.0	0	1.5558
Poly[oxy(2,6-dimethyl-1,4-phenylene)]	1.5750	4.0	0	1.5939
Poly(vinyl benzoate)	1.5775	5.0	0	1.5721
Poly[2,2-propane bis{4-(2-methylphenyl)}carbonate]	1.5783	10.0	0	1.5787
Poly[1,1-butane bis(4-phenyl)carbonate]	1.5792	9.0	0	1.5789
Poly(1,2-diphenylethyl methacrylate)	1.5816	9.0	0	1.5810
Poly(o-chlorobenzyl methacrylate)	1.5823	6.0	-3	1.5743
Poly[2,2-butane bis(4-phenyl)carbonate]	1.5827	9.0	0	1.5800
Poly(m-nitrobenzyl methacrylate)	1.5845	8.0	0	1.5536
Bisphenol-A polycarbonate	1.5850	8.0	0	1.5870
Poly[N-(2-phenylethyl)methacrylamide]	1.5857	8.0	12	1.5737
Poly(1,1-cyclohexane bis{4-(2,6-dichlorophenyl)}carbonate]	1.5858	9.0	-12	1.6040
Poly(o-methyl styrene)	1.5874	4.0	0	1.5921
Poly[1,1-cyclohexane bis(4-phenyl)carbonate]	1.5900	9.0	0	1.5896
Poly[2,2-propane bis{4-(2-chlorophenyl)}carbonate]	1.5900	8.0	-6	1.5971
Polystyrene	1.5920	3.0	0	1.6037
Poly(o-methoxy styrene)	1.5932	5.0	0	1.5783
Poly(diphenylmethyl methacrylate)	1.5933	8.0	0	1.5867
Poly[1,1-ethane bis(4-phenyl)carbonate]	1.5937	7.0	0	1.5922
Poly(propylene sulfide)	1.5960	4.0	18	1.5852
Poly(p-bromophenyl methacrylate)	1.5964	6.0	0	1.5974
Poly(N-benzyl methacrylamide)	1.5965	7.0	12	1.5821
Poly(p-methoxy styrene)	1.5967	5.0	0	1.5785
Poly[1,1-cyclopentane bis(4-phenyl)carbonate]	1.5993	8.5	0	1.5903
Poly(vinylidene chloride)	1.6000	2.0	0	1.6080
Poly[2,2-propane bis{4-(2,6-dichlorophenyl)}carbonate]	1.6056	8.0	-12	1.6037
Poly(pentachlorophenyl methacrylate)	1.6080	6.0	-15	1.5997
Poly(o-chloro styrene)	1.6098	3.0	-3	1.6125

☞ ☞ ☞   **TABLE 8.1 IS CONTINUED IN THE NEXT PAGE.** ☞ ☞ ☞

Table 8.1. CONTINUED FROM THE PREVIOUS PAGE.

Polymer	n(exp)	N_{rot}	N_{ref}	n(fit)
Arylef™ U100 *(equimolar random copolymer of bisphenol-A*				
terephthalate and bisphenol-A isophthalate)	1.6100	10.0	0	1.5955
Poly(phenyl α-bromoacrylate)	1.6120	5.0	0	1.6054
Poly[1,1-(1-phenylethane) *bis*(4-phenyl)carbonate]	1.6130	8.0	0	1.6066
Poly[2,2-propane *bis*{4-(2,6-dibromophenyl)}carbonate]	1.6147	8.0	0	1.6555
Poly(N-vinyl phthalimide)	1.6200	3.0	18	1.6520
Poly(2,6-dichlorostyrene)	1.6248	3.0	-6	1.6196
Poly(chloro-*p*-xylylene)	1.6290	3.0	-3	1.6145
Poly(β-naphthyl methacrylate)	1.6298	6.0	18	1.6244
Poly(α-naphthyl carbinyl methacrylate)	1.6300	7.0	18	1.6160
Poly(phenylmethylsilane)	1.6300	3.0	32	1.6271
Poly[4,4'-isopropylidene diphenoxy di(4-phenylene)sulfone]	1.6330	10.0	18	1.6230
Poly(2-vinylthiophene)	1.6376	3.0	18	1.6651
Poly[oxy(2,6-diphenyl-1,4-phenylene)]	1.6400	4.0	0	1.6397
Poly(α naphthyl methacrylate)	1.6410	6.0	18	1.6243
Poly[4,4'-sulfone diphenoxy di(4-phenylene)sulfone]	1.6500	8.0	36	1.6402
Poly[diphenylmethane *bis*(4-phenyl)carbonate]	1.6539	8.0	0	1.6194
Poly(styrene sulfide)	1.6568	4.0	18	1.6397
Poly(*p*-xylylene)	1.6690	3.0	0	1.6060
Poly(α-vinyl naphthalene)	1.6818	3.0	18	1.6696
Poly(N-vinyl carbazole)	1.6830	3.0	27	1.6893
Poly(pentabromophenyl methacrylate)	1.7100	6.0	0	1.6874

The dataset used in developing the correlation for n(298K) did not include any polyimides. See Rich *et al* [20] for a recent study of the refractive indices of aromatic polyimides. These authors showed that the decrease of the refractive indices of polyimides with increasing fluorine content is predicted successfully, while the measured refractive indices of the non-fluorinated polyimides systematically exceed the predicted values. They explained the disparities in terms of crystallinity and interchain interactions, both of which are expected to be more prevalent in the non-fluorinated polyimides. In a future statistical analysis on an extended dataset including the polyimides, a refinement similar to the term accounting for hydrogen bonding can be made to Equation 8.12 by introducing a term with a positive coefficient proportional to the number of imide rings to account for the effect of the strong interchain interactions in non-fluorinated polyimides.

8.D.   Example of Application: Specific Refractive Index Increments of Solutions

Light scattering experiments in dilute solution are often used to determine the *weight-average molecular weights* (M_w) of polymers. See van Krevelen [5] for a review of the fundamentals of light scattering. It suffices, for our purposes, to recall that optical inhomogeneities scatter light. A polymer dissolved in a solvent is an optical inhomogeneity. Some of the light scattering in solutions therefore arises from the fluctuations of refractive index caused by fluctuations in composition. The rate of change of the refractive index n of a dilute solution with changing concentration c of the polymer plays an important role in the extraction of information from light scattering experiments. For such dilute solutions, this *specific refractive index increment* (dn/dc) is a constant whose value only depends on the polymer, solvent, and temperature:

$$\frac{dn}{dc} \approx \frac{(n_P - n_S)}{\rho_P} \tag{8.14}$$

In Equation 8.14, n_P, ρ_P and n_S are the refractive index and the density of the polymer, and the refractive index of the solvent, respectively. Since $1/\rho_p$ is the specific volume of the polymer (see equations 3.1 and 3.2) in cm^3/g, Equation 8.14 indicates that the rate of change of the refractive index of a dilute solution with increasing polymer concentration near the limit of c→0 is proportional to the product of the difference between the refractive indices of the polymer and solvent with the sizes of the optical inhomogeneities (dissolved polymer chains).

The ability to estimate (dn/dc) with reasonable accuracy without performing detailed measurements of n as a function of c can simplify the determination of M_w by light scattering experiments, and is obviously a very useful application of any new correlation for n_P. In addition, since the difference ($n_P - n_S$) is much smaller than either n_P or n_S, the comparison of calculated and experimental values of (dn/dc) for various polymer-solvent combinations is an especially stringent test of the quality of a new correlation for n_P.

The refractive indices of a vast number of solvents, as well as the specific refractive index increments of a vast number of polymer-solvent combinations, have been measured and listed in standard reference publications. The (dn/dc) values of 100 polymer-solvent combinations, involving a diverse set of 16 polymers, will now be calculated at room temperature, and compared with the experimental values.

Most of the experimental (dn/dc) values were taken from the *Polymer Handbook* [21], with some (dn/dc) values being taken from van Krevelen's book [5].

Equation 8.14 was used to calculate (dn/dc) at room temperature. Values provided in the *Polymer Handbook* [22] were used for the n_S of most solvents. Other references [5,23,24] were used if n_S was not listed in the *Polymer Handbook*. For solvents which have two or more isomers of slightly different n_S, the n_S of the most common isomer was used in the calculations if the references for (dn/dc) did not indicate which isomer was used in the measurements.

The conformations of polymer chains in dilute solutions under theta conditions are essentially the same as the random coil conformations of chains in amorphous polymers [25], where the interactions of polymer chains with solvent molecules are replaced by interactions between the polymer chains. The density and the refractive index in the amorphous limit of a polymer are therefore the appropriate values of ρ_P and n_P to use in calculations of the specific refractive index increment via Equation 8.14. The correlation developed for V(298K) in Section 3.C was hence used to calculate ρ_P, and the correlation developed in Section 8.C was used to calculate n_P.

The results of the calculations of (dn/dc) for 100 polymer/solvent combinations at or near room temperature are listed in Table 8.2 and depicted in Figure 8.2. The calculated values have a standard deviation of 0.0158 relative to, and a correlation coefficient of 0.9796 with, the experimental values measured at room temperature and standard wavelengths of light. Equation 8.14 can thus be combined with refractive indices predicted by using Equation 8.13, and molar volumes predicted by using Equation 3.13, to obtain reliable predictions of the specific refractive index increment at room temperature without being limited by the lack of group contributions.

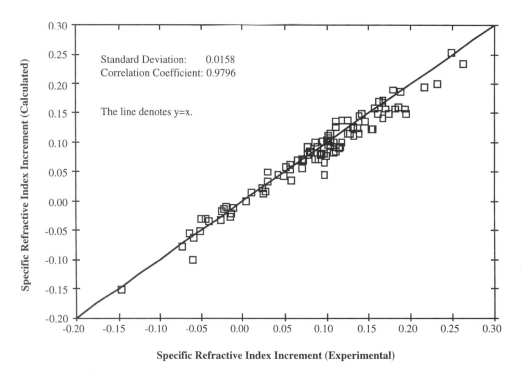

Figure 8.2. Comparison of the experimental values of the specific refractive index increment measured at or near room temperature using standard frequencies of light with calculated values, for 100 polymer-solvent combinations.

Table 8.2. Comparison of experimental specific refractive index increments measured for 100 polymer-solvent combinations (16 different polymers) at or near 298K using standard frequencies of light with calculated values. The calculated refractive indices n_P and densities ρ_P of the polymers, and experimental refractive indices n_S of the solvents, are also listed.

Polymer	Solvent	dn/dc(exp)	n_S	dn/dc(calc)
Poly(propylene oxide)	Benzene	-0.045	1.498	-0.030
(n_P=1.467, ρ_P=1.02)	Chlorobenzene	-0.064	1.523	-0.055
	n-Hexane	0.078	1.372	0.093
	2-Methylheptane	0.066	1.397	0.069
	Methanol	0.118	1.326	0.138
Polypropylene	Benzene	-0.050	1.498	-0.031
(n_P=1.471, ρ_P=0.86)	Bromobenzene	-0.060	1.557	-0.100
	n-Butyl chloride	0.108	1.400	0.083
	n-Butyl propionate	0.110	1.398	0.085
	Carbon tetrachloride	0.011	1.459	0.014
	Cyclohexane	0.057	1.424	0.055
Polystyrene	Benzene	0.108	1.498	0.099
(n_P=1.604, ρ_P=1.07)	Bromobenzene	0.042	1.557	0.044
	Bromonaphthalene	-0.051	1.660	-0.052
	Carbon tetrachloride	0.146	1.459	0.136
	Chlorobenzene	0.099	1.523	0.076
	Chloroform	0.195	1.444	0.150
	Cyclohexane	0.167	1.424	0.168
	p-Chlorotoluene	0.093	1.521	0.078
	Cis-decalin	0.128	1.481	0.115
	1,2-Dichloroethane	0.161	1.444	0.150
	p-Dioxane	0.168	1.420	0.172
	Tetrahydrofuran	0.189	1.404	0.187
	Toluene	0.104	1.494	0.103
Poly(vinyl chloride)	Acetone	0.138	1.357	0.144
(n_P=1.556, ρ_P=1.38)	Cyclohexanone	0.078	1.448	0.078
	p-Dioxane	0.086	1.420	0.099
	Tetrahydrofuran	0.102	1.404	0.110
Poly(α-vinyl naphthalene)	Benzene	0.175	1.498	0.150
(n_P=1.670, ρ_P=1.15)	Cis-1,2-dichloroethylene	0.217	1.445	0.196
	Ethylene glycol dimethyl ether	0.248	1.380	0.252

☞ ☞ ☞ **TABLE 8.2 IS CONTINUED IN THE NEXT PAGE.** ☞ ☞ ☞

Table 8.2. CONTINUED FROM THE PREVIOUS PAGE.

Polymer	Solvent	dn/dc(exp)	n_S	dn/dc(calc)
Poly(N-vinyl carbazole)	Benzene	0.193	1.498	0.157
(n_P=1.689, ρ_P=1.22)	Chloroform	0.232	1.444	0.201
	1,1,2,2-Tetrachlorooethane	0.185	1.493	0.161
	Tetrahydrofuran	0.262	1.404	0.234
Poly(vinyl cyclohexane)	Cyclohexane	0.101	1.424	0.103
(n_P=1.524, ρ_P=0.97)	n-Hexane	0.170	1.372	0.157
	Tetrahydrofuran	0.131	1.404	0.124
Poly(vinyl acetate)	Acetone	0.105	1.357	0.102
(n_P=1.481, ρ_P=1.22)	Acetonitrile	0.104	1.342	0.114
	Benzene	-0.023	1.498	-0.014
	Chlorobenzene	-0.040	1.523	-0.034
	p-Dioxane	0.030	1.420	0.050
	n-Butyl acetate	0.072	1.392	0.073
	Ethyl formate	0.095	1.358	0.101
	Methanol	0.132	1.326	0.127
	2-Butanone	0.080	1.377	0.085
	Tetrahydrofuran	0.058	1.404	0.063
	Toluene	-0.020	1.494	-0.011
Poly(n-butyl methacrylate)	Acetone	0.124	1.357	0.114
(n_P=1.476, ρ_P=1.04)	Benzene	-0.014	1.498	-0.021
	Bromobenzene	-0.073	1.557	-0.078
	Carbon tetrachloride	0.027	1.459	0.016
	Isopropanol	0.102	1.375	0.097
	2-Butanone	0.104	1.377	0.095
	Toluene	-0.024	1.494	-0.017
Poly(methyl methacrylate)	Acetone	0.132	1.357	0.110
(n_P=1.485, ρ_P=1.16)	Acetonitrile	0.137	1.342	0.123
	Benzene	-0.010	1.498	-0.011
	Bromobenzene	-0.058	1.557	-0.062
	1-Bromonaphthalene	-0.147	1.660	-0.151
	2-Butanone	0.112	1.377	0.093
	n-Butyl acetate	0.097	1.392	0.080
	n-Butyl chloride	0.090	1.400	0.073

☞ ☞ ☞   TABLE 8.2 IS CONTINUED IN THE NEXT PAGE. ☞ ☞ ☞

Table 8.2. CONTINUED FROM THE PREVIOUS PAGE.

Polymer	Solvent	dn/dc(exp)	n_S	dn/dc(calc)
Poly(methyl methacrylate)	Carbon tetrachloride	0.023	1.459	0.022
(continued)	Chlorobenzene	-0.026	1.523	-0.033
	Chloroform	0.059	1.444	0.035
	p-Dioxane	0.071	1.420	0.056
	Ethyl acetate	0.118	1.370	0.099
	Isoamyl acetate	0.091	1.403	0.071
	Nitroethane	0.094	1.390	0.082
	1,1,2-Trichloroethane	0.025	1.471	0.012
	Tetrahydrofuran	0.087	1.404	0.070
Poly(vinyl alcohol)				
(n_P=1.524, ρ_P=1.20)	Water	0.157	1.333	0.159
Poly(hexamethylene adipamide)	m-Cresol	-0.016	1.542	-0.027
(n_P=1.513, ρ_P=1.08)	Dichloroacetic acid	0.098	1.466	0.044
	95% Sulfuric acid	0.082	1.422	0.084
Poly[oxy(2,6-dimethyl-				
1,4-phenylene)]	Benzene	0.114	1.498	0.089
(n_P=1.594, ρ_P=1.08)	Chlorobenzene	0.098	1.523	0.066
	Chloroform	0.124	1.444	0.139
	Toluene	0.116	1.494	0.093
	1,1,2-Trichloroethane	0.139	1.471	0.114
	p-Xylene	0.114	1.493	0.094
Bisphenol-A polycarbonate	Bromoform	0.004	1.587	0.000
(n_P=1.587, ρ_P=1.17)	Chloroform	0.155	1.444	0.122
	1,2-Dibromoethane	0.050	1.538	0.042
	p-Dioxane	0.168	1.420	0.143
	1,2-Dichloroethane	0.154	1.444	0.122
	Pyridine	0.073	1.507	0.068
	1,1,2,2-Tetrachloroethane	0.094	1.493	0.080
	Tetrahydrofuran	0.182	1.404	0.156
Cellulose (n_P=1.551, ρ_P=1.45)	Acetone	0.111	1.357	0.134
Poly[oxy(methylphenylsilylene)]	Acetone	0.179	1.357	0.190
(n_P=1.558, ρ_P=1.06)	Benzene	0.053	1.498	0.057
	n-Butyl chloride	0.142	1.400	0.149

☞ ☞ ☞   **TABLE 8.2 IS CONTINUED IN THE NEXT PAGE.** ☞ ☞ ☞

Table 8.2. CONTINUED FROM THE PREVIOUS PAGE.

Polymer	Solvent	dn/dc(exp)	n_S	dn/dc(calc)
Poly[oxy(methylphenylsilylene)]	Carbon tetrachloride	0.089	1.459	0.093
(continued)	Chlorobenzene	0.030	1.523	0.033
	Chloroform	0.102	1.444	0.108
	Cyclohexane	0.111	1.424	0.126
	2-Butanone	0.163	1.377	0.171

Dilute solution properties in water are very important for water-soluble polymers [26], not only in their characterization but also in determining their performance characteristics under actual use conditions. These polymers include the acrylamide polymers, namely polyacrylamide and its structural variants, whose range of applications encompasses their use as soil modifiers. The results of calculations [27] of the specific refractive index increments of random copolymers of acrylamide with N-benzyl methacrylamide and with N-methyl methacrylamide are shown in Figure 8.3. (See Section 17.E for the method used to calculate the properties of random copolymers.) The details of these calculations (not shown) indicate that: (a) the specific volumes of the homopolymers are in the order polyacrylamide < poly(N-benzyl methacrylamide) < poly(N-methyl methacrylamide), and (b) the refractive indices are in the order of polyacrylamide ≈ poly(N-methyl methacrylamide) << poly(N-benzyl methacrylamide). The results shown for the copolymers in Figure 8.3 arise from the multiplicative combination of these trends.

For polyacrylamide in water, the five reported observed values of dn/dc range from 0.149 to 0.187 [25]. Inspection of the point for a comonomer mole fraction of 0.0 in Figure 8.3 shows that the predicted value is within this range, very close to its low end. It was suggested [26] that the low end of the range may correspond to polymer samples which were not completely dry to begin with, and the high end to dry initial polymer samples, so that the high end may be preferable although three of the five observed values are near the low end.

The specific refractive index increment (dn/dc) can be predicted at temperatures other than room temperature by utilizing Equation 8.5 or Equation 8.6 with R_{LL} or R_{GD} values calculated as will be discussed in Section 8.E to estimate $n_P(T)$, and the equations listed in Section 3.D to estimate $\rho_P(T)$, for use in Equation 8.14. Since n_P and n_S both change in the same manner (i.e., they both decrease) with increasing temperature, small changes in the measurement temperature only have a minor effect on the value of (dn/dc).

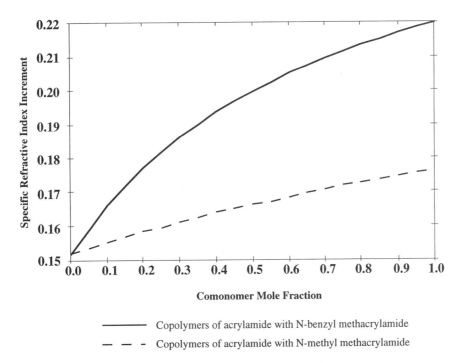

Figure 8.3. Calculated specific refractive index increments, in water at room temperature, of random copolymers of acrylamide with N-benzyl methacrylamide and N-methyl methacrylamide.

8.E. Calculation of the Molar Refraction

The correlation developed for n(298K) in Section 8.C did not require the prior calculation of the molar refraction; however, the calculation of the effects of crystallinity or temperature variation on the value of n via Equation 8.5 or Equation 8.6 still requires the estimation of the molar refraction. Furthermore, R_{LL} and R_{GD} quantify the intrinsic refractive power of a material, so that, in addition to their predictive value, their calculation can provide valuable physical insights.

R_{LL} and R_{GD} can both be calculated for a polymer if n and V are both known, or can both be predicted, at a given temperature and percent crystallinity. General correlations were derived for the amorphous molar volume at room temperature in Section 3.C, and for the amorphous refractive index at room temperature in Section 8.C. V(298K) and n(298K) values calculated by using these correlations can simply be substituted into equations 8.7 and 8.8, to estimate R_{LL} and R_{GD}, respectively, for polymers of interest.

The results of calculations of R_{LL} by using this procedure are summarized in Table 8.3 and depicted in Figure 8.4 for 100 polymers. The standard deviation between R_{LL} values calculated in this manner and R_{LL} values calculated for the same polymers by using group contributions originally developed for liquid organic molecules is only 1.0. This standard deviation is only 2.6% of the average R_{LL}(group) value of 38.9 for the same set of polymers. The correlation coefficient between the two sets of R_{LL} values is 0.9989.

These comparisons are not independent checks of the accuracy of the new correlations for V(298K) and n(298K). R_{LL} and R_{GD} values calculated from the correlations for V(298K) and n(298K), which were developed specifically for polymers, should describe the molar refraction of polymers better than R_{LL} and R_{GD} values calculated by group contributions developed for organic molecules. These calculations are, however, quite instructive since they show the transferability of the values of the molar refraction from liquid organic compounds to high polymers.

Finally, since the correlations developed for V(298K) and n(298K) in this book do not use group contributions, R_{LL} and R_{GD} can now be estimated for all polymers without requiring any tables of group contributions.

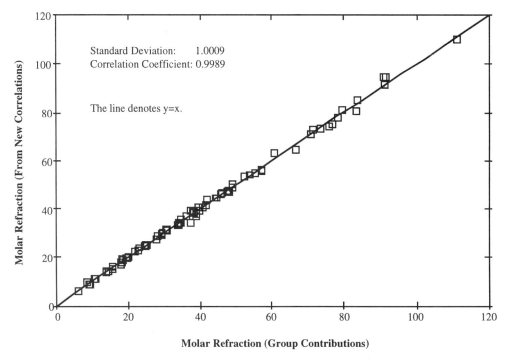

Figure 8.4. Comparison of molar refraction according to Lorentz and Lorenz (R_{LL}) calculated by using group contributions, with R_{LL} calculated from the new correlations developed in this book for the molar volume and refractive index of polymers at room temperature, for 100 polymers. R_{LL}(new) can be calculated for all polymers, without any need for group contributions.

Table 8.3. Comparison of the molar refraction according to Lorentz and Lorenz (R_{LL}) calculated in the usual manner by using group contributions for organic liquids, with values of R_{LL} calculated from the new correlations developed in this manuscript for the molar volume and the refractive index of polymers at room temperature, for 100 polymers. V(new) and n(new) are the molar volumes and refractive indices, respectively, calculated by using the new correlations. R_{LL}(new) can be calculated for all polymers, without any need for group contributions.

Polymer	R_{LL}(group)	V(new)	n(new)	R_{LL}(new)
Polyoxymethylene	6.290	23.0	1.4647	6.355
Polytetrafluoroethylene	8.634	48.7	1.3353	10.081
Poly(vinylidene fluoride)	9.025	39.9	1.3709	9.044
Polyethylene	9.298	32.2	1.4648	8.899
Poly(vinyl alcohol)	10.723	36.7	1.5239	11.228
Polyoxyethylene	10.939	39.1	1.4648	10.805
Polychlorotrifluoroethylene	13.759	57.5	1.4144	14.381
Polyacrylonitrile	13.793	45.1	1.5425	14.204
Polypropylene	13.909	49.0	1.4713	13.703
Poly(vinyl chloride)	14.288	45.2	1.5560	14.529
Poly(acrylic acid)	15.477	53.5	1.5182	16.216
Poly(vinyl methyl ether)	15.496	55.4	1.4704	15.468
Poly(propylene oxide)	15.550	56.8	1.4667	15.753
Poly(1,2-butadiene)	18.180	58.3	1.5055	17.308
Poly(1,4-butadiene)	18.180	59.1	1.5188	17.933
Polymethacrylonitrile	18.401	62.3	1.5325	19.321
Polyisobutylene	18.517	65.6	1.4795	18.617
Poly(1-butene)	18.558	65.6	1.4669	18.197
Poly(vinyl methyl ketone)	18.696	66.7	1.4843	19.093
Poly(vinylidene chloride)	19.275	57.4	1.6080	19.849
Poly(methyl acrylate)	20.146	69.3	1.4833	19.802
Poly(vinyl ethyl ether)	20.199	73.4	1.4641	20.259
Poly(vinyl acetate)	20.215	70.6	1.4814	20.106
Poly(propylene sulfide)	21.979	67.0	1.5852	22.461
Polyisoprene	22.788	76.6	1.5158	23.131
Polychloroprene	23.167	72.5	1.5673	23.694
Poly(methyl methacrylate)	24.754	86.4	1.4846	24.746
Poly(vinyl propionate)	24.864	87.7	1.4765	24.757
Poly(ethyl acrylate)	24.933	87.3	1.4770	24.666
Poly(methyl α-chloroacrylate)	25.039	82.3	1.5214	25.075

☞ ☞ ☞   **TABLE 8.3 IS CONTINUED IN THE NEXT PAGE.** ☞ ☞ ☞

Table 8.3. CONTINUED FROM THE PREVIOUS PAGE.

Polymer	R$_{LL}$(group)	V(new)	n(new)	R$_{LL}$(new)
Poly(vinyl chloroacetate)	25.265	84.2	1.5120	25.268
Poly(4-methyl-1-pentene)	27.818	98.1	1.4713	27.433
Poly(methyl α-bromoacrylate)	28.144	90.7	1.5534	29.042
Poly(ε-caprolactone)	29.451	103.6	1.4754	29.189
Poly(vinyl *sec*-butyl ether)	29.459	107.2	1.4655	29.661
Poly(vinyl isobutyl ether)	29.459	105.8	1.4687	29.448
Poly(vinyl *n*-butyl ether)	29.497	105.5	1.4643	29.129
Poly(ethyl methacrylate)	29.541	104.4	1.4789	29.601
Poly(ethyl α-chloroacrylate)	29.826	100.3	1.5111	30.051
Poly(ε-caprolactam)	30.475	105.0	1.5130	31.558
Poly(2-hydroxyethyl methacrylate)	30.928	106.6	1.4973	31.207
Poly(ethylene succinate)	31.008	110.7	1.4818	31.548
Polystyrene	33.679	97.0	1.6037	33.351
Poly(*n*-propyl methacrylate)	34.021	120.5	1.4774	34.070
Poly(*p* xylylcnc)	34.030	97.9	1.6060	33.763
Poly(isopropyl methacrylate)	34.051	121.6	1.4795	34.512
Poly(*n*-butyl acrylate)	34.062	119.5	1.4743	33.604
Poly(vinyl *n*-pentyl ether)	34.146	121.6	1.4644	33.579
Poly(2-chloroethyl methacrylate)	34.422	117.0	1.5050	34.704
Poly(oxy-2,2-dichloromethyltrimethylene)	34.907	115.3	1.5299	35.610
Poly(vinyl cyclohexane)	34.951	113.9	1.5237	34.833
Poly[oxy(2,6-dimethyl-1,4-phenylene)]	36.560	110.8	1.5939	37.592
Poly(ethylmercaptyl methacrylate)	37.222	111.4	1.5344	34.650
Poly(2-bromoethyl methacrylate)	37.274	129.0	1.5243	39.490
Poly(α-methyl styrene)	38.093	114.1	1.5914	38.580
Poly(*o*-methyl styrene)	38.359	114.7	1.5921	38.821
Poly(*o*-chloro styrene)	38.489	110.6	1.6125	38.472
Poly(*t*-butyl methacrylate)	38.591	138.3	1.4819	39.421
Poly(isobutyl methacrylate)	38.632	136.9	1.4792	38.832
Poly(*n*-butyl methacrylate)	38.670	136.6	1.4761	38.536
Poly(vinyl *n*-hexyl ether)	38.795	137.7	1.4644	38.028
Polyoxyoctamethylene	38.833	135.6	1.4648	37.474
Poly(*sec*-butyl α-chloroacrylate)	38.917	134.1	1.5026	39.616

☞☞☞   **TABLE 8.3 IS CONTINUED IN THE NEXT PAGE.** ☞☞☞

Table 8.3. CONTINUED FROM THE PREVIOUS PAGE.

Polymer	R_{LL}(group)	V(new)	n(new)	R_{LL}(new)
Poly(8-aminocaprylic acid)	39.773	137.1	1.5034	40.555
Poly(vinyl benzoate)	40.081	119.4	1.5721	39.292
Poly(p-methoxy styrene)	40.613	122.1	1.5785	40.550
Poly(styrene sulfide)	41.749	114.9	1.6397	41.380
Poly(sec-butyl α-bromoacrylate)	42.022	142.4	1.5250	43.642
Poly(2-nitro-2-methylpropyl methacrylate)	44.258	154.5	1.4917	44.803
Poly(phenyl methacrylate)	44.589	136.7	1.5653	44.545
Poly(cyclohexyl methacrylate)	45.765	153.3	1.5123	46.022
Poly(cyclohexyl α-chloroacrylate)	46.050	149.3	1.5337	46.388
Poly(ethylene terephthalate)	47.748	145.9	1.5558	46.885
Poly(2-ethylbutyl methacrylate)	47.930	170.0	1.4749	47.854
Poly(n-hexyl methacrylate)	47.968	168.7	1.4742	47.431
Poly(vinyl 2-ethylhexyl ether)	48.055	171.2	1.4653	47.351
Poly(vinyl n-octyl ether)	48.093	169.9	1.4645	46.927
Poly(benzyl methacrylate)	49.089	151.4	1.5590	48.880
Poly(cyclohexyl α-bromoacrylate)	49.155	157.6	1.5524	50.388
Poly(p-t-butyl styrene)	52.421	165.1	1.5643	53.722
Poly(1-phenylethyl methacrylate)	53.753	169.6	1.5522	54.211
Poly(m-nitrobenzyl methacrylate)	55.241	170.9	1.5536	54.740
Poly(tetramethylene isophthalate)	57.043	178.1	1.5444	56.258
Poly(n-octyl methacrylate)	57.266	200.9	1.4729	56.348
Poly(vinyl n-decyl ether)	57.391	202.0	1.4646	55.798
Poly(hexamethylene adipamide)	60.950	209.9	1.5130	63.086
Poly(vinyl n-dodecyl ether)	66.689	234.2	1.4646	64.697
Poly(p-cyclohexylphenyl methacrylate)	70.795	220.0	1.5584	70.968
Bisphenol-A polycarbonate	71.378	216.5	1.5870	72.762
Poly(diphenylmethyl methacrylate)	73.619	217.5	1.5867	73.066
Poly(dodecyl methacrylate)	75.862	265.2	1.4711	74.140
Poly[oxy(2,6-diphenyl-1,4-phenylene)]	76.640	208.4	1.6397	75.056
Poly(1,2-diphenylethyl methacrylate)	78.119	233.1	1.5810	77.686
Poly(hexamethylene sebacamide)	79.546	274.2	1.5034	81.105
Poly(1-octadecene)	83.644	290.7	1.4653	80.408
Poly(2,2,2'-trimethylhexamethylene terephthalamide)	83.754	263.2	1.5578	84.828

☞ ☞ ☞   **TABLE 8.3 IS CONTINUED IN THE NEXT PAGE.** ☞ ☞ ☞

Table 8.3. CONTINUED FROM THE PREVIOUS PAGE.

Polymer	R_{LL}(group)	V(new)	n(new)	R_{LL}(new)
Poly[2,2-propane *bis*{4-(2,6-dimethylphenyl)}carbonate]	90.898	287.5	1.5718	94.571
Poly[1,1-(1-phenylethane) *bis*(4-phenyl)carbonate]	91.244	265.0	1.6066	91.462
Poly[2,2-propane *bis*{4-(2,6-dichlorophenyl)}carbonate]	91.418	271.2	1.6037	93.242
Poly[diphenylmethane *bis*(4-phenyl)carbonate]	111.110	313.5	1.6194	110.038

References and Notes for Chapter 8

1. H. A. Lorentz, *Wied. Ann. Phys.*, *9*, 641-665 (1880).

2. L. V. Lorenz, *Wied. Ann. Phys.*, *11*, 70-103 (1880).

3. J. H. Gladstone and T. P. Dale, *Transactions of the Royal Society (London)*, *A148*, 887-894 (1858).

4. J. C. Seferis, in *Polymer Handbook*, edited by J. Brandrup and E. H. Immergut, Wiley, New York, third edition (1989), VI/451-VI/461.

5. D. W. van Krevelen, *Properties of Polymers*, second edition, Elsevier, Amsterdam (1976). Chapter 10 of this book deals with optical properties. The refractive indices of a large number of polymers and solvents are listed in Table V and Table VI, respectively, of Part VII at the end of the book.

6. D. W. van Krevelen, in *Computational Modeling of Polymers*, edited by J. Bicerano, Marcel Dekker, New York (1992), Chapter 1.

7. D. J. Goedhart, *Communication Gel Permeation Chromatography International Seminar*, Monaco, October 12-15, 1969.

8. W. Groh, *Makromol. Chem.*, *189*, 2861-2874 (1988).

9. Y. Takezawa, S. Tanno, N. Taketani, S. Ohara and H. Asano, *J. Appl. Polym. Sci. 42*, 2811-2817 (1991).

10. Y. Takezawa, S. Tanno, N. Taketani, S. Ohara and H. Asano, *J. Appl. Polym. Sci.*, *42*, 3195-3203 (1991).

11. Y. Takezawa, N. Taketani, S. Tanno and S. Ohara, *J. Polym. Sci., Polym. Phys. Ed.*, *30*, 879-885 (1992).

12. Y. Takezawa, N. Taketani, S. Tanno and S. Ohara, *J. Appl. Polym. Sci.*, *46*, 1835-1841 (1992).

13. Y. Takezawa, N. Taketani, S. Tanno and S. Ohara, *J. Appl. Polym. Sci.*, *46*, 2033-2037 (1992).

14. Y. Takezawa and S. Ohara, *J. Appl. Polym. Sci.*, *49*, 169-173 (1993).

15. N. B. Colthup, L. H. Daly and S. E. Wiberly, *Introduction to Infrared and Raman Spectroscopy*, second edition, Academic Press, New York (1975).

16. *Silicon Compounds: Register and Review*, Petrarch Systems, Silanes and Silicones Group, Bristol, Pennsylvania (1987).

17. R. West, *Pure & Applied Chemistry*, *54*, 1041-1050 (1982).

18. P. John, I. M. Odeh and J. Wood, *J. Chem. Soc., Chem. Commun.*, 1496-1497 (1983).

19. P. Trefonas, article titled *"Polysilanes and Polycarbosilanes"*, *Encyclopedia of Polymer Science and Engineering*, *13*, Wiley-Interscience, New York (1988).

20. D. C. Rich, P. Cebe and A. K. St. Clair, in *High-Temperature Properties and Applications of Polymeric Materials*, edited by M. R. Tant, J. W. Connell and H. L. N. McManus, *ACS Symposium Series*, *603*, American Chemical Society, Washington, D. C. (1995), 238-246.

21. M. B. Huglin, in *Polymer Handbook*, edited by J. Brandrup and E. H. Immergut, Wiley, New York, third edition (1989), VII/409-VII/484.

22. H.-G. Elias, in *Polymer Handbook*, edited by J. Brandrup and E. H. Immergut, Wiley, New York, third edition (1989), III/25-III/28.

23. *Lange's Handbook of Chemistry*, thirteenth edition, edited by J. A. Dean, Mc-Graw Hill, New York (1985).

24. *The Merck Index*, tenth edition, edited by M. Windholz, S. Budavari, R. F. Blumetti and E. S. Otterbein, Merck & Co., Inc., Rahway, New Jersey (1983).

25. P. J. Flory, *Pure & Applied Chemistry*, *56*, 305-312 (1984).

26. P. Molyneux, *CRC Handbook of Water-Soluble Synthetic Polymers: Properties and Behavior*, CRC Press, Boca Raton, Florida (1984).

27. J. Bicerano, *Soil Science*, *158*, 255-266 (1994).

CHAPTER 9

ELECTRICAL PROPERTIES

9.A. Background Information

The electrical properties of polymers are important in many applications [1]. The most widespread electrical application of polymers is the insulation of cables. In recent years, high-performance polymers have become important in the electronics industry, as encapsulants for electronic components, interlayer dielectrics, and printed wiring board materials. The *dielectric constant* (or *permittivity*) ε and the *dissipation factor* (or *power factor* or *electrical loss tangent*) tan δ, which are dimensionless quantities, are the key electrical properties of polymers.

The *dielectric constant* ε is a measure of the polarization of the medium between two charges when this medium is subjected to an electric field. A larger value of ε implies greater polarization of the medium between the two charges. The vacuum contains nothing that could be polarized, and therefore has ε=1. All materials have ε>1. The dielectric constant of a nonconducting material is generally defined as the ratio of the capacities of a parallel plate condenser with and without the material placed between the plates.

The dielectric constant of a polymer is a function of the following variables:

1. *Temperature* of measurement. The most common temperature at which the electrical properties are reported in standard reference tables is room temperature (298K±5K, with some variation).
2. *Rate (frequency)* of measurement. The most common two frequencies at which the electrical properties are reported in standard reference tables are one kilohertz (1 kHz = 1000 Hertz) and one megahertz (1 MHz = 1000000 Hertz).
3. *Structure and composition* of polymer, and especially the presence of any polar groups.
4. *Morphology* of specimens, and especially any *crystallinity* and/or *orientation*.
5. *Impurities, fillers, plasticizers, other additives*, and *moisture* (water molecules) in the polymer.

The temperature and frequency dependences of ε depend drastically upon whether a polymer is polar or nonpolar, i.e., whether it contains permanent dipoles or does not contain them.

Nonpolar polymers (such as the hydrocarbon polymers which by definition contain only carbon and hydrogen atoms) have low dielectric constants which change very little with temperature or frequency. For a nonpolar material, ε equals the square of the refractive index n:

$$\varepsilon(\text{nonpolar insulator}) = n^2 \qquad (9.1)$$

The small temperature dependence of the dielectric constant of a nonpolar insulator can generally be approximated in terms of the molar polarization according to Lorentz and Lorenz

240

(P_{LL}, in units of cc/mole), by Equation 9.2. Comparison of equations 9.1, 9.2 and 8.1 shows that Equation 9.3 holds for nonpolar materials.

$$\varepsilon(T, \text{nonpolar insulator}) \approx \frac{[V(T) + 2P_{LL}]}{[V(T) - P_{LL}]} \tag{9.2}$$

$$P_{LL}(\text{nonpolar insulator}) = R_{LL} \tag{9.3}$$

Polar materials behave very differently from nonpolar materials. They have $\varepsilon > n^2$ and $P_{LL} > R_{LL}$, with ($\varepsilon - n^2$) and ($P_{LL} - R_{LL}$) both increasing with increasing amount of polarity. The temperature dependence of the dielectric constant of a polar polymer is not controlled by volumetric effects. It is, instead, determined by the ability of the permanent dipoles in the material to align in the direction of an applied electric field. Equation 9.2 therefore *cannot* be used to calculate this temperature dependence.

The dielectric constant of a polar solid typically increases with increasing temperature. The permanent dipoles become able to align more easily with the applied electric field as molecular mobility increases. When the temperature increases above the glass transition temperature of an amorphous material or the melting temperature of a crystalline material, ε typically peaks out and decreases with further increase of temperature. This type of behavior observed in the "liquid" state occurs because the energies and thus the amplitudes of the random thermal motions increase, making it more difficult for the dipoles to align in the direction of the applied electric field.

The frequency dependence of the ε of a polar polymer is also strong. A peak is often observed between 1 Hz and 1 MHz in measurements made at room temperature. The precise location and width of this peak depends on the detailed nature of the atomic motions taking place, and the resulting relaxation times.

The dielectric constants of polar polymers are also extremely sensitive to moisture. Most polar polymers, and especially polymers containing hydrogen-bonding moieties, are susceptible to moisture. Water molecules can become incorporated into them upon exposure to humidity. Water has $\varepsilon \approx 80.4$ at room temperature. The incorporation of even a small amount of moisture can therefore drastically increase ε. In addition, ε is often sensitive to the presence of even small quantities of many common plasticizers, stabilizers and other additives. Very different values are therefore often quoted for the ε of the same polar polymer. These values may all be "correct", in the sense that, even though they may not describe the "intrinsic properties" of the polymer, they describe the properties of the specimen being tested.

The same types of electrical forces which determine the values of the molar polarization and the dielectric constant also determine the cohesive energy density [2]. An approximate correlation, given by Equation 9.4, has thus been found [3] between ε at room temperature and the solubility parameter δ for polymers. This equation can be used to provide a rough estimate of the value of the dielectric constant *at room temperature*.

$$\varepsilon(298K) \approx \frac{\delta(298K)}{7} \tag{9.4}$$

Equation 9.5, which is obtained by setting T=298K in Equation 9.2, is another relationship commonly used to estimate ε at room temperature. Group contributions for P_{LL} [3], if available, can be used to estimate P_{LL} for substitution into Equation 9.5. As discussed above, the complicated temperature dependence of the dielectric constants of polar polymers precludes the use of Equation 9.2 itself to predict the temperature dependence.

$$\varepsilon(298K) \approx \frac{[V(298K) + 2P_{LL}]}{[V(298K) - P_{LL}]} \tag{9.5}$$

If ε(298K) and V(298K) are both known, Equation 9.6 which is obtained by inverting Equation 9.5 and expressing P_{LL} in terms of ε(298K) and V(298K) can be used to calculate P_{LL}:

$$P_{LL} = V(298K)\frac{[\varepsilon(298K) - 1]}{[\varepsilon(298K) + 2]} \tag{9.6}$$

The dielectric constant can be used to estimate many other electrical properties of a polymer. For example, let R(298K) denote the volume resistivity in ohm·cm measured at room temperature, and let $\log_{10}$ denote the logarithm to the base ten. As shown by Equation 9.7, the electrical resistance of polymers decreases exponentially, so that the electrical conductivity increases exponentially, with increasing ε(298K) [3].

$$\log_{10}[R(298K)] \approx 23 - 2 \cdot \varepsilon(298K) \tag{9.7}$$

Capacitive interference, the signal delay caused by the permittivity of insulating layers, the minimum required interlayer dielectric thicknesses, and the power dissipation, can all be decreased by designing polymers with lower values of ε. The synthesis of polymers which simultaneously have low values of ε and outstanding thermal and mechanical properties is therefore crucial for applications in the electronics industry.

A new correlation will be developed for ε(298K) in Section 9.B.

In Section 9.C, P_{LL} values calculated for a set of polymers via Equation 9.6, by using the new correlation for ε(298K) and the predicted values of V(298K), will be compared with P_{LL} values calculated by using the observed values of ε(298K) and V(298K).

The values of $(\varepsilon - n^2)$ and $(P_{LL} - R_{LL})$ both quantify the polarity of a material, although each quantity provides the optimum description of the polarity for different purposes. For example, the *effective dipole moment* μ of a solid polymer in Debye units at room temperature can be very roughly estimated [3] by using the following equation:

$$\mu(298K) \approx 0.220326\sqrt{(P_{LL} - R_{LL})} \tag{9.8}$$

A polar group is equally polar whether it is located on a polymer chain dissolved in a solvent, or on a polymer chain in a solid polymer. On the other hand, the constraints of the solid phase cause a drastic reduction of the mobilities (i.e., the degrees of freedom) of the structural units in polymer chains, compared with the same structural units in either polymer chains in dilute solutions or simple molecular liquids. Consequently, the *effective* dipole moments of solid polymers *appear to be* much smaller than the μ's of the same polymers measured [4] in dilute solutions or the μ's of simple molecular liquids containing the same types of polar groups. The effective dipole moments of polymers, calculated via Equation 9.8 from the "experimental" and predicted values of P_{LL} and R_{LL}, will be compared in Section 9.D.

The dielectric constant has, thus far, been treated as a single number. Such a treatment is equivalent to assuming that ε is a static property of a material. Some of the energy of an applied electric field is, however, dissipated. Dissipation occurs when energy is lost to the "internal motions" of the material, which are defined as motions of the atoms from which the material is built [5,6]. This "lossy" component of the response of a material to an electric field is usually expressed in terms of the imaginary component of the complex quantity ε^*:

$$\varepsilon^* \equiv \varepsilon' - i \cdot \varepsilon'' \tag{9.9}$$

The real component ε' of ε^* has the same significance as the ordinary dielectric constant ε discussed above. It quantifies the energy from the applied alternating electric field which is elastically stored in the material during each cycle, to be returned to the electric field at the end of the cycle. The imaginary component ε'' of ε^* quantifies the energy lost to the internal motions. The *dissipation factor* $\tan \delta$ is defined as follows:

$$\tan \delta \equiv \frac{\varepsilon''}{\varepsilon'} \tag{9.10}$$

The frequency and temperature dependences of ε'' and $\tan \delta$ resemble the behavior of ε' in many respects. These quantities are also small for nonpolar polymers, increase with increasing polarity, and manifest peaks at certain combinations of temperature and frequency. The detailed analysis of the frequency and temperature dependences of ε', ε'' and $\tan \delta$ requires the consideration of *relaxation times*. Many useful general functional forms have been developed empirically and fitted to experimental data [7-13]. On the other hand, the *a priori* prediction of ε', ε'' and tan δ as functions of the temperature and frequency prior to running any experiments is not possible at this time.

The dissipation factor measured at room temperature, in the frequency range of 50 Hz to 10 MHz, will be considered in Section 9.E. It will be shown that tan δ correlates, to within one order of magnitude, with the difference $(\varepsilon - n^2)$. The use of an additional factor related to the rotational degrees of freedom will be shown to improve the quality of the correlation at the lowest frequencies (50 Hz to 100 Hz) in this range.

The dielectric strength is an important property which cannot yet be predicted in a quantitative manner from the polymeric structure. Dielectric strength will be discussed briefly in Section 9.F.

See Section 8.A for additional background information on the electrical properties, since these properties are intimately related to the optical properties which also describe the behavior of a material in its interactions with electromagnetic fields.

9.B. Correlation for the Dielectric Constant at Room Temperature

A dataset containing the dielectric constants of 61 polymers measured at room temperature was prepared by careful comparison and combination of the data provided by many sources [3,14-20]. For polar polymers, special care was exercised to select values of ε which represented, whenever possible, the "intrinsic" properties of the polymers, rather than the effects of the additives and fillers used. It will be seen in Section 9.D that the dielectric constants of typical commercial grades of many polar polymers, which contain significant amounts of additives, are considerably higher than the "intrinsic" values used in the present section.

A new correlation similar to Equation 9.4 was sought, mindful of the fact that there are important differences as well as similarities between the physical factors which determine the solubility parameter and the dielectric constant of a polymer. The following observations were made as a result of the analysis of the dataset:

1. The cohesive energy E_{coh1} (Section 5.B) gave a better correlation than E_{coh2} (Section 5.C).

2. The cohesive energy density E_{coh1}/V resulted in a slight improvement over $\delta = \sqrt{(E_{coh1}/V)}$.

3. The van der Waals volume V_w gave slightly improved results compared to the molar volume V.

4. The use of $(c \cdot E_{coh1} + N_{dc})$ instead of E_{coh1} by itself resulted in a major further improvement. N_{dc} is defined by Equation 9.11, and c is a fitting parameter. $(E_{coh1} + N_{dc}/c)$ is essentially an "effective cohesive energy" which emphasizes those structural features which contribute more significantly to ε than to E_{coh1}. All terms in N_{dc} are either structural parameters or atomic correction terms as defined in Section 2.C, so that N_{dc} does not contain group correction terms.

$$N_{dc} \equiv 19N_N + 7N_{(backbone\ O,\ -S-)} + 12N_{(side\ group\ O,\ -S-)} + 52N_{sulfone} - 2N_F + 8N_{ClBrasym}$$
$$+ 20N_{Si} - 14N_{cyc} \qquad (9.11)$$

The final correlation is given by Equation 9.12. It has a standard deviation of 0.0871 and a correlation coefficient of 0.9788. The standard deviation corresponds to 3% of the average experimental $\varepsilon(298K)$ value of 2.881 for the polymers in the dataset. This correlation has been used extensively and successfully in our practical work, and therefore has a greater reliability than is apparent from the size of the dataset initially used in developing it.

$$\varepsilon(298K) \approx 1.412014 + \frac{(0.001887E_{coh1} + N_{dc})}{V_w} \qquad (9.12)$$

In Equation 9.11, N_N is the number of nitrogen atoms in the repeat unit, regardless of the bonding environment. The $19N_N$ term corrects for the nearly uniform underestimation of the contribution of all moieties containing nitrogen atoms in the polymers in our dataset to $\varepsilon(298K)$ when E_{coh1}/V_w is used as the main fitting variable. Differences between various bonding environments containing nitrogen atoms are accounted for by E_{coh1} itself.

The second and third terms refer to the total numbers of oxygen and divalent sulfur atoms, regardless of bonding environment. They distinguish between such atoms in the chain backbone and in side groups as defined in Section 2.D. It is often more difficult for polar groups to align in the direction of an applied electric field when they are incorporated into and "tied down" by the backbone than when they are in side groups. Polar groups therefore often make smaller contributions to ε when they are in the chain backbone than when they are pendant to the chain.

$N_{ClBrasym}$ is the number of chlorine and/or bromine atoms which are *not* symmetrically attached to the same backbone atom. The symmetry of the substitution on backbone atoms affects the value of ε. There is considerable mutual cancellation of dipole moment vectors when substitution is symmetrical, resulting in smaller net dipole moment vectors of structural units consisting of the backbone carbon atoms and their substituents. This fact explains why the term proportional to $N_{(Cl\ or\ Br)}$ is only added if Cl and/or Br atoms are not attached symmetrically to the same backbone atom. See Figure 9.1 for three examples, explained in the next paragraph.

The contribution to N_{dc} of the Cl atom in poly(vinyl chloride), where a backbone carbon atom is asymmetrically substituted by a Cl atom, is 8. The contribution to N_{dc} of the two Cl atoms symmetrically substituted to the same backbone carbon atom in poly(vinylidene chloride) is zero. Since the two Cl atoms are not attached to the backbone atom but located on pendant chloromethyl groups, the contribution of the Cl atoms in poly(oxy-2,2-dichloromethyltrimethylene) is 16 despite the fact that the backbone carbon atom is symmetrically substituted.

The results of the calculations of $\varepsilon(298K)$ are listed in Table 9.1 and shown in Figure 9.2.

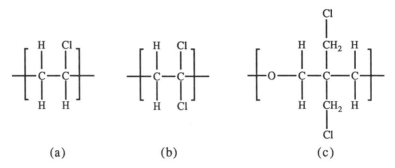

Figure 9.1. The structures of three of the chlorine-containing polymers considered in developing the correlation for the dielectric constant ε at room temperature. (a) Poly(vinyl chloride). (b) Poly(vinylidene chloride). (c) Poly(oxy-2,2-dichloromethyltrimethylene).

The physical significance of the terms in Equation 9.11 can be summarized as follows:

1. Heteroatoms usually found in highly polar moieties, such as nitrogen, oxygen, sulfur and chlorine, are more important in determining the dielectric constant than in determining E_{coh1}.

2. The small and highly electronegative fluorine atoms, which hold their electrons very tightly and thus contribute relatively little to E_{coh1}, generally contribute even less to $\varepsilon(298K)$. There are, however, some important and very dramatic exceptions to this rule, as discussed briefly below.

3. The two observations summarized above indicate that the *polar component* of E_{coh1} (Chapter 5) is weighted more heavily in the determination of $\varepsilon(298K)$ than in the value of E_{coh1} itself.

4. The often greater mobility of polar units in side groups than in backbones is accounted for.

5. The effects of the symmetry of substitution are more important in determining $\varepsilon(298K)$ than in determining E_{coh1}/V_w, and they are accounted for.

The E_{coh1} values listed in Table 9.1 were calculated by using equations 5.8-5.10. V_w was calculated by using equations 3.10 and 3.11. Note that there are two copolymers in this dataset. The E_{coh1} and V_w of the random copolymer of bisphenol-A terephthalate and bisphenol-A isophthalate are identical to the E_{coh1} and V_w of the homopolymer of either repeat unit. On the other hand, the twelve weight percent residual poly(vinyl alcohol), which remains even in the most completely reacted grade of poly(vinyl butyral), affects $\varepsilon(298K)$ significantly because of the much greater cohesive energy density of poly(vinyl alcohol) and the tendency of its exposed hydroxyl groups to absorb moisture. The properties of random copolymers can be calculated by taking weighted averages, as discussed in Section 17.E where the calculation of $\varepsilon(298K)$ for poly(vinyl butyral) is used as the example.

Table 9.1. Experimental dielectric constants $\varepsilon(exp)$, the quantities (cohesive energy E_{coh1} in J/mole, correction index N_{dc}, and van der Waals volume V_w in cc/mole) used in the correlation for ε, and fitted dielectric constants $\varepsilon(fit)$, for 61 polymers. E_{coh1} is calculated from equations 5.8-5.10, and V_w is calculated from equations 3.10 and 3.11, for use in this correlation. Calculations on random copolymers are discussed in Section 17.E.

Polymer	$\varepsilon(exp)$	E_{coh1}	N_{dc}	V_w	$\varepsilon(fit)$
Polytetrafluoroethylene	2.10	16097	-8.0	32.96	2.09
Poly(4-methyl-1-pentene)	2.13	27549	0.0	60.97	2.26
Polypropylene	2.20	13774	0.0	30.48	2.26
Polyisobutylene	2.23	16870	0.0	39.33	2.22
Poly(vinyl cyclohexane)	2.25	39197	-14.0	74.85	2.21
Poly(1-butene)	2.27	19092	0.0	41.32	2.28
Polyethylene	2.30	9883	0.0	20.37	2.33
Poly($\alpha,\alpha,\alpha',\alpha'$-tetrafluoro-$p$-xylylene)	2.35	45353	-8.0	84.43	2.33
Polyisoprene	2.37	21863	0.0	47.71	2.28
Poly(o-methyl styrene)	2.49	43256	0.0	73.85	2.52
Poly(1,4-butadiene)	2.51	17972	0.0	37.49	2.32
Poly(β-vinyl naphthalene)	2.51	58630	0.0	87.14	2.68
Polystyrene	2.55	39197	0.0	64.04	2.57
Poly(α-methyl styrene)	2.57	42616	0.0	73.55	2.51
Poly(cyclohexyl methacrylate)	2.58	59978	10.0	99.86	2.65
Polychlorotrifluoroethylene	2.60	24706	2.0	39.44	2.64
Poly(α-vinyl naphthalene)	2.60	58796	0.0	87.25	2.68
Poly[oxy(2,6-dimethyl-1,4-phenylene)]	2.60	42207	7.0	68.58	2.68
Poly[1,1-cyclohexane bis(4-phenyl)carbonate]	2.60	109931	12.0	159.61	2.79
Poly(p-xylylene)	2.65	39031	0.0	64.25	2.56
Poly(p-chloro styrene)	2.65	49546	8.0	74.21	2.78
Poly(vinyl butyral) *[has 12 weight percent poly(vinyl alcohol)]*	2.69	41789	3.7	65.73	2.67
Ethyl cellulose	2.70	73180	36.0	136.89	2.68
Poly(isobutyl methacrylate)	2.70	48496	24.0	85.60	2.76
Arylef™ U100 *(equimolar random copolymer of bisphenol-A terephthalate and bisphenol-A isophthalate)*	2.73	134962	38.0	195.16	2.91
Poly(dimethyl siloxane)	2.75	16870	27.0	43.79	2.76
Poly[oxy(2,6-diphenyl-1,4-phenylene)]	2.80	93052	7.0	138.38	2.73
Poly(m-chloro styrene)	2.80	49546	8.0	74.21	2.78
Poly(n-butyl methacrylate)	2.82	49921	24.0	86.33	2.78

☞☞☞ **TABLE 9.1 IS CONTINUED IN THE NEXT PAGE.** ☞☞☞

Table 9.1. CONTINUED FROM THE PREVIOUS PAGE.

Polymer	ε(exp)	E_{coh1}	N_{dc}	V_w	ε(fit)
Poly(vinylidene chloride)	2.85	29784	0.0	38.89	2.86
Bisphenol-A polycarbonate	2.90	93060	26.0	135.05	2.90
Poly(N-vinyl carbazole)	2.90	73620	19.0	105.22	2.91
Poly[1,1-ethane *bis*(4-phenyl)carbonate]	2.90	89765	26.0	125.65	2.97
Poly(3,4-dichlorostyrene)	2.94	60062	16.0	84.49	2.94
Poly(chloro-*p*-xylylene)	2.95	49546	8.0	74.53	2.77
Poly(vinyl chloride)	2.95	20231	8.0	30.44	2.93
Poly(1,4-cyclohexylidene dimethylene terephthalate)	3.00	103236	24.0	153.89	2.83
Poly(ethyl methacrylate)	3.00	40039	24.0	65.96	2.92
Poly(oxy-2,2-dichloromethyltrimethylene)	3.00	48954	23.0	75.42	2.94
Poly(*p*-methoxy-*o*-chloro styrene)	3.08	58922	20.0	88.26	2.90
Poly(methyl methacrylate)	3.10	35097	24.0	54.27	3.07
Poly[thio(*p*-phenylene)]	3.10	42698	7.0	51.98	3.10
Polyoxymethylene	3.10	8089	7.0	13.13	3.11
Poly(tetramethylene terephthalate)	3.10	83970	38.0	115.42	3.11
Poly(ethyl α-chloroacrylate)	3.10	46495	32.0	65.73	3.23
Ultem 1000	3.15	242995	100.0	307.10	3.23
Poly[4,4'-isopropylidene diphenoxy di(4-phenylene)sulfone]	3.18	177223	90.0	237.49	3.20
Poly(ether ether ketone)	3.20	119048	26.0	153.89	3.04
Poly(hexamethylene sebacamide) *(measured dry)*	3.20	139771	62.0	181.22	3.21
Poly(vinyl acetate)	3.25	31136	24.0	45.42	3.23
Poly(ethylene terephthalate)	3.25	74087	38.0	95.04	3.28
Poly(*p*-hydroxybenzoate)	3.28	46677	19.0	60.33	3.19
Poly[2,2'-(*m*-phenylene)-5,5'-bibenzimidazole]	3.30	124100	76.0	158.28	3.37
Poly(methyl α-chloroacrylate)	3.40	41554	32.0	54.05	3.45
Poly[4,4'-diphenoxy di(4-phenylene)sulfone]	3.44	164930	90.0	204.30	3.38
Poly(ε-caprolactam) *(measured dry)*	3.50	60003	31.0	70.23	3.47
Poly(hexamethylene adipamide) *(measured dry)*	3.50	120006	62.0	140.47	3.47
Poly[N,N'-(*p,p*'-oxydiphenylene)pyromellitimide]	3.50	167297	93.0	187.08	3.60
Torlon 2000 *(basic unfilled grade, measured dry)*	3.70	165305	74.0	183.33	3.52
Poly[4,4'-sulfone diphenoxy di(4-phenylene)sulfone]	3.80	203050	166.0	228.11	3.82
Polyacrylonitrile	4.00	34157	19.0	32.36	3.99

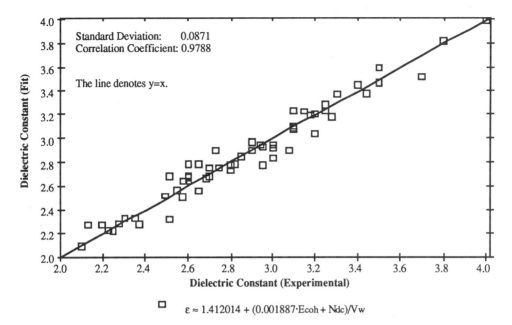

$$\varepsilon \approx 1.412014 + (0.001887 \cdot E_{coh} + N_{dc})/V_w$$

Figure 9.2. Correlation for the dielectric constants of 61 polymers measured at room temperature, in terms of the cohesive energy, the van der Waals volume, and a correction index.

Although fluorination normally reduces ε, asymmetric fluorination can sometimes create very large net dipole moment vectors, and result in polymers which have very large ε, and sometimes also have other unusual electrical properties which cannot be predicted by the existing structure-property relationships. Examples include poly(vinyl fluoride) ($\varepsilon \approx 6.8$ to 8.5 at 1 kHz [16], see Figure 2.3), poly(vinylidene fluoride) ($\varepsilon \approx 8.4$ at 60 Hz and 6.6 at 1 MHz [16], see Figure 9.3), and poly[oxy(methyl γ-trifluoropropylsilylene)] ($\varepsilon \approx 7.35$ [20], see Figure 9.3). The development of the ability to predict such special physical effects with accuracy is a remaining challenge.

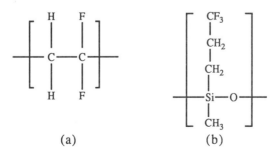

(a)   (b)

Figure 9.3. Polymers with very large dielectric constants caused by asymmetric fluorination. (a) Poly(vinylidene fluoride), where the fluorine substitution is symmetric at a given C atom, but asymmetric (alternating) along the chain. (b) Poly[oxy(methyl γ-trifluoropropylsilylene)].

9.C. Calculation of the Molar Polarization

An experimental value was available for the molar volume V(298K) at room temperature for 55 of the polymers listed in Table 9.1, enabling the calculation of an "experimental" value of P_{LL} by using Equation 9.6. The "fitted" values of P_{LL} were calculated by using ε(298K) and V(298K) obtained form the correlations in sections 9.B and 3.C, respectively. The P_{LL}(fit) values have a standard deviation of 2.2 relative to, and a correlation coefficient of 0.9988 with, the P_{LL}(exp) values. The standard deviation is 3.8% of the average P_{LL}(exp) of 57.2 for this set of polymers. The results of these calculations are listed in Table 9.2 and shown in Figure 9.4.

Table 9.2. Experimental molar volumes V(exp) at 298K in cc/mole, predicted molar volumes V(pred) at 298K calculated by using equations 3.13 and 3.14, and experimental and predicted values of the molar polarization P_{LL} in cc/mole, for 61 polymers. The calculations also utilize the experimental and fitted values of the dielectric constant, which are listed in Table 9.1. The V(exp) values listed for semicrystalline polymers are extrapolations to the amorphous limit. P_{LL}(exp) was not calculated for six polymers because V(exp) was not known.

Polymer	V(exp)	V(pred)	P_{LL}(exp)	P_{LL}(pred)
Polyoxymethylene	24.0	23.0	9.88	9.49
Polyethylene	33.1	32.2	10.01	9.88
Polytetrafluoroethylene	50.0	48.7	13.41	12.99
Polypropylene	49.5	49.0	14.14	14.53
Poly(vinyl chloride)	45.1	45.2	17.77	17.69
Poly(1-butene)	65.2	65.6	19.39	19.66
Polyisobutylene	66.8	65.6	19.42	18.98
Poly(1,4-butadiene)	60.7	59.1	20.32	18.03
Polychlorotrifluoroethylene	60.7	57.5	21.11	20.36
Poly(vinylidene chloride)	58.4	57.4	22.28	21.95
Polyacrylonitrile	44.8	45.1	22.40	22.52
Polyisoprene	75.2	76.6	23.58	22.87
Poly(4-methyl-1-pentene)	100.5	98.1	27.50	29.09
Poly(dimethyl siloxane)	75.6	79.6	27.85	29.39
Poly(vinyl acetate)	72.4	70.6	31.03	30.13
Poly[thio(p-phenylene)]	80.7	80.5	33.23	33.12
Polystyrene	99.1	97.0	33.76	33.28
Poly(vinyl cyclohexane)	116.0	113.9	34.12	32.80
Poly(p-xylylene)	96.8	97.9	34.35	33.47

☞ ☞ ☞   **TABLE 9.2 IS CONTINUED IN THE NEXT PAGE.** ☞ ☞ ☞

Table 9.2. CONTINUED FROM THE PREVIOUS PAGE.

Polymer	V(exp)	V(pred)	P_{LL}(exp)	P_{LL}(pred)
Poly(methyl methacrylate)	85.6	86.4	35.25	35.32
Poly($\alpha,\alpha,\alpha',\alpha'$-tetrafluoro-$p$-xylylene)	116.9	124.0	36.28	38.11
Poly(methyl α-chloroacrylate)	82.0	82.3	36.44	37.04
Poly(p-hydroxybenzoate)	86.9	91.1	37.53	38.41
Poly(α-methyl styrene)	111.0	114.1	38.13	38.12
Poly(o-methyl styrene)	115.1	114.7	38.20	38.53
Poly[oxy(2,6-dimethyl-1,4-phenylene)]	112.2	110.8	39.03	39.70
Poly(ethyl α-chloroacrylate)	96.8	100.3	39.86	42.81
Poly(ethyl methacrylate)	102.0	104.4	40.80	40.76
Poly(p-chloro styrene)	------	110.6	------	41.20
Poly(m-chloro styrene)	------	110.6	------	41.20
Poly(chloro-p-xylylene)	108.3	111.6	42.66	41.47
Poly(oxy-2,2-dichloromethyltrimethylene)	111.8	115.3	44.72	45.31
Poly(β-vinyl naphthalene)	140.0	134.3	46.87	48.24
Poly(vinyl butyral)	131.3	132.2	47.31	47.21
Poly(ϵ-caprolactam)	104.4	105.0	47.45	47.37
Poly(α-vinyl naphthalene)	137.7	134.3	47.90	48.28
Poly(3,4-dichlorostyrene)	------	124.3	------	48.86
Poly(isobutyl methacrylate)	136.1	136.9	49.23	50.64
Poly(n-butyl methacrylate)	134.8	136.6	50.90	50.89
Poly(o-chloro-p-methoxy styrene)	------	135.7	------	52.59
Poly(cyclohexyl methacrylate)	153.2	153.3	52.85	54.30
Poly(ethylene terephthalate)	144.0	145.9	61.71	63.05
Poly(N-vinyl carbazole)	161.0	158.0	62.43	61.52
Poly(tetramethylene terephthalate)	177.1	178.1	72.92	73.62
Ethyl cellulose	217.0	222.5	78.49	79.99
Poly[1,1-ethane bis(4-phenyl)carbonate]	------	199.5	------	79.01
Poly[oxy(2,6-diphenyl-1,4-phenylene)]	214.3	208.4	80.36	76.26
Bisphenol-A polycarbonate	211.9	216.5	82.17	84.08
Poly(1,4-cyclohexylidene dimethylene terephthalate)	229.3	227.5	91.72	86.31
Poly[1,1-cyclohexane bis(4-phenyl)carbonate]	------	248.9	------	92.91
Poly(hexamethylene adipamide)	211.5	209.9	96.14	94.69
Poly(ether ether ketone)	227.9	233.9	96.42	94.69

☞ ☞ ☞ **TABLE 9.2 IS CONTINUED IN THE NEXT PAGE.** ☞ ☞ ☞

Table 9.2. CONTINUED FROM THE PREVIOUS PAGE.

Polymer	V(exp)	V(pred)	P_{LL}(exp)	P_{LL}(pred)
Poly[2,2'-(*m*-phenylene)-5,5'-bibenzimidazole]	237.2	231.9	102.94	102.39
Arylef U100	296.2	298.8	108.34	116.30
Poly(hexamethylene sebacamide)	271.5	274.2	114.87	116.30
Torlon 2000	251.3	275.3	119.04	125.60
Poly[N,N'-(*p,p*'-oxydiphenylene)pyromellitimide]	269.2	278.2	122.36	129.07
Poly[4,4'-diphenoxy di(4-phenylene)sulfone]	310.5	310.1	139.27	137.05
Poly[4,4'-isopropylidene diphenoxy di(4-phenylene)sulfone]	356.9	360.2	150.20	152.36
Poly[4,4'-sulfone diphenoxy di(4-phenylene)sulfone]	339.1	336.4	163.70	162.98
Ultem 1000	466.6	463.3	194.79	197.58

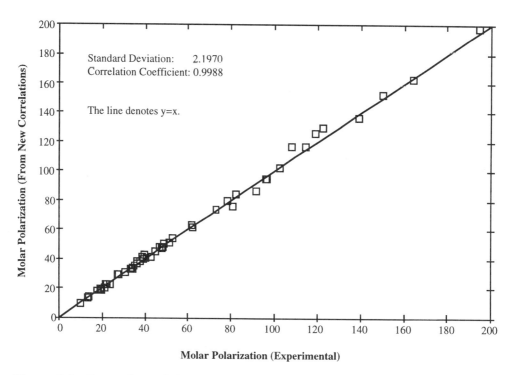

Figure 9.4. Comparison of the molar polarization according to Lorentz and Lorenz (P_{LL}) calculated from measured dielectric constants and molar volumes, with the molar polarization calculated from dielectric constants and molar volumes predicted by using the new correlations developed in this book, for 55 polymers.

9.D. Calculation of the Effective Dipole Moment

Available experimental values of the refractive index were used to calculate "experimental" values of the molar refraction R_{LL} by using Equation 8.3. Equation 9.8 was then used to calculate the "experimental" effective dipole moments at room temperature for the 40 polymers from the dataset for which all of the necessary information was available to calculate both P_{LL} and R_{LL} from experimental data. "Fitted" values of R_{LL}, calculated as described in Section 8.E, were then combined with the fitted values of P_{LL}, to estimate the effective dipole moment.

The μ(fit) values have a standard deviation of 0.1076 debyes relative to, and a correlation coefficient of 0.9592 with, the μ(exp) values. The standard deviation is 20.6% of the average μ(exp) value of 0.5226 debyes for this set of polymers. The results of these calculations are summarized in Table 9.3 and shown in Figure 9.5.

An important issue needs to be clarified regarding the calculations on nonpolar polymers. The theory of electromagnetism indicates that $\varepsilon \geq n^2$ and $P_{LL} \geq R_{LL}$ for all materials. There are, nonetheless, some nonpolar polymers whose tabulated experimental values of ε and n make it appear as if these fundamental inequalities are being violated. For example, the experimental dielectric constant [14] and refractive index of poly(α-vinyl naphthalene) [3] have been listed as 2.60 and 1.6818, respectively, so that $n^2 \approx 2.83 > \varepsilon$. Such apparent discrepancies arise only because different tests are used to measure ε and n. A slight overestimation of n and/or a slight underestimation of ε can make n^2 appear to be somewhat larger than ε if $\varepsilon \approx n^2$. In all calculations involving the difference $(P_{LL} - R_{LL})$, or the difference $(\varepsilon - n^2)$, these quantities should therefore simply be set equal to zero if they appear to be negative. This procedure was followed in the calculation of μ(298K).

Note, from inspection of Table 9.3, that the largest differences between μ(exp) and μ(fit) occur for the nonpolar hydrocarbon polymers which have very small or zero dipole moments. Any errors in measured and/or fitted ε and n values are amplified when the very small difference $(P_{LL} - R_{LL})$ is calculated for such polymers. The agreement between μ(exp) and μ(fit) is far better for polymers containing polar groups, whose $(P_{LL} - R_{LL})$ is larger. For example, if the 11 hydrocarbon polymers are excluded from the comparison between μ(exp) and μ(fit), then for the 29 remaining polymers, all of which contain heteroatoms, the standard deviation is only 0.0790 debyes, and the correlation coefficient is 0.9684. The standard deviation is only 11.4% of the average μ(exp) value of 0.6903 debyes for these polymers, which really should manifest polarity and have nonzero dipole moments.

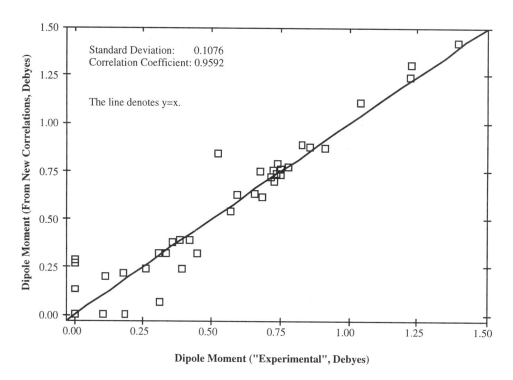

Figure 9.5. Comparison of the "experimental" effective dipole moment μ(298K) at room temperature, calculated from observed dielectric constants, refractive indices and molar volumes, with μ(298K) calculated from the values predicted for the same properties by using the new correlations developed in this book, for 40 polymers.

Table 9.3. Experimental and predicted values of the molar polarization P_{LL} and the molar refraction R_{LL} in cc/mole, and the effective dipole moment μ at room temperature in debyes, for 61 polymers. Experimental values of μ were not calculated for 21 of the polymers because some of the required information was not available to calculate P_{LL} and/or R_{LL} for these polymers.

Polymer	Experimental			Predicted		
	P_{LL}	R_{LL}	μ	P_{LL}	R_{LL}	μ
Poly(o-methyl styrene)	38.20	38.703	0.00	38.53	38.821	0.00
Poly(α-vinyl naphthalene)	47.90	52.145	0.00	48.28	50.144	0.00
Poly(p-xylylene)	34.35	36.117	0.00	33.47	33.763	0.00
Poly(vinyl cyclohexane)	34.12	--------	-----	32.80	34.833	0.00
Poly(β-vinyl naphthalene)	46.87	--------	-----	48.24	50.150	0.00
Poly(α−methyl styrene)	38.13	--------	-----	38.12	38.580	0.00

☞ ☞ ☞   **TABLE 9.3 IS CONTINUED IN THE NEXT PAGE.** ☞ ☞ ☞

Table 9.3. CONTINUED FROM THE PREVIOUS PAGE.

Polymer	Experimental			Predicted		
	P_{LL}	R_{LL}	μ	P_{LL}	R_{LL}	μ
Polyisobutylene	19.42	19.813	0.00	18.98	18.617	0.13
Poly(1-butene)	19.39	19.581	0.00	19.66	18.197	0.27
Poly(4-methyl-1-pentene)	27.50	27.783	0.00	29.09	27.433	0.28
Polystyrene	33.76	33.535	0.10	33.28	33.351	0.00
Polypropylene	14.14	13.899	0.11	14.53	13.703	0.20
Polyisoprene	23.58	22.898	0.18	22.87	23.131	0.00
Polyethylene	10.01	9.336	0.18	9.88	8.899	0.22
Poly(N-vinyl carbazole)	62.43	61.052	0.26	61.52	60.343	0.24
Poly(1,4-butadiene)	20.32	18.334	0.31	18.03	17.933	0.07
Poly[oxy(2,6-dimethyl-1,4-phenylene)]	39.03	37.077	0.31	39.70	37.592	0.32
Poly[thio(*p*-phenylene)]	33.23	--------	-----	33.12	31.059	0.32
Poly(vinylidene chloride)	22.28	19.979	0.33	21.95	19.849	0.32
Poly(*p*-chloro styrene)	------	--------	-----	41.20	38.483	0.36
Poly(*m*-chloro styrene)	------	--------	-----	41.20	38.483	0.36
Poly($\alpha,\alpha,\alpha',\alpha'$-tetrafluoro-*p*-xylylene)	36.28	--------	-----	38.11	35.432	0.36
Polytetrafluoroethylene	13.41	10.759	0.36	12.99	10.081	0.38
Poly[oxy(2,6-diphenyl-1,4-phenylene)]	80.36	77.209	0.39	76.26	75.056	0.24
Polyoxymethylene	9.88	6.818	0.39	9.49	6.355	0.39
Poly(vinyl chloride)	17.77	4.128	0.42	17.69	14.529	0.39
Poly(chloro-*p*-xylylene)	42.66	38.484	0.45	41.47	38.922	0.35
Poly(3,4-dichlorostyrene)	------	--------	-----	48.86	43.647	0.50
Arylef U100	108.34	102.694	0.52	116.30	101.599	0.84
Polychlorotrifluoroethylene	21.11	14.389	0.57	20.36	14.381	0.54
Poly(*o*-chloro-*p*-methoxy styrene)	------	--------	-----	52.59	45.668	0.58
Poly(cyclohexyl methacrylate)	52.85	45.561	0.59	54.30	46.022	0.63
Poly(*p*-hydroxybenzoate)	37.53	--------	-----	38.41	31.010	0.60
Polyacrylonitrile	22.40	13.620	0.65	22.52	14.204	0.64
Poly[1,1-cyclohexane *bis*(4-phenyl)carbonate]	------	-------	-----	92.91	83.950	0.66
Poly(dimethyl siloxane)	27.85	18.469	0.67	29.39	17.878	0.75
Poly(vinyl butyral)	47.31	37.631	0.69	47.21	39.447	0.61
Poly(oxy-2,2-dichloromethyltrimethylene)	44.72	--------	-----	45.31	35.610	0.69
Poly(methyl methacrylate)	35.25	24.718	0.71	35.32	24.746	0.72

☞☞☞ **TABLE 9.3 IS CONTINUED IN THE NEXT PAGE.** ☞☞☞

Table 9.3. CONTINUED FROM THE PREVIOUS PAGE.

Polymer	Experimental			Predicted		
	P_{LL}	R_{LL}	μ	P_{LL}	R_{LL}	μ
Poly(isobutyl methacrylate)	49.23	38.456	0.72	50.64	38.832	0.76
Poly(vinyl acetate)	31.03	20.089	0.73	30.13	20.106	0.70
Bisphenol-A polycarbonate	82.17	71.016	0.74	84.08	72.762	0.74
Poly(ethyl α-chloroacrylate)	39.86	28.567	0.74	42.81	30.051	0.79
Poly(ethyl methacrylate)	40.80	29.233	0.75	40.76	29.601	0.74
Poly[1,1-ethane *bis*(4-phenyl)carbonate]	------	--------	-----	79.01	67.529	0.75
Poly(methyl α-chloroacrylate)	36.44	24.808	0.75	37.04	25.075	0.76
Poly(*n*-butyl methacrylate)	50.90	38.498	0.78	50.89	38.536	0.77
Poly(ether ether ketone)	96.42	--------	-----	94.69	82.055	0.78
Poly[2,2'-(*m*-phenylene)-5,5'-bibenzimidazole]	102.94	--------	-----	102.39	89.293	0.80
Poly(1,4-cyclohexylidene dimethylene terephthalate)	91.72	--------	-----	86.31	72.725	0.81
Poly(ethylene terephthalate)	61.71	47.585	0.83	63.05	46.885	0.89
Poly(ε-caprolactam)	47.45	32.249	0.86	47.37	31.558	0.88
Ethyl cellulose	78.49	61.535	0.91	79.99	64.217	0.87
Poly(tetramethylene terephthalate)	72.92	--------	-----	73.62	56.258	0.92
Torlon 2000	119.04	--------	-----	125.60	100.670	1.10
Poly[4,4'-isopropylidene diphenoxy di(4-phenylene)sulfone]	150.20	127.465	1.05	152.36	127.017	1.11
Poly[4,4'-diphenoxy di(4-phenylene)sulfone]	139.27	--------	-----	137.05	111.419	1.12
Poly[N,N'-(*p,p*'-oxydiphenylene)pyromellitimide]	122.36	--------	-----	129.07	103.069	1.12
Ultem 1000	194.79	--------	-----	197.58	168.491	1.19
Poly(hexamethylene adipamide)	96.14	65.332	1.22	94.69	63.086	1.24
Poly(hexamethylene sebacamide)	114.87	83.866	1.23	116.30	81.105	1.31
Poly[4,4'-sulfone diphenoxy di(4-phenylene)sulfone]	163.70	123.684	1.39	162.98	121.231	1.42

9.E. Dissipation Factor

It is impossible to predict the complete dissipation factor (tan δ) curves of polymers as functions of temperature and frequency without detailed consideration of relaxation times. At present, the *a priori* estimation of relaxation times requires detailed computer-intensive calculations, such as force field or quantum mechanical methods to estimate rotation barriers, and molecular dynamics simulations. A much less ambitious goal was therefore pursued. A simple variable was sought, for use as a rough "order of magnitude" estimate of the "lossiness" of a polymer, at room temperature, over the most important frequency range for typical applications.

A dataset of 206 points, containing the experimental values [14-18,21] of tan δ measured by dielectric relaxation experiments at room temperature (298±5K) for a large number of polymers, was prepared. The test frequencies included in the dataset ranged from 50 Hz to 10 MHz.

The following types of polymers were *not* included in this dataset:

1. Cellulose derivatives, whose electrical properties are extremely dependent upon formulation.

2. The asymmetrically fluorinated poly(vinyl fluoride) and poly(vinylidene fluoride), whose electrical properties differ qualitatively from the general behavior of polymers (Section 9.B).

3. Polymers which were identified only by an acronym in the sources of data, and whose structures could not be determined from the acronym.

4. Polymers for which the frequency of measurement of tan δ was not provided in the sources.

The difference ($\varepsilon - n^2$) was used as the main fitting variable. As discussed in Section 9.D, ($\varepsilon - n^2$) was set equal to zero for nonpolar polymers when experimental errors resulted in an unphysical apparent slightly negative value.

The experimental values used for ε in these calculations were generally those obtained for the specimens whose tan δ was reported, so that they do not describe "intrinsic" properties of polar polymers but rather the properties of samples often containing considerable amounts of additives.

Roughly speaking, structural and/or compositional modifications change the dielectric constant and the dissipation factor of a polymer in the same direction. For example, moisture and polar plasticizers increase both ε and tan δ, although the magnitude of the change is often much larger for tan δ than for ε. If an approximate general relationship is found for tan δ by using the available data, such a relationship can then be used to estimate the "intrinsic" lossiness of a polymer by substituting the value of ε calculated by Equation 9.12 into the correlation instead of using the value of ε measured for a polymer containing additives, fillers, and/or moisture.

Experimental refractive indices were used whenever available in these calculations. For semicrystalline polymers such as polyethylene, the typical refractive indices of the semicrystalline specimens were used instead of taking the amorphous limit as was done whenever possible in Section 8.C. Whenever experimental refractive indices were not available, the best possible estimate was made for the refractive index, in most cases by using Equation 8.6.

The behavior for 50 Hz to 100 Hz was qualitatively different from the behavior for 1 kHz to 10 MHz. Inclusion of the degrees of freedom defined in Chapter 4 improved the quality of the fit for tan δ measured at 50 Hz to 100 Hz. The more numerous the degrees of freedom (especially of the side groups) were, the more a polymer was able to dissipate energy in proportion to its apparent polarity as manifested by the value of ($\varepsilon - n^2$). On the other hand, inclusion of the degrees of freedom had no effect, and the apparent polarity as quantified by ($\varepsilon - n^2$) was the only important physical factor in determining tan δ, at higher frequencies (1 kHz to 10 Mhz).

Equation 9.13 was found to fit the 43 data points in the frequency range of 50 Hz to 100 Hz with a correlation coefficient of 0.9707 and a standard deviation of 0.0154. N_{BBrot} and N_{SGrot} are the rotational degrees of freedom parameters of the backbone and the side groups, calculated as described in Section 4.C. N is the number of non-hydrogen atoms in the repeat unit.

$$\tan \delta \approx 0.029694(\varepsilon - n^2)\frac{(N_{BBrot} + 1.5 N_{SGrot})}{N} \tag{9.13}$$

Equation 9.14 was found to fit the 163 data points in the frequency range of 1 kHz to 10 MHz with a correlation coefficient of 0.9080 and a standard deviation of 0.0101.

$$\tan \delta \approx 0.022599 \cdot (\varepsilon - n^2) \tag{9.14}$$

The fit, obtained by using equations 9.13 and 9.14 as appropriate, for the complete set of 206 data points, has a standard deviation of 0.0114 and a correlation coefficient of 0.9493.

The results of these calculations are summarized in Table 9.4 and depicted in Figure 9.6. The degrees of freedom and the refractive indices used in the calculations have not been listed in Table 9.4, to avoid overcrowding this table.

The typical magnitude of the deviation between $\tan \delta$(exp) and $\tan \delta$(fit) changes slowly with changing $\tan \delta$(exp); however, the relative (fractional) deviation increases dramatically with decreasing $\tan \delta$(exp). Most of the $\tan \delta$(fit) values are within an order of magnitude of $\tan \delta$(exp). Equations 9.13 and 9.14 can thus be used to provide very rough order of magnitude estimates of the intrinsic lossiness of polymers, by utilizing ε values calculated via Equation 9.12. The effects of structural, compositional, and/or morphological changes on $\tan \delta$ can be estimated by modifying ε to reflect these changes.

Only a very rough order of magnitude estimate is obtained for $\tan \delta$ from equations 9.13 and 9.14, useful for qualitative comparisons but not as input into more quantitative calculations.

Most of the remaining systematic portion of the deviation between experimental and calculated values is caused by the facts that (a) equations 9.13 and 9.14 do not account for the variation of $\tan \delta$ with temperature or frequency, and (b) $\tan \delta$ is much more sensitive than ε to many factors. Another important source of deviation is the uncertainty of the tabulated experimental values compiled from the results of measurements made over many decades, by many different workers, in different laboratories, on different instruments. Because of all of these serious limitations, this index of lossiness should only be used to estimate the order of magnitude of $\tan \delta$ at frequencies falling within the range used in measuring the data included in its development, namely, (50 Hz)$\leq$frequency$\leq$(10 MHz).

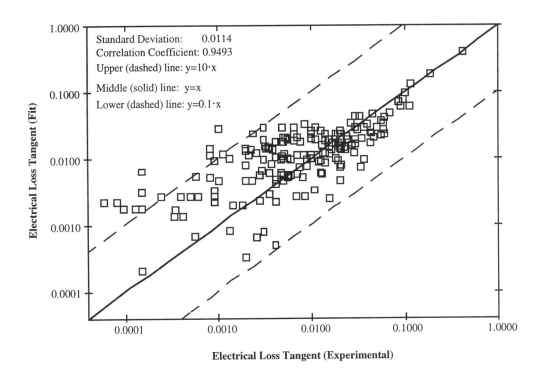

Figure 9.6. Comparison of observed and fitted electrical loss tangent (tan δ) at room temperature, with 206 data points for a wide variety of polymers, at frequencies ranging from 50 Hz to 10 MHz. The logarithmic scale is used for the axes, so that data points at very low tan δ(exp) or tan δ(fit) are not seen. The typical absolute deviation between tan δ(exp) and tan δ(fit) does not change much with changing tan δ(exp); however, the relative (fractional) deviation increases with decreasing tan δ(exp). Most of the tan δ(fit) values fall within an order of magnitude of the tan δ(exp) values.

Table 9.4. Frequency (ν) of measurement in Hertz, dielectric dissipation factor (tan δ) and dielectric constant (ε) observed at room temperature at the specified frequency of measurement, and the fitted value of tan δ calculated as described in the text, for 206 data points covering a wide variety of polymers. Most measurements were made on commercial samples containing various plasticizers, fillers and other additives, so that the ε(exp) values listed for many polar polymers differ significantly from (i.e., exceed) the more "intrinsic" ε(exp) values listed in Table 9.1.

Polymer	ν	tan δ(exp)	ε(exp)	tan δ(fit)
Poly(1,4-butadiene) *(99% trans)*	50	0.00200	2.51	0.00472
Poly[oxy(2,6-dimethyl-1,4-phenylene)]	60	0.00035	2.58	0.00164

☞ ☞ ☞   **TABLE 9.4 IS CONTINUED IN THE NEXT PAGE.**   ☞ ☞ ☞

Table 9.4. CONTINUED FROM THE PREVIOUS PAGE.

Polymer	ν	tan δ(exp)	ε(exp)	tan δ(fit)
Poly[4,4'-diphenoxy di(4-phenylene)sulfone]	60	0.00058	3.44	0.00543
Poly[4,4'-isopropylidene diphenoxy di(4-phenylene)sulfone]	60	0.00080	3.18	0.00524
Bisphenol-A polycarbonate	60	0.00090	3.17	0.00925
Poly(ether ether ketone)	60	0.00100	3.20	0.00469
Phenoxy resin	60	0.00100	4.10	0.02831
Poly[4,4'-sulfone diphenoxy di(4-phenylene)sulfone]	60	0.00200	3.80	0.00800
Poly(ethylene terephthalate)	60	0.00250	3.30	0.01217
Poly(dichloro-*p*-xylylene)	60	0.00400	2.84	0.00049
Poly(oxy-2,2-dichloromethyltrimethylene)	60	0.01100	3.50	0.03012
Poly(chloro-*p*-xylylene)	60	0.02000	3.15	0.00459
Polyacrylonitrile	60	0.11300	6.50	0.06220
Polystyrene	100	0.00015	2.55	0.00020
Polypropylene	100	0.00015	2.26	0.00308
Poly(*p*-xylylene)	100	0.00020	2.65	0.00000
Polytetrafluoroethylene	100	0.00025	2.10	0.00275
Polyisobutylene	100	0.00040	2.23	0.00000
Poly(*p*-chloro styrene)	100	0.00056	2.65	0.00068
Poly(N-vinyl carbazole)	100	0.00130	2.95	0.00081
Poly(vinyl cyclohexane)	100	0.00150	2.25	0.00000
Polyisoprene	100	0.00280	2.40	0.00231
Poly(*p*-hydroxybenzoate)	100	0.00289	3.22	0.00665
Poly(vinyl butyral)	100	0.00380	2.69	0.01368
Poly(cyclohexyl methacrylate)	100	0.00460	2.58	0.00959
Poly(vinyl acetate)	100	0.00490	3.09	0.03017
Poly(vinyl formal)	100	0.00540	3.16	0.01930
Poly(vinyl acetal)	100	0.00650	3.17	0.02292
Poly(ε-caprolactam)	100	0.00650	3.50	0.03012
Poly(3,4-dichlorostyrene)	100	0.00850	2.94	0.00279
Poly(hexamethylene adipamide)	100	0.00850	3.60	0.03271
Poly(hexamethylene sebacamide)	100	0.01000	3.20	0.02296
Poly(isobutyl methacrylate)	100	0.01110	2.70	0.01924
Poly(vinyl chloride)	100	0.01300	3.18	0.01606
Poly(*m*-chloro styrene)	100	0.01570	2.80	0.00241

☞ ☞ ☞ TABLE 9.4 IS CONTINUED IN THE NEXT PAGE. ☞ ☞ ☞

Table 9.4. CONTINUED FROM THE PREVIOUS PAGE.

Polymer	ν	tan δ(exp)	ε(exp)	tan δ(fit)
Polychlorotrifluoroethylene	100	0.02100	2.72	0.00780
(85% vinylidene chloride)/(15% vinyl chloride) copolymer	100	0.03000	5.30	0.04282
Poly(*o*-vinyl pyridine)	100	0.03600	4.91	0.03044
Poly(ethyl methacrylate)	100	0.04200	2.90	0.02450
Poly(*n*-butyl methacrylate)	100	0.06050	2.82	0.02304
Poly(methyl methacrylate) *(Plexiglas™)*	100	0.06050	3.40	0.04011
Poly(12-aminododecanoic acid)	100	0.07000	4.00	0.04888
Poly(vinyl alcohol)	100	0.40500	14.00	0.40706
Poly[2,2'-(*m*-phenylene)-5,5'-bibenzimidazole]	1000	0.00000	3.30	0.00953
Polyethylene	1000	0.00006	2.28	0.00229
Polyisobutylene	1000	0.00010	2.23	0.00000
Polytetrafluoroethylene	1000	0.00015	2.10	0.00627
Poly(vinyl cyclohexane)	1000	0.00020	2.25	0.00000
Poly(*p*-chloro styrene)	1000	0.00034	2.65	0.00132
Poly(*p*-xylylene)	1000	0.00034	2.65	0.00000
Poly[4,4'-isopropylidene diphenoxy di(4-phenylene)sulfone]	1000	0.00080	3.17	0.01137
Alternating copolymer of ethylene and tetrafluoroethylene	1000	0.00080	2.60	0.01427
Poly(N-vinyl carbazole)	1000	0.00090	2.95	0.00266
Poly(butadiene-co-styrene) *(23.5% bound styrene)*	1000	0.00090	2.50	0.00328
Poly(α-vinyl naphthalene)	1000	0.00095	2.60	0.00000
Bisphenol-A polycarbonate	1000	0.00110	3.02	0.01148
Ultem	1000	0.00130	3.15	0.00984
Polyisoprene	1000	0.00180	2.40	0.00196
Poly(tetramethylene terephthalate)	1000	0.00200	3.00	0.01390
Poly[4,4'-sulfone diphenoxy di(4-phenylene)sulfone]	1000	0.00240	3.80	0.02435
Poly(dichloro-*p*-xylylene)	1000	0.00300	2.82	0.00078
Polyoxymethylene	1000	0.00300	3.60	0.02983
Poly(*p*-hydroxybenzoate)	1000	0.00316	3.21	0.01496
Poly[thio(*p*-phenylene)]	1000	0.00400	3.10	0.00487
Poly(vinyl butyral)	1000	0.00400	2.67	0.01050
Poly(cyclohexyl methacrylate)	1000	0.00475	2.52	0.00565
Poly(ethylene terephthalate)	1000	0.00500	3.25	0.01739
Poly(vinyl acetate)	1000	0.00500	3.07	0.02074

☞ ☞ ☞   TABLE 9.4 IS CONTINUED IN THE NEXT PAGE.   ☞ ☞ ☞

Table 9.4. CONTINUED FROM THE PREVIOUS PAGE.

Polymer	ν	tan δ(exp)	ε(exp)	tan δ(fit)
Poly(3,4-dichlorostyrene)	1000	0.00700	2.93	0.00584
Poly(isobutyl methacrylate)	1000	0.00700	2.68	0.01127
Poly(vinyl acetal)	1000	0.00700	3.14	0.02079
Poly(vinyl formal)	1000	0.01000	3.12	0.01966
Poly(ε-caprolactam)	1000	0.01000	3.50	0.02620
Poly(*m*-chloro styrene)	1000	0.01072	2.74	0.00336
Poly(hexamethylene adipamide)	1000	0.01250	3.50	0.02620
Poly(hexamethylene sebacamide)	1000	0.01500	3.20	0.01942
Poly(vinyl chloride)	1000	0.01850	3.10	0.01653
Poly(chloro-*p*-xylylene)	1000	0.01900	3.10	0.00935
Torlon 4203L	1000	0.02600	4.20	0.03323
Polychlorotrifluoroethylene	1000	0.02700	2.63	0.01577
Poly(ethyl methacrylate)	1000	0.02940	2.75	0.01231
Poly(*n*-butyl methacrylate)	1000	0.03600	2.62	0.00951
Poly(*o*-vinyl pyridine)	1000	0.04600	4.64	0.04685
Poly(methyl methacrylate) *(Plexiglas)*	1000	0.04650	3.12	0.02038
Poly(12-aminododecanoic acid)	1000	0.05000	3.50	0.02876
Poly(11-aminoundecanoic acid)	1000	0.06000	3.70	0.03314
Polyacrylonitrile	1000	0.08500	5.50	0.07208
Poly(vinyl alcohol)	1000	0.18500	10.40	0.18418
Poly(*p*-xylylene)	10000	0.00039	2.65	0.00000
Poly(*p*-chloro styrene)	10000	0.00042	2.65	0.00132
Poly(N-vinyl carbazole)	10000	0.00060	2.95	0.00266
Poly(α-vinyl naphthalene)	10000	0.00067	2.60	0.00000
Polyisoprene	10000	0.00140	2.40	0.00196
Poly(ether ether ketone)	10000	0.00200	3.20	0.01308
Bisphenol-A polycarbonate	10000	0.00210	3.00	0.01102
Poly[2,2'-(*m*-phenylene)-5,5'-bibenzimidazole]	10000	0.00300	3.30	0.00953
Poly(*p*-hydroxybenzoate)	10000	0.00336	3.19	0.01451
Poly(isobutyl methacrylate)	10000	0.00500	2.63	0.01014
Poly(vinyl acetate)	10000	0.00520	3.05	0.02029
Poly(vinyl butyral)	10000	0.00560	2.65	0.01005
Poly(3,4-dichlorostyrene)	10000	0.00570	2.91	0.00539

☞ ☞ ☞   TABLE 9.4 IS CONTINUED IN THE NEXT PAGE.   ☞ ☞ ☞

Table 9.4. CONTINUED FROM THE PREVIOUS PAGE.

Polymer	ν	tan δ(exp)	ε(exp)	tan δ(fit)
Poly(cyclohexyl methacrylate)	10000	0.00600	2.52	0.00565
Poly(m-chloro styrene)	10000	0.00676	2.71	0.00268
Poly(vinyl acetal)	10000	0.01000	3.09	0.01966
Poly(ε-caprolactam)	10000	0.01400	3.40	0.02394
Poly(vinyl formal)	10000	0.01540	3.08	0.01876
Poly(hexamethylene sebacamide)	10000	0.01700	3.10	0.01716
Poly(hexamethylene adipamide)	10000	0.01800	3.40	0.02396
Poly(ethyl methacrylate)	10000	0.01850	2.65	0.01005
Poly(n-butyl methacrylate)	10000	0.02000	2.52	0.00725
Poly(vinyl chloride)	10000	0.02250	3.02	0.01472
Polychlorotrifluoroethylene	10000	0.02300	2.53	0.01351
Poly(methyl methacrylate) *(Plexiglas)*	10000	0.03000	2.95	0.01654
Poly(12-aminododecanoic acid)	10000	0.04500	3.30	0.02424
Poly(o-vinyl pyridine)	10000	0.05600	4.26	0.03827
Poly(vinyl alcohol)	10000	0.11500	8.00	0.12994
Polypropylene	100000	0.00015	2.25	0.00178
Poly(p-xylylene)	100000	0.00020	2.65	0.00000
Poly(p-chloro styrene)	100000	0.00042	2.65	0.00132
Poly(N-vinyl carbazole)	100000	0.00050	2.95	0.00266
Poly(α-vinyl naphthalene)	100000	0.00075	2.60	0.00000
Polyisoprene	100000	0.00140	2.40	0.00196
Poly(p-hydroxybenzoate)	100000	0.00348	3.17	0.01405
Poly(isobutyl methacrylate)	100000	0.00370	2.55	0.00833
Poly(m-chloro styrene)	100000	0.00410	2.69	0.00223
Bisphenol-A polycarbonate	100000	0.00490	2.99	0.01080
Poly(3,4-dichlorostyrene)	100000	0.00500	2.88	0.00471
Poly(cyclohexyl methacrylate)	100000	0.00500	2.52	0.00565
Poly(vinyl acetate)	100000	0.00560	3.02	0.01961
Poly(vinyl butyral)	100000	0.00880	2.63	0.00960
Poly(ethyl methacrylate)	100000	0.01180	2.60	0.00892
Poly(n-butyl methacrylate)	100000	0.01250	2.47	0.00612
Polychlorotrifluoroethylene	100000	0.01350	2.46	0.01193
Poly(vinyl acetal)	100000	0.01400	3.03	0.01830

☞☞☞   **TABLE 9.4 IS CONTINUED IN THE NEXT PAGE.** ☞☞☞

Table 9.4. CONTINUED FROM THE PREVIOUS PAGE.

Polymer	ν	tan δ(exp)	ε(exp)	tan δ(fit)
Poly(vinyl formal)	100000	0.01900	3.00	0.01695
Poly(ε-caprolactam)	100000	0.01900	3.40	0.02394
Poly(methyl methacrylate) *(Plexiglas)*	100000	0.02000	2.84	0.01406
Poly(hexamethylene sebacamide)	100000	0.02000	3.00	0.01490
Poly(vinyl chloride)	100000	0.02100	2.96	0.01337
Poly(hexamethylene adipamide)	100000	0.02150	3.40	0.02394
Poly[2,2'-(*m*-phenylene)-5,5'-bibenzimidazole]	100000	0.03400	3.20	0.00727
Poly(12-aminododecanoic acid)	100000	0.03500	3.10	0.01972
(85% vinylidene chloride)/(15% vinyl chloride) copolymer	100000	0.04000	4.00	0.03254
Poly(*o*-vinyl pyridine)	100000	0.06000	3.77	0.02719
Poly(vinyl alcohol)	100000	0.10000	6.60	0.09831
Poly(*p*-xylylene)	1000000	0.00007	2.65	0.00000
Polyethylene	1000000	0.00008	2.28	0.00229
Polypropylene	1000000	0.00010	2.25	0.00178
Poly(N-vinyl carbazole)	1000000	0.00040	2.95	0.00266
Poly(α-vinyl naphthalene)	1000000	0.00087	2.60	0.00000
Poly[oxy(2,6-dimethyl-1,4-phenylene)]	1000000	0.00090	2.58	0.00225
Polyisoprene	1000000	0.00180	2.40	0.00196
Poly(dichloro-*p*-xylylene)	1000000	0.00200	2.80	0.00033
Poly(*m*-chloro styrene)	1000000	0.00252	2.62	0.00065
Poly(cyclohexyl methacrylate)	1000000	0.00280	2.52	0.00565
Poly(dimethyl siloxane)	1000000	0.00300	2.80	0.01876
Poly(*p*-hydroxybenzoate)	1000000	0.00325	3.16	0.01383
Poly(isobutyl methacrylate)	1000000	0.00350	2.45	0.00607
Poly[4,4'-sulfone diphenoxy di(4-phenylene)sulfone]	1000000	0.00400	3.50	0.01757
Poly(3,4-dichlorostyrene)	1000000	0.00420	2.85	0.00403
Poly[4,4'-isopropylidene diphenoxy di(4-phenylene)sulfone]	1000000	0.00510	3.19	0.01183
Poly(vinyl acetate)	1000000	0.00650	2.98	0.01871
Poly[4,4'-diphenoxy di(4-phenylene)sulfone]	1000000	0.00760	3.45	0.01735
Poly(*n*-butyl methacrylate)	1000000	0.00800	2.43	0.00521
Polychlorotrifluoroethylene	1000000	0.00820	2.43	0.01125
Poly(ethyl methacrylate)	1000000	0.00900	2.55	0.00779
Bisphenol-A polycarbonate	1000000	0.01000	2.96	0.01012

☞☞☞ **TABLE 9.4 IS CONTINUED IN THE NEXT PAGE.** ☞☞☞

Table 9.4. CONTINUED FROM THE PREVIOUS PAGE.

Polymer	ν	tan δ(exp)	ε(exp)	tan δ(fit)
Poly(vinyl butyral)	1000000	0.01240	2.61	0.00915
Poly(chloro-*p*-xylylene)	1000000	0.01300	2.95	0.00596
Poly(methyl methacrylate) *(Plexiglas)*	1000000	0.01400	2.76	0.01225
Poly(vinyl chloride)	1000000	0.01600	2.88	0.01156
Poly(ethylene terephthalate)	1000000	0.01600	3.00	0.01174
Poly(vinyl acetal)	1000000	0.01800	2.96	0.01672
Poly(vinyl formal)	1000000	0.01900	2.92	0.01514
Poly(hexamethylene azelamide)	1000000	0.02000	3.20	0.01942
Nylon 6,12	1000000	0.02000	3.50	0.02620
Poly(hexamethylene sebacamide)	1000000	0.02100	3.00	0.01490
Poly(ε-caprolactam)	1000000	0.02400	3.30	0.02168
Poly(hexamethylene adipamide)	1000000	0.02500	3.20	0.01942
Phenoxy resin	1000000	0.03000	3.80	0.02941
Torlon 4203L	1000000	0.03100	3.90	0.02645
Polyacrylonitrile	1000000	0.03300	4.20	0.04270
Poly(*o*-vinyl pyridine)	1000000	0.05600	3.56	0.02245
Poly(vinyl alcohol)	1000000	0.09500	5.70	0.07797
Polypropylene	10000000	0.00013	2.25	0.00178
Poly(*p*-xylylene)	10000000	0.00037	2.65	0.00000
Poly(N-vinyl carbazole)	10000000	0.00050	2.95	0.00266
Poly(α-vinyl naphthalene)	10000000	0.00130	2.60	0.00000
Poly(*p*-hydroxybenzoate)	10000000	0.00220	3.15	0.01360
Poly(cyclohexyl methacrylate)	10000000	0.00230	2.48	0.00475
Poly(3,4-dichlorostyrene)	10000000	0.00350	2.80	0.00290
Poly(isobutyl methacrylate)	10000000	0.00460	2.45	0.00607
Poly(vinyl acetate)	10000000	0.00490	2.94	0.01781
Poly(*n*-butyl methacrylate)	10000000	0.00550	2.42	0.00499
Polychlorotrifluoroethylene	10000000	0.00600	2.35	0.00944
Poly(ethyl methacrylate)	10000000	0.00750	2.53	0.00734
Poly(methyl methacrylate) *(Plexiglas)*	10000000	0.01000	2.71	0.01112
Poly(vinyl chloride)	10000000	0.01150	2.87	0.01133
Poly(vinyl butyral)	10000000	0.01380	2.59	0.00870
Poly(vinyl acetal)	10000000	0.01600	2.90	0.01537

☞ ☞ ☞   **TABLE 9.4 IS CONTINUED IN THE NEXT PAGE.**   ☞ ☞ ☞

Table 9.4. CONTINUED FROM THE PREVIOUS PAGE.

Polymer	ν	tan δ(exp)	ε(exp)	tan δ(fit)
Poly(vinyl formal)	10000000	0.01650	2.85	0.01356
Poly(hexamethylene sebacamide)	10000000	0.02100	3.00	0.01490
Poly(hexamethylene adipamide)	10000000	0.02550	3.10	0.01715
Poly(ε-caprolactam)	10000000	0.03000	3.20	0.01941
Poly(o-vinyl pyridine)	10000000	0.04200	3.33	0.01725
Poly(vinyl alcohol)	10000000	0.09000	5.00	0.06215

9.F. Dielectric Strength

Electrical degradation and breakdown phenomena [22] limit the utility of polymers in many applications ranging from electrical cable insulation to utilization in advanced electronic devices. The key property of a polymer in relation to its electrical degradation and breakdown behavior is its *dielectric strength*. This property cannot be predicted quantitatively as a function of the polymeric structure at this time. However, because of its importance, we will discuss it briefly.

The dielectric strength can be defined as the voltage that will produce a catastrophic decrease in the resistance of an insulator, divided by the thickness of the specimen. Consequently, it has units of volts/length. It is affected by a variety of factors, including the polymeric structure, the specimen geometry, and the testing conditions.

The types of electrical tests used for insulating resins (tests for dielectric strength, dissipation factor and dielectric constant) were reviewed by Winkeler [23]. The conditioning of a resin and the specimen thickness are both seen to be very important in determining the dielectric strength. The dielectric strength varies roughly as the reciprocal square root of the specimen thickness, so that it decreases monotonically with increasing thickness. Consequently, accurate comparisons of dielectric strength can only be made if the specimen thickness is known. Mason [24] studied the effects of both the area and the thickness of the specimens on the dielectric strength of polymers, also reaching the conclusion that dielectric strength is more dependent on the thickness than it is on the area. Another important factor related to the specimen geometry is encapsulation. For example, it has been shown [25] that polyethylene specimens have significantly greater dielectric strength, and that their dielectric strength drops off more slowly with increasing temperature, when they are fully encapsulated than when they are merely recessed.

There are many published discussions of both the effects of polymeric structure and testing conditions on the dielectric strength (for example, see [22,25,26]) and the underlying mechanisms of electrical breakdown (for example, see [22,25-27]). At the risk of greatly oversimplifying these complicated effects and mechanisms, they will now be reviewed briefly.

One factor affecting the dielectric strength is the electronic structure of the polymer, and in particular its *band gap*. In quantum mechanics [28], each electron in a molecule can only occupy one of a discrete set of allowed energy levels. In solids, the overlaps between different repeating units of the material (for example, the repeat units in quasi-one-dimensional systems such as polymer chains [28-30]) cause these discrete energy levels to broaden into bands. The band gap is the energy difference between the top of the valence band and the bottom of the conduction band. (In terms which are equivalent but more familiar to chemists, the band gap is the energy difference between the highest occupied and the lowest unoccupied molecular orbitals of a solid). The band gaps of polymers range from very large values characteristic of insulators for polymers only containing σ electrons, to small values for polymers built from discrete repeat units with conjugated π-electron backbones, to zero for some unique aromatic ladder polymers [31-33]. The dielectric strength should have some correlation with the band gap, since it requires more energy to excite an electron from an occupied energy level to an unoccupied level (and thus increase its mobility) if the band gap is larger. As will be seen below, however, this is not a simple correlation, and its prominence also depends on other factors such as the temperature.

There is much *qualitative* similarity between mechanical strength (to be discussed in Chapter 11) and dielectric strength, so that all of the factors which affect the mechanical strength also affect the dielectric strength. For example, defects, which act as stress concentrators and reduce the mechanical strength, also usually reduce the dielectric strength. Consequently, the factors determining the abundance of defects, such as the specimen thickness, the presence of impurities, and the fabrication methods, often have similar effects of the mechanical and dielectric strength.

The qualitative similarity between mechanical and dielectric strength extends beyond just the similarities of the effects of specimen geometry, defects and fabrication conditions. The effects of temperature are also similar. Dielectric strength generally decreases with increasing temperature, and manifests a large drop near the glass transition temperature or the melting temperature, just like mechanical strength. The effects of crystallinity, crosslinking and plasticization *at high temperatures* also qualitatively resemble their effects on the mechanical strength.

The polarity of a polymer affects its dielectric strength. The dielectric strength of an insulating resin typically increases with increasing cohesive energy density, since the greater cohesive forces hold the polymer chains together more tightly. In particular, the dielectric strength at low temperatures is strongly dependent on the cohesive energy density. It decreases continuously and relatively rapidly for highly polar glassy polymers with increasing temperature due to the breakup of dipole-dipole interactions, while it is much more constant for nonpolar glassy polymers.

Differences are often found between mechanical and dielectric strength at *low temperatures*. These differences include (a) the relatively rapid drop of the dielectric strength (mentioned above) with increasing temperature in polar glassy polymers, (b) the slight *increase* of dielectric strength with increasing temperature in some nonpolar glassy polymers, (c) the decrease of dielectric strength sometimes observed with increasing crystallinity, and (d) the increase of dielectric strength by the presence of impurities rich in π electrons which allow increased energy absorption while mechanical strength is almost invariably reduced by the presence of impurities.

The combined consideration of three different general types of mechanisms can explain both the qualitative similarities and the differences between dielectric and mechanical strength, including why the differences are most prominent at low temperatures while the similarities are most obvious at high temperatures. *Electronic breakdown* (caused by the effect of the applied voltage on the electronic structure of the polymer) is considered to be the most important type of mechanism at low temperatures. *Thermal* and *electromechanical* mechanisms gain importance, and eventually become dominant, with increasing temperature. Thermal mechanisms involve the effects of the applied voltage in causing failure by local heating, decomposition or melting. Electromechanical effects involve the deformation of a compliant material such as a polymer by the mechanical pressure exerted by the large electrostatic fields in a stressed dielectric.

References and Notes for Chapter 9

1. *Encyclopedia of Polymer Science and Engineering*, *5*, Wiley-Interscience, New York (1986), contains excellent articles reviewing the electrical properties of polymers, the measurement of these properties, and the electrical, electronic and electrooptical applications of polymers.
2. J. R. Darby, N. W. Touchette and K. Sears, *Polymer Engineering and Science*, 7, 295-309 (1967).
3. D. W. van Krevelen, *Properties of Polymers*, second edition, Elsevier, Amsterdam (1976). Chapter 11 of this book deals with electrical properties.
4. W. R. Krigbaum and J. V. Dawkins, in *Polymer Handbook*, edited by J. Brandrup and E. H. Immergut, Wiley, New York, third edition (1989), VII/493-VII/496.
5. H. Fröhlich, *Theory of Dielectrics*, second edition, Clarendon Press, Oxford (1958).
6. N. G. McCrum, B. E. Read and G. Williams, *Anelastic and Dielectric Effects in Polymeric Solids*, John Wiley & Sons, New York (1967).
7. K. S. Cole and R. H. Cole, *J. Chem. Phys.*, *9*, 341-351 (1941).
8. D. W. Davidson and R. H. Cole, *J. Chem. Phys.*, *18*, 1417 (1950).
9. D. W. Davidson and R. H. Cole, *J. Chem. Phys.*, *19*, 1484-1490 (1951).
10. S. Havriliak and S. Negami, *J. Polym. Sci., C, No. 14*, 99-117 (1966).
11. S. Havriliak and S. Negami, *Polymer*, *8*, 161-210 (1967).
12. G. Williams and D. C. Watts, *Transactions of the Faraday Society*, *66*, 80-85 (1971).
13. S. Havriliak and D. G. Watts, *Polymer*, *27*, 1509-1512 (1986).

14. *Tables of Dielectric Materials*, Volume IV: Technical Report No. 57, Laboratory for Insulation Research, Massachusetts Institute of Technology (January 1953).

15. W. B. Westphal and A. Sils, *Dielectric Constant and Loss Data*, Technical Report AFML-TR-72-39, Air Force Materials Laboratory, Air Force Systems Command, Wright-Patterson Air Force Base, Ohio (April 1972).

16. J. A. Brydson, *Plastics Materials*, fifth edition, Butterworths, London (1989).

17. A. J. Curtis, *SPE Transactions*, 82-85 (January 1962).

18. Articles on special types of polymers in various editions and volumes of the *Encyclopedia of Polymer Science and Engineering*, tabulations in the *Polymer Handbook*, and product brochures and catalogs on commercial polymers, were among the sources of data used for the dielectric constant and dissipation factor.

19. A. J. Polak and R. C. Sundahl, *Polymer*, *30*, 1314-1318 (1989).

20. *Silicon Compounds: Register and Review*, Petrarch Systems, Silanes and Silicones Group, Bristol, Pennsylvania (1987).

21. *Plastics, Desk-Top Data Bank, Edition 5, Book A*, The International Plastics Selector, Inc., San Diego, California (1980).

22. L. A. Dissado and J. C. Fothergill, *Electrical Degradation and Breakdown in Polymers*, Peter Peregrinus Ltd., London (1992).

23. M. Winkeler, *Proceedings of the 19th Electrical/Electronics Insulation Conference*, held by the Institute of Electrical and Electronics Engineers (IEEE) in Chicago, 191-194 (1989).

24. J. H. Mason, *IEEE Transactions on Electrical Insulation*, *26*, 318-322 (1991).

25. J. K. Nelson, in *Electrical Properties of Solid Insulating Materials* (from Engineering Dielectrics Series), edited by R. Bartnikas and R. M. Eichhorn, ASTM Press, American Society for Testing and Materials, Philadelphia (1983), 445-520.

26. M. Ieda, *IEEE Transactions on Electrical Insulation*, *EI-15*, 206-224 (1980).

27. R. M. Eichhorn, *Conference Record of the 1990 IEEE International Symposium on Electrical Insulation*, held in Toronto, 2-3 (1990).

28. J.-M. André, in *Computational Modeling of Polymers*, edited by J. Bicerano, Marcel Dekker, New York (1992), Chapter 10.

29. R. Hoffmann, *Angew. Chem. Int. Ed. Engl.*, *26*, 846-878 (1987).

30. M.-H. Whangbo, in *Crystal Chemistry and Properties of Materials With Quasi-One-Dimensional Structures*, edited by J. Rouxel, D. Reidel Publishing Company, Boston (1986), 27-85.

31. *Conjugated Conducting Polymers*, edited by H. G. Kiess, Springer-Verlag, Heidelberg (1992).

32. M. Pomerantz, R. Cardona and P. Rooney, *Macromolecules*, *22*, 304-308 (1989).

33. J.-L. Brédas, in *Handbook of Conducting Polymers*, edited by T. Skotheim, Marcel Dekker, New York (1986), Chapter 25.

CHAPTER 10

MAGNETIC PROPERTIES

10.A. Background Information

The magnetic properties of polymers have historically not received nearly as much attention as their optical and electrical properties. This state of affairs is quite understandable. Polymers with certain optical and/or electrical properties are often required, for a wide variety of technological applications. On the other hand, there are no major applications where any of the magnetic properties of a polymer is a key performance requirement. Magnetic properties are therefore far more often studied for the information and the insights that they provide about the structure and dynamics of polymers, instead of being utilized to screen polymers for specific applications.

Materials have two types of magnetic properties:

1. The *magnetic susceptibility* ζ is the key magnetic property of a material as a whole [1]. (The Greek letter χ is usually used to denote the magnetic susceptibility. We are instead using the Greek letter ζ to avoid confusion with the connectivity indices which are also denoted by χ.) ζ is defined as follows:

$$\zeta \equiv \frac{\text{(Intensity of magnetization)}}{\text{(Strength of magnetic field)}} \qquad (10.1)$$

The more "susceptible" a material is to magnetic fields, the more intense the magnetization will be when the material is subjected to a magnetic field of a given strength. Materials fall into three general classes, namely, *diamagnetic*, *paramagnetic* and *ferromagnetic*, in the order of increasing magnitude of ζ. The difference between the ζ's of a diamagnetic and ferromagnetic material can be up to eleven orders of magnitude.

2. *Magnetic resonance* involves interactions of electrons or nuclei in certain configurations with a magnetic field, and has very important applications in the characterization of polymers.

(a) *Nuclear magnetic resonance* (NMR) spectroscopy is used to study the interactions of certain types of nuclei with a magnetic field. NMR provides detailed information concerning the structures (i.e., stereochemical configuration, geometrical isomerism, regioregularity of chain propagation during polymerization, branching of polymer chains, and copolymer sequence distribution), morphologies (i.e., orientation) and dynamic processes (i.e., segmental and side group motions, relaxations and transitions) of polymers at a local (atomic) level [2,3]. NMR chemical shifts are best calculated by means of sophisticated *ab initio* quantum mechanical calculations, especially for small molecules. However, the analysis of data on the NMR spectra of specific classes of materials with specific magnetically active nuclei has also led to empirical correlations of limited applicability which can be used for those combinations of material type and nucleus. The classic example is the study of the ^{13}C NMR spectra of saturated hydrocarbons, which led to an additive linear empirical equation for the ^{13}C NMR chemical shifts of this family of compounds [4]. Extensions of such studies to other types of chemical moieties and the incorporation of the effects of chain conformation statistics [5] within the framework of rotational isomeric state theory [6,7] have allowed similar calculations to become possible for the ^{13}C NMR spectra of many polymers.

(b) *Electron spin resonance* (ESR) spectroscopy is used to study the interaction of electrons with a magnetic field. ESR is especially useful in studying paramagnetic materials, such as materials containing unpaired electrons. For example, "free radicals" with unpaired electrons, which are often created during polymerization or thermooxidative degradation reactions, as well as during fracture processes induced by mechanical deformation, can be studied by ESR [8,9], providing valuable information concerning the processes and polymers.

Organic polymers generally only manifest paramagnetism under two circumstances:
1. If they contain transition metals.
2. If they contain unpaired electrons, as in the free radical or the triplet states.

Organic polymers are also generally believed to manifest ferromagnetism only when impurities (especially transition metals) are present, or when transition metals are deliberately incorporated into them. There is, however, some relatively recent evidence [10] of ferromagnetism in organic polymers containing some specialized types of structural and electronic features but not containing appreciable amounts of transition metal atoms.

Both paramagnetism and ferromagnetism involve orientation (alignment) of the spin axes of a substantial fraction of the electrons in a material upon application of a magnetic field. Elaborate theoretical models exist for ferromagnetism [11,12], which is a prototype problem of the modern theory of "critical phenomena" consisting of phase transitions and similar processes. It becomes clear, upon examination of the physical factors governing magnetization in these theories, why an ordinary organic polymer not containing either any unpaired electrons or any transition metal atoms should not normally be expected to manifest either paramagnetism or ferromagnetism.

Diamagnetism only involves a change in the paths of electrons under the influence of a magnetic field, without any need for spin alignment. The paths of electrons in all materials, including organic polymers, are affected (albeit often by a very small amount) when a magnetic field is applied. Diamagnetism is therefore a universal property of all materials. It is thus most appropriate to calculate the *molar diamagnetic susceptibility* for ordinary polymers. The symbol ζ will denote the molar diamagnetic susceptibility of polymers in the remainder of this chapter.

The magnetic susceptibility ζ and the molar diamagnetic susceptibility ζ_m (see Equation 10.2) are nearly independent of the temperature. They are usually expressed in units of 10^{-6} cc/g and 10^{-6} cc/mole, respectively, and ζ_m is predicted by using group contributions.

$$\zeta = \frac{\zeta_m}{M} \tag{10.2}$$

Van Krevelen summarized alternative group contribution tables for ζ_m, and suggested sets of recommended group contributions [1,13] for polymers. A correlation in terms of connectivity indices will be developed in Section 10.B for ζ_m values calculated from the recommended values of the group contributions. This new correlation will enable the prediction of ζ even when a polymer contains structural units for which the group contribution to ζ_m is not known.

10.B.   Correlation for the Molar Diamagnetic Susceptibility

Group contributions [1,13] were used to calculate the molar diamagnetic susceptibility ζ_m of a set of 110 polymers. The correlation between ζ_m(group) and all four of the zeroth-order and first-order connectivity indices was very strong. The largest correlation coefficient (0.9965) was obtained with $^0\chi^v$. This strong correlation indicates that the zeroth-order index $^0\chi^v$ accounts for 99.3% of the variation of ζ_m among the polymers in the dataset. Weighted fits were used, with weight factors of $100/\zeta_m$(group). The following equation has a standard deviation of only 4.1:

$$\zeta_m \approx 14.451591 \cdot {}^0\chi^v \tag{10.3}$$

A correction index N_{ds} was defined by Equation 10.4, to improve the correlation.

$$N_{ds} \equiv -24N_{Si} + 2N_{(carbon\ atoms\ with\ \delta=\delta^v=2)} + 2N_F + 2N_{Cl} - 5N_{Br} - 5N_{cyanide} - 7N_{C=C}$$
$$+ 3N_{(six\text{-}membered\ aromatic\ rings)} - 12N_{[-(C=O)-\ groups,\ except\ in\ carboxylic\ acid\ moieties]} \tag{10.4}$$

The first six terms in Equation 10.4 are atomic correction terms. Their use can be made unnecessary by optimizing the value of δ^v for use in the calculation of $^0\chi^v$, as illustrated in Section 3.B.2. For example, $N_{cyanide}$ equals the number of nitrogen atoms which have $\delta=1$ and $\delta^v=5$ in a standard calculation of $^0\chi^v$. Carbon atoms with $\delta=\delta^v=2$ refer to C atoms in methylene groups, namely in the bonding configuration $R-CH_2-R'$, where $R\neq H$ and $R'\neq H$. The last three terms in Equation 10.4 are group correction terms. $N_{C=C}$ was defined in Section 3.B.1.

The use of both $^0\chi^v$ and N_{ds} in the linear regression procedure for ζ_m resulted in the correlation given by Equation 10.5. This equation has a standard deviation of only 1.5, and a correlation coefficient of 0.9995. The standard deviation equals 1.6% of the average ζ_m(group) value of 95.6 for the polymers in the dataset.

$$\zeta_m \approx 14.646919 \cdot {^0\chi^v} + 0.498884 \cdot N_{ds} \tag{10.5}$$

The results of these calculations are summarized in Table 10.1 and shown in Figure 10.1. The predictive power of Equation 10.5 is also demonstrated in Table 10.1, where predicted values of ζ_m are listed for 15 additional polymers whose ζ_m(group) cannot be calculated as a result of the lack of some of the necessary group contribution values.

Table 10.1. The molar diamagnetic susceptibility ζ_m calculated by group contributions, the correction index N_{ds} used in the correlation equation, and the fitted or predicted (fit/pre) values of ζ_m, for 125 polymers. For 15 of these polymers, ζ_m cannot be calculated via group contributions, but can be calculated by using the new correlation. The values of $^0\chi^v$ used in the correlation equation are all listed in Table 2.2. ζ_m is expressed in units of 10^{-6} cc/mole.

Polymer	ζ_m(group)	N_{ds}	ζ_m(fit/pre)
Polyoxymethylene	16.35	2	17.33
Polyethylene	22.70	4	22.71
Poly(vinyl fluoride)	26.95	4	26.34
Polyoxyethylene	27.70	4	28.69
Poly(vinyl alcohol)	27.85	2	26.36
Polyacrylonitrile	31.35	-3	31.19
Poly(vinylidene fluoride)	31.55	6	31.75
Polypropylene	34.85	2	34.46
Polytrifluoroethylene	35.80	6	35.38
Poly(1,4-butadiene)	35.90	-3	36.13
Poly(1,2-butadiene)	35.95	-5	35.13
Poly(thiocarbonyl fluoride)	36.20	4	38.33
Poly(β-alanine)	36.70	-8	37.35
Polyacrylamide	37.35	-10	35.58
Poly(maleic anhydride)	--------	-24	37.53
Poly(ethylene sulfide)	38.70	4	40.65
Poly(vinyl chloride)	38.85	4	37.42

☞ ☞ ☞ **TABLE 10.1 IS CONTINUED IN THE NEXT PAGE.** ☞ ☞ ☞

Table 10.1. CONTINUED FROM THE PREVIOUS PAGE.

Polymer	ζ_m(group)	N_{ds}	ζ_m(fit/pre)
Polyoxytrimethylene	39.05	6	40.04
Poly(acrylic acid)	39.35	2	39.66
Poly(vinyl methyl ether)	39.85	2	40.44
Poly(propylene oxide)	39.85	2	40.44
Polyfumaronitrile	40.00	-10	39.67
Polytetrafluoroethylene	40.40	8	40.78
Polymethacrylonitrile	43.85	-3	44.70
Poly(dimethyl siloxane)	45.00	-24	45.27
Poly(1-butene)	46.20	4	45.81
Polyisobutylene	47.35	2	47.97
Poly(vinyl bromide)	47.85	-3	46.08
Polyisoprene	48.30	-3	49.64
Poly(methyl acrylate)	48.85	-10	47.75
Poly(vinyl acetate)	48.85	-10	47.75
Poly(p-phenylene)	50.00	3	49.97
Polyoxytetramethylene	50.40	8	51.40
Poly(vinyl ethyl ether)	51.20	4	51.79
Poly(vinylidene chloride)	51.35	6	53.89
Poly(methacrylic acid)	51.85	2	53.18
Polychlorotrifluoroethylene	52.30	8	51.85
Polychloroprene	52.30	-1	52.60
Polyepichlorohydrin	55.35	6	54.75
Poly(propylene sulfone)	--------	2	55.39
Poly(methyl α-cyanoacrylate)	57.85	-15	58.00
Poly(ethyl acrylate)	60.20	-8	59.11
Poly(vinyl propionate)	60.20	-8	59.11
Poly(methyl methacrylate)	61.35	-10	61.27
Poly(methyl α-chloroacrylate)	65.35	-8	64.23
Poly(1-butene sulfone)	--------	4	66.74
Poly(N-vinyl pyrrolidone)	--------	-4	67.74
Poly(2,5-benzoxazole)	--------	3	68.69
Poly(p-vinyl pyridine)	--------	5	69.01
Poly(vinylidene bromide)	69.35	-8	71.22
Poly(4-methyl-1-pentene)	70.05	4	69.68

☞ ☞ ☞   TABLE 10.1 IS CONTINUED IN THE NEXT PAGE. ☞ ☞ ☞

Table 10.1. CONTINUED FROM THE PREVIOUS PAGE.

Polymer	ζ_m(group)	N_{ds}	ζ_m(fit/pre)
Poly(ε-caprolactone)	70.75	-2	70.07
Poly(ε-caprolactam)	70.75	-2	71.41
Poly(p-xylylene)	72.70	7	72.68
Poly(ethyl methacrylate)	72.70	-8	72.62
Poly(methyl ethacrylate)	72.70	-8	72.62
Polystyrene	73.35	5	70.91
Poly(vinyl n-butyl ether)	73.90	8	74.50
Poly(vinyl sec-butyl ether)	74.70	4	74.90
Poly(2-hydroxyethyl methacrylate)	77.05	-6	75.88
Poly[oxy(methyl γ-trifluoropropylsilylene)]	78.20	-14	80.26
Poly[oxy(2,6-dimethyl-1,4-phenylene)]	80.00	3	82.98
Poly(n-butyl acrylate)	82.90	-4	81.83
Poly[oxy(methylphenylsilylene)]	83.50	-21	81.73
Poly(o-methyl styrene)	84.85	5	84.43
Poly(p-methyl styrene)	84.85	5	84.43
Poly(vinyl pivalate)	84.85	-10	84.37
Poly(α-methyl styrene)	85.85	5	84.43
Poly(vinyl cyclohexane)	85.85	12	85.04
Poly(vinyl benzoate)	87.35	-7	84.21
Poly(o-chloro styrene)	88.85	7	87.39
Poly(p-chloro styrene)	88.85	7	87.39
Poly(N-phenyl maleimide)	--------	-21	89.20
Poly(styrene sulfide)	89.35	5	88.85
Poly(1-hexene sulfone)	--------	8	89.45
Poly(p-methoxy styrene)	89.85	5	90.41
Poly[oxy(dipropylsilylene)]	90.40	-16	90.69
Poly(8-aminocaprylic acid)	93.45	2	94.12
Poly(oxy-2,2-dichloromethyltrimethylene)	94.70	12	93.93
Poly(oxy-1,1-dichloromethyltrimethylene)	94.70	12	93.93
Poly(n-butyl methacrylate)	95.40	-4	95.33
Poly(sec-butyl methacrylate)	96.20	-8	95.73
Poly(t-butyl methacrylate)	97.35	-10	97.89
Poly(vinyl butyral)	--------	8	97.39
Poly(p-bromo styrene)	97.85	0	96.05

☞ ☞ ☞ **TABLE 10.1 IS CONTINUED IN THE NEXT PAGE.** ☞ ☞ ☞

Table 10.1. CONTINUED FROM THE PREVIOUS PAGE.

Polymer	ζ_m(group)	N_{ds}	ζ_m(fit/pre)
Poly(ethylene terephthalate)	100.70	-17	99.27
Poly(N-methyl glutarimide)	---------	-20	102.48
Poly(α-vinyl naphthalene)	---------	8	103.97
Poly(benzyl methacrylate)	111.20	-5	109.08
Poly(cyclohexyl methacrylate)	112.35	0	111.85
Poly(n-hexyl methacrylate)	118.10	0	118.04
Poly(2-ethylbutyl methacrylate)	118.90	-4	118.44
Poly(p-t-butyl styrene)	120.85	5	121.05
Poly(tetramethylene terephthalate)	123.40	-13	121.98
Poly(N-vinyl carbazole)	---------	8	126.30
Poly(p-phenylene terephthalamide)	128.00	-18	129.22
Poly[thio bis(4-phenyl)carbonate]	135.00	-6	137.15
Poly[3,5-(4-phenyl-1,2,4-triazole)-1,4-phenylene]	---------	6	135.37
Poly(ethylene-2,6-naphthalenedicarboxylate)	138.70	-14	132.33
Polybenzobisoxazole	140.00	6	137.38
Poly(n-octyl methacrylate)	140.80	4	140.75
Poly(hexamethylene adipamide)	141.50	-4	142.83
Poly(hexamethylene isophthalamide)	146.10	-9	147.38
Bisphenol-A polycarbonate	155.00	-6	155.83
Polybenzobisthiazole	160.00	6	161.30
Poly(1,4-cyclohexylidene dimethylene terephthalate)	164.20	-9	161.60
Poly(oxy-1,4-phenylene-oxy-1,4-phenylene-carbonyl-1,4-phenylene)	166.50	-3	169.18
Poly[2,2-hexafluoropropane bis(4-phenyl)carbonate]	176.00	6	180.39
Phenoxy resin	185.20	10	186.23
Poly(hexamethylene sebacamide)	186.90	4	188.25
Poly[N,N'-(p,p'-oxydiphenylene)pyromellitimide]	190.00	-39	195.99
Torlon	193.00	-25	195.95
Poly[1,1-(1-phenylethane) bis(4-phenyl)carbonate]	193.50	-3	192.29
Poly[2,2'-(m-phenylene)-5,5'-bibenzimidazole]	200.00	9	190.04
Poly[1,1-(1-phenyltrifluoroethane) bis(4-phenyl)carbonate]	204.00	3	204.56
Poly[2,2-propane bis{4-(2,6-dimethylphenyl)}carbonate]	205.00	-6	209.89
Resin F	211.25	-8	212.96
Poly[2,2-propane bis{4-(2,6-dichlorophenyl)}carbonate]	221.00	2	221.72
Poly[diphenylmethane bis(4-phenyl)carbonate]	232.00	0	228.74

☞ ☞ ☞   **TABLE 10.1 IS CONTINUED IN THE NEXT PAGE.** ☞ ☞ ☞

Table 10.1. CONTINUED FROM THE PREVIOUS PAGE.

Polymer	ζ_m(group)	N_{ds}	ζ_m(fit/pre)
Poly[4,4'-diphenoxy di(4-phenylene)sulfone]	---------	12	232.76
Poly[1,1-biphenylethane *bis*(4-phenyl)carbonate]	243.50	0	242.26
Poly[2,2-propane *bis*{4-(2,6-dibromophenyl)}carbonate]	257.00	-26	256.38
Resin G	260.60	18	259.76
Poly[4,4'-isopropylidene diphenoxy di(4-phenylene)sulfone]	---------	12	269.38
Ultem	332.00	-33	338.52

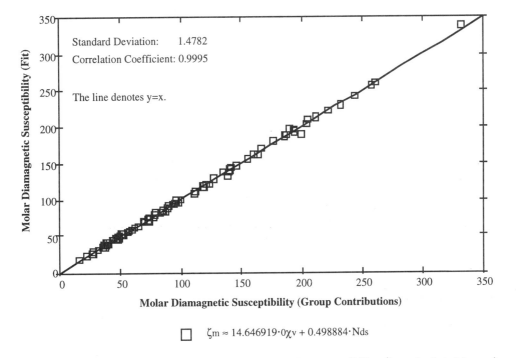

Standard Deviation: 1.4782
Correlation Coefficient: 0.9995

The line denotes y=x.

$$\zeta_m \approx 14.646919 \cdot 0\chi v + 0.498884 \cdot N_{ds}$$

Figure 10.1. A correlation for the molar diamagnetic susceptibility ζ_m calculated by using group contributions for 110 polymers. ζ_m is expressed in 10^{-6} cc/mole.

The observed magnetic susceptibilities ζ of 14 polymers, gathered from a variety of literature sources, were tabulated by van Krevelen [1]. These values are compared in Table 10.2 with the results predicted by using equations 10.4 and 10.5. Van Krevelen [1] indicated that two of the experimental values were questionable, by placing a question mark in parentheses (?) after them. The experimental and predicted values of ζ are compared in Figure 10.2 for the dataset of 12 polymers obtained after removing these two outliers. A correlation coefficient of 0.9526 and a standard deviation of $0.0394 \cdot 10^{-6}$ cc/g is found between the observed and predicted values of ζ. The correlation coefficient indicates that equations 10.4 and 10.5 account for 90.7% of the variation of the ζ values in the dataset. The standard deviation is equal to 6.2% of the average value of $0.6342 \cdot 10^{-6}$ cc/g for ζ in this dataset. The agreement is again very good, but the correlation coefficient is lower and the relative standard deviation is higher than was found above for ζ_m. This is not surprising, since the much wider range of values of the extensive (molar) property ζ_m includes the effects of the repeat unit molecular weight which varies greatly among polymers, while this superimposed effect has been removed from the intensive property ζ.

Table 10.2. Observed magnetic susceptibilities (ζ) of polymers [1], and the ζ values predicted by using equations 10.4 and 10.5. ζ is in units of 10^{-6} cc/g. Experimental values followed by a question mark in parentheses (?) have been designated in this manner in the source.

Polymer	ζ(exp)	ζ(pre)	Polymer	ζ(exp)	ζ(pre)
Polytetrafluoroethylene	0.38	0.41	Polyoxyethylene	0.63	0.65
Poly[oxy(2,6-dimethyl-1,4-phenylene)]	0.47 (?)	0.69	Polystyrene	0.705	0.681
Poly(ethylene terephthalate)	0.505	0.516	Poly(2,3-dimethyl-1,4-butadiene)	0.72	0.77
Polyoxymethylene	0.52	0.58	Polycyclopentadiene	0.72	0.63
Poly(methyl methacrylate)	0.59	0.61	Poly(hexamethylene adipamide)	0.76 (?)	0.63
Poly[oxy(methylphenylsilylene)]	0.60	0.60	Polypropylene	0.80	0.82
Poly(dimethyl siloxane)	0.62	0.61	Polyethylene	0.82	0.81

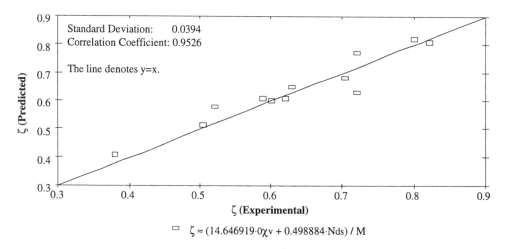

$$\zeta \approx (14.646919 \cdot 0\chi v + 0.498884 \cdot Nds) / M$$

Figure 10.2. Comparison between observed and predicted magnetic susceptibilities ζ, in units of 10^{-6} cc/g, for 12 polymers. The two outlying data points for which the experimental values have been indicated to be questionable in the source ([1], see Table 10.2) are not shown.

References and Notes for Chapter 10

1. D. W. van Krevelen, *Properties of Polymers*, second edition, Elsevier, Amsterdam (1976). Chapter 12 of this book deals with magnetic properties.

2. F. A. Bovey and L. W. Jelinski, article titled *"Nuclear Magnetic Resonance"*, *Encyclopedia of Polymer Science and Engineering*, *10*, Wiley-Interscience, New York (1987).

3. *Solid State NMR of Polymers*, edited by L. J. Mathias, Plenum Press, New York (1991).

4. D. M. Grant and E. G. Paul, *J. Am. Chem. Soc.*, *86*, 2984-2990 (1964).

5. A. E. Tonelli, *NMR Spectroscopy and Polymer Microstructure: The Conformational Connection*, VCH Publishers, New York (1989).

6. P. J. Flory, *The Principles of Polymer Chemistry*, Cornell University Press, Ithaca, New York (1953).

7. P. J. Flory, *Statistical Mechanics of Chain Molecules*, Interscience Publishers, New York (1969).

8. K. DeVries and D. Roylance, article titled *"Electron-Spin Resonance"*, *Encyclopedia of Polymer Science and Engineering*, *5*, Wiley-Interscience, New York (1986).

9. S. N. Zhurkov, *International Journal of Fracture Mechanics*, *1*, 311-323 (1965).

10. H. Ishida, article titled *"Magnetic Polymers"*, *Encyclopedia of Polymer Science and Engineering, Supplement Volume*, Wiley-Interscience, New York (1989).

11. J. M. Ziman, *Principles of the Theory of Solids*, second edition, Cambridge University Press, London (1972).

12. S.-K. Ma, *Modern Theory of Critical Phenomena*, W. A. Benjamin, Inc., Reading, Massachusetts (1976).

13. D. W. van Krevelen, in *Computational Modeling of Polymers*, edited by J. Bicerano, Marcel Dekker, New York (1992), Chapter 1.

CHAPTER 11

MECHANICAL PROPERTIES

11.A. Stress-Strain Behavior of Polymers

The mechanical properties of materials are of great importance in engineering applications [1]. When a mechanical force is applied to a specimen, the deformation of the specimen is described in terms of its *stress-strain behavior*. The stress-strain behavior quantifies the *stress* (mechanical load) σ required to achieve a certain amount of *strain* (deformation or displacement) ε, as a function of ε and variables such as the temperature T and the strain rate $\dot{\varepsilon}$.

The stress has dimensions of force per unit area, i.e., negative pressure. In this chapter, all quantities with dimensions of stress will be expressed in megapascals (MPa). The strain is always a dimensionless quantity. For a tensile deformation, it is defined simply as the fractional change of the length of the specimen as a result of the deformation.

Many different types of testing modes can be used to measure the mechanical properties of polymers. Uniaxial tension, uniaxial compression, plane strain compression, and simple shear, are among the most important testing modes. Each type of testing mode creates a different stress state along the three principal axes of the specimen during deformation. Several general types of stress-strain behavior are exhibited by specimens, depending on the intrinsic material properties, the preparation and processing conditions of the specimens, and the test conditions. For example, the stress-strain behavior of rubbery amorphous polymers (above T_g at the testing temperature) is drastically different from the stress-strain behavior of glassy amorphous polymers (below T_g at the testing temperature). The types of tests used to measure the mechanical properties of polymers, and the general types of stress-strain behavior which are observed, are discussed in many standard references [2-6].

For our purposes, it suffices to recall that there are significant differences between the *small-strain* (i.e., small deformation) and the *large-strain* behavior of polymers. The small-strain behavior is discussed in Section 11.B. It is mainly described by the moduli (or compliances) and Poisson's ratio. The large-strain behavior is discussed in Section 11.C. It refers to failure mechanisms observed in specimens, such as brittle fracture, shear yielding and crazing, resulting either in their complete breakage or in a catastrophic deterioration of their mechanical properties.

The correlations reviewed in sections 11.B and 11.C estimate the mechanical properties of amorphous polymers in terms of several material parameters of a more fundamental nature. Creep, stress relaxaion and fatigue will also be discussed briefly in Section 11.D. As discussed in Section 11.E, the key contribution of our work has been the development of the general correlations presented in this book to calculate these material parameters, thus significantly extending the range and structural diversity of polymers for which the existing structure-property relationships for the mechanical properties can be applied.

All mechanical properties depend on crystallinity, orientation and crosslinking. These three factors fall outside the main focus of this book, which is on the properties of uncrosslinked isotropic amorphous polymers. See Kinloch and Young [2] and van Krevelen [7] for the effects of crystallinity and orientation on the mechanical properties.

11.B. Small-Strain Behavior: Moduli, Compliances, and Poisson's Ratio

11.B.1. Definitions and Phenomenology

The *modulus* is the most important small-strain mechanical property. It is the key indicator of the "stiffness" or "rigidity" of specimens made from a material. It quantifies the resistance of the specimens to mechanical deformation, in the limit of infinitesimally small deformation. There are three major types of moduli. The *bulk modulus* B is the resistance of a specimen to isotropic compression (pressure). The *Young's modulus* E is its resistance to uniaxial tension (being stretched). The *shear modulus* G is its resistance to simple shear deformation (being twisted).

Each type of modulus is defined in terms of the stress σ required to deform a specimen by a strain of ε, in the limit of an infinitesimally small deformation of the type quantified by that modulus. For example, Young's modulus is defined by Equation 11.1, in the limit of $\varepsilon \to 0$ under uniaxial tension. This equation shows that the stress σ required to achieve a small strain of ε under uniaxial tension is proportional to E.

$$\sigma = E \cdot \varepsilon \tag{11.1}$$

As a rule of thumb, the "stiffer" or "more rigid" an uncrosslinked material is, the greater will the resistance of its specimens be to any type of deformation. Consequently, the larger the moduli of such a material, the less will its specimens "comply" with a deformation. For example, a pendant weight will cause much more "creep" (extension with time, as a result of the force exerted by the hanging weight) when hung at the end of a low-modulus fiber than when hung at the end of a high-modulus fiber.

The *compliance* is the reciprocal of the modulus. It indicates the extent to which the specimens of a given material are expected to comply with a deformation. The *shear compliance* J, the *bulk compliance* (or *compressibility*) κ and the *tensile compliance* D are defined as follows:

$$J \equiv \frac{1}{G} \tag{11.2}$$

$$\kappa \equiv \frac{1}{B} \tag{11.3}$$

$$D \equiv \frac{1}{E} \tag{11.4}$$

Poisson's ratio ν is defined by Equation 11.5 for an isotropic (unoriented) specimen. It describes the effect of the application of a deformation (strain) in one direction (i.e., along the x axis) on the dimensions of the specimen along the other two directions (i.e., the y and z axes) perpendicular to the direction of the applied deformation. The fractional change of volume dV/V of the specimen is given by Equation 11.6 in terms of the strains $d\varepsilon_x$, $d\varepsilon_y$ and $d\varepsilon_z$ along the three axes. If $\nu=0.5$, the strains along the y and z axes will each be opposite in sign and of exactly half the magnitude of the strain applied along the x axis, so that the total volume of the specimen will not change. The value of ν is very close to 0.5 for rubbery polymers. When $\nu<0.5$, as is the case for glassy polymers, the strains along the y and z axes will not be sufficient to counter the strain applied along the x axis, so that the total volume of the specimen will change.

$$\nu \equiv -\frac{d\varepsilon_y}{d\varepsilon_x} = -\frac{d\varepsilon_z}{d\varepsilon_x} \tag{11.5}$$

$$\frac{dV}{V} = d\varepsilon_x + d\varepsilon_y + d\varepsilon_z \tag{11.6}$$

The value of ν provides the fundamental relationships, which are given by Equation 11.7, between the three types of moduli. It is therefore only necessary to know the value of one of the moduli, and the value of ν, to estimate the remaining two moduli by using Equation 11.7, and all three compliances by using equations 11.2-11.4.

$$E = 2(1 + \nu)G = 3(1 - 2\nu)B \tag{11.7}$$

E, G, B and ν are functions of both the temperature and the frequency (rate) of measurement. They are often treated as complex (dynamic) properties. The real portion quantifies the energy which is reversibly stored by the "elastic" component of the deformation. The imaginary portion quantifies the energy lost (i.e., dissipated) by the "viscous" component of the deformation. For example, equations 11.8 and 11.9 define the complex Young's modulus E^*, its real and imaginary components E' and E", and the mechanical loss tangent $\tan \delta_E$ under uniaxial tension.

$$E^* \equiv E' - i \cdot E" \tag{11.8}$$

$$\tan \delta_E \equiv \frac{E"}{E'} \tag{11.9}$$

Equations 11.8 and 11.9 are isomorphous to equations 9.9 and 9.10, which define the storage and loss components of the complex dielectric constant ε^*. Similar equations are also used to define the complex bulk modulus B^*, the complex shear modulus G^*, and the complex Poisson's ratio ν^*, in terms of their elastic and viscous components. The physical mechanism giving rise to the viscous portion of the mechanical properties is often called "damping" or "internal friction", and has important implications in terms of the performance of materials [8-15].

There are fundamental interrelationships, as well as significant differences, between the physical factors which determine the magnitudes of the elastic and viscous components of the three types of mechanical moduli, and of the corresponding electrical quantities. The definitive identification of these interrelationships and differences, and their embodiment in simple and reliable predictive equations, are areas of ongoing research in fundamental polymer physics. We will, therefore, only deal with the real-valued properties E, G, B and ν, which are equivalent to the elastic (real) components E', G', B' and ν' of the corresponding complex quantities. Furthermore, only the temperature dependences of these properties will be discussed. The values calculated for these properties will therefore correspond to measurements made under "typical" testing conditions, with commonly used strain rates for each type of test.

The moduli are approximately proportional both to the strengths of the links between the atoms in a material and to the number of links per unit of cross-sectional area. (The use of the term "link" instead of "bond" is deliberate, to encompass both the covalent chemical bonds in polymer chains, and the nonbonded interchain attractions such as hydrogen bonds and van der Waals interactions.) Each "link" can be viewed as a spring, with a certain value of its "spring constant" or "force constant". The moduli of a polymer (i.e., the rigidities of macroscopic specimens made from the polymer) thus generally increase with increasing chain stiffness and with increasing cohesive energy density. When a stress is applied, the weakest links (i.e., the nonbonded interchain interactions) deform much more easily than the strong covalent bonds along the individual chains. The "network" of nonbonded interchain interactions therefore plays an especially crucial role in determining the magnitudes of the moduli of a polymer.

E and G are both three to four orders of magnitude lower in rubbery polymers than in glassy polymers as a result of the breakdown of most of the interchain "links" at the glass transition. Low but nonzero values of E and G remain, however, in the "rubbery plateau" region above T_g, as a result of the restoring forces of entropic origin which oppose the deformation of the entangled polymer chains.

While most of the translational constraints to local motion disappear at T_g, a volumetric deformation under pressure does not invoke translational constraints. The reduction of B at the glass transition is therefore much less drastic than the reductions of E and G. Consequently, while all three moduli are of the same order of magnitude below T_g, E and G are of the same order of magnitude but B is much larger than both E and G above T_g.

The general trends for the $\nu(T)$ of amorphous polymers [16] can be summarized as follows:

1. For glassy polymers below (T_g - 20K), $v(T)$ typically ranges from 0.32 to 0.44. It decreases with increasing cross-sectional area of the polymer chains. It is largest (0.40 to 0.44) for very "thin" polymer chains, and smallest (0.32 to 0.36) for very "fat" polymer chains with large side groups.

2. $v(T)$ increases very slowly up to about 20K below T_g. This increase is essentially monotonic. The details of the usually small variations from a simple and smooth increase are mainly caused by factors, such as the dynamics of the polymer (for example, its secondary relaxations) and the thermal history of a given specimen, which cannot be easily accounted for by simple and general structure-property relationships.

3. $v(T)$ approaches 0.5 rapidly as the specimen softens rapidly between $T=(T_g$ - 20K) and $T=T_g$.

4. $v(T)$ is very close to 0.5 for $T>T_g$. It is, however, always very slightly lower than 0.5, and usually between 0.499 and 0.5. There is a physical reason for why $v(T)$ is not exactly equal to 0.5 for $T>T_g$. If $v(T)$ were exactly equal to 0.5, $B(T)$ would become infinite according to Equation 11.7 since $E(T)>0$, and the polymer would become totally incompressible above T_g according to Equation 11.3, in contradiction to the observed behavior.

11.B.2.   Structure-Property Relationships for Glassy Polymers

11.B.2.a.   Introductory Remarks

Because of the great practical importance of the small-strain mechanical properties of polymers, many different structure-property relationships have been developed for these properties. In our practical work, we have found a set of correlations developed by Seitz [16] to have the greatest utility. Consequently, these correlations, which are summarized in Section 11.B.2.b, have been implemented in the software package which automates the use of our predictive schemes. It will be seen that the input parameters needed to use the equations of Seitz for a polymer of arbitrary structure can all be estimated by using correlations developed in earlier chapters of this monograph.

An alternative correlation for the bulk modulus, in terms of the molar Rao function [7], is of historical interest. In addition, a correlation of similar nature has been developed [7b] for the shear modulus, in terms of the molar Hartmann function. These two relationships will be discussed in sections 11.B.2.c and 11.B.2.d, respectively. Correlations will also be developed in these sections to enable the estimation of the molar Rao function and the molar Hartmann function for a polymer of arbitrary structure.

11.B.2.b. Correlations by Seitz for the Elastic Moduli

Seitz [16] developed structure-property relationships for the mechanical properties, up to $(T_g - 20K)$, of isotropic amorphous polymers which are glassy at room temperature, i.e., which have $T_g > 298K$. If only the values of the properties at room temperature are of interest, equations 11.10 and 11.13 can be utilized to estimate $v(298K)$ and $B(298K)$, respectively, and equations 11.2, 11.3, 11.4 and 11.7 can then be used to estimate the compliances and the other two moduli. If the temperature dependence of the mechanical properties is of interest, equations 11.10-11.12 must all be used to estimate $v(T)$, for substitution into Equation 11.7. *These equations by Seitz are our preferred method for predicting the elastic properties of glassy polymers.*

$$v(298K) \approx 0.513 - 3.054 \cdot 10^{-6} \cdot \sqrt{\frac{V_w}{l_m}} \tag{11.10}$$

$$v_0 \approx v(298K) - \frac{14900}{T_g}\{0.00163 + \exp[0.459(285 - T_g)]\} \tag{11.11}$$

$$v(T) \approx v_0 + \frac{50T}{T_g}\{0.00163 + \exp[0.459(T - T_g - 13)]\} \tag{11.12}$$

$$B(T) \approx 8.23333E_{cohl}\left[\frac{5V(0K)^4}{V(T)^5} - \frac{3V(0K)^2}{V(T)^3}\right] \tag{11.13}$$

The standard deviations of predictions made by using these equations at $T=298K$ are roughly 12% (relative) for E and 0.019 (absolute) for v. In Equation 11.10, l_m denotes the length of a repeat unit of the polymer in its fully extended conformation, in centimeters (1 Å = 10^{-8} cm). For example, for a vinylic polymer such as poly(methyl methacrylate) (PMMA, Figure 11.1), the fully extended conformation corresponds to the all-*trans* conformation. It is shown in Figure 11.1 that l_m can be estimated easily and quite accurately for PMMA, by assuming ideal tetrahedral bonding around the carbon atoms in the backbone, and then using simple trigonometry. The same estimate is also valid for other polymers with vinylic backbones, such as polystyrene (Figure 1.1), poly(vinyl fluoride) (Figure 2.3), and the four polymers shown in Figure 4.1.

For polymers with more complicated repeat units, it is generally both much more efficient and much more accurate to use an interactive molecular modeling software package to estimate l_m. When molecular modeling is used to estimate l_m, caution must be exercised to ensure that the model molecule containing the repeat unit is built in its fully extended conformation. Depending on the complexity of the repeat unit, and on the quality of the default values of the bond lengths and bond angles supplied by the software package, it may or may not be necessary to optimize the geometry of the model molecule via force field energy minimization.

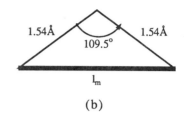

$$l_m = 1.54 \cdot \sin(109.5°)/\sin(35.25°) \approx 2.52\text{Å}.$$

(a)   (b)

Figure 11.1. Calculation of a rough estimate for the length of a polymeric repeat unit in its fully extended conformation, with poly(methyl methacrylate) as the example. (a) The repeat unit structure. The distance defined as l_m is shown. (b) Use of simple trigonometry (i.e., the "sine rule"), and the assumption of ideal tetrahedral bonding (bond angles of 109.5°) with a carbon-carbon bond length of 1.54 Å, to estimate that $l_m \approx 2.52$ Å$=2.52 \cdot 10^{-8}$ cm. This rough estimate is also valid for other polymers with vinylic backbones, such as polystyrene (Figure 1.1). *A much better estimate ($l_m=2.32$Å) is obtained for polystyrene by using Biosym's Polymer Version 6.0 molecular modeling program. This difference has a large effect on some predicted properties.*

V(0K) is the molar volume at absolute zero temperature. *The Fedors-type cohesive energy E_{cohl} (see Section 5.B) must be used in Equation 11.13.* The substitution of l_m in cm, E_{cohl} in J/mole, and V_w, V(T) and V(0K) in cc/mole, into equations 11.10-11.13, provides the calculated values of the moduli in MPa. For example, equations 3.15 [to estimate V(0K)], 11.10, 11.13 and 11.7, with $T_g \approx 373$K, $l_m \approx 2.52 \cdot 10^{-8}$ cm, $V_w \approx 63.25$ cc/mole, V(298K)≈ 99.1 cc/mole, V(0K)≈ 91.4 cc/mole, and $E_{coh} \approx 40310$ J/mole, give the predicted values of v(298K)≈ 0.360, E(298K)≈ 3000 MPa, G(298K)≈ 1100 MPa, and B(298K)≈ 3570 MPa, for polystyrene. (The calculated moduli are rounded off to the nearest 10 MPa, since the moduli generally cannot be measured to greater accuracy.) The best available experimental results (also rounded off) are v(298K)≈ 0.354, E(298K)≈ 3300 MPa, G(298K)≈ 1220 MPa and B(298K)≈ 3770 MPa [16], and B(298K)≈ 3500 MPa ([17], measured directly by pressure-volume-temperature experiments).

Equation 11.13, when it is combined with Equation 11.7, can be used to estimate E(T) and G(T) from T=0K up to approximately (T_g - 20K). It cannot be used for T>(T_g - 20K), since v(T)$\rightarrow 0.5$, so that E(T) and G(T) estimated by combining equations 11.13 and 11.7 rapidly approach zero, between (T_g - 20K) and T_g. As mentioned earlier, the experimental values of E(T) and G(T) do not approach zero as T$\rightarrow T_g$, but gradually decrease, and then rapidly drop between T_g and approximately (T_g + 30K), to the much lower range of values for the "rubbery plateau" region which will be discussed in Section 11.B.3.

The data for E and v at room temperature, assembled by Seitz [16] and used in developing his correlations, are listed in Table 11.1. His paper also contains additional data, as a function of T. Note that the range of variation of E at room temperature (from 2100 MPa to 4000 MPa) is quite small for this very diverse set of glassy amorphous polymers, all of which happen to be significantly below their glass transition temperatures at room temperature. The Poisson's ratios also occupy a distinct range (0.320≤v≤0.441), well below the range for liquids and for rubbery amorphous polymers (0.499<v<0.5) and above that for rigid inorganic solids (0.1≤v≤0.3). Bulk and shear moduli calculated from these data by using Equation 11.7 are also listed in Table 11.1, and correspond to indirectly-measured experimental values since Equation 11.7 is exact.

Table 11.1. Data for Young's modulus (E) and Poisson's ratio (v) at room temperature [16]. The bulk (B) and shear (G) moduli were calculated by substituting E and v into Equation 11.7, and rounded off to the nearest 10 MPa. The moduli (E, B and G) are all listed in MPa. See Table 11.6 and Table 11.7 for additional and/or alternative measured values of E.

Polymer	E	v	G	B
Equimolar copolymers of bisphenol-A terephthalate and isophthalate	2100	0.433	730	5220
Bisphenol-A polycarbonate	2300	0.401	820	3870
Phenoxy resin	2300	0.402	820	3910
Poly[oxy(2,6-dimethyl-1,4-phenylene)]	2300	0.410	820	4260
Equimolar copolymers of bisphenol-A based ester and carbonate	2300	0.440	800	6390
Poly[4,4'-isopropylidene diphenoxy di(4-phenylene)sulfone]	2500	0.441	870	7060
Poly(vinyl chloride)	2600	0.385	940	3770
Poly[4,4'-sulfone diphenoxy di(4-phenylene)sulfone]	2600	0.420	920	5420
Poly(p-t-butyl styrene)	3000	0.330	1130	2940
Poly(ethylene terephthalate)	3000	0.430	1050	7140
Poly(α-methyl styrene)	3100	0.320	1170	2870
Poly(o-methyl styrene)	3100	0.345	1150	3330
Poly(methyl methacrylate)	3200	0.371	1170	4130
Polystyrene	3300	0.354	1220	3770
Poly(styrene-co-methyl methacrylate) (35/65 by weight)	3500	0.361	1290	4200
Poly(styrene-co-α-methyl styrene) (52/48 by weight)	3800	0.330	1430	3730
Poly(styrene-co-acrylonitrile) (76/24 by weight)	3800	0.366	1390	4730
Poly(o-chloro styrene)	4000	0.320	1520	3700
Poly(p-methyl styrene)	-------	0.341	-------	-------
Torlon	-------	0.380	-------	-------

The correlations summarized above provide a general method to estimate the moduli, compliances and Poisson's ratio for glassy amorphous polymers, as functions of the structure of the polymeric repeat unit and the temperature of measurement. Independent correlations for B(T) and G(T) will now be discussed, and compared with the correlations presented above.

11.B.2.c.   Bulk Modulus via Molar Rao Function

A method used to calculate B(T) [7] utilizes relationships between the bulk modulus B, the density ρ, and the velocity of acoustic (sound) waves in materials. B(T) is approximately equal to the product of the density with the sixth power of the ratio (U_R/V), where U_R is the *molar Rao function* (or the *molar sound velocity function*). U_R is independent of the temperature, and is also useful in predicting the thermal conductivity (Chapter 14).

$$B(T) \approx \rho(T)\left[\frac{U_R}{V(T)}\right]^6$$

(11.14)

A unit conversion factor of 10^{-7} must be used to convert the result of Equation 11.14, with $\rho(T)$ in grams/cc, V(T) in cc/mole and U_R in $cm^{10/3}/(sec^{1/3}\cdot mole)$, into B(T) in MPa. For example, for polystyrene, which has $U_R=5980$ $cm^{10/3}/(sec^{1/3}\cdot mole)$, $\rho(298K)=1.05$ grams/cc and V(298K)=99.1 cc/mole, Equation 11.14 gives $B(298K) \approx 10^{-7}\cdot 1.05\cdot(5980/99.1)^6 \approx 5070$ MPa, comparing poorly with the best available experimental values of 3770 MPa [16] and 3500 MPa [17]. By contrast, Equation 11.13 gave excellent agreement with these experimental values.

Based on such comparisons, Equation 11.13 in our judgement provides a better description of B(T) for $T \leq T_g$. Consequently, only Equation 11.13 was implemented for the calculation of B(T) for glassy polymers in the software package automating the use of the preferred version of our work. A correlation was also developed, however, for U_R, and will be presented below.

The bulk modulus has a very strong (sixth power) dependence on the molar Rao function U_R. It is, therefore, essential to develop an accurate correlation for U_R in terms of connectivity indices for the correlation to be useful. U_R was calculated for 129 polymers from group contributions. The group contributions of a structural unit to U_R are different in liquids and in solid polymers. Group contributions provided by van Krevelen [7a,18] for polymers were used for most groups.

The group contribution of 1280 used for bromine (-Br), 1460 used for carboxylic acid (-COOH), 4035 used for *meta*-phenylene rings, 3975 used for *ortho*-phenylene rings, 5100 used for 2,5-benzoxazole groups (Figure 11.2), 6100 used for *cis*-benzobisoxazole groups (Figure 11.2), and 7300 used for *trans*-benzobisthiazole groups (Figure 11.2), were our own estimates.

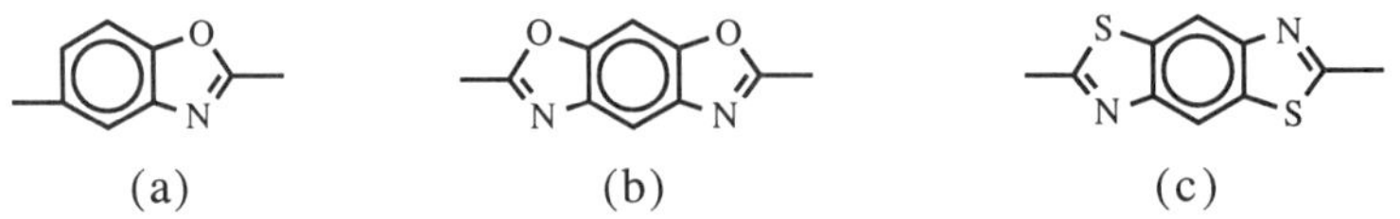

(a)   (b)   (c)

Figure 11.2. Schematic illustration of three structural units. (a) 2,5-Benzoxazole is the repeat unit of poly(2,5-benzoxazole) (AB-PBO). (b) *Cis*-benzobisoxazole. (c) *Trans*-benzobisthiazole.

Most of the latest set of values of the group contributions published for U_R [7b] only differ by a small amount from earlier values [7a,18]. Our correlation for U_R, which was developed prior to the publication of this latest set of group contributions, was not revised to incorporate these changes in the group contributions. On the other hand, the earlier sources did not contain group contributions for silicon-containing moieties. The values provided in the latest edition of van Krevelen's book [7b] were therefore used for silicon-containing structural units.

The zeroth-order and first-order connectivity indices correlated strongly with U_R(group). The strongest correlation for U_R, with a correlation coefficient of 0.9868, was found with $^0\chi^v$. The use of additional terms proportional to $^1\chi$ and N_H, where N_H denotes the total number of hydrogen atoms in the repeat unit, significantly improved the quality of the fit. In graph theoretical terms, N_H is the number of vertices removed from the "complete graph" of a repeat unit, containing all atoms as vertices, to obtain the hydrogen-suppressed graph (see Chapter 2) which forms the basis of the connectivity index method. Weighted fits with weight factors of 10000/U(group) were used in the linear regression procedure. The best fit found completely within the formalism of zeroth-order and first-order connectivity indices is given by Equation 11.15. Equation 11.15 has a standard deviation of 281.2, and a correlation coefficient of 0.9983 which indicates that it accounts for 99.7% of the variation of U_R(group).

$$U_R \approx 407.75 \cdot {}^0\chi^v + 674.60 \cdot {}^1\chi + 132.62 \cdot N_H \tag{11.15}$$

Most of the remaining deviation in the correlation for U_R is corrected when a term proportional to N_{UR}, which is defined by Equation 11.16, is included in the linear regression procedure. The final correlation for U_R is given by Equation 11.17. It has a standard deviation of only 100.7, and a correlation coefficient of 0.9997 which indicates that Equation 11.30 accounts for 99.94% of the variation of the U_R(group) values. The standard deviation is only 1.3% of the average U_R(group) value of 7693 for the polymers in this dataset.

$$N_{UR} \equiv - N_F + 2N_{Cl} - 2N_{(-S-)} - 2N_{OH} + N_{cyanide} - 3N_{cyc}$$
$$+ 4N_{(six-membered\ aromatic\ rings)} + N_{Si} \tag{11.16}$$

$$U_R \approx 441.30 \cdot {}^0\chi^v + 540.62 \cdot {}^1\chi + 154.69 \cdot N_H + 105.91 \cdot N_{UR} \tag{11.17}$$

When using Equation 11.15, or using equations 11.16 and 11.17, $^0\chi^v$ should be evaluated with all silicon atoms replaced by carbon atoms in the repeat unit. For example, $^0\chi^v$ values listed in Table 2.3 should be used instead of the values listed in Table 2.2 for silicon-containing polymers whose connectivity indices were listed in Chapter 2.

The first five terms in Equation 11.16 are atomic correction terms, whose use can be made unnecessary, as discussed in Section 3.B.2, by optimizing the δ^v values of the corresponding atomic configurations. N_{cyc} (defined in Section 3.B.1) and $N_{(six\text{-membered aromatic rings})}$ are the only two terms in N_{UR} which are not simple atomic correction terms. They refer to general types of structural features, and therefore allow all of the different types of groups containing those structural features to be treated in exactly the same manner.

The results of these calculations are summarized in Table 11.2 and shown in Figure 11.3. The predictive power of Equation 11.17 is also demonstrated in Table 11.2, where predicted values of U_R are listed for 21 polymers whose U_R(group) cannot be calculated because of the lack of some of the group contributions.

Table 11.2. Molar Rao function U_R calculated by group contributions, the number of hydrogen atoms N_H in the repeat unit, the correction index N_{UR} used in the correlation equation, and the fitted or predicted (fit/pre) values of U_R, for 150 polymers. For 21 of these polymers, U_R(group) cannot be calculated because of the lack of some of the group contributions, but U_R(fit/pre) can be calculated by using the new correlation. The values of $^0\chi^v$ and $^1\chi$ used in the correlation equation are listed in Table 2.2, with the exception of silicon-containing polymers for which the alternative $^0\chi^v$ values listed in Table 2.3 are used. U_R is in $cm^{10/3}/(sec^{1/3}\cdot mole)$.

Polymer	U_R(group)	N_H	N_{UR}	U_R(fit/pre)
Polyoxymethylene	1280	2	0	1342
Poly(thiocarbonyl fluoride)	1600	0	-4	1594
Polyethylene	1760	4	0	1783
Poly(vinyl fluoride)	1830	3	-1	1845
Poly(vinylidene fluoride)	1930	2	-2	1887
Poly(vinyl alcohol)	1930	4	-2	1925
Polytrifluoroethylene	2000	1	-3	1978
Polytetrafluoroethylene	2100	0	-4	2036
Polyoxyethylene	2160	4	0	2234
Poly(ethylene sulfide)	2310	4	-2	2382
Poly(vinyl chloride)	2480	3	2	2497
Poly(vinyl bromide)	2610	3	0	2651

☞ ☞ ☞   **TABLE 11.2 IS CONTINUED IN THE NEXT PAGE.** ☞ ☞ ☞

Table 11.2. CONTINUED FROM THE PREVIOUS PAGE.

Polymer	U_R(group)	N_H	N_{UR}	U_R(fit/pre)
Polyacrylonitrile	2630	3	1	2599
Polypropylene	2730	6	0	2690
Polychlorotrifluoroethylene	2750	0	-1	2688
Poly(acrylic acid)	2790	4	-2	2818
Polyoxytrimethylene	3040	6	0	3126
Poly(1,2-butadiene)	-------	6	0	3106
Poly(propylene oxide)	3130	6	0	3140
Poly(vinyl methyl ether)	3130	6	0	3161
Poly(1,4-butadiene)	3160	6	0	3143
Poly(vinyl sulfonic acid)	-------	4	-2	3210
Poly(vinylidene chloride)	3230	2	4	3189
Poly(maleic anhydride)	-------	2	-3	3269
Poly(vinyl methyl sulfide)	3280	6	-2	3309
Poly(dimethyl siloxane)	3300	6	1	3240
Poly(β-alanine)	3460	5	0	3313
Poly(vinylidene bromide)	3490	2	0	3498
Polyfumaronitrile	3500	2	2	3433
Polymethacrylonitrile	3600	5	1	3497
Poly(1-butene)	3610	8	0	3602
Poly(vinyl methyl ketone)	3630	6	0	3583
Polychloroprene	3660	5	2	3879
Polyisobutylene	3730	8	0	3576
Polyepichlorohydrin	3760	5	2	3859
Poly(methacrylic acid)	3790	6	-2	3721
Polyisoprene	3910	8	0	4072
Polyoxytetramethylene	3920	8	0	4017
Poly(propylene sulfone)	3980	6	0	3987
Poly(vinyl acetate)	3980	6	0	4024
Poly(methyl acrylate)	3980	6	0	4054
Poly(vinyl ethyl ether)	4010	8	0	4053
Poly(p-phenylene)	4100	4	4	4106
Poly(ethylene oxalate)	4260	4	0	4462
Poly[oxy(p-phenylene)]	4500	4	4	4548

☞☞☞   **TABLE 11.2 IS CONTINUED IN THE NEXT PAGE.** ☞☞☞

Table 11.2. CONTINUED FROM THE PREVIOUS PAGE.

Polymer	U_R(group)	N_H	N_{UR}	U_R(fit/pre)
Poly[thio(p-phenylene)]	4650	4	2	4696
Poly(methyl α-chloroacrylate)	4730	5	2	4764
Poly(1-butene sulfone)	4860	8	0	4900
Poly(vinyl propionate)	4860	8	0	4937
Poly(ethyl acrylate)	4860	8	0	4946
Poly(methyl α-cyanoacrylate)	4880	5	1	4879
Poly(methyl methacrylate)	4980	8	0	4958
Poly(2,5-benzoxazole)	5100	3	4	5309
Poly(N-vinyl pyrrolidone)	-------	9	-3	5272
Poly(p-hydroxybenzoate)	5350	4	4	5441
Poly(4-methyl-1-pentene)	5460	12	0	5380
Poly(vinyl cyclopentane)	5480	12	-3	5482
Poly(vinyl trimethylsilane)	5640	12	1	5482
Poly(ϵ-caprolactone)	5650	10	0	5793
Poly(p-vinyl pyridine)	-------	7	4	5655
Poly(o-vinyl pyridine)	-------	7	4	5655
Poly(vinyl n-butyl ether)	5770	12	0	5836
Poly(ethyl methacrylate)	5860	10	0	5849
Poly(vinyl sec-butyl ether)	5860	12	0	5851
Poly(p-xylylene)	5860	8	4	5881
Poly(methyl ethacrylate)	5860	10	0	5882
Poly(vinyl n-butyl sulfide)	5920	12	-2	5985
Poly(p-fluorostyrene)	5930	7	3	5952
Poly(2-hydroxyethyl methacrylate)	5940	10	-2	5976
Cellulose	-------	10	-9	5963
Polystyrene	5980	8	4	5867
Poly(ϵ-caprolactam)	6100	11	0	5988
Poly($\alpha,\alpha,\alpha',\alpha'$-tetrafluoro-$p$-xylylene)	6200	4	0	6140
Poly(oxy-2,2-dichloromethyltrimethylene)	6270	8	4	6381
Poly(oxy-1,1-dichloromethyltrimethylene)	6270	8	4	6381
Poly(vinyl cyclohexane)	6380	14	-3	6374
Poly[oxy(methylphenylsilylene)]	6400	8	5	6435
Poly(o-chloro styrene)	6455	7	6	6612

☞☞☞   **TABLE 11.2 IS CONTINUED IN THE NEXT PAGE.** ☞☞☞

Table 11.2. CONTINUED FROM THE PREVIOUS PAGE.

Polymer	U_R(group)	N_H	N_{UR}	U_R(fit/pre)
Poly(styrene sulfide)	6530	8	2	6466
Poly[oxy(2,6-dimethyl-1,4-phenylene)]	6550	8	4	6425
Poly(p-chloro styrene)	6580	7	6	6603
Poly(5-vinyl-2-methylpyridine)	-------	9	4	6584
Poly(1-hexene sulfone)	6620	12	0	6683
Poly(n-butyl acrylate)	6620	12	0	6730
Poly(o-methyl styrene)	6705	10	4	6805
Poly(p-bromo styrene)	6710	7	4	6758
Poly(chloro-p-xylylene)	6725	7	6	6626
Poly[oxy(dipropylsilylene)]	6820	14	1	6873
Poly(vinyl pivalate)	6830	12	0	6711
Poly(p-methyl styrene)	6830	10	4	6796
Poly(α-methyl styrene)	6980	10	4	6770
Poly(vinyl benzoate)	7230	8	4	7201
Poly(p-methoxy styrene)	7230	10	4	7267
Poly(vinyl butyral)	-------	14	-3	7301
Poly(N-phenyl maleimide)	-------	7	1	7600
Poly(n-butyl methacrylate)	7620	14	0	7633
Poly(sec-butyl methacrylate)	7710	14	0	7647
Poly(t-butyl methacrylate)	7830	14	0	7609
Poly(8-aminocaprylic acid)	7860	15	0	7772
Poly(N-methyl glutarimide)	-------	13	-3	7996
Polyoxynaphthoate	-------	6	8	8188
Poly(phenyl methacrylate)	8230	10	4	8105
Poly(ethylene terephthalate)	8360	8	4	8568
Poly(cyclohexyl α-chloroacrylate)	8380	13	-1	8419
Poly(α-vinyl naphthalene)	-------	10	8	8623
Poly(cyclohexyl methacrylate)	8630	16	-3	8612
Poly(benzyl methacrylate)	9110	12	4	8997
Poly(n-hexyl methacrylate)	9380	18	0	9416
Poly(2-ethylbutyl methacrylate)	9470	18	0	9451
Poly(p-t-butyl styrene)	9680	16	4	9483
Poly(tetramethylene terephthalate)	10120	12	4	10352

☞☞☞   **TABLE 11.2 IS CONTINUED IN THE NEXT PAGE.** ☞☞☞

Table 11.2. CONTINUED FROM THE PREVIOUS PAGE.

Polymer	U_R(group)	N_H	N_{UR}	U_R(fit/pre)
Polybenzobisoxazole	10200	6	8	10617
Poly(N-vinyl carbazole)	--------	11	8	10261
Poly[thio *bis*(4-phenyl)carbonate]	10350	8	6	10579
Poly[3,5-(4-phenyl-1,2,4-triazole)-1,4-phenylene]	--------	9	8	10787
Poly(*n*-octyl methacrylate)	11140	22	0	11200
Poly(ethylene-2,6-naphthalenedicarboxylate)	--------	10	8	11315
Polybenzobisthiazole	11400	6	4	10914
Poly(*p*-phenylene terephthalamide)	11600	10	8	11272
Poly(hexamethylene adipamide)	12200	22	0	11977
Poly(pentabromophenyl methacrylate)	--------	5	4	12614
Bisphenol-A polycarbonate	12650	14	8	12693
Poly(hexamethylene isophthalamide)	12715	18	4	12525
Poly[2,2-hexafluoropropane *bis*(4-phenyl)carbonate]	12950	8	2	13040
Poly[oxy(2,6-diphenyl-1,4-phenylene)]	13050	12	12	12779
Poly[2,2-propane *bis*{4-(2,6-difluorophenyl)}carbonate]	13150	10	4	13069
Poly(oxy-1,4-phenylene-oxy-1,4-phenylene-carbonyl-1,4-phenylene)	14000	12	12	14095
Poly[1,1-cyclohexane *bis*(4-phenyl)carbonate]	--------	18	5	14594
Phenoxy resin	14660	20	6	14625
Poly[2,2'-(*m*-phenylene)-5,5'-bibenzimidazole]	--------	12	12	15114
Poly(hexamethylene sebacamide)	15720	30	0	15544
Poly[2,2-propane *bis*{4-(2,6-dichlorophenyl)}carbonate]	15750	10	16	15674
Poly[1,1-(1-phenylethane) *bis*(4-phenyl)carbonate]	15900	16	12	15888
Poly[1,1-(1-phenyltrifluoroethane) *bis*(4-phenyl)carbonate]	16050	13	9	16061
Poly[2,2-propane *bis*{4-(2,6-dibromophenyl)}carbonate]	16270	10	8	16292
Poly[2,2-hexafluoropropane *bis*{4-(2,6-dibromophenyl)}carbonate]	16570	4	2	16640
Poly[2,2-propane *bis*{4-(2,6-dimethylphenyl)}carbonate]	16750	22	8	16447
Torlon	17080	14	12	16797
Poly[N,N'-(*p,p*'-oxydiphenylene)pyromellitimide]	17100	10	12	16891
Resin F	17720	22	6	17647
Poly(dicyclooctyl itaconate)	--------	34	-6	18093
Poly[4,4'-diphenoxy di(4-phenylene)sulfone]	18450	16	16	18612
Poly[*o*-biphenylenemethane *bis*(4-phenyl)carbonate]	--------	16	16	18705
Poly[diphenylmethane *bis*(4-phenyl)carbonate]	19150	18	16	19082

☞☞☞   **TABLE 11.2 IS CONTINUED IN THE NEXT PAGE.** ☞☞☞

Table 11.2. CONTINUED FROM THE PREVIOUS PAGE.

Polymer	U_R(group)	N_H	N_{UR}	U_R(fit/pre)
Poly[di(*n*-octyl) itaconate]	19430	38	0	19701
Poly[4,4'-sulfone diphenoxy di(4-phenylene)sulfone]	19700	16	16	19915
Resin G	19930	31	2	20066
Poly[1,1-biphenylethane *bis*(4-phenyl)carbonate]	20000	20	16	19994
Poly[4,4'-isopropylidene diphenoxy di(4-phenylene)sulfone]	21300	22	16	21316
Ultem	28485	24	20	28249

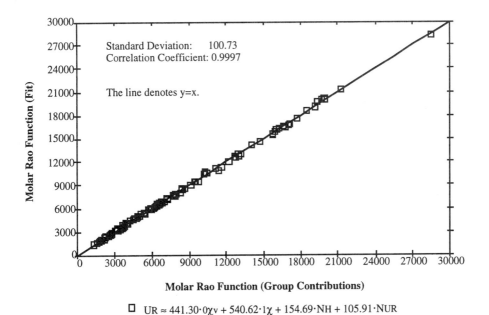

Standard Deviation: 100.73
Correlation Coefficient: 0.9997

The line denotes y=x.

$\square$ $U_R \approx 441.30 \cdot {}^0\chi v + 540.62 \cdot {}^1\chi + 154.69 \cdot N_H + 105.91 \cdot N_{UR}$

Figure 11.3. A fit using connectivity indices, to the molar Rao function U_R calculated by group contributions, for 129 polymers. U_R is expressed in units of $cm^{10/3}/(sec^{1/3} \cdot mole)$.

11.B.2.d. Shear Modulus via Molar Hartmann Function

Hartmann and Lee [19] extended the types of calculations possible by using U_R to crosslinked polymers. They then developed [20] an alternative additive quantity similar to U_R, to calculate the shear modulus. Although the additive quantity for the shear modulus was initially developed [20] by using crosslinked epoxies as test cases, it is also useful for the much simpler uncrosslinked polymers. In the latest edition of his book, van Krevelen [7b] provided tentative values for the group contributions to this new additive quantity, which he named the *molar Hartmann function* U_H. U_H has the same units as U_R. Its value can be used in Equation 11.18 to estimate the shear modulus G, again with a unit conversion factor of 10^{-7} to convert the calculated G(T) to MPa. For example, for polystyrene, U_H=4725 $cm^{10/3}$/($sec^{1/3}$·mole), ρ(298K)=1.05 grams/cc and V(298K)=99.1 cc/mole. Equation 11.15 gives G(298K) $\approx$ 10^{-7}·1.05·$(4725/99.1)^6$ $\approx$ 1230 MPa, compared with the experimental value ([16], also see Table 11.1) of 1220 MPa.

$$G(T) \approx \rho(T)\left[\frac{U_H}{V(T)}\right]^6 \qquad (11.18)$$

Caution must be exercised in using the molar Hartmann function U_H as a predictive tool:

1. The group contributions for U_H have been indicated [7b] to be tentative at this time.
2. These group contributions are not strictly additive. For example, if PMMA (Figure 11.1) is broken down into its smallest subunits (a methylene group, a carbon atom with four non-hydrogen nearest neighbors, an acrylic ester group, and two methyl groups) and the group contributions are added, U_H(group)=3870. On the other hand, if the tetravalent backbone carbon atom and all of the side groups attached to it are combined into one unit, and the group contribution for this larger subunit is used, U_H(group)=4325. Similarly, in calculating U_H(group) for many other polymers with sizeable repeat units, it is possible to combine group contributions in different but equally reasonable ways to calculate somewhat different values.
3. Acoustic (sound propagation) measurements in materials, rather than merely conventional types of mechanical tests, played an important role in developing U_H.

With these cautionary remarks in mind, a correlation was developed for U_H. The following argument shows that, to a reasonable first approximation, U_H should be proportional to U_R:

1. The shear and the bulk moduli are related to each other by Equation 11.7.
2. Poisson's ratios of glassy polymers fall into the relatively narrow range of 0.32 to 0.44.
3. According to equations 11.14 and 11.18, (U_H/U_R) is proportional to $(G/B)^{1/6}$, so that the differences between the (G/B) ratios result from small differences between the (U_H/U_R) ratios.

The calculations were started with the dataset of 150 polymers listed in Table 11.2. The group contributions provided in Table 14.2 of the latest edition of van Krevelen's book [7b] were used to calculate U_H(group) for the 118 polymers in this dataset for which all of the required group contributions were available. Weight factors of 10000/U_H(group) were used in the linear regression. Equation 11.19 resulted in a standard deviation of 142.0, and a correlation coefficient of 0.9992 which indicates that it accounts for 99.84% of the variation of U_H(group).

$$U_H \approx 0.8017 \cdot U_R \qquad (11.19)$$

Most of the remaining deviation in the correlation for U_H is corrected when a term proportional to N_{UH}, which is defined by Equation 11.32, is included. In Equation 11.20, an α-substituted acrylate is any acrylate which has a non-hydrogen substituent attached to the α-carbon atom, namely the backbone carbon atom to which the acrylate unit is attached. The polymers which require this correction term are all structural variants of PMMA (Figure 11.1). A backbone ester unit is simply an ester (-COO-) linkage in the backbone of the repeat unit, as in poly(glycolic acid) (Figure 2.4), in contrast to an ester unit in a side group, as in PMMA.

$$N_{UH} \equiv 14N_{(\alpha\text{-substituted acrylate})} + 5N_{cyanide} + 3N_{Cl} + 5N_{C=C} + 5N_{sulfone}$$
$$- 4N_{(backbone\ ester)} \tag{11.20}$$

The final correlation for U_H is given by Equation 11.21. It has a standard deviation of only 77.2, and a correlation coefficient of 0.9998 which indicates that it accounts for 99.96% of the variation of the U_H(group) values. The standard deviation is only 1.3% of the average U_H(group) value of 5968 for the polymers in this dataset. The results of these calculations are summarized in Table 11.3 and shown in Figure 11.4. The predictive power of Equation 11.21 is also demonstrated in Table 11.3, where predicted values of U_H are listed for 32 polymers whose U_H(group) cannot be calculated because of the lack of some of the group contributions.

$$U_H \approx 0.7931 \cdot U_R + 24.7545 \cdot N_{UH} \tag{11.21}$$

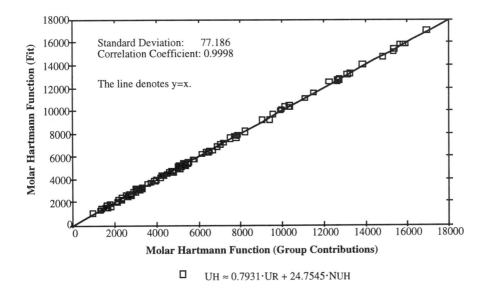

Figure 11.4. A correlation for the molar Hartmann function U_H calculated by group contributions, for a set of 118 polymers. U_H is in units of $cm^{10/3}/(sec^{1/3} \cdot mole)$.

Table 11.3. Molar Hartmann function U_H calculated by group contributions, the molar Rao function U_R and the correction index N_{UH} used in the correlation for U_H, and the fitted or predicted (fit/pre) values of U_H, for 150 polymers. For 32 of these polymers, U_H(group) cannot be calculated because of the lack of some of the group contributions, but U_H(fit/pre) can be calculated by using the new correlation. U_H is in $cm^{10/3}/(sec^{1/3}\cdot mole)$.

Polymer	U_H(group)	U_R	N_{UH}	U_H(fit)
Polyoxymethylene	975	1280	0	1015
Poly(thiocarbonyl fluoride)	1340	1600	0	1269
Polyethylene	1350	1760	0	1396
Poly(vinyl fluoride)	1445	1830	0	1451
Poly(vinyl alcohol)	1545	1930	0	1531
Poly(vinylidene fluoride)	1575	1930	0	1531
Polyoxyethylene	1650	2160	0	1713
Polytrifluoroethylene	1670	2000	0	1586
Poly(ethylene sulfide)	1790	2310	0	1832
Polytetrafluoroethylene	1800	2100	0	1666
Poly(vinyl chloride)	2125	2480	3	2041
Poly(vinyl bromide)	2125	2610	0	2070
Polyacrylonitrile	2195	2630	5	2210
Poly(acrylic acid)	-------	2790	0	2213
Polypropylene	2325	2730	0	2165
Polyoxytrimethylene	2325	3040	0	2411
Polychlorotrifluoroethylene	2335	2750	3	2255
Poly(1,4-butadiene)	2550	3160	5	2630
Poly(1,2-butadiene)	-------	3106	5	2587
Poly(maleic anhydride)	-------	3269	0	2593
Poly(propylene oxide)	2625	3130	0	2482
Poly(vinyl methyl ether)	2625	3130	0	2482
Poly(dimethyl siloxane)	2640	3300	0	2617
Poly(vinyl sulfonic acid)	-------	3210	5	2670
Poly(vinylidene chloride)	2710	3230	6	2710
Poly(vinylidene bromide)	2710	3490	0	2768
Poly(β-alanine)	2750	3460	0	2744
Poly(vinyl methyl sulfide)	2765	3280	0	2601
Poly(vinyl methyl ketone)	2925	3630	0	2879
Polymethacrylonitrile	2990	3600	5	2979

☞ ☞ ☞   **TABLE 11.3 IS CONTINUED IN THE NEXT PAGE.** ☞ ☞ ☞

Table 11.3. CONTINUED FROM THE PREVIOUS PAGE.

Polymer	U_H(group)	U_R	N_{UH}	U_H(fit)
Poly(1-butene)	3000	3610	0	2863
Polyoxytetramethylene	3000	3920	0	3109
Poly(methacrylic acid)	-------	3790	0	3006
Polyisobutylene	3025	3730	0	2958
Polyfumaronitrile	3040	3500	10	3023
Polyepichlorohydrin	3100	3760	3	3056
Polychloroprene	3150	3660	8	3101
Poly(ethylene oxalate)	3150	4260	-8	3181
Poly(vinyl acetate)	3225	3980	0	3157
Poly(methyl acrylate)	3225	3980	0	3157
Polyisoprene	3280	3910	5	3225
Poly(vinyl ethyl ether)	3300	4010	0	3180
Poly(p-phenylene)	3300	4100	0	3252
Poly(propylene sulfone)	3325	3980	5	3280
Poly[oxy(p-phenylene)]	3600	4500	0	3569
Poly[thio(p-phenylene)]	3740	4650	0	3688
Poly(vinyl propionate)	3900	4860	0	3854
Poly(ethyl acrylate)	3900	4860	0	3854
Poly(1-butene sulfone)	4000	4860	5	3978
Poly(2,5-benzoxazole)	-------	5100	0	4045
Poly(N-vinyl pyrrolidone)	-------	5272	0	4181
Poly(methyl α-chloroacrylate)	4195	4730	17	4172
Poly(p-hydroxybenzoate)	4200	5350	-4	4144
Poly(ε-caprolactone)	4275	5650	-4	4382
Poly(methyl methacrylate)	4325	4980	14	4296
Poly(methyl α-cyanoacrylate)	4345	4880	19	4341
Poly(vinyl cyclopentane)	-------	5480	0	4346
Poly(4-methyl-1-pentene)	4350	5460	0	4330
Poly(p-vinyl pyridine)	-------	5655	0	4485
Poly(o-vinyl pyridine)	-------	5655	0	4485
Poly(vinyl trimethylsilane)	4515	5640	0	4473
Poly(vinyl n-butyl ether)	4650	5770	0	4576
Poly(p-xylylene)	4650	5860	0	4648

☞ ☞ ☞   **TABLE 11.3 IS CONTINUED IN THE NEXT PAGE.** ☞ ☞ ☞

Table 11.3. CONTINUED FROM THE PREVIOUS PAGE.

Polymer	U_H(group)	U_R	N_{UH}	U_H(fit)
Polystyrene	4725	5980	0	4743
Cellulose	-------	5963	0	4729
Poly(p-fluorostyrene)	4745	5930	0	4703
Poly(ε-caprolactam)	4775	6100	0	4838
Poly(vinyl n-butyl sulfide)	4790	5920	0	4695
Poly(vinyl sec-butyl ether)	4800	5860	0	4648
Poly(ethyl methacrylate)	5000	5860	14	4994
Poly(methyl ethacrylate)	5000	5860	14	4994
Poly(oxy-2,2-dichloromethyltrimethylene)	5035	6270	6	5121
Poly(oxy-1,1-dichloromethyltrimethylene)	5035	6270	6	5121
Poly(2-hydroxyethyl methacrylate)	5045	5940	14	5058
Poly(vinyl cyclohexane)	5045	6380	0	5060
Poly(α,α,α',α'-tetrafluoro-p-xylylene)	5100	6200	0	4917
Poly[oxy(2,6-dimethyl-1,4-phenylene)]	5100	6550	0	5195
Poly[oxy(methylphenylsilylene)]	5160	6400	0	5076
Poly(styrene sulfide)	5165	6530	0	5179
Poly(o-chloro styrene)	-------	6455	3	5194
Poly(5-vinyl-2-methylpyridine)	-------	6584	0	5222
Poly(n-butyl acrylate)	5250	6620	0	5250
Poly(chloro-p-xylylene)	5250	6725	3	5408
Poly(o-methyl styrene)	-------	6705	0	5318
Poly[oxy(dipropylsilylene)]	5340	6820	0	5409
Poly(p-chloro styrene)	5345	6580	3	5293
Poly(p-bromo styrene)	5345	6710	0	5322
Poly(1-hexene sulfone)	5350	6620	5	5374
Poly(vinyl pivalate)	5370	6830	0	5417
Poly(p-methyl styrene)	5475	6830	0	5417
Poly(α-methyl styrene)	5490	6980	0	5536
Poly(vinyl benzoate)	5745	7230	0	5734
Poly(p-methoxy styrene)	5775	7230	0	5734
Poly(vinyl butyral)	-------	7301	0	5790
Poly(N-phenyl maleimide)	-------	7600	0	6028
Poly(8-aminocaprylic acid)	6125	7860	0	6234

☞ ☞ ☞   **TABLE 11.3 IS CONTINUED IN THE NEXT PAGE.** ☞ ☞ ☞

Table 11.3. CONTINUED FROM THE PREVIOUS PAGE.

Polymer	U_H(group)	U_R	N_{UH}	U_H(fit)
Poly(N-methyl glutarimide)	-------	7996	0	6341
Poly(*n*-butyl methacrylate)	6350	7620	14	6390
Polyoxynaphthoate	-------	8188	-4	6395
Poly(ethylene terephthalate)	6450	8360	-8	6432
Poly(*sec*-butyl methacrylate)	6500	7710	14	6461
Poly(*t*-butyl methacrylate)	6620	7830	14	6557
Poly(α-vinyl naphthalene)	-------	8623	0	6839
Poly(phenyl methacrylate)	6845	8230	14	6874
Poly(cyclohexyl α-chloroacrylate)	7065	8380	17	7067
Poly(cyclohexyl methacrylate)	7195	8630	14	7191
Poly(benzyl methacrylate)	7520	9110	14	7572
Poly(*n*-hexyl methacrylate)	7700	9380	14	7786
Poly(*p-t*-butyl styrene)	7770	9680	0	7677
Poly(tetramethylene terephthalate)	7800	10120	-8	7828
Poly(2-ethylbutyl methacrylate)	7850	9470	14	7857
Polybenzobisoxazole	-------	10200	0	8090
Poly(N-vinyl carbazole)	-------	10261	0	8138
Poly[thio *bis*(4-phenyl)carbonate]	8240	10350	0	8209
Poly[3,5-(4-phenyl-1,2,4-triazole)-1,4-phenylene]	-------	10787	0	8555
Poly(ethylene-2,6-naphthalenedicarboxylate)	-------	11315	-8	8776
Polybenzobisthiazole	-------	11400	0	9041
Poly(*n*-octyl methacrylate)	9050	11140	14	9182
Poly(*p*-phenylene terephthalamide)	9400	11600	0	9200
Poly(hexamethylene adipamide)	9550	12200	0	9676
Bisphenol-A polycarbonate	9900	12650	0	10033
Poly(hexamethylene isophthalamide)	9950	12715	0	10084
Poly[oxy(2,6-diphenyl-1,4-phenylene)]	10140	13050	0	10350
Poly[2,2-hexafluoropropane *bis*(4-phenyl)carbonate]	10305	12950	0	10271
Poly(pentabromophenyl methacrylate)	--------	12614	14	10350
Poly[2,2-propane *bis*{4-(2,6-difluorophenyl)}carbonate]	10350	13150	0	10429
Poly(oxy-1,4-phenylene-oxy-1,4-phenylene-carbonyl-1,4-phenylene)	11100	14000	0	11103
Phenoxy resin	11520	14660	0	11627
Poly[1,1-cyclohexane *bis*(4-phenyl)carbonate]	--------	14594	0	11575

☞ ☞ ☞   **TABLE 11.3 IS CONTINUED IN THE NEXT PAGE.** ☞ ☞ ☞

Table 11.3. CONTINUED FROM THE PREVIOUS PAGE.

Polymer	U_H(group)	U_R	N_{UH}	U_H(fit)
Poly[2,2'-(*m*-phenylene)-5,5'-bibenzimidazole]	--------	15114	0	11987
Poly(hexamethylene sebacamide)	12250	15720	0	12468
Poly[1,1-(1-phenylethane) *bis*(4-phenyl)carbonate]	12615	15900	0	12610
Poly[1,1-(1-phenyltrifluoroethane) *bis*(4-phenyl)carbonate]	12720	16050	0	12729
Poly[2,2-propane *bis*{4-(2,6-dichlorophenyl)}carbonate]	12750	15750	12	12788
Poly[2,2-propane *bis*{4-(2,6-dibromophenyl)}carbonate]	12750	16270	0	12904
Poly[2,2-hexafluoropropane *bis*{4-(2,6-dibromophenyl)}carbonate]	13155	16570	0	13142
Poly[2,2-propane *bis*{4-(2,6-dimethylphenyl)}carbonate]	13270	16750	0	13284
Torlon	--------	17080	0	13546
Kapton	--------	17100	0	13562
Resin F	13845	17720	0	14054
Poly(dicyclooctyl itaconate)	--------	18093	0	14349
Poly[4,4'-diphenoxy di(4-phenylene)sulfone]	14800	18450	5	14756
Poly[*o*-biphenylenemethane *bis*(4-phenyl)carbonate]	--------	18705	0	14835
Poly[diphenylmethane *bis*(4-phenyl)carbonate]	15330	19150	0	15188
Poly[di(*n*-octyl) itaconate]	15350	19430	0	15410
Resin G	15640	19930	0	15806
Poly[4,4'-sulfone diphenoxy di(4-phenylene)sulfone]	15800	19700	10	15872
Poly[1,1-biphenylethane *bis*(4-phenyl)carbonate]	15915	20000	0	15862
Poly[4,4'-isopropylidene diphenoxy di(4-phenylene)sulfone]	16900	21300	5	17017
Ultem	--------	28485	0	22591

11.B.3.   Structure-Property Relationships for Rubbery Polymers

11.B.3.a.   Shear Modulus

According to the theory of rubber elasticity [21], the "equilibrium" shear modulus $G_E^o(T)$, above T_g, of a polymer crosslinked beyond its gel point, is determined by the average molecular weight M_c of the chain segments between the chemical crosslinks. By analogy, $G_N^o(T)$, i.e., the shear modulus of an uncrosslinked polymer in the "rubbery plateau", is usually assumed to be determined by the physical interactions caused by "entanglements" between polymer chains [22]. Equation 11.16 can therefore be used to define an average molecular weight between these entanglements, i.e., the "entanglement molecular weight" M_e. The utilization of R≈8.31451

J/(mole·K) for the gas constant, units of g/cc for the density ρ, and units of g/mole for M_e, results in the expression of $G_N^0(T)$ in MPa by Equation 11.22.

$$G_N^0(T) \approx \frac{\rho(T) \cdot RT}{M_e}$$ (11.22)

In a crosslinked polymer, $G_E^0(T)$ increases monotonically with increasing T for a given M_c, until T becomes high enough to cause the dissociation of chemical bonds. Physical interactions resulting in the rubbery plateau region in an uncrosslinked polymer are much weaker, and they gradually break as the temperature is increased in this regime. Consequently, $G_N^0(T)$ does not normally increase with increasing temperature in the rubbery plateau region. Equation 11.22 should therefore be used to predict $G_N^0(T)$ only at the onset temperature of the rubbery plateau [often roughly around $(T_g + 30K)$], since $G_N^0(T)$ is *not* simply proportional to the absolute temperature at higher temperatures.

Data for the M_e of polymers [16] are listed in Table 11.4. M_e can be predicted by using two different correlations, as discussed below. Both of these predictions can differ significantly from the "observed" values of M_e deduced from measured values of $G_N^0(T)$ by using Equation 11.22.

1. M_e can be assumed to be equal to half of the "critical molecular weight" M_{cr}, which is discussed in Chapter 13.

$$M_e \approx \frac{M_{cr}}{2}$$ (11.23)

2. Equation 11.24 (with a relative standard deviation of 21%) often provides a better estimate for M_e. This equation is a slightly modified version of an equation by Seitz [16], with the backbone rotational degrees of freedom parameter N_{BBrot} defined in Chapter 4 being used instead of the corresponding parameter in Ref. [16] which was defined in a different manner.

$$M_e \approx 1039.7 + 1.36411 \cdot 10^{-23} \cdot \frac{N_{BBrot} \cdot M \cdot V_w}{l_m^3}$$ (11.24)

For polystyrene, if we use $N_{BBrot}=2$, M=104.1, $V_w \approx 63.25$ cc/mole and $l_m \approx 2.52 \cdot 10^{-8}$ cm, the result is $M_e \approx 12265$ according to Equation 11.24. Two "measured" values of M_e for polystyrene are 18700 [23] and 17851 [16]. The use of the Polymer Version 6.0 software package of Biosym Technologies (an advanced molecular modeling program), gives $l_m=2.32$Å as indicated in Figure 11.1, resulting in the prediction of $M_e=15536$ which improves the agreement with the observed values very significantly. It also shows that greater accuracy (in addition to convenience) may be an advantage of doing the calculations on a computer.

Table 11.4. Data for the entanglement molecular weight M_e [16].

Polymer	M_e (g/mole)
Polyethylene	1422
Poly(ethylene terephthalate)	1450
Polyisoprene	1833
Polybutadiene (*trans*-1,4: 51%; *cis*-1,4: 37%; 1,2: 12%)	1844
Poly(hexamethylene adipamide)	2000
Poly(hexamethylene isophthalamide)	2040
Polyoxyethylene	2200
Poly[4,4'-isopropylidene diphenoxy di(4-phenylene)sulfone]	2250
Poly(vinylidene fluoride)	2400
Bisphenol-A poly(ester-*co*-carbonate) (1:1 by mole)	2402
Bisphenol-A poly(ester-*co*-carbonate) (1:2 by mole)	2429
Poly(ε-caprolactam)	2490
Bisphenol-A polycarbonate	2495
Polyoxymethylene	2540
Phenoxy resin	2670
Poly(1,4-cyclohexylene dimethylene terephthalate-*co*-isophthalate) (1:1 by mole)	2880
Poly(1,4-butadiene) *(cis)*	2936
Poly(1,2-butadiene)	3529
Poly[oxy(2,6-dimethyl-1,4-phenylene)]	3620
Poly(styrene-*co*-acrylonitrile) (50/50 by weight)	5030
Polytetrafluoroethylene	5580
Poly(styrene-*co*-acrylonitrile) (63/37 by weight)	7005
Poly(styrene-*co*-methyl methacrylate) (35/65 by weight)	7624
Poly(dimethyl siloxane)	8160
Poly(ethyl methacrylate)	8590
Poly(vinyl acetate)	8667
Poly(styrene-*co*-acrylonitrile) (76/24 by weight)	8716
Polyisobutylene	8818
Poly(methyl acrylate)	9070
Poly(styrene-*co*-acrylonitrile) (71/29 by weight)	9154
Poly(methyl methacrylate)	9200
Poly(styrene-*co*-acrylonitrile) (78/22 by weight)	9536
Poly(α-methyl styrene)	12800

☞ ☞ ☞   TABLE 11.4 IS CONTINUED IN THE NEXT PAGE. ☞ ☞ ☞

Table 11.4. CONTINUED FROM THE PREVIOUS PAGE.

Polymer	M_e (g/mole)
Poly(styrene-*co*-maleic anhydride) (67/33 by weight)	14522
Poly(styrene-*co*-acrylic acid) (92/8 by weight)	14916
Poly(styrene-*co*-maleic anhydride) (91/9 by weight)	16462
Poly(styrene-*co*-acrylic acid) (87/13 by weight)	16680
Poly(styrene-*co*-maleic anhydride) (71/29 by weight)	17750
Polystyrene	17851
Poly(2-ethylbutyl methacrylate)	22026
Poly(*p*-methyl styrene)	24714
Poly(*p*-bromo styrene)	29845
Poly(*n*-hexyl methacrylate)	33800
Poly(*p-t*-butyl styrene)	37669

Since estimates of M_e are subject to large errors, while $\rho(T)$ varies slowly with increasing temperature, the use of $\rho(298K)$ rather than $\rho(T)$ in Equation 11.22 to simplify the estimation of $G_N^o(T)$ usually does not introduce any significant errors. Substitution of $\rho(298K) \approx 1.05$ grams/cc, $M_e \approx 15536$ g/mole, $T_g=373K$, and $T=403K=(T_g + 30K)$, into Equation 11.16, gives $G_N^o(403K) \approx 0.226$ MPa for polystyrene, in comparison with an experimental value [22] of 0.200 MPa. On the other hand, Equation 3.16 with $V(298K)=99.1$ cc/mole and $T_g=373K$ gives $V(403K) \approx 102.6$ cc/mole and $\rho(403K) \approx 1.015$ grams/cc. Substitution of $\rho(403K) \approx 1.015$ instead of 1.05 g/cc into Equation 11.22 gives $G_N^o(403K) \approx 0.219$ MPa, which differs only negligibly from 0.226 MPa. Of course, such a simplification does not have any advantages in terms of convenience when the calculations are performed on a computer, and is therefore not needed.

The fact that the computational estimates and the indirectly deduced "experimental" values of M_e are both subject to substantial uncertainties does not diminish the usefulness of equations 11.22-11.24 in predicting the general trends for the $G_N^o(T)$ of rubbery polymers. Reported experimental values of M_e (see Table 11.4, also see references [16,23]) range from 1390 for polyethylene to the somewhat questionable enormous value of 110000 for poly(*t*-butyl methacrylate). The largest reported value of M_e is therefore 79 times as large as the smallest reported value. It follows from Equation 11.16 that the observed values [22] of $G_N^o(T)$ span a range of almost two orders of magnitude. Consequently, even an error of 50% in the predicted value of M_e, which is considerably larger than the typical magnitude of the error in the prediction of this property, only results in a fairly small relative error when the entire range of $G_N^o(T)$ values is considered. Trends between the $G_N^o(T)$ values of members of a structurally diverse set of polymers are thus usually predicted with reasonable accuracy.

11.B.3.b.   Bulk Modulus and Young's Modulus

B(T) does not decrease nearly as drastically as E(T) or G(T) above T_g, but often shows a significant drop (up to a factor of three) when T increases above T_g. Equation 11.25 (developed by Arends [17] by analysis of a large amount of pressure-volume-temperature data) provides a reasonable prediction of B(T) for $T \geq (T_g + 30K)$. In the software package automating the use of our correlations, Equation 11.13 is used to predict B(T) for $T \leq (T_g - 20K)$, Equation 11.25 is used for $T \geq (T_g + 30K)$, and an interpolation is performed for $(T_g - 20K) < T < (T_g + 30K)$.

$$B(T) \approx \frac{\dfrac{205V(T)}{V_w}}{\left[\dfrac{V(T)}{V_w} - 1.27\right]^2} - 2329\left[\frac{V_w}{V(T)}\right]^2 \tag{11.25}$$

For example, for polystyrene at T=403K, the substitution of V(403K)≈102.6 cc/mole and V_w≈63.25 cc/mole into Equation 11.25 gives B(403K) ≈ 1797 MPa.

Equation 11.25 was developed independently of the methods developed to estimate B(T) for $T \leq T_g$, such as Equation 11.13. Consequently, qualitatively significant inconsistencies can sometimes occur between B(T) values calculated for $T \leq T_g$ and $T > T_g$ since B(T) is not being calculated by a single internally consistent procedure over the two temperature regimes. Fortunately, however, since both equations were derived from physically sound considerations and have reasonably good accuracy, we have only rarely encountered such inconsistencies.

ν(T) can be estimated for rubbery polymers by rearranging Equation 11.7 into the form given by Equation 11.26, and then substituting the values of B(T) estimated from Equation 11.25 and G(T) estimated from Equation 11.22. Such an estimate, however, is of questionable usefulness and significance, since the ν(T) of a rubbery polymer *cannot* be measured to great accuracy.

$$\nu(T) \approx \frac{[3B(T) - 2G(T)]}{[6B(T) + 2G(T)]} \tag{11.26}$$

For example, it was predicted above that G(403K)≈0.277 MPa and B(403K)≈1797 MPa for polystyrene. The insertion of these values into Equation 11.26 gives ν(403K)≈0.499923. A Poisson's ratio very close to but just below 0.5, usually somewhere between 0.499 and 0.499999, is typical for rubbery polymers. The approximation ν(T)≈0.5 in Equation 11.7, to relate E(T) to $G_N^o(T)$ for a rubbery polymer, results in the following simple expression for E(T):

$$E(T) \approx 3 \cdot G_N^o(T) \tag{11.27}$$

11.C. Large-Strain Behavior: Failure Mechanisms

11.C.1. Phenomenology

The usefulness of a polymer in many applications is largely determined by its predominant failure mechanism under the conditions of the application. Other factors being equal, a polymer whose failure requires the application of a large stress will be more useful than a polymer which fails under less rigorous conditions. It is, therefore, very important to be able to predict the failure mechanism of polymeric specimens as a function of the structure of the polymer, the processing conditions used in the manufacture of the specimens, and the test conditions. See the book by Ashby [1] for a broad perspective, encompassing all engineering materials, of how parameters quantifying resistance to failure can be utilized in practical guidelines for the optimum selection of materials for use in applications where specimens of various shapes are subjected to various modes of mechanical loading.

The general trends for the mechanical failure of uncrosslinked amorphous specimens as a function of the material parameters and test conditions will now be summarized. Effects due to (a) variations of the specimen geometry, (b) thermal history, (c) crystallinity, (d) orientation, (e) crosslinking, and (f) defects incorporated in the specimens during processing or use, will not be discussed. It should be kept in mind that the predominant failure mechanism is strongly affected by these additional factors. The "material" failure properties calculated in this section are only rough indicators of the intrinsic proclivities of different polymers. Trends predicted from these material properties can be contravened by differences in the preparation and quality of the specimens of the polymers being compared.

Three general modes of failure have been identified in amorphous polymers, namely *brittle fracture* [24], *crazing* [25-31] and *shear yielding* [2-6,32]. When it is not clear which mechanism is dominant, this situation is usually caused by the fact that, under the given test or application conditions, two of these mechanisms are competing, and neither mechanism is overwhelmingly favored. For example, in measurements of tensile strength or fatigue under tension, it is possible to get mixed failure modes, with crazes and shear deformation zones occurring in the same specimen. It is, therefore, only necessary to consider the three main mechanisms.

In *brittle fracture*, which is illustrated schematically in Figure 11.5, failure occurs in a *brittle* fashion both at a *microscopic (local)* and a *macroscopic (bulk)* level. A very familiar example of brittle fracture is the shattering of ordinary (silica-based) glass (used in windows, etc.) when hit by a stone. Although most polymeric materials are less susceptible to brittle fracture than ordinary glass, they can all be made to undergo brittle fracture under the right set of conditions. For example, at extremely low temperatures (below -100°C) natural rubber can shatter and undergo brittle fracture just like glass. When a specimen undergoes brittle fracture, the bonds [whether primary (covalent) or secondary (van der Waals)] crossing the fracture surface break, and the specimen fractures as a result. There is no plastic flow during this cleavage process, since the mobilities of the subunits of the polymer are far too low. Brittle fracture has been explained in terms of (a) a defect mechanism involving a "characteristic flaw size" [2,33], and (b) considerations of the energy required for bond breaking [24]. The defect mechanism provides a better description of the fundamental physics of brittle fracture than the simple bond breakage models. On the other hand, it has proven easier to develop a simple predictive structure-property relationship for the brittle fracture stresses of polymers in terms of bond breakage.

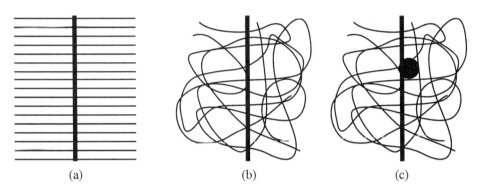

(a)   (b)   (c)

Figure 11.5. Schematic illustrations of *brittle fracture*. (a) Idealized limiting case of perfectly uniaxially oriented polymer chains (horizontal lines), with a fracture surface (thick vertical line) resulting from the scission of the chain backbone bonds crossing these chains and perpendicular to them. This limit is approached, but not reached, in fracture transverse to the direction of orientation of highly oriented fibers. (b) Isotropic amorphous polymer with a typical random coil type of chain structure. Much fewer bonds cross the fracture surface (thick vertical line), and therefore much fewer bonds have to break, than was the case in the brittle fracture of a polymer whose chains are perfectly aligned and perpendicular to the fracture surface. (c) Illustration of a defect, such as a tiny dust particle (shown as a filled circle), which was incorporated into the specimen during fabrication and which can act as a stress concentrator, facilitating brittle fracture.

Unlike brittle fracture, crazing (Figure 11.6) and shear yielding (Figure 11.7) both require sufficient mobility of the chain segments of a polymer for plastic flow to occur at a local (molecular) level. Both of these processes can therefore be considered to be ductile at a *local* level. On the other hand, at a *macroscopic* level, shear yielding by the homogeneous and continuous plastic deformation of a specimen is much more ductile than the heterogeneous processes of cavitation, craze nucleation, propagation and breakdown, and crack propagation, which are the stages of failure via crazing.

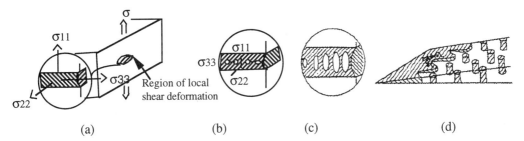

(a) Formation of a localized plastic zone and buildup of significant lateral stresses.

(b) Nucleation of voids in the plastic zone (very often at or near a defect close to the surface of the specimen) to relieve the triaxial constraints.

(c) Further deformation of polymer ligaments between voids and coalescence of individual voids to form a void network. The columnar structures remaining between voids are craze fibrils.

(d) Schematic view of the wedge of deformed polymer at the craze tip, showing a polymer chain about to be drawn into two different fibrils, either via chain scission or via disentanglement.

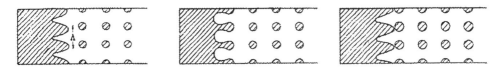

(e) The advance of the craze tip (shown from the left to the right for an advance of one fibril spacing) is believed to occur via the so-called meniscus instability mechanism, in which void fingers advance into the wedge of deformed polymer, leaving behind them some trailing fibrils. The void fingers and craze fibrils observed experimentally are not normally nearly as regular as those shown in this schematic illustration.

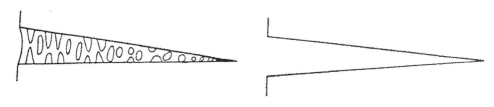

(f) The difference between a craze (left) and a crack (right) is shown. As the craze tip continues to propagate, some of the craze fibrils start to break down. Once a few neighboring fibrils have broken down, a large void is formed in the craze. If the stress is high enough, these voids may grow slowly by slow fibril breakdown at their edges until a crack of critical size has formed within a craze. The crack then propagates rapidly, breaking craze fibrils as it grows, and eventually resulting in fracture.

Figure 11.6. Schematic illustration of the stages of failure via crazing. Figures (a) to (e) have been reproduced from Ref. [27], while Figure (f) has been reproduced from Ref. [25].

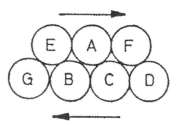

Intermolecular Shear: "Atoms" E, A and F cooperatively shear relative to atoms G, B, C and D.

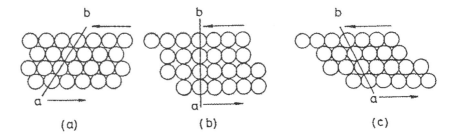

Intramolecular Shear: The motion proceeds from (a) to (c). The difference from intermolecular shear (see above) is that the atoms connected by the line labeled a------b represent a chain "caught" at an angle to the direction of intermolecular shear, so that intermolecular shear is accompanied by a rotation of this line (chain) as it attempts to keep up with the intermolecular shear motion.

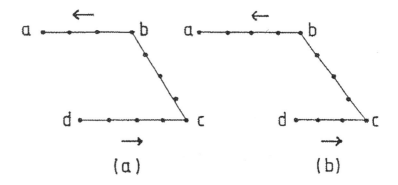

Tubular Motion: The motion proceeds from (a) to (b). It is combined with intermolecular shear. Note the resemblance of this motion to the movements of a snake, which is approximately tubular in shape, as it glides around an obstacle.

Figure 11.7. A highly stylized schematic illustration, instructive but not intended to be interpreted literally, of some of the types of plastic flow processes believed to play an important role in shear yielding [32].

A rule of thumb is that, *all other factors being equal, the polymer with the most ductile mode of failure will be the most useful one.* Failure by either crazing or shear yielding is, therefore, to be favored over failure by brittle fracture. The desired mechanism among crazing and shear yielding is not always as clearcut, and depends on additional factors such as whether the specimens are notched or unnotched, but shear yielding is preferable in most cases.

Many brittle polymers are "toughened" by modification, for example by the incorporation of rubber particles. Depending on the nature of a polymer, toughening [2,34,35] can occur by imparting either the ability to craze more effectively (as in polystyrene [27]) or the ability to undergo shear yielding (as in nylon [36]).

The most common trends which determine the preferred mode of failure of a polymer are schematically depicted in Figure 11.8, which shows the stresses required for brittle fracture (σ_f), crazing (σ_c) and shear yielding (σ_y) as functions of the temperature. Each failure stress will, of course, vary significantly among different polymers, so that this illustration is only a rather general schematic representation of the phenomenology of the failure of polymers.

Because of the existence of relationships between temperature and rate dependences in mechanical and rheological tests, such as the time-temperature superposition principle [37] which often holds, similar curves can also be drawn for the failure stresses as functions of the strain rate. For example, σ_y increases with increasing rate of deformation. Unfortunately, simple and reasonably reliable quantitative structure-property relationships are only available for the temperature dependence of the failure stresses at the present time. Such relationships are therefore only intended for comparison with the results of measurements at "typical" strain rates.

As depicted in Figure 11.8, σ_f generally has a very weak dependence on the temperature, since it is related to the scission of chemical bonds. The number of bonds crossing a unit cross-sectional area of a specimen decreases very slowly with thermal expansion, resulting in a very slow decrease of σ_f. The only major exception to this trend occurs when the temperature of measurement is sufficiently high for significant thermal and/or thermooxidative degradation to take place in the time scale of the test. In that case, the thermal energy added to the specimen can supply most of the energy needed to break chemical bonds, and σ_f can rapidly decrease with increasing temperature.

As the temperature increases, both σ_c and σ_y decrease. The rate of decrease of σ_y with increasing temperature is usually considerably greater than the rate of decrease of σ_c because of the greater role played by plastic deformation in shear yielding. Shear yielding can be treated as a simple activated flow process, while crazing occurs by the superposition of plastic flow and surface separation processes.

At the lowest temperatures (Regime I), σ_f is always the lowest failure stress, i.e., specimens are brittle. At an intermediate temperature range (Regime II), σ_c often becomes the lowest failure stress. At the highest temperatures (Regime III), σ_y often becomes the lowest failure stress.

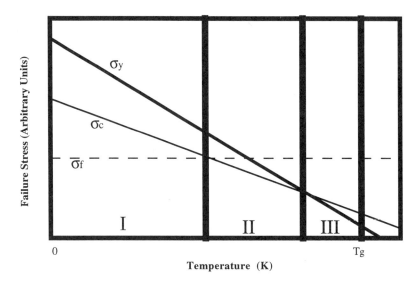

Figure 11.8. Schematic illustration of the stresses required for *brittle fracture* (σ_f), *crazing* (σ_c) and *shear yielding* (σ_y), as functions of the absolute temperature. Each Roman numeral refers to the temperature regime circumscribed by the two thick vertical lines closest to it on either side. The failure stresses have been approximated by straight lines as functions of the temperature for simplicity. The experimental failure stress curves are not straight lines. Deviations from linearity are smallest (negligible for testing temperatures well below the temperature for the onset of thermal and/or thermooxidative degradation) for σ_f, and largest for σ_y.

The mechanical properties also depend on the ratio M_w/M_e, where M_w denotes the weight-average molecular weight. The predominant mode of failure therefore also depends on M_w. An "optimum" M_w of $10M_e$ to $15M_e$ often results in the best balance of properties [16]. Polymers in this optimum range of M_w usually have chains which are long enough to approach the "high polymer" limit for the mechanical properties, but short enough for the melt viscosity (Chapter 13) to be sufficiently low not to cause great difficulty in melt processing.

When the failure stress for the mechanism requiring the smallest amount of applied stress under a given set of test or application conditions is reached, that mechanism is triggered, while competing mechanisms are not triggered. Specimens thus preferentially fail via the mechanism which requires the least amount of applied stress, with mixed failure modes or a statistical distribution of the different failure modes over a set of specimens being observed in borderline situations. Amorphous polymers therefore usually go through regimes where brittle fracture, crazing and shear yielding are predominant (I, II and III, respectively, in Figure 11.8), with increasing temperature.

Whether a certain failure regime is observed, and the width of the temperature range over which it is predominant, vary greatly among polymers. Many polymers are so brittle that they only undergo brittle fracture all the way up to T_g, or only have regimes of brittle fracture and crazing. For example, isotropic (unoriented) polystyrene does not undergo shear yielding under uniaxial tension at any temperature below T_g, but only manifests brittle fracture and crazing. Another example is that many polymers with extremely high glass transition temperatures are brittle at room temperature, and often even at elevated temperatures which are substantially lower than T_g. Great chain stiffness, which is the most important factor resulting in extremely high values of T_g (Chapter 6), can make it very difficult for a polymer to dissipate a significant amount of the applied mechanical energy via plastic deformation.

While Figure 11.8 summarizes much of the phenomenology of polymer failure, there are some exceptions to the trends it depicts. For example, the mechanism of crazing in *high entanglement density polymers*, including such thermoplastics as bisphenol-A polycarbonate and poly[4,4'-isopropylidene diphenoxy di(4-phenylene)sulfone], is somewhat different from the mechanism in polymers with a low entanglement density such as polystyrene [27,28]. In polymers with a high entanglement density, the predominant failure mechanisms in Regimes II and III are reversed, with specimens predominantly undergoing shear yielding in Regime II and crazing in Regime III [38,39].

Determination of the preferred mode of failure has been shown above to require the estimation of which one of the three major types of failure stresses will be the lowest one for specimens of a given material, being tested under a given set of conditions. These ideas are depicted in a flow chart in Figure 11.9, assuming that a specimen is being subjected to uniaxial tension.

If a small deformation is applied to the specimen, it can break via brittle fracture when the stress σ reaches σ_f, if σ_f is the lowest failure stress. (The use of the term "small deformation" for the brittle fracture process may appear to be an oxymoron, given the fact that the specimen falls apart into two or more pieces. It is, however, an accurate usage. The *local* deformations in brittle fracture, which does not involve plastic flow but only involves bond scission, are indeed small, in spite of their catastrophic effect [40,41] on the structural integrity of the specimen.) If σ_y or σ_c is the lowest failure stress, the specimen survives the small deformation, manifesting Hookean (spring-like) behavior, with σ equal to the product of E and ε initially, and deviating from Hookean behavior with increasing ε. Finally, as the deformation becomes large, failure via shear yielding or crazing takes place, depending on whether σ_y or σ_c is lower.

These considerations show that the *toughness* can be quantified by the total amount of energy required to induce failure, or equivalently by the total area under the stress-strain curve. Failure mechanisms which involve plastic flow under a given set of test or use conditions result in the requirement of more energy for failure, and thus result in the manifestation of greater toughness.

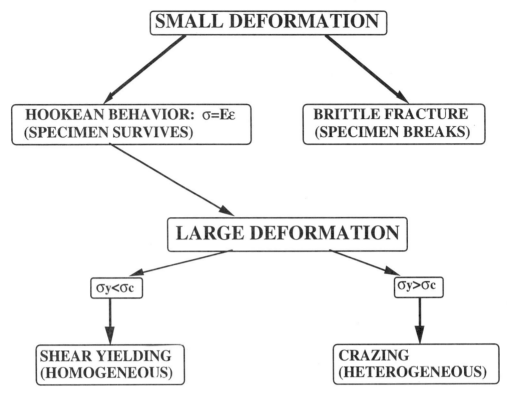

Figure 11.9. Flow chart for the determination of the preferred failure mechanism, assuming that a specimen is being subjected to uniaxial tension.

In addition to models which deal with the failure of specific types of materials, or with specific modes of failure, there are general models and computational methods which deal with universal aspects of fracture shared by all materials and all modes of failure. Such models and computational methods often provide the most appropriate description of the physical processes occurring in bulk specimens. The statistical theory of fracture kinetics [41] for the time-dependent fracture strength of solids is an example of a general model attempting to incorporate the universal aspects of fracture. The numerical methods of continuum mechanics [42,43], such as the finite element method, are leading techniques used in design engineering to predict the mechanical properties of bulk specimens.

The bridge between the molecular and macroscopic levels of treatment consists of the use of the structure-property relationships to estimate the material parameters used as input parameters in models describing the bulk behavior. The "intrinsic" material mechanical and thermal properties predicted by the correlations provided in this book can be used as input parameters in such "bulk specimen" approaches, which consider factors such as the sizes and the shapes of the specimens. Since the predominant failure mechanism of a specimen depends strongly on many such factors besides intrinsic material properties, such a use of structure-property relationships is a key step towards developing a truly predictive capability for the preferred failure mechanisms of polymers.

11.C.2. Structure-Property Relationships for the Brittle Fracture Stress

The brittle fracture stress σ_f is proportional to the product of the average number of chain backbone bonds crossing a unit area of the specimen with the average strength of a backbone bond. This correlation was first recognized by Vincent [24]. It was later refined by Seitz [16], who augmented the dataset with data on additional polymers (see Table 11.6 for the full dataset [16]) and obtained the correlation given by Equation 11.28 (with a relative standard deviation of 16%) which provides $\sigma_f(T)$ in units of MPa if l_m is in cm and V(T) is in cc/mole:

$$\sigma_f(T) \approx \frac{2.288424 \cdot 10^{11} \cdot l_m}{V(T)} \tag{11.28}$$

For example, for polystyrene at T=298K, $l_m \approx 2.52 \cdot 10^{-8}$ cm, $V(298K) \approx 99.1$ cc/mole, and $\sigma_f(298K) \approx 58$ MPa from Equation 11.28. The experimental value [16] of $\sigma_f(298K)$ is 41 MPa.

It is instructive to note that $\sigma_f(T)$ as calculated for an isotropic amorphous polymer from Equation 11.28 is roughly two orders of magnitude smaller than what would be expected if all polymer chains were perfectly aligned and if defects did not play a role in initiating brittle fracture. On the other hand, the chains are *not* perfectly aligned, and defects *do* play a key role, as shown schematically in Figure 11.5. Breakage of many chemical bonds along chain backbones is needed to fracture the specimen in the direction of alignment of the perfectly oriented polymer shown in Figure 11.5(a), while fracture perpendicular to the direction of alignment only requires breakage of the much weaker nonbonded (physical) interchain interactions. See Adams *et al* [44] for a compendium of recent research on "rigid rod" polymers, whose tensile properties can be made to approach the theoretical limit for perfectly aligned chains by using sophisticated fabrication methods. In the isotropic (unoriented) system shown in Figure 11.5(b), the average number of chemical bonds which must be broken to fracture the specimen (and thus also the value of σ_f) is the same in all directions and intermediate between what would be expected for the longitudinal and the transverse directions of the perfectly oriented system. For the system depicted in Figure 11.5(c), σ_f is still isotropic, but much smaller than it would be for the system in Figure 11.5(b).

Seitz also showed [16] that σ_f decreases with decreasing number-average molecular weight M_n, and that its dependence on M_n can be approximated by Equation 11.29 for $M_n > 3.4 M_e$.

$$\sigma_f(M_n) \approx \sigma_f(\infty) \cdot \left(1 - \frac{3.4 \cdot M_e}{M_n}\right) \tag{11.29}$$

Table 11.5. Brittle fracture stress σ_f just below the ductile-brittle transition temperature [16].

Polymer	σ_f (MPa)
Poly(*p-t*-butyl styrene)	31
Polystyrene	41
Poly(*p*-chloro styrene)	41
Poly(*o*-methyl styrene)	46
Poly(4-methyl-1-pentene)	53
Poly(1-pentene)	58
Poly(styrene-*co*-acrylonitrile) (76/24 by weight)	62
Poly(methyl methacrylate)	68
Poly(1-butene)	81
Polypropylene	98
Polytetrafluoroethylene	117
Poly(vinyl chloride)	142
Bisphenol-A polycarbonate	145
Poly[4,4'-sulfone diphenoxy di(4-phenylene)sulfone]	148
Poly(ethylene terephthalate)	155
Polyethylene	160
Poly(hexamethylene adipamide)	179
Polyoxymethylene	216

11.C.3. Structure-Property Relationships for the Yield Stress

The yield stress in uniaxial tension is roughly proportional to Young's modulus. Equation 11.30 (with a relative standard deviation of 22% at T=298K) is a slight modification of the original equation by Seitz [16], who used a proportionality constant of 0.025 instead of 0.028 based on the analysis of the data listed in Table 11.6.

$$\sigma_y(T) \approx 0.028 \cdot E(T) \tag{11.30}$$

It is interesting to note that Equation 11.30 is valid for semicrystalline polymers as well as for amorphous polymers. This is the reason for the presence of some highly crystalline polymers such as high-density polyethylene in Table 11.6.

For polystyrene at room temperature, E(298K) was predicted to be 3000 MPa from equations 11.13, 11.10 and 11.7. Equation 11.30 gives $\sigma_y(298K) \approx 84$ MPa, in reasonable agreement with the observed value [16] of 76 MPa. Even more importantly, the predicted values of $\sigma_y(298K)$

and σ_f(298K) suggest that σ_y(298K)>σ_f(298K), i.e., that brittle fracture is favored over shear yielding, for polystyrene at room temperature and commonly used strain rates. This prediction agrees with the experimental results, which show that σ_y(298K)>σ_f(298K) for polystyrene.

Table 11.6. Tensile yield stress σ_y and Young's modulus E at room temperature [16]. See Table 11.7 for additional and/or alternative measured values of σ_y.

Polymer	σ_y (MPa)	E (MPa)
Low-density polyethylene	8	200
Polytetrafluoroethylene	13	400
High-density polyethylene	30	1000
Polypropylene	32	1400
Cellulose acetate	42	2000
Poly(hexamethylene sebacamide)	45	1200
Poly(vinyl chloride)	48	2600
Poly(ε-caprolactam)	50	1900
Poly(hexamethylene adipamide)	57	2000
Bisphenol-A polycarbonate	62	2300
Phenoxy resin	66	2300
Poly[4,4'-isopropylidene diphenoxy di(4-phenylene)sulfone]	69	2500
Poly[oxy(2,6-dimethyl-1,4-phenylene)]	72	2300
Poly(ethylene terephthalate)	72	3000
Polystyrene	76	3300
Poly(styrene-co-acrylonitrile) (76/24 by weight)	83	3800
Poly[4,4'-sulfone diphenoxy di(4-phenylene)sulfone]	84	2600
Poly(methyl methacrylate)	90	3200
Poly(o-chloro styrene)	90	4000

Another structure-property relationship (Equation 11.31) for σ_y(T) was provided by Wu [22], where C_∞ is the characteristic ratio (see Chapter 12), and δ^2 is the cohesive energy density expressed as the square of the solubility parameter δ (see Chapter 5).

$$\log_{10}\left\{\frac{\sigma_y(T)}{\left[(T_g - T)\delta^2\right]}\right\} \approx -3.36 + \log_{10}(C_\infty) \tag{11.31}$$

A standard yield criterion, such as the modified von Mises criterion or the modified Tresca criterion, can be used to predict the yield stress in other modes of testing (such as uniaxial compression, plane strain compression and simple shear), from the value of $\sigma_y(T)$ in uniaxial tension calculated by using Equation 11.30 or Equation 11.31. These two criteria are discussed in standard references [2-4], and are generally of comparable quality.

In predicting whether or not a candidate polymer will be ductile, it is recommended to err on the conservative side, and not to propose that the polymer will prefer shear yielding to brittle fracture unless $\sigma_f(T) > 1.2 \cdot \sigma_y(T)$, instead of using the criterion $\sigma_f(T) > \sigma_y(T)$ suggested by Figure 11.8 for preference for shear yielding. There are several reasons for such caution:

1. When $\sigma_f(T)$ and $\sigma_y(T)$ are very similar at a given T, mixed ductile/brittle failure modes and statistical variations between ductile and brittle behavior over a set of specimens are often seen.

2. Many important factors which were not explicitly considered in developing the correlations for the failure stresses, such as the effects of deformation rate (as in impact phenomena where the deformation rate is very high), the presence of defects which act as stress concentrators, and the effects of physical aging, favor brittle fracture over shear yielding.

3. Finished polymeric parts often encounter more severe conditions than initially anticipated.

11.C.4. An Attempt to Account for Rate Dependence of the Yield Stress

A major attempt was made to account simultaneously for the temperature and the strain rate $\dot{\varepsilon}$ (i.e., the measurement frequency) dependendence of the yield stress [45]. This attempt was unsuccessful in producing a reliable predictive relationship for the rate dependence. It will, nonetheless, be briefly discussed, to help future workers in this field. The following procedure was used to analyze a vast amount of data on the yield stresses of glassy polymers:

1. Begin with experimental σ_y data [46-65] obtained in various testing modes as a function of T and/or $\dot{\varepsilon}$, with emphasis on data obtained as a function of both variables in the same test on the same set of specimens since the largest amount of internally consistent data is then obtained.

2. Apply the modified von Mises yield criterion to convert these data into the simple shear mode, to estimate the yield stress τ_y in simple shear, and thus to provide a common working framework to compare all of the data.

3. Approximate the dependence of τ_y on T and $\dot{\varepsilon}$ by using a correlation suggested by Brown [66], who combined the kinetic theory of deformation [67] with the empirical observation that τ_y is approximately proportional to the shear modulus (G) just as $\sigma_y(T)$ in uniaxial tension is roughly proportional to Young's modulus (E).

$$\tau_y(T,\dot{\varepsilon}) \approx aG(T,\dot{\varepsilon}) + \left(\frac{kT}{v_s}\right)\ln\left(\frac{\dot{\varepsilon}}{\dot{\varepsilon}_0}\right)$$

$$(11.32)$$

In Equation 11.32, k is Boltzmann's constant. The "activation volume" [67,68] v_s, $\dot{\varepsilon}_0$ and a are material parameters. The value of a is estimated by extrapolating $\tau_y(T,\dot{\varepsilon})$ and $G(T,\dot{\varepsilon})$ to T=0K so that the second term in Equation 11.32 becomes zero. Brown showed [66] that, on the average, a≈0.076±0.030 for a number of polymers.

4. Fit the experimental $\tau_y(T,\dot{\varepsilon})$ data to Equation 11.32, to obtain the values of the adjustable parameters v_s, $\dot{\varepsilon}_0$ and a for as many polymers as possible.

5. Correlate the values found for v_s, $\dot{\varepsilon}_0$ and a with the structure of the polymeric repeat unit.

6. Use the correlations that will be developed for v_s, $\dot{\varepsilon}_0$ and a, in order to estimate the values of these parameters for new polymers, thus endowing Equation 11.31 with predictive powers.

7. Apply modified von Mises criterion to convert predictions from τ_y to σ_y in other testing modes.

The most important observations made as a result of this effort can be summarized as follows:

1. Contrary to expectations based on qualitative explanations commonly found in textbooks, that $\dot{\varepsilon}_0$ represents a fundamental (molecular level) flow rate parameter, the values of $\dot{\varepsilon}_0$ show an enormous variation, not only between different polymers, but often also for data collected by different workers on the same polymer. For example, the values of $\dot{\varepsilon}_0$ obtained for PMMA by analyzing four sets of data as a function of the strain rate ranged from $4.9 \cdot 10^{14}$ to $2.2 \cdot 10^{39}$.

2. The standard treatment of v_s for shear yielding as a constant activation volume independent of temperature for a polymer is inadequate. This oversimplification often causes large deviations between observed values of $\tau_y(T,\dot{\varepsilon})$ and the results of fitting these data with Equation 11.32.

3. The best simple functional form of $v_s(T)$ was found to be given by Equation 11.33:

$$v_s(T) \approx b \cdot \exp\left(\frac{cT}{T_g}\right) \qquad (11.33)$$

In Equation 11.33, b and c are fitting constants. The value of c was typically found to range from 1 to 2.75. Equation 11.33 implies that v_s, which is essentially the size of a typical volume element over which plastic flow takes place during shear yielding, increases exponentially with increasing temperature. This result is consistent with the increase in the mobility of the chain segments of polymers with increasing temperature.

4. The use of $v_s(T)$ as given by Equation 11.33 instead of a constant value of v_s in Equation 11.32 results in fits of outstanding quality for *individual sets* of experimental $\tau_y(T,\dot{\varepsilon})$ data.

5. Nonetheless, it proved to be impossible to develop reliable quantitative structure-property relationships for the parameters of equations 11.32 and 11.33 from available data, preventing the use of these equations as predictive tools. This failure was traced back to two causes:

(a) The existence of data for σ_y simultaneously as a function of T and $\dot{\varepsilon}$, over a sufficiently wide range of values of both variables, for only a few polymers.

(b) The large variations observed in some cases for the values of the adjustable parameters giving the best fit for measurements made by different workers on the same polymer.

Computer simulations [69-72] of the deformation of microstructural models based on the kinetic theory of deformation appears to be a more promising approach than the development of simple structure property-relationships, to predict the rate dependence of tensile deformation. The tradeoffs in using such simulation models, however, are that large amounts of computational resources (both computer time and computer memory) are needed, and furthermore that the use of these sophisticated methods requires users to have a significant amount of advanced knowledge.

11.C.5. Structure-Property Relationships for the Crazing Stress

Development of simple but reliable correlations for the crazing stress is a major challenge for future work. The complicated heterogeneous mechanism of crazing has made the development of such relationships very difficult. The following may be some of the most promising approaches:

1. Theories were developed to deal with crazing under different circumstances. Future work may lead to simple structure-property relationships for the crazing stress σ_c based on these theories.

(a) The theory of Andrews *et al* [25,26] deals with environmental crazing. It uses a weighted arithmetic mean of the yield stress and the surface tension to calculate σ_c.

(b) The theory of Kramer [27,28] treats ordinary crazing in air. It uses a complicated expression, involving both the yield stress and a modified type of surface tension term and roughly proportional to the geometric mean of these two terms, to calculate σ_c.

2. Kambour [29,30] developed a correlation for the minimum stress required for the inception of crazing, in terms of the cohesive energy density, T_g, and E (or σ_y). Since failure with crazing as the primary mechanism requires the growth and propagation of crazes after their inception, this correlation describes only one of the key aspects crazing. It is, however, a major step in the development of a complete structure-property relationship. Some of the data of Kambour [29] for the craze inception strain (ε_{ci}) and craze inception stress (σ_{ci}) are listed in Table 11.7.

3. Seitz [16] showed that the tendency to craze increases with increasing average length of the chain segment between entanglements. For example, if the contour length is greater than approximately 200Å, the material is likely to craze. On the other hand, if the contour length is less than 200Å, shear yielding is expected to be preferred over crazing.

4. Wu [23] correlated craze inception stress data [29] with the density of entanglements, as shown in Equation 11.34 where σ_{ci} is in MPa and ρ/M_e (entanglement density) is in millimoles/cc:

$$\log_{10}\left(\sigma_{ci}\right) \approx 1.83 + 0.5 \cdot \log_{10}\left(\frac{\rho}{M_e}\right) \qquad (11.34)$$

Table 11.7. Craze inception strain (ε_{ci}) and craze inception stress (σ_{ci}) for polymers [29]. The σ_{ci} values were calculated from ε_{ci} and Young's modulus (E) values by assuming linear elasticity up to the inception of crazing and thus using Equation 11.1, and rounded off to the nearest integer. E and yield stress (σ_y) values from Ref. [29] are also listed, and differ (in some cases by a significant amount) from the values of these properties listed in Tables 11.1 and 11.6 for many of the same polymers since they are from different experiments. E, σ_{ci} and σ_y are in MPa. The stress for failure via crazing (σ_c) is usually significantly larger than σ_{ci} since it also includes the effects of craze growth and propagation, but it is not yet as well-quantified as σ_{ci}.

Polymer	ε_{ci}	σ_{ci}	E	σ_y
Polystyrene	0.0035	11	3280	82.8
Polystyrene/Poly[oxy(2,6-dimethyl-1,4-phenylene)] blend (75/25 by weight)	0.0073	23	3130	86.2
Polystyrene/Poly[oxy(2,6-dimethyl-1,4-phenylene)] blend (50/50 by weight)	0.0087	25	2830	91.0
Polystyrene/Poly[oxy(2,6-dimethyl-1,4-phenylene)] blend (25/75 by weight)	0.0111	30	2680	83.5
Poly[oxy(2,6-dimethyl-1,4-phenylene)]	0.015	35	2350	72.4
Polystyrene/o-Dichlorobenzene blend (85/15 by weight)	0.0013	---	-------	53.8
Polystyrene/o-Dichlorobenzene blend (90/10 by weight)	0.0014	---	-------	69.7
Polystyrene/o-Dichlorobenzene blend (92.5/7.5 by weight)	0.0017	---	-------	73.1
Polystyrene/o-Dichlorobenzene blend (95/5 by weight)	0.0020	---	-------	79.3
Polystyrene/o-Dichlorobenzene blend (97.5/2.5 by weight)	0.0030	---	-------	81.4
Poly(styrene-co-acrylonitrile) (73/27 by weight)	0.005	15	3060	82.8
Poly(vinyl cyclohexane)	0.0035	10	2900	>138.0
Poly[2,2-propane bis(4-phenyl)oxymethyleneoxy]	0.005	13	2520	48.3
Poly(methyl methacrylate)	0.0105	31	2980	59.3
Amorphous poly(ethylene terephthalate)	0.0115	23	1970	47.6
Poly(vinyl chloride)	0.012	42	3520	41.4
Poly[2,2-propane bis{4-(2,6-dimethylphenyl)}carbonate]	0.0177	39	2180	93.8
Bisphenol-A polycarbonate	0.018	45	2520	58.6
Poly[1,1-dichloroethylene bis(4-phenyl)carbonate]	0.02	51	2570	69.0
Poly[dicyanomethane bis(4-phenyl)carbonate]	0.026	---	-------	84.8
Poly(dixylenylsulfone-co-bisphenol-A carbonate) (1/1 by mole)	0.024	55	2280	81.4
Poly[4,4'-isopropylidene diphenoxy di(4-phenylene)sulfone]	0.025	62	2480	69.0
Poly[4,4'-diphenoxy (4-phenylene)sulfone]	0.0255	75	2930	82.8
Poly[4,4'-phenoxy (4-phenylene)sulfone]	0.04	89	2230	71.7
Ultem/Poly[4,4'-phenoxy (4-phenylene)sulfone] blend (50/50 by weight)	0.03	---	-------	82.8
Ultem	0.022	68	3100	103.5
Bisphenol-A poly(carbonate-co-terephthalate-co-isophthalate) (7/11/2 by mole)	0.029	---	-------	59.8
Bisphenol-A terephthalate-co-isophthalate (1/1)	0.032	---	-------	63.2

The approaches of Seitz [16] and Wu [23] are both consistent with the overall physical picture of crazing provided by Kramer's theory [26,27], since they both show that a more dense network of entanglements should lead to a higher σ_c.

11.C.6. Stress-Strain Curves of Elastomers

The stress-strain relationships of elastomeric (rubbery) networks at low extension (or draw) ratios (λ, which is simply the length of the deformed specimen divided by the length of the initial undeformed specimen) can be described in terms of Equation 11.35 [21], which is the simplest possible constitutive equation for the deformation of an isotropic incompressible medium.

$$\sigma \approx G \cdot \left(\lambda - \frac{1}{\lambda^2} \right) \qquad\qquad\qquad (11.35)$$

Since Equation 11.35 was historically first derived from considerations of entropy, it is often called the "entropy spring" model. Here, G is the shear modulus given by Equation 11.22, where the entanglement molecular weight M_e can be replaced by the average molecular weight between the chemical crosslinks (M_c) if the polymer is chemically crosslinked, by an average molecular weight based on the existence of both entanglements in the amorphous regions and of crystallites as additional physical entanglement junctions if the polymer is rubbery and semicrystalline (has its amorphous regions above T_g and its crystalline regions below T_m), or by some other appropriate average molecular weight which takes into account the different types of chemical and physical entanglement junctions which are elastically active at the temperature and deformation state of a given experiment. The apparent vagueness of this last statement is intentional. While a chemical crosslink is a very well-defined entity, the various types of weaker physical network junctions are not as well-defined, and their abundance and elastic activity depend on the test conditions.

The criterion of low draw ratio is $\lambda \ll \sqrt{n}$, where n is the average number of statistical chain segments (Kuhn segments) between the entanglement junctions which are elastically active under the testing conditions. A Kuhn segment is the shortest chain segment which can be used to describe the conformation of a real polymer chain, where the bond angles are constrained to narrow ranges of values, in terms of a "freely jointed" chain model where the angles between the directions of successive random walk steps are unconstrained and can have any value [73,74].

Several types of events may occur at higher draw ratios. Weak physical entanglements may slip, and overstretched chain segments may break (see Figure 11.10), reducing the number of elastically-active network junctions. However, in some elastomers, strain-induced crystallization [21,75-100] may create additional elastically-active network junctions with increasing λ. Whether because of the presence of chemical crosslinks, or strain-induced crystallization, or a combination of these factors, a strong network may survive up to high draw ratios. The resistance to further deformation then increases rapidly because of the resistance of this network. However, since even these junctions only have a finite strength, they also continue to break. A fracture surface eventually traverses the entire macroscopic specimen, which then breaks at a finite draw ratio.

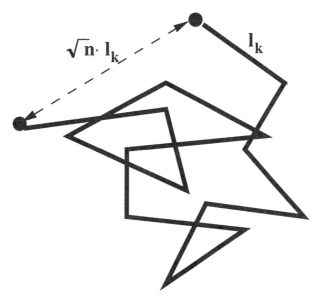

Figure 11.10. Schematic illustration of finite chain extensibility, for a chain segment between two entanglement junctions. A random walk of n steps of length l_k is shown, where n is the number of Kuhn segments and l_k is the length of each Kuhn segment. The average end-to-end distance of such chain segments in the undeformed elastomer is $\sqrt{n} \cdot l_k$. The total chain segment length is $n \cdot l_k$. The maximum extensibility of the average chain segment is, consequently, $\sqrt{n}$.

The simple entropy spring model (Equation 11.35) becomes invalid at high draw ratios, since it fails to consider the finite extensibilities of the chains. A more general theoretical model [21], which results in Equation 11.36, provides a better description of an elastomeric stress-strain curve all the way up to fracture, by including the finite extensibilities of the chains. In this equation, $£^{-1}$ is the inverse Langevin function. $£^{-1}$ is a transcendental function which is defined by Equation 11.37, where "coth" is the hyperbolic cotangent function and the superscript of -1 represents the functional inversion (and not merely the simple reciprocal) of the function in square brackets.

$$\sigma \approx G \cdot \frac{\sqrt{n}}{3} \cdot \left[£^{-1}\left(\frac{\lambda}{\sqrt{n}}\right) - \lambda^{-3/2} \cdot £^{-1}\left(\frac{1}{\sqrt{n} \cdot \lambda}\right) \right] \qquad (11.36)$$

$$\text{£}^{-1}(x) = \left[\coth(x) - \frac{1}{x} \right]^{-1} \tag{11.37}$$

The stress-strain behaviors predicted from Equation 11.35 and Equation 11.36 are compared in Figure 11.11, for n=100. The two models produce similar results at low draw ratios. In fact, it can be shown that Equation 11.36 simplifies to Equation 11.35 in the limit of $\lambda \rightarrow 1$. On the other hand, the entropy spring model becomes inadequate at large λ, and fails to predict the very rapid increase of σ as $\lambda \rightarrow \sqrt{n}$ (which is equal to 10 in this example). Equation 11.36 predicts this *strain hardening* effect correctly, and gives a finite draw ratio of less than $\sqrt{n}$ since σ exceeds the strength of the surviving network junctions during its rapid climb towards infinity as $\lambda \rightarrow \sqrt{n}$.

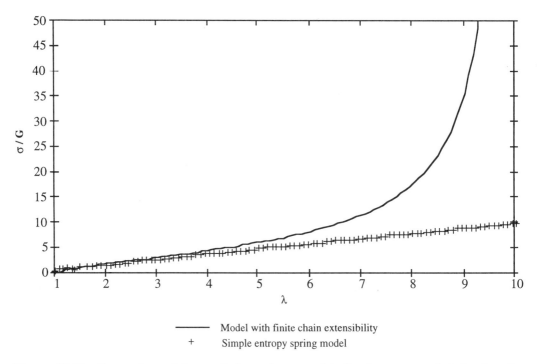

Figure 11.11. Comparison of the predictions of two models for the stress-strain behavior of elastomeric networks. There are 100 Kuhn segments between network junctions in this particular example. The stress is denoted by σ, the shear modulus by G, and the draw ratio by λ.

It can be shown that the inverse Langevin function $\pounds^{-1}(x)$, with $0 \le x < 1$, can be estimated very accurately in terms of a simple closed-form expression called a Padé approximant [101]:

$$\pounds^{-1}(x) \approx x \cdot \frac{\left(3 - x^2\right)}{\left(1 - x^2\right)} \tag{11.38}$$

Insertion of Equation 11.38 into Equation 11.36, followed by some algebraic rearrangement [102], gives a much-simplified version of the equation for the rubber elasticity model with finite chain extensibility, representing the stress-strain behavior of an elastomer over the entire range of draw ratios as successfully as the original expression involving the transcendental $\pounds^{-1}$ function.

$$\sigma \approx G \cdot \left[\frac{\lambda}{3} \cdot \left(\frac{3n - \lambda^2}{n - \lambda^2} \right) - \frac{1}{3\lambda^2} \cdot \left(\frac{3n\lambda - 1}{n\lambda - 1} \right) \right] \tag{11.39}$$

11.C.7. Ductile Thermoplastics at Large Extension Ratios

The stress-strain curves of ductile thermoplastics (including both glassy amorphous polymers such as bisphenol-A polycarbonate and semicrystalline polymers such as polyethylene at room temperature) have the general shapes shown in Figure 11.12.

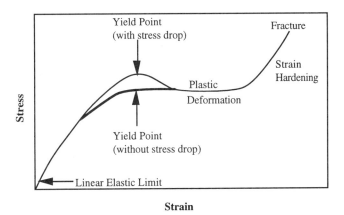

Figure 11.12. General shapes of the stress-strain curves of ductile thermoplastics. Some of these polymers manifest a very distinct post-yield stress drop, while many others do not.

The specimens exhibit "irreversible" plastic deformation with increasing strain after the yield strain is exceeded. (In other words, the plastic deformation can only be reversed by melting and reshaping the specimens, but not by merely removing the load which is being applied.) The stress in the plastic deformation region varies little with the strain and is often lower than the yield stress. This reduction of stress is referred to as the *post-yield stress drop*. The magnitude of the post-yield stress drop depends on factors such as the thermal history of the specimens and the propensity of the polymer to form shear bands, and may be zero. As the extension ratio increases even further, a strain is eventually reached where the stress again begins to increase rapidly with increasing strain, eventually resulting in fracture.

The *strain hardening* behavior shown in Figure 11.12 resembles that shown in Figure 11.11 for elastomeric networks. The recognition of this similarity has stimulated the development of models based on rubber elasticity theory to describe the strain hardening of ductile thermoplastics. In other words, it has been proposed that the deformation behavior of a ductile thermoplastic at very high strains is rubberlike even in the absence of such dominant attributes of ordinary rubbers as the presence of chemical crosslinks or the occurrence of strain-induced crystallization. This is a very active area of current research [103], which may eventually result in the development of simple but reliable quantitative structure-property relationships for the ultimate mechanical properties (fracture strain and stress) of ductile thermoplastics. There are no existing relationships of sufficient accuracy to be used with a reasonable degree of confidence for these properties.

11.C.8. Thermoset Resins

Unlike elastomers, whose stress-strain behavior is well-described, the effects of crosslinking on the thermal and mechanical properties of thermosets (crosslinked glassy polymers) are not well-understood. For example, a considerable amount of data can be found [19,20,104-113] for properties such as the density, coefficient of thermal expansion, and elastic moduli of thermosets. Any trends which may exist in these data are obscured by the differences in specimen preparation conditions, and perhaps especially by the different amounts of "free volume" frozen into different systems due to kinetic reasons as the segmental mobilities of chains decrease with increasing crosslinking during a curing reaction. It is clear, from an analysis of these data, that they do not allow the development of any statistically significant correlations for the effects of crosslinking on the thermoelastic properties of thermosets.

The only generalization which can be made, at a qualitative level and with some reservations, involves the effect of crosslinking on the "toughness" of thermosets. When a glassy polymer is crosslinked, the number of degrees of freedom available for the local motions of chain segments is reduced by the replacement of some of the weak nonbonded (physical) interactions with much stronger chemical bonds. Large-scale plastic flow processes, such as those occurring in either shear yielding or crazing, therefore become more difficult. Consequently, when a polymer becomes crosslinked very densely, the usual result is a catastrophic amount of embrittlement (reduction of toughness). On the other hand, the effects of the reduction of the number of degrees of freedom are often relatively minor at low crosslink densities, and the brittle fracture stress may even go up a little as a result of the increase of the number of backbone bonds crossing the cross-sectional area of the specimen until the opposing effect of the reduction of the number of degrees of freedom becomes more important and begins to dominate. These general trends are shown for styrene-divinylbenzene copolymers in Figure 11.13, which was redrawn from information given in Ref. [114]. Similar trends have also been observed for other polymers. The available data, however, are insufficient in quantity and in quality to allow the development of a statistically significant quantitative structure-property relationship which could be used with some confidence.

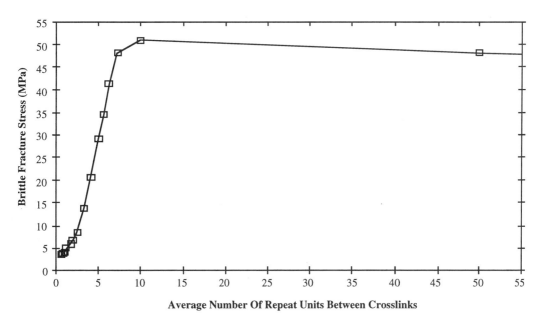

Figure 11.13. Effects of crosslinking on the brittle fracture stress of styrene-divinylbenzene copolymers. Note the catastrophic embrittlement at very high crosslink densities (i.e., at an average number of repeat units between crosslinks less than the Kuhn segment length of 8 repeat units). The data point (not shown) for the uncrosslinked polymer limit (polystyrene) is at (∞,41).

11.D. Creep, Stress Relaxation and Fatigue

The mechanical properties discussed above all have a major common feature, since they are all measured by deforming a specimen continuously at a given rate until it fails. While the mode of deformation (uniaxial tension, uniaxial compression, plane strain compression, or simple shear) may vary, these differences can be roughly taken into account by imposing criteria (such as yield criteria) to describe the stress states imposed on specimens by different deformation geometries.

There are, however, three additional types of important and qualitatively different mechanical tests representing deformation modes often occurring in the practical use of engineering materials. These tests, which will be discussed briefly below, can be used to anticipate the performance of a plastic part as a function of the time under some deformation states which are encountered very often in technological applications; namely, under constant stress for a long time, under constant strain for a long time, and under periodic stress or periodic strain for a long time. More detailed discussions of time-dependent phenomena are provided in many references [2-6,67,115-119].

In *creep*, a constant load (stress or force) is imposed on a specimen, and its response (strain) is measured over time. Depending on the temperature, the relaxation times of the viscoelastic processes occurring in the specimen, and the magnitude of the load, the observed behavior can range from very little perceptible strain over long periods of time, to large elongations, to rupture. *Linear viscoelastic creep* occurs when the stress is sufficiently low that the strains remain below roughly 0.005 by the time the buildup of a back stress in the polymer because of its long and entangled chains counterbalances the applied stress and the polymer ceases to creep. Linear viscoelastic creep is reversible, and the polymer recovers completely when the stress is removed. At the other extreme is *creep rupture*, which is also often referred to as *static fatigue*.

Creep is intimately related to *physical aging* which is the slow structural relaxation of a glassy polymer at a temperature below T_g but sufficiently high to allow some mobility of the chain segments. Creep and physical aging have in common the fact that a slow relaxation occurs over a long period of time. The difference between them is that, by definition, creep occurs under a constant applied load which may result in overall deformation (strain), while physical aging occurs without an applied load and thus without an overall strain on the specimen. Physical aging may, however, result in volumetric relaxation (densification), enthalpy relaxation (reduction of enthalpy), and changes in the subsequent deformation and yielding behavior of the specimen [119-132]. Not surprisingly, the theoretical treatments of creep and physical aging are related.

In *stress relaxation*, a given deformation (strain) is imposed on the specimen which is then allowed to relax at constant strain and temperature. The relaxation times of the viscoelastic processes taking place in the material are the key determinants of the rate and the extent of stress relaxation. There is a quantitative similarity between the relaxation processes taking place during stress relaxation and during creep at small deformations. However, creep at large deformations is dominated by large-scale viscous flow processes (which are suppressed in stress relaxation by the imposition of a constant strain on the specimen) and thus differs from stress relaxation.

Molecular dynamics simulations of stress relaxation on idealized models [133] have shown, in agreement with experimental trends, similarities between the simulated stress relaxation (stress versus time) curves of metals and polymers despite the existence of essential differences between their stress-strain curves. The simulations also show the existence of domains in which motions are highly correlated, suggesting that cooperative motions are important in stress relaxation.

In *fatigue* [116] (also referred to as *dynamic fatigue* to distinguish it from creep rupture), the specimen is subjected to repeated (oscillatory) cycles of stress or strain until it fails. The results are usually reported in terms of the number of cycles to failure as a function of the stress level and the measurement temperature. The *fatigue limit* or the *endurance limit* is the limiting stress level below which the specimen is not observed to fail. The temperature of measurement and the damping (energy dissipation by "internal friction" [8]) characteristics of a polymer play a major role in determining both the predominant mechanism of fatigue failure [2] and the number of cycles to failure. *Mechanical fatigue failure* occurs in materials with relatively low energy dissipation characteristics and/or at low measurement temperatures, and involves the initiation and subsequent propagation of a crack. Heat buildup becomes important and can lead to *thermal fatigue failure* [10-12] if a material has a large amount of damping and/or the temperature of measurement is close to T_g or T_m so that the specimen temperature rises rapidly and exceeds T_g or T_m instead of stabilizing at some lower value. As a result of these many complications, the correlation or prediction of the fatigue performance of a polymer in terms of both the polymeric structure and the many possible experimental variables under testing conditions is even more difficult than the correlation or prediction of its creep or stress relaxation behavior.

The complexity of the interplay between various factors (time, temperature, loading level, and material properties) involved in determining the creep, stress relaxation and fatigue behavior has made it impossible, thus far, to develop quantitative structure-property relationships of predictive value. The best that can be done is to fit experimental data to empirical functional forms, without being able to relate the values of the fitting parameters quantitatively to polymeric structure. It is also possible (with considerable trepidation) to use these empirical fits to extrapolate the expected behavior to times and/or test conditions beyond those available by merely using the relationships to interpolate within the range of the available data. Since the main focus of this book is on the *prediction* of polymer properties, creep, stress relaxation and fatigue will not be discussed further.

11.E.   Improvements in the Ability to Predict the Mechanical Properties

The quantitative structure-property relationships summarized earlier in this chapter necessarily treat the mechanical properties of polymers as "derived" properties, in the sense defined in Section 1.B.2. In other words, the mechanical properties are expressed by equations in terms of material parameters of a more "fundamental" nature, instead of being correlated directly with either group contributions or with connectivity indices. This necessity to treat the mechanical properties as "derived" properties is a direct consequence of their great complexity.

The following material properties are used as input parameters in one or more of the correlations discussed in this chapter for the mechanical properties: molecular weight, length and number of backbone degrees of freedom of the polymeric repeat unit; volumetric properties (molar volume and density as functions of temperature, and van der Waals volume), cohesive energy, glass transition temperature, entanglement (or critical) molecular weight, molar Rao function, molar Hartmann function, characteristic ratio, and surface tension.

The molecular weight of a repeat unit is determined exactly from its composition. The length of a repeat unit can be measured by using an interactive molecular modeling program. The rotational degrees of freedom of the backbone can be counted by using the rules provided in Section 4.C. The new quantitative structure-property relationships developed in this book for the volumetric properties (Chapter 3), cohesive energy (Chapter 5), glass transition temperature (Chapter 6), critical molecular weight (Chapter 13), molar Rao function (Section 11.B), molar Hartmann function (Section 11.B), characteristic ratio (Chapter 12), and surface tension (Chapter 7), allow the application of the correlations for the mechanical properties to all polymers built from the nine elements (C, N, O, H, F, Si, S, Cl and Br) included in the scope of our work.

The main contribution to the ability to predict the mechanical properties of polymers, by the work presented in this book, is therefore the new set of quantitative structure-property relationships, not subject to the limitations traditionally imposed by the need for group contributions, for the many input parameters entering the correlations developed for the mechanical properties. This development has allowed the prediction of the mechanical properties of many novel polymers whose mechanical properties could not be estimated previously.

References and Notes for Chapter 11

1. M. F. Ashby, *Materials Selection in Mechanical Design*, Pergamon Press, Oxford (1992).

2. A. J. Kinloch and R. J. Young, *Fracture Behaviour of Polymers*, Elsevier Applied Science Publishers, London (1983).

3. *The Physics of Glassy Polymers*, edited by R. N. Haward, Applied Science Publishers Ltd, London (1973).

4. *Failure of Plastics*, edited by W. Brostow and R. D. Corneliussen, Hanser Publishers, Munich (1986).

5. N. G. McCrum, C. P. Buckley and C. B. Bucknall, *Principles of Polymer Engineering*, Oxford University Press, New York (1988).

6. J. G. Williams, *Fracture Mechanics of Polymers*, Ellis Horwood Ltd., West Sussex, England (1984).

7. D. W. van Krevelen, *Properties of Polymers*. (a) Second edition, Elsevier, Amsterdam (1976). (b) Third edition, Elsevier, Amsterdam (1990).

8. C. A. Wert, *J. Appl. Phys.*, *60*, 1888-1895 (1986).

9. A. Hiltner and E. Baer, *Polymer*, *15*, 805-813 (1974).

10. R. J. Crawford and P. P. Benham, *J. Materials Science*, *9*, 18-28 (1974).

11. R. J. Crawford and P. P. Benham, *J. Materials Science*, *9*, 1297-1304 (1974).

12. R. J. Crawford and P. P. Benham, *Polymer*, *16*, 908-914 (1975).

13. J. Bicerano, *J. Polym. Sci., Polym. Phys. Ed.*, *29*, 1329-1343 (1991).

14. J. Bicerano, *J. Polym. Sci., Polym. Phys. Ed.*, *29*, 1345-1359 (1991).

15. N. G. McCrum, B. E. Read and G. Williams, *Anelastic and Dielectric Effects in Polymeric Solids*, Wiley, New York (1967).

16. J. T. Seitz, *J. Appl. Polym. Sci.*, *49*, 1331-1351 (1993).

17. See C. B. Arends, *J. Appl. Polym. Sci.*, *49*, 1931-1938 (1993) and *51*, 711-719 (1994), for the discussion of the PVT behavior of polymers and the methods used in obtaining the data.

18. D. W. van Krevelen, in *Computational Modeling of Polymers*, edited by J. Bicerano, Marcel Dekker, New York (1992), Chapter 1.

19. B. Hartmann and G. F. Lee, *J. Appl. Phys.*, *51*, 5140-5144 (1980).

20. B. Hartmann and G. F. Lee, *J. Polym. Sci., Polym. Phys. Ed.*, *20*, 1269-1278 (1982).

21. L. R. G. Treloar, *The Physics of Rubber Elasticity*, third edition, Clarendon Press, Oxford (1975).

22. W. W. Graessley and S. F. Edwards, *Polymer*, *22*, 1329-1334 (1981).

23. S. Wu, *Polymer Engineering and Science*, *30*, 753-761 (1990).

24. P. I. Vincent, *Polymer*, *13*, 558-560 (1972).

25. E. H. Andrews and L. Bevan, *Polymer*, *13*, 337-346 (1972).

26. E. H. Andrews, G. M. Levy and J. Willis, *J. Materials Science*, *8*, 1000-1008 (1973).

27. E. J. Kramer, *Advances in Polymer Science*, *52/53*, 1-56 (1983).

28. E. J. Kramer and L. L. Berger, *Advances in Polymer Science*, *91/92*, 1-68 (1990).

29. R. P. Kambour, *Polymer Communications*, *24*, 292-296 (1983).

30. M. T. Takemori, R. P. Kambour and D. S. Matsumoto, *Polymer Communications*, *24*, 297-299 (1983).

31. G. H. Michler, *J. Materials Science*, *25*, 2321-2334 (1990).

32. N. Brown, *J. Materials Science*, *18*, 2241-2254 (1983).

33. A. A. Griffith, *Phil. Trans. Roy. Soc.*, *A221*, 163-198 (1920).

34. M. A. Maxwell and A. F. Yee, *Polymer Engineering and Science*, *21*, 205-211 (1981).

35. *Polymer Toughening*, edited by C. B. Arends, Marcel Dekker, New York (1994).

36. A. Margolina and S. Wu, *Polymer*, *29*, 2170-2173 (1988).

37. J. D. Ferry, *Viscoelastic Properties of Polymers*, John Wiley & Sons, New York (1961).

38. C. J. G. Plummer and A. M. Donald, *J. Polym. Sci., Polym. Phys. Ed.*, *27*, 325-336 (1989).

39. T. C. B. McLeish, C. J. G. Plummer and A. M. Donald, *Polymer*, *30*, 1651-1655 (1989).

40. *Fracture Processes in Polymeric Solids*, edited by B. Rosen, Interscience Publishers, New York (1964).

41. O. M. Ettouney and C. C. Hsiao, *J. Appl. Phys.*, *64*, 4884-4888 (1988).

42. C. L. Dym, *Computers & Structures*, *16*, 101-107 (1983).

43. T. H. H. Pian, *RCA Review*, *39*, 648-664 (1978).

44. *The Materials Science and Engineering of Rigid-Rod Polymers*, edited by W. W. Adams, R. K. Eby and D. E. McLemore, Materials Research Society, Pittsburgh (1989).

45. L. L. Beecroft, J. Bicerano and J. T. Seitz, unpublished calculations. Some of this work was a part of L. L. Beecroft's project as a summer intern at Dow, and was presented by her as a course report at the Massachussetts Institute of Technology (late 1989).

46. C. Bauwens-Crowet, J.-C. Bauwens and G. Homes, *J. Materials Science*, *7*, 176-183 (1972).

47. J. Haussy, J. P. Cavrot, B. Escaig and J. M. Lefebvre, *J. Polym. Sci., Polym. Phys. Ed.*, *18*, 311-325 (1980).

48. P. Beardmore, *Philosophical Magazine*, *19*, 389-401 (1969).

49. A. Thierry, R. J. Oxborough and P. B. Bowden, *Philosophical Magazine*, *30*, 527-536 (1974).

50. C. Bauwens-Crowet, *J. Materials Science*, *8*, 968-979 (1973).

51. P. I. Vincent, *Plastics*, *27 (August)*, 105-110 (1962).

52. R. E. Robertson, *Appl. Polym. Symp.*, *7*, 201-213 (1968).

53. P. B. Bowden and S. Raha, *Philosophical Magazine*, *22*, 463-482 (1970).

54. R. D. Andrews and W. Whitney, *Report No. TD-123-64*, Textile Division, Department of Mechanical Engineering, Massachusetts Institute of Technology, Cambridge, Massachusetts (May 1, 1964).

55. Y. Imai and N. Brown, *Polymer*, *18*, 298-304 (1977).

56. A. S. Argon and M. I. Bessonov, *Philosophical Magazine*, *35*, 917-933 (1977).

57. S. Yamini and R. J. Young, *J. Materials Science*, *15*, 1814-1822 (1980).

58. C. Bauwens-Crowet, J.-C. Bauwens and G. Homes, *J. Polym. Sci.*, A-2, *7*, 735-742 (1969).

59. W. Wu and A. P. L. Turner, *J. Polym. Sci., Polym. Phys. Ed.*, *13*, 19-34 (1975).

60. J. T. Ryan, *Polymer Engineering and Science*, *18*, 264-267 (1978).

61. S. S. Sternstein, L. Ongchin and A. Silverman, *Appl. Polym. Symp.*, *7*, 175-199 (1968).

62. J. A. Roetling, *Polymer*, *6*, 311-317 (1965).

63. J. A. Roetling, *Polymer*, *6*, 615-619 (1965).

64. R. N. Haward, B. M. Murphy and E. F. T. White, *J. Polym. Sci.*, A-2, *9*, 801-814 (1971).

65. Measured by A. Ahn, L. L. Beecroft, C. Broomall, J. McPeak, T. Tomczak and C. A. Wedelstaedt (1989-1991).

66. N. Brown, Ref. [4], Chapter 6.

67. A. S. Krausz and H. Eyring, *Deformation Kinetics*, John Wiley & Sons, New York (1975).

68. J. C. M. Li, C. A. Pampillo and L. A. Davis, in *Deformation and Fracture of High Polymers*, edited by H. H. Kausch, J. A. Hassell and R. I. Jaffee, Plenum Press, New York (1974), 239-258.

69. Y. Termonia and P. Smith, *Macromolecules*, 20, 835-838 (1987).

70. Y. Termonia and P. Smith, *Macromolecules*, 21, 2184-2189 (1988).

71. Y. Termonia, S. R. Allen and P. Smith, *Macromolecules*, 21, 3485-3489 (1988).

72. Y. Termonia and P. Smith, *Colloid & Polymer Science*, 270, 1085-1090 (1992).

73. P. J. Flory, *The Principles of Polymer Chemistry*, Cornell University Press, Ithaca, New York (1953).

74. P. J. Flory, *Statistical Mechanics of Chain Molecules*, Interscience Publishers, New York (1969).

75. B. Wunderlich, *Macromolecular Physics, Volume 2, Crystal Nucleation, Growth, Annealing*, Academic Press, New York (1976).

76. A. N. Gent, *Trans. Faraday Soc.*, 50, 521-533 (1954).

77. A. N. Gent, *J. Polym. Sci., A*, 3, 3787-3801 (1965).

78. A. N. Gent, *J. Polym. Sci., A-2*, 4, 447-464 (1966).

79. G. S. Y. Yeh and P. H. Geil, *J. Macromol. Sci., Phys.*, B1, 251-277 (1967).

80. J. J. Klement and P. H. Geil, *J. Macromol. Sci., Phys.*, B6, 31-56 (1972).

81. G. S. Y. Yeh and S. L. Lambert, *J. Appl. Phys.*, 42, 4614-4621 (1971).

82. J. H. Southern and R. S. Porter, *J. Appl. Polym. Sci.*, 14, 2305-2317 (1970).

83. J. H. Southern and G. L. Wilkes, *J. Polym. Sci., Polym. Lett. Ed.*, 11, 555-562 (1973).

84. K. Kobayashi and T. Nagasawa, *J. Macromol. Sci., Phys.*, B4, 331-345 (1970).

85. E. H. Andrews, P. J. Owen and A. Singh, *Proc. Roy. Soc. London A*, 324, 79-97 (1971).

86. M. R. Mackley and A. Keller, *Polymer*, 14, 16-20 (1973).

87. R. Oono, K. Miyasaka and K. Ishikawa, *J. Polym. Sci., Polym. Phys. Ed.*, 11, 1477-1488 (1973).

88. T. Takahashi, H. Iwamoto, K. Inoue and I. Tsujimoto, *J. Polym. Sci., Polym. Phys. Ed.*, 17, 115-122 (1979).

89. J. W. C. van Bogart, A. Lilaonitkul and S. L. Cooper, in *Multiphase Polymers*, edited by S. L. Cooper and G. M. Estes, *Advances in Chemistry Series*, 176, American Chemical Society, Washington, D. C. (1979), Chapter 1.

90. C. K. L. Davies, S. V. Wolfe, I. R. Gelling and A. G. Thomas, *Polymer*, 24, 107-113 (1983).

91. B. J. R. Scholtens, *Rubber Chemistry and Technology*, 57, 703-724 (1984).

92. D. Goeritz and M. Kiss, *Rubber Chemistry and Technology*, 59, 40-45 (1986).

93. T. A. Speckhard and S. L. Cooper, *Rubber Chemistry and Technology*, 59, 405-431 (1986).

94. E. Riande, J. Guzman and J. E. Mark, *Polymer Engineering and Science*, 26, 297-303 (1986).

95. Y.-H. Hsu and J. E. Mark, *Polymer Engineering and Science*, 27, 1203-1208 (1987).

96. C.-C. Sun and J. E. Mark, *J. Polym. Sci., Polym. Phys. Ed.*, 25, 2073-2083 (1987).

97. M. C. Chien and R. A. Weiss, *Polymer Engineering and Science*, 28, 6-12 (1988).

98. K. Sakurai and T. Takahashi, *J. Appl. Polym. Sci.*, 38, 1191-1194 (1989).

99. W. J. Orts, R. H. Marchessault, T. L. Bluhm and G. K. Hamer, *Macromolecules*, 23, 5368-5370 (1990).

100. J. C. Wittmann and P. Smith, *Nature*, 352, 414-417 (1991).

101. A. Cohen, *Rheologica Acta*, 30, 270-273 (1991).

102. J. T. Seitz, unpublished calculations.

103. R. N. Haward, *Macromolecules*, 26, 5860-5869 (1993).

104. V. V. Kolokol'chikov and L. P. Li, *Mekhanika Polimerov, No. 4*, 597-602 (1976).

105. V. V. Shapovalenko, V. I. Pakhomov, E. I. Gol'dshtein, L. I. Voitenko and M. V. Pomoshnikova, *Strength Mater. (USA)*, 10, 112-114 (1978).

106. G. C. Martin and M. Shen, *ACS Polymer Preprints*, 20, 786-789 (1979).

107. A. Kanno and K. Kurashiki, *J. Soc. Mater. Sci. Japan*, 33, 102-108 (1984).

108. V. V. Bulatov, O. B. Salamatina, A. A. Gusev, G. A. Vorobjeva and E. F. Oleinik, *Polymer Bulletin*, 13, 21-27 (1985).

109. M. Ogata, T. Kawata and N. Kinjo, *Kobunshi Ronbunshu*, 44, 193-199 (1987).

110. T. S. Chow, *ACS Symposium Series*, 367, *Crosslinked Polymers* (1988), Chapter 10.

111. B. L. Burton, *SAMPE Journal*, 27-30 (May/June 1988).

112. S. Murakami, O. Watanabe, H. Inoue, M. Ochi and M. Shimbo, *Proc. Japan. Congr. Mater. Res.*, 32, 249-253 (1989).

113. H. J. Bixler, A. S. Michaels and M. Salame, *J. Polym. Sci., Part A*, 1, 895-919 (1963).

114. J. L. Amos, L. C. Rubens and H. G. Hornbacher, in *Styrene: Its Polymers, Copolymers and Derivatives*, edited by R. H. Boundy and R. F. Boyer, Reinhold Publishing Corporation, New York (1952), 723-729.

115. W. N. Findlay, J. S. Lai and K. Onaran, *Creep and Relaxation of Nonlinear Viscoelastic Materials*, Dover, New York (1989).

116. R. M. Hertzberg and J. A. Manson, *Fatigue of Engineering Plastics*, Academic Press, New York (1980).

117. I. M. Ward, *Mechanical Properties of Solid Polymers*, second edition, Wiley, New York (1983).

118. L. C. E. Struik, *Internal Stresses, Dimensional Instabilities and Molecular Orientations in Plastics*, Wiley, New York (1990).

119. L. C. E. Struik, *Physical Aging in Amorphous Polymers and Other Materials*, Elsevier, Amsterdam (1978).

120. M. R. Tant and G. L. Wilkes, *Polymer Engineering and Science*, *21*, 874-895 (1981).

121. M. Washer, *Polymer*, *26*, 1546-1548 (1985).

122. R. A. Bubeck, S. E. Bales and H.-D. Lee, *Polymer Engineering and Science*, *24*, 1142-1148 (1984).

123. R. A. Bubeck, P. B. Smith and S. E. Bales, from *Order in the Amorphous "State" of Polymers*, edited by S. K. Keinath, R. L. Miler and J. K. Rieke, Plenum Press, New York (1987), 347-358.

124. R. A. Bubeck, H. Y. Yasar and B. H. Hammouda, *Polymer Communications*, *30*, 25-27 (1989).

125. G. R. Mitchell and A. H. Windle, *Colloid & Polymer Science*, *263*, 280-285 (1985).

126. J. A. Zurimendi, F. Biddlestone, J. N. Hay and R. N. Haward, *J. Materials Science*, *17*, 199-203 (1982).

127. R. N. Haward, *Colloid & Polymer Science*, *258*, 643-662 (1980).

128. G. A. Adam, A. Cross and R. N. Haward, *J. Materials Science*, *10*, 1582-1590 (1975).

129. J. R. Flick and S. E. B. Petrie, in *Studies in Physical and Theoretical Chemistry, 10, Structure and Properties of Amorphous Polymers*, edited by A. G. Walton, Elsevier, Amsterdam (1978), 145-163.

130. K. Neki and P. H. Geil, *J. Macromol. Sci.-Phys.*, *B8*, 295-341 (1973).

131. G. L. Pitman, I. M. Ward and R. A. Duckett, *J. Materials Science*, *13*, 2092-2104 (1978).

132. C. Bauwens-Crowet and J.-C. Bauwens, *Polymer*, *23*, 1599-1604 (1982).

133. S. Blonski, W. Brostow and J. Kubat, *Makromol. Chem., Macromol. Symp.*, *65*, 109-121 (1993).

CHAPTER 12

PROPERTIES OF POLYMERS
IN DILUTE SOLUTIONS

12.A.   Background Information

12.A.1.   General Considerations

The *conformational properties* of polymer chains [1-4] are usually determined in *dilute solutions* of the polymers. Viscosity, light scattering, small-angle x-ray scattering, and osmotic pressure, are the main types of measurements made in solution.

The so-called Θ *(theta) conditions*, in weak solvents, are usually preferred in solution viscosity measurements. Polymer chains are believed to manifest their "unperturbed dimensions" under Θ conditions. This is a result of the nearly perfect balancing of the "excluded volume" (a consequence of the self-avoidance of the random walk path of a polymer chain in a random coil configuration) by unfavorable interactions with the solvent molecules.

X-ray or neutron scattering measurements can be used to probe the conformations of polymer chains in the solid state, instead of in solution. In these measurements, a small fraction of the polymer chains are labeled with a less common isotope of one of the elements in the chemical composition of the polymer, to distinguish them from the other chains surrounding them. For example, hydrogen can be replaced by deuterium in a small fraction of the chains, and neutron scattering can be used. Such measurements, however, do not directly probe the conformations of isolated polymer chains, but of chains surrounded by a dense medium of the unlabeled chains.

The conformational properties of polymer chains are very important for several reasons:
1. They play a key role in determining the properties of polymer solutions, and therefore in both synthesis (i.e., polymerization in solution) and processing (i.e., solvent casting of thin films).
2. The conformations of polymer chains in solution under Θ conditions are essentially the same as the random coil conformations of chains in glassy and rubbery amorphous polymers [5,6], where interactions of polymer chains with solvent molecules are replaced by interactions between the polymer chains. Knowledge of the "unperturbed dimensions" of polymer chains in solution therefore also plays an important role in predicting the physical properties of glassy and rubbery amorphous polymers. For example, the characteristic ratio C_∞ was used as an input parameter in a structure-property relationship [7] (Equation 11.24) for the yield stress of glassy polymers in uniaxial tension.
3. It will be shown in Chapter 13 that information gained from *solution* viscosity measurements under Θ conditions can also be used to predict the critical molecular weight M_{cr}. M_{cr} is important in determining the dependence of the zero-shear viscosity η_0 of polymer *melts* on the average molecular weight of the polymer.

4. Solution viscosity or light scattering measurements can also be used to estimate the weight-average molecular weight M_w of polymer chains. The specific refractive index increment, which is used as an input parameter in determining M_w from the results of light scattering measurements, was discussed in Section 8.D.

As mentioned in Section 1.C.2 and listed among the references for Chapter 1, many articles applying the formalism of graph theory to the properties of polymer chains that can be studied in solutions, including viscoelastic behavior and chain configurations, have been published. This work, however, has almost exclusively been of academic interest. It has mainly focused on the development of elaborate and rigorous mathematical formalisms, and has not provided simple predictive techniques and correlations that could be used on a routine basis in industrial research and development. In this section, we will develop simple correlations based on variations of the formalism of connectivity indices, for selected dilute solution properties of polymers.

Several parameters, most of which are interrelated and can be estimated in terms of each other, are utilized to describe the conformational properties of polymer chains [1,2]. These quantities include the *steric hindrance parameter* σ, the *characteristic ratio* C_∞, the *persistence length*, the *statistical chain segment* (or *Kuhn segment*) *length*, and the *molar stiffness function* K which is directly related to the intrinsic viscosity. These parameters are discussed in many textbooks. For example, see Elias [8] and Flory [2] for excellent and detailed discussions. It is, therefore, outside the scope of this research monograph to describe these parameters in detail. In addition, see deGennes [9] and des Cloizeaux and Jannink [10] for fundamental theoretical considerations.

Three key properties of polymer chains in dilute solutions (σ, C_∞ and K) will be discussed, and then new correlations will be developed for these properties, in the remainder of this chapter.

12.A.2. Steric Hindrance Parameter

The *steric hindrance parameter* σ is the simplest conformational property of a polymer chain. Let $\langle r^2 \rangle_0$ denote the mean-square end-to-end distance of an unperturbed linear chain molecule in solution, and $\langle r^2 \rangle_{0f}$ denote the mean-square end-to-end distance of the idealized and hypothetical "freely rotating" state of the chain. The formal definition of σ is then given by Equation 12.1:

$$\sigma \equiv \sqrt{\frac{\langle r^2 \rangle_0}{\langle r^2 \rangle_{0f}}} \tag{12.1}$$

It can be seen from this definition that the following general trends may be expected for σ:

1. Any hindrance of the torsional motions of polymer chains, regardless of how small this hindrance is, will cause the chains to "expand" relative to their "freely rotating" states, increasing the mean-square end-to-end distance. Consequently, $\sigma > 1$ for all polymers.

2. Structural features (such as bulky side groups) which increase the steric hindrance to torsional motions will increase σ. For example, the value of σ for a series of vinylic polymers will generally increase with increasing size(s) of the side group(s) attached to the vinylic backbone.

3. Polymers with rigid (for example, aromatic) rings as major portions of their chain backbones will generally have small values of σ. Such rings reduce the number of torsional degrees of freedom (Section 4.C), and introduce subunits of approximately fixed length into the polymer chains. The relative effect of the remaining torsional motions on the chain dimensions thus becomes smaller than in chains not containing such rings, so that the difference between the end-to-end distances of the "freely rotating" and the sterically hindered chains decreases.

It will be shown in Section 12.B that these trends indeed hold for the measured σ values of polymers, so that a simple correlation can be derived by elaborating on them further and expressing them in a quantitative manner.

12.A.3.   Characteristic Ratio

The formal definition of the *characteristic ratio* C_∞, which is a more popular descriptor of polymer chain conformation than σ, is given by Equation 12.2, where "lim" denotes "limit", n is the number of bonds along the shortest path across the chain backbone; and n_i is the number of times the i'th type of bond, which has a bond length of l_i, occurs along this shortest path.

$$C_\infty \equiv \lim_{n \to \infty} \left[\frac{\langle r^2 \rangle_0}{\Sigma(n_i l_i^2)} \right] \tag{12.2}$$

C_∞ is proportional to σ^2. Values calculated for σ by using the correlation developed in Section 12.B can therefore be squared, and multiplied by the appropriate proportionality constant, to predict C_∞ at a level of accuracy comparable to the accuracy of the predicted value of σ.

It will be shown below that the proportionality constant between σ and C_∞ depends on the geometry of the chain backbone in a simple manner.

C_∞ and σ are generally both determined from the same measured value of the unperturbed end-to-end distance of the polymer chains, by assuming that the bond lengths and the bond angles across the chain backbone are all constant. For example, if the bond angles in the chain backbone are all equal to τ degrees, then the relationship between σ^2 and C_∞ is given by Equation 12.3 [8].

Examples of the dependence of the proportionality constant between σ^2 and C_∞ on the geometry of the chain backbone will be given in Section 12.C.

$$C_\infty = \frac{[1 - \cos(\tau)]\, \sigma^2}{[1 + \cos(\tau)]} \tag{12.3}$$

Experimental values of $\sigma(\text{exp})$ and $C_\infty(\text{exp})$ listed in the *Polymer Handbook* [11] will be used, to develop the correlation for σ in Section 12.B, and to provide the examples of the relationship between C_∞ and σ^2 in Section 12.C. Values of σ and C_∞ observed in viscosity measurements under Θ conditions will be used in most (but not all) cases, in preference to values of σ and C_∞ observed in other types of tests. When several measured values are available for a polymer, the average of the measured values will be used in most (but not all) cases.

12.A.4. Intrinsic Viscosity Under Theta Conditions

Let η_{ss}, η_s and c denote the solution viscosity, solvent viscosity and polymer concentration, respectively. The *intrinsic viscosity*, represented by the symbol $[\eta]$, is defined as follows:

$$[\eta] \equiv \lim_{c \to 0} \left[\frac{(\eta_{ss} - \eta_s)}{c \cdot \eta_s} \right] \tag{12.4}$$

The intrinsic viscosity depends on the average molecular weight of the polymer by a power law behavior. It is given by Equation 12.5 for a dilute polymer solution under Θ conditions.

$$[\eta]_\Theta = K_\Theta \cdot M_v^{0.5} \tag{12.5}$$

In Equation 12.5, M_v is the *viscosity-average molecular weight*. In monodisperse polymers (i.e., polymers whose chains all have the same degree of polymerization), M_v is equal to the number-average molecular weight M_n. $M_v > M_n$ for polydisperse polymers, and increases with increasing polydispersity. There is no simple general expression for the M_v of polymers with a high degree of polydispersity. The lack of such a general expression is a source of potential uncertainty in predicting the viscosities of dilute solutions of polymers of high polydispersity.

The K_Θ parameter in Equation 12.5 is related, by definition [12a], to the *molar stiffness function* (also called the *molar limiting viscosity number function*) K and the molecular weight M per repeat unit (Equation 12.6), so that Equation 12.5 can be rewritten as Equation 12.7. If units of cc/gram are used for $[\eta]_\Theta$, then K is in units of grams$^{0.25}$·cm$^{1.5}$/mole$^{0.75}$.

$$K \equiv M \cdot K_\Theta{}^{0.5} \qquad\qquad\qquad (12.6)$$

$$[\eta]_\Theta = \left(\frac{K}{M}\right)^2 \cdot M_v{}^{0.5} \qquad\qquad\qquad (12.7)$$

K can be expressed as the sum of two types of terms [12]:

$$K = J + 4.2 \cdot N_{SP} \qquad\qquad\qquad (12.8)$$

The *molar intrinsic viscosity function* [12b] J is estimated as the sum of group contributions [12,13] from the structural units in the repeat unit. N_{SP} is the number of atoms located along the shortest path (Section 2.E) between the two ends of the repeat unit.

Since N_{SP} is exactly determined from the structure of the repeat unit, it is only necessary to develop a new correlation for J. The fixed term $4.2N_{SP}$ is a usually a large portion (sometimes the dominant portion) of K, which is the quantity of real physical interest for estimating $[\eta]_\Theta$. Any correlation developed for J will thus automatically result in a much better correlation for K when the same $4.2N_{SP}$ values are added to both the values of J calculated by group contributions and the values of J estimated from the fitting procedure. The final correlation for K will thus have the same standard deviation as the correlation for J, but a much smaller relative error [(standard deviation)/(average value) ratio], and a much larger correlation coefficient.

A new correlation in terms of connectivity indices will be developed for J, and thus also for K, in Section 12.D, by using van Krevelen's group contributions [12,13] for a large and structurally diverse dataset of polymers.

12.A.5. Intrinsic Viscosity Away from Theta Conditions

If the measurement of $[\eta]$ is not performed under Θ conditions, the power law exponent (a), which is usually called the *Mark-Houwink exponent*, becomes different from 0.5, and its value depends on the magnitude of the deviation from Θ conditions. Experimental data on intrinsic viscosities can be correlated very roughly by using the following empirical expressions [12b]:

$$a \approx 0.8 - 0.1 \cdot abs(\delta_{solvent} - \delta_{polymer}) \qquad \textit{[for abs(}\delta_{solvent} - \delta_{polymer}\textit{)}\leq 3] \qquad (12.9)$$

$$a \approx 0.5 \qquad\qquad\qquad\qquad\qquad \textit{[for abs(}\delta_{solvent} - \delta_{polymer}\textit{)}>3] \qquad (12.10)$$

The solubility parameters (δ) are in $\sqrt{(J/cc)}$. Equations 12.6 and 12.7 work reasonably well unless the polymer has a tendency to be highly crystalline in the bulk (such as polyethylene) and/or the solvent and polymer differ very significantly in their hydrogen bonding capacities [12].

It can be shown, by combining the equations provided above and in [12b] in a straightforward manner, and fitting some relevant information presented only graphically in [12b] by algebraic expressions for convenience, that the final expression for [η] is as follows:

$$[\eta] \approx 0.99328 \cdot \left(\frac{K}{M}\right)^2 \cdot \frac{\exp\left(8.5 \cdot a^{10.3}\right) \cdot M_v^a}{M_{cr}^{(a-0.5)}} \tag{12.11}$$

The critical molecular weight (M_{cr}) will be discussed further in Chapter 13. Note that Equation 12.11 reduces correctly to Equation 12.7 when a=0.5.

An estimate of the expected accuracy of Equation 12.11 can be obtained from van Krevelen's calculations [12b] of [η] for 65 polymer-solvent combinations involving ten different polymers with $M_v=2.5 \cdot 10^5$, and his comparison of the results with the observed values in terms of the logarithms to the base of ten of both the calculated and the experimental values. These results are listed in Table 12.1 and depicted in Figure 12.1. Whenever a range of experimental values was indicated for the logarithm of [η] by van Krevelen [12b], we used the midpoint of this range to facilitate the statistical analysis of the quality of the correlation, and we also used [η] itself instead of its logarithm. Since van Krevelen [12b] had indicated that the calculation of the Mark-Houwink exponent via equations 12.9 and 12.10 is unreliable for systems in which the polymer and the solvent differ significantly in their hydrogen bonding capabilities, two separate correlations are indicated in Figure 12.1; namely, one for all 65 data points and one for only the 58 data points for those polymer-solvent combinations where the polymer and the solvent have similar hydrogen-bonding capabilities. It is seen from Figure 12.1 that the agreement between theory and experiment is very rough and at best semi-quantitative.

The difficulty in the accurate estimation of the value of [η] is not surprising, since the measured values of this quantity for the same combination of polymer and solvent in different laboratories often differ widely from each other. It is very difficult to develop an accurate quantitative structure-property relationship for a property whose measured values themselves are not reproducible with reasonable accuracy.

12.A.6.   Solution Viscosity At Small But Finite Concentrations

Van Krevelen [12b] examined various equations for solution viscosity, and recommended the use of the following two equations [14,15] for the η_{ss} of dilute polymer solutions (at polymer concentrations above the limit of c→0 but with low values of c, especially c≤0.01 g/cc). In these equations, Φ_{solv} denotes the solvated volume fraction of the polymer, and ρ is the polymer bulk density.

Table 12.1. Comparison of calculated and observed intrinsic viscosities (in units of cc/g) of 65 polymer-solvent combinations for 10 different polymers with $M_v=2.5 \cdot 10^5$. The logarithmic average is listed for systems where a range was provided in data source [12b]. The solvents marked by an asterisk (*) differ significantly from the polymer in hydrogen bonding capability.

Polymer	Solvent	[η] (exp)	[η] (pre)
Polypropylene	Cyclohexane	295	309
	Toluene	182	151
	Benzene	160	120
Polyisobutylene	Cyclohexane	209	186
	Carbon tetrachloride	135	95
	Toluene	87	74
	Benzene	59	63
Polystyrene	Cyclohexane	42	49
	n-Butyl chloride	55	74
	Ethylbenzene	83	81
	Decalin	44	83
	Toluene	104	129
	Benzene	114	132
	Chloroform (*)	94	182
	Butanone (*)	52	195
	Chlorobenzene	81	120
	Dioxane	85	79
Poly(vinyl acetate)	Methyl isobutyl ketone	78	46
	Toluene	78	62
	3-Heptanone	44	72
	Benzene	94	76
	Chloroform	158	91
	Butanone	82	93
	Ethyl formate (*)	102	145
	Chlorobenzene	98	166
	Dioxane	115	87
	Acetone	94	85
	Acetonitrile	101	36
	Methanol	61	36
Poly(propylene oxide)	Toluene	145	166
	Benzene	162	316

☞ ☞ ☞   **TABLE 12.1 IS CONTINUED IN THE NEXT PAGE.** ☞ ☞ ☞

Table 12.1. CONTINUED FROM THE PREVIOUS PAGE.

Polymer	Solvent	[η] (exp)	[η] (pre)
Poly(ethylene oxide)	Cyclohexane	186	89
	Carbon tetrachloride	135	145
	Benzene	120	269
	Chloroform	102	363
	Dioxane	127	219
	Acetone (*)	78	214
	Dimethyl formamide (*)	209	78
	Methanol (*)	257	78
Poly(methyl methacrylate)	n-Butyl chloride	25	35
	Methyl isobutyrate	42	36
	Methyl methacrylate	52	44
	Toluene	101	52
	Heptanone	27	60
	Ethyl acetate	60	65
	Benzene	70	68
	Chloroform	124	93
	Butanone	49	98
	Dichloroethane	60	65
	Tetrachloroethane	112	47
	Acetone	48	43
	Nitroethane	58	27
	Acetonitrile	26	27
Polyacrylonitrile	Dimethyl acetamide	398	129
	Dimethyl formamide	343	479
	Dimethyl sulfoxide	363	437
	Butyrolactone	248	129
Polybutadiene	Cyclohexane	126	257
	Isobutyl acetate (*)	93	372
	Toluene	211	269
	Benzene	184	204
Polyisoprene	Hexane	91	72
	Isooctane	110	110
	Toluene	184	174
	Benzene	186	135

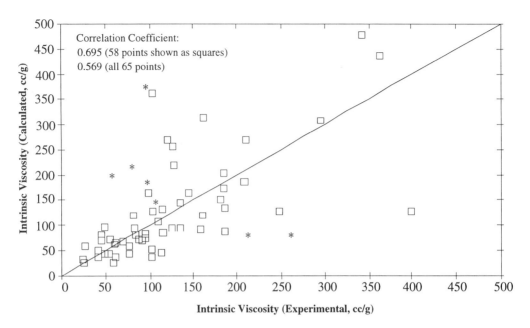

Figure 12.1. Comparison of calculated and observed intrinsic viscosities [12b] of 65 polymer-solvent combinations involving 10 different polymers with $M_v=2.5 \cdot 10^5$.

$$\Phi_{solv} = \frac{0.4 \cdot [\eta] \cdot c}{\left\{ 1 + 0.765 \cdot [\eta] \cdot c - 1.91 \cdot \dfrac{c}{\rho} \right\}} \tag{12.12}$$

$$\eta_{ss} = \frac{\eta_s}{\left(1 - 2.5 \cdot \Phi_{solv} + 11 \cdot \Phi_{solv}^5 - 11.5 \cdot \Phi_{solv}^7 \right)} \tag{12.13}$$

For example, consider a polymer with M=100 g/mole, M_{cr}=30000 g/mole, $M_v=2 \cdot 10^5$, $\delta_p=20$ $\sqrt{(J/cc)}$, K=30 $g^{0.25} \cdot cm^{1.5}/mole^{0.75}$ and ρ=1 g/cc, dissolved at a concentration of c=0.01 g/cc in a solvent with δ_s=18 $\sqrt{(J/cc)}$ and $\eta_s=1.000 \cdot 10^{-3}$ N·sec/m². The difference between δ_p and δ_s is 2 $\sqrt{(J/cc)}$, so that Equation 12.9 gives a≈0.6. Substitution of the values of the input parameters into Equation 12.11 gives [η]≈50.5 cc/g. Equation 12.12 then gives Φ_{solv}≈0.148 and Equation 12.13 gives η_{ss}≈1.585·10⁻³ N·sec/m².

The value of η_{ss} estimated by Equation 12.13 goes to the correct limit of η_s at zero polymer concentration (c=0). The use of equations 12.12 and 12.13 results in the prediction that the viscosity of a dilute polymer solution (with c≤0.05 g/cc as the most generous possible upper limit) increases slowly with increasing polymer concentration until the concentration becomes sufficiently large for the dilute solution equations to become inapplicable. Since these two equations depend strongly on [η], their accuracy depends on the accuracy of the measured or predicted value of [η] inserted into them. We have found these equations to work quite well in practice provided that the value of [η] used in them is accurate.

The effect of the temperature on η_{ss} for a dilute polymer solution can be described in terms of an activated process [12b], as in Equation 12.14. However, given the difficulty of predicting η_{ss} accurately even at a single reference temperature (T_{ref}), the attempted prediction of the temperature dependence of η_{ss} in this manner often results in a compounding of the errors and is, therefore, at best of only qualitative validity. Equation 12.14 is probably more useful for correlating experimental data than it is for predicting it with reasonable accuracy.

$$\eta_{ss}(T) = \eta_{ss}(T_{ref}) \cdot \exp\left[\frac{E_{\eta ss} \cdot (T_{ref} - T)}{R \cdot T \cdot T_{ref}}\right] \tag{12.14}$$

The situation is especially complicated since the activation energy depends on the concentration and may vary according to different mixing rules depending on the nature and strength of the interactions between the polymer and the solvent. For example, if we obtain the volume fraction Φ_p of the polymer from Equation 12.15, examples can be found where the activation energy for viscous flow of the solution ($E_{\eta ss}$) is closer to the upper-bound average of the activation energies for the polymer ($E_{\eta p}$) and the solvent ($E_{\eta s}$) given by Equation 12.16, or closer to the lower-bound average given by Equation 12.17 (shown in an algebraically rearranged form in Equation 12.18), or even changing gradually in the dominant form of its dependence on Φ_p with increasing Φ_p. The situation is extremely complex, and an accurate and reliable quantitative resolution of these complexities is difficult to anticipate at this time.

$$\Phi_p = \frac{c}{\rho} \tag{12.15}$$

$$E_{\eta ss} \approx E_{\eta s} + \Phi_p \cdot (E_{\eta p} - E_{\eta s}) \tag{12.16}$$

$$\frac{1}{E_{\eta ss}} = \frac{(1 - \Phi_p)}{E_{\eta s}} + \frac{\Phi_p}{E_{\eta p}} \tag{12.17}$$

$$E_{\eta ss} = \frac{E_{\eta s} \cdot E_{\eta p}}{\left[\Phi_p \cdot E_{\eta s} + (1 - \Phi_p) \cdot E_{\eta p}\right]} \tag{12.18}$$

12.B.   Correlation for the Steric Hindrance Parameter

12.B.1.   Definitions of the Fitting Variables

A dataset consisting of the experimental $\sigma(exp)$ values of 54 polymers was prepared, by using data provided in the *Polymer Handbook* [11]. This dataset was used to develop a direct correlation, utilizing four adjustable parameters, between the structures of polymers and their steric hindrance parameters. The intuitively expected general trends summarized in Section 12.A.2 were used as guidelines in developing this correlation.

A detailed analysis showed that the value of σ depends strongly on two types of structural features, namely, (a) the nature of the connectivity and conformations of the chain backbone, and (b) the relative size of the side group portion of the hydrogen-suppressed graph of the repeat unit.

The dependence of σ on these structural factors is not described adequately by the parameters defined and used in earlier chapters. Four specialized indices were therefore defined, and used in the correlation for σ. First of all, two backbone indices, ξ^{BB1} and ξ^{BB2}, were defined in terms of averages of functions of the atomic indices δ and δ^v over the N_{BB} backbone atoms:

$$\xi^{BB1} \equiv \frac{\displaystyle\sum_{BB\ atoms}^{'} \left(\frac{\delta}{\delta^{v'}}\right)}{N_{BB}} \tag{12.19}$$

$$\xi^{BB2} \equiv 3 - \frac{\displaystyle\sum_{BB\ atoms} (\delta)}{N_{BB}} \tag{12.20}$$

In Equation 12.19, $\delta^{v'}$ denotes:

1. The valence index δ^v for atoms from the first row of the periodic table (C, N, O and F).
2. The δ^v of the isoelectronic first-row atom for an atom from a lower row of the periodic table in its standard (lowest) oxidation state. The following are two examples:

     (a) $\delta^{v'}=4$, which is the δ^v of carbon, instead of being equal to 4/9, which is the δ^v of silicon itself, for a silicon atom with sp^3 hybridization bonded to four non-hydrogen atoms.

     (b) $\delta^{v'}=6$, which is the δ^v of oxygen, instead of being equal to 2/3, which is the δ^v of sulfur itself, for a sulfur atom with sp^3 hybridization bonded to two non-hydrogen atoms.

3. The δ^v of the atom itself for an atom from a lower row of the periodic table in a higher oxidation state. For example, $\delta^{v'}=\delta^v=(8/3)$ for a sulfur atom in its highest oxidation state, in the bonding configuration R-SO$_2$-R'.

A third backbone index ξ^{trans} was defined by Equation 12.21, as the number of carbon-carbon double bonds along the chain backbone which are in a *trans* configuration divided by the number N_{BB} of backbone atoms. If there is a mixture of *cis* and *trans* isomerization around the C=C bonds along the chain backbone, only the *trans* double bonds are counted. For example, all-*trans* polyisoprene has one *trans* C=C backbone bond per repeat unit with $N_{BB}=4$, so that $\xi^{trans}=(1/4)=0.25$. On the other hand, 85%-*trans* polychloroprene has $\xi^{trans}=(0.85/4)=0.2125$.

$$\xi^{trans} \equiv \frac{N_{(trans\ C=C\ along\ backbone)}}{N_{BB}} \qquad (12.21)$$

Finally, a side group index ξ^{SG} was defined as the fourth power of the fraction $(N - N_{BB})/N$ of the vertices of the hydrogen-suppressed graph of the repeat unit that are in its side groups:

$$\xi^{SG} \equiv \frac{(N - N_{BB})^4}{N^4} \equiv \left(\frac{N_{SG}}{N}\right)^4 \qquad (12.22)$$

12.B.2. Development of the Correlation

One polymer, namely poly[oxy(dipropylsilylene)], was found to be an outlier in preliminary calculations. It was therefore removed from the dataset, leaving 53 polymers in the final dataset.

The single most important index was found to be ξ^{SG}. The correlation coefficient between σ and ξ^{SG} is 0.8337. There is, therefore, a fairly strong correlation between σ and increasing size of the side groups, since ξ^{SG} is the *fourth power* of the fraction of the vertices in the hydrogen-suppressed graph that are in side groups.

The correlation coefficient between $\sigma(exp)$ and ξ^{BB1}, which is the second most important index, is 0.7980. The ease of rotational or oscillational motions around a backbone bond emanating from an atom which is singly bonded to all of its neighbors increases with decreasing $(\delta/\delta^{v'})$, as shown in Figure 12.2 for the first-row atoms C, N and O in bonding environments in which these atoms are often found in the backbones of polymers. ξ^{BB1} is thus, in general, an indicator of the extent of the "hinge-like" behavior (i.e., the flexibility) of backbone bonds, so that σ increases with increasing stiffness and decreases with increasing flexibility of backbone bonds.

As shown in Figure 12.3, the ratio $(\delta/\delta^{v'})$ also contributes a value smaller than one to the sum in Equation 12.19 for backbone atoms that are multiply bonded. Obviously, a double bond in a chain backbone is much *less* flexible than a single bond. Nevertheless, as discussed in Section 12.A.2, because of the manner in which σ is related to the bond lengths, bond angles and number of rotational degrees of freedom along the shortest path across the chain backbone, σ decreases by the presence, in the backbone, of structural features such as aromatic rings, which contain multiply-bonded atoms. The ratio $(\delta/\delta^{v'})$ therefore also represents, quite adequately, the contribution of structural features such as double bonds and aromatic rings along the backbone to

ξ^{BB1}. The only such structural features which need to be treated in a different manner are *trans* carbon-carbon double bonds along the chain backbone.

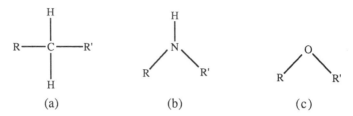

(a)   (b)   (c)

Figure 12.2. Let R and R' denote the structural units surrounding a backbone C, N or O atom with only two non-hydrogen neighbors, and therefore with δ=2. The relative ease of rotational or oscillational motions around a backbone bond emanating from an atom which is singly bonded to all of its neighbors generally increases with decreasing value of the ratio $(\delta/\delta^{v'})$. For example: (a) C has $\delta^{v'}$=2, so that $(\delta/\delta^{v'})$=1. (b) N has $\delta^{v'}$=4, so that $(\delta/\delta^{v'})$=(2/4)=0.5. (c) O has $\delta^{v'}$=6, so that $(\delta/\delta^{v'})$=(2/6)$\approx$0.33333.

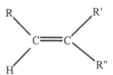

Figure 12.3. $(\delta/\delta^{v'})$<1 for multiply-bonded backbone atoms. Let R, R' and R" denote non-hydrogen structural units. [It makes no difference whether the two C atoms are in a linear chain segment as in polyisoprene, or in a ring structure such as a backbone phenyl ring of bisphenol-A polycarbonate. (See Figure 5.1 for the structures of these two polymers.)] The carbon atom on the left has δ=2 and $\delta^{v'}$=3, so that $(\delta/\delta^{v'})$=(2/3)$\approx$0.66667. The carbon atom on the right has δ=3 and $\delta^{v'}$=4, so that $(\delta/\delta^{v'})$=(3/4)=0.75.

Carbon-carbon double bonds (C=C) in a *trans* configuration along the chain backbone cause a significant reduction of σ. This reduction is accounted for by the parameter ξ^{trans}. The correlation coefficient between σ and ξ^{trans} for the entire dataset is only -0.4212. This correlation coefficient is small because only three of the polymers in the dataset have C=C bonds in a *trans* configuration. ξ^{trans} is, nonetheless, an important parameter. Its use in the linear regression systematically corrects for most of the large difference between the σ values of the same polymer with *trans* versus *cis* isomerization for C=C bonds in the chain backbone.

ξ^{BB2} is used to distinguish between backbone atoms with equal values of $(\delta/\delta^{v'})$ but different values of δ, resulting from the presence of different numbers of non-hydrogen neighbors of a given backbone atom. (See Figure 12.4 for several examples.) The correlation between σ and ξ^{BB2} is weak, with a correlation coefficient of merely -0.4084; however, in a three-parameter fit utilizing only ξ^{BB1}, ξ^{trans} and ξ^{SG}, the deviation between $\sigma(exp)$ and $\sigma(fit)$ correlates fairly well with ξ^{BB2}. The use of ξ^{BB2} therefore results in an improved correlation.

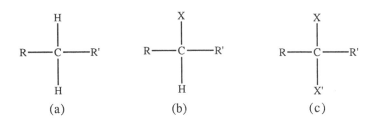

Figure 12.4. Examples of backbone atoms with equal values of $(\delta/\delta^{v'})$ but different δ, resulting from the presence of different numbers of non-hydrogen neighbors of a given backbone atom. Let R, R', X and X' all be different from H. The C atoms in all three illustrations have $\delta = \delta^{v'}$, and therefore $(\delta/\delta^{v'}) = 1$. On the other hand, $\delta = 2$ in (a), $\delta = 3$ in (b), and $\delta = 4$ in (c).

The final quantitative structure-property relationship for σ (Equation 12.23) has a standard deviation of 0.0889, a correlation coefficient of 0.9729, and no statistically significant trend in its deviations. The results of these calculations are listed in Table 12.2 and depicted in Figure 12.5.

$$\sigma \approx 0.244978 \cdot (7 \cdot \xi^{BB1} + \xi^{BB2} + 5 \cdot \xi^{SG} - 6 \cdot \xi^{trans}) \tag{12.23}$$

If additional σ values are measured with great accuracy in the future, a refined relationship may emerge. For example, polypropylene, poly(vinyl alcohol), poly(vinyl fluoride), poly(vinyl chloride) and poly(vinyl bromide) all have equal ξ^{BB1}, ξ^{BB2}, ξ^{SG} and ξ^{trans}, and therefore the same σ according to Equation 12.23. The existing data do not show trends of sufficient statistical significance to justify the introduction of additional parameters to distinguish between them.

$\xi^{BB1} \leq 1$ and $\xi^{BB2} \leq 1$ for all uncrosslinked polymers constructed from the nine elements (C, N, O, H, F, Si, S, Cl and Br) of interest in this book. Furthermore, all polymers have backbones, i.e., $N_{BB} \geq 1$ for all polymers. ξ^{SG} can thus only asymptotically approach its maximum value of 1 in the limit of infinitely large side groups. Inspection of Table 12.2 shows that $\xi^{SG} \ll 1$ for all ordinary polymers. Equations 12.19-12.23 thus result in a predicted upper limit for σ, namely,

0.244978·(7+1+5)≈3.185. As discussed in Section 12.A.2, σ>1 for all polymers. Combining these upper and lower limits, the σ values of all polymers are predicted to fall into the range of 1<σ<3.185 when deduced from the results of viscosity measurements under Θ conditions.

Table 12.2. Experimental values [11] σ(exp) of the steric hindrance parameter, the quantities used in the correlation equation for σ, and the fitted values σ(fit), for 53 polymers.

Polymer	σ(exp)	N	N_{BB}	ξ^{BB1}	ξ^{BB2}	ξ^{SG}	ξ^{trans}	σ(fit)
Bisphenol-A polycarbonate	1.18	19	16	0.6719	0.5625	0.0006	0.0000	1.29
Poly(1,4-butadiene) *(trans)*	1.23	4	4	0.8333	1.0000	0.0000	0.2500	1.31
Polyisoprene *(trans)*	1.30	5	4	0.8542	0.7500	0.0016	0.2500	1.28
Poly(ethylene terephthalate)	1.33	14	12	0.6944	0.6667	0.0004	0.0000	1.35
Poly[oxy(2-methyl-6-phenyl-1,4-phenylene)]	1.36	14	7	0.6667	0.4286	0.0625	0.0000	1.32
Poly(dimethyl siloxane)	1.37	4	2	0.6667	0.0000	0.0625	0.0000	1.22
Polychloroprene *(85% trans)*	1.40	5	4	0.8542	0.7500	0.0016	0.2125	1.34
Polyoxyethylene	1.43	3	3	0.7778	1.0000	0.0000	0.0000	1.58
Poly(propylene oxide)	1.56	4	3	0.7778	0.6667	0.0039	0.0000	1.50
Poly[oxy(methylphenylsilylene)]	1.58	9	2	0.6667	0.0000	0.3660	0.0000	1.59
Poly[oxy(methyl γ-trifluoropropylsilylene)]	1.61	9	2	0.6667	0.0000	0.3660	0.0000	1.59
Polyoxyethylethylene	1.66	5	3	0.7778	0.6667	0.0256	0.0000	1.53
Poly(1,4-butadiene) *(cis)*	1.66	4	4	0.8333	1.0000	0.0000	0.0000	1.67
Polyisoprene *(cis)*	1.67	5	4	0.8542	0.7500	0.0016	0.0000	1.65
Polyoxytetramethylene	1.69	5	5	0.8667	1.0000	0.0000	0.0000	1.73
Polypropylene	1.76	3	2	1.0000	0.5000	0.0123	0.0000	1.85
Poly(hexamethylene adipamide)	1.78	16	14	0.8929	0.8571	0.0002	0.0000	1.74
Poly(ε-caprolactam)	1.78	8	7	0.8929	0.8571	0.0002	0.0000	1.74
Polyisobutylene	1.80	4	2	1.0000	0.0000	0.0625	0.0000	1.79
Poly(vinyl bromide)	1.82	3	2	1.0000	0.5000	0.0123	0.0000	1.85
Poly(1-butene)	1.82	4	2	1.0000	0.5000	0.0625	0.0000	1.91
Poly(acrylic acid)	1.83	5	2	1.0000	0.5000	0.1296	0.0000	2.00
Polyethylene	1.87	2	2	1.0000	1.0000	0.0000	0.0000	1.96
Poly(vinyl chloride)	1.92	3	2	1.0000	0.5000	0.0123	0.0000	1.85
Poly(ethyl methacrylate)	2.00	8	2	1.0000	0.0000	0.3164	0.0000	2.10

☞ ☞ ☞   **TABLE 12.2 IS CONTINUED IN THE NEXT PAGE.** ☞ ☞ ☞

Table 12.2. CONTINUED FROM THE PREVIOUS PAGE.

Polymer	σ(exp)	N	N_{BB}	ξ^{BB1}	ξ^{BB2}	ξ^{SG}	ξ^{trans}	σ(fit)
Poly(methyl methacrylate)	2.01	7	2	1.0000	0.0000	0.2603	0.0000	2.03
Poly(methyl acrylate)	2.03	6	2	1.0000	0.5000	0.1975	0.0000	2.08
Poly(n-butyl methacrylate)	2.06	10	2	1.0000	0.0000	0.4096	0.0000	2.22
Poly(vinyl acetate)	2.08	6	2	1.0000	0.5000	0.1975	0.0000	2.08
Poly(1-pentene)	2.14	5	2	1.0000	0.5000	0.1296	0.0000	2.00
Poly(1-octene)	2.14	8	2	1.0000	0.5000	0.3164	0.0000	2.22
Poly(methyl ethacrylate)	2.15	8	2	1.0000	0.0000	0.3164	0.0000	2.10
Poly(2-ethylbutyl methacrylate)	2.16	12	2	1.0000	0.0000	0.4823	0.0000	2.31
Polystyrene	2.18	8	2	1.0000	0.5000	0.3164	0.0000	2.22
Poly(2,5-dichlorostyrene)	2.18	10	2	1.0000	0.5000	0.4096	0.0000	2.34
Poly(N-vinyl pyrrolidone)	2.21	8	2	1.0000	0.5000	0.3164	0.0000	2.22
Poly(o-chloro styrene)	2.22	9	2	1.0000	0.5000	0.3660	0.0000	2.29
Poly(α-methyl styrene)	2.25	9	2	1.0000	0.0000	0.3660	0.0000	2.16
Poly(o-methoxy styrene)	2.26	10	2	1.0000	0.5000	0.4096	0.0000	2.34
Poly(n-hexyl methacrylate)	2.27	12	2	1.0000	0.0000	0.4823	0.0000	2.31
Poly(methyl butacrylate)	2.28	10	2	1.0000	0.0000	0.4096	0.0000	2.22
Poly(n-octyl methacrylate)	2.28	14	2	1.0000	0.0000	0.5398	0.0000	2.38
Poly(vinyl pivalate)	2.29	9	2	1.0000	0.5000	0.3660	0.0000	2.29
Poly(p-methyl styrene)	2.31	9	2	1.0000	0.5000	0.3660	0.0000	2.29
Poly(p-bromo styrene)	2.33	9	2	1.0000	0.5000	0.3660	0.0000	2.29
Poly(cyclohexyl methacrylate)	2.33	12	2	1.0000	0.0000	0.4823	0.0000	2.31
Poly(5-vinyl-2-methylpyridine)	2.35	9	2	1.0000	0.5000	0.3660	0.0000	2.29
Poly(p-methoxy styrene)	2.37	10	2	1.0000	0.5000	0.4096	0.0000	2.34
Poly(1-methoxycarbonyl-1-phenylethylene)	2.42	12	2	1.0000	0.0000	0.4823	0.0000	2.31
Poly(3,4-dichlorostyrene)	2.44	10	2	1.0000	0.5000	0.4096	0.0000	2.34
Poly(vinyl benzoate)	2.46	11	2	1.0000	0.5000	0.4481	0.0000	2.39
Poly(dodecyl methacrylate)	2.57	18	2	1.0000	0.0000	0.6243	0.0000	2.48
Poly(N-vinyl carbazole)	2.78	15	2	1.0000	0.5000	0.5642	0.0000	2.53

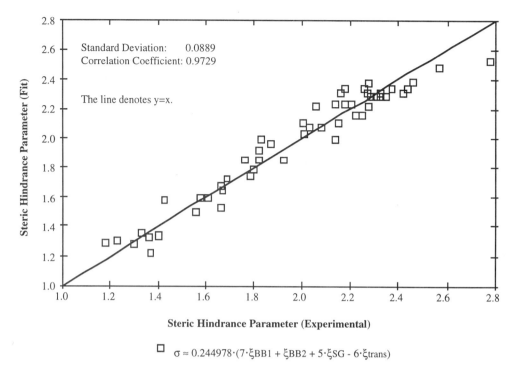

$$\sigma \approx 0.244978 \cdot (7 \cdot \xi_{BB1} + \xi_{BB2} + 5 \cdot \xi_{SG} - 6 \cdot \xi_{trans})$$

Figure 12.5. Comparison of the experimental values [11] of the steric hindrance parameters σ of 53 polymers with the calculated values obtained by a four-parameter linear regression.

12.C. Calculation of the Characteristic Ratio

As discussed in Section 12.A.3, C_∞ is related to σ^2 in a simple manner [8]. C_∞ can be predicted with reasonable accuracy by assuming that it is proportional to σ^2. The proportionality constant depends on the average value of the angles between successive bonds in the shortest path across the chain backbone. Values of σ estimated by using the new correlation developed in Section 12.B can therefore be used to predict C_∞.

The relationship between C_∞ and σ^2 will now be illustrated for the most common types of chain backbone geometries. The results to be presented below are not new correlations. They are, instead, some of the most useful examples of the application of Equation 12.3, which itself is a direct consequence [8] of the definitions of C_∞ and σ.

If a chain backbone were made up of bonds of equal length, with the ideal tetrahedral angle of 109.5° for all bond angles, the substitution of $\tau=109.5°$ into Equation 12.3 would result in a proportionality constant of $2.00213 \approx 2$ between σ^2 and C_∞. The backbones of most polymer chains are constructed from atoms from the first row of the periodic table (C, N and O), with bond angles in a narrow range of values (100° to 120°) approximately centered around 109.5°. The relationship between σ and C_∞ (Table 12.3 and Figure 12.6) is thus given by Equation 12.24 for most polymers, whose backbones have neither ordinary double bonds around which *cis* and *trans* isomerization is possible, nor silicone-type (-Si-O-) bonding configurations.

$$C_\infty \approx 2 \cdot \sigma^2 \qquad \qquad \textit{(for most polymers)} \qquad \qquad (12.24)$$

As can be seen from the data points for poly(1,4-butadiene) and polyisoprene in Figure 12.6, the C_∞ of polymers which have all-*cis* double bonds in their backbones deviates only slightly (about 10%) from Equation 12.19, while the C_∞ of polymers with all-*trans* double bonds shows a very large but systematic (positive) deviation from Equation 12.24. These results are also consequences of Equation 12.3.

The correlation between σ^2 and C_∞ for silicone-type polymers (see Figure 12.6) is given by Equation 12.25. The increase of the proportionality constant relative to its value in Equation 12.24 is a consequence of Equation 12.3. The bond angles in the backbones of silicones are substantially larger than the bond angles in the backbones of typical polymers. For example, Si-O-Si bond angles in molecules are larger than C-O-C and C-O-Si bond angles in similar molecules by at least 20° [16]. The use of $\tau \approx 121°$ (considering both the Si-O-Si and the O-Si-O bond angles in the chain backbone) in Equation 12.3 would result in a proportionality constant of 3.1240, which is very similar to the proportionality constant of 3.1285 in Equation 12.25.

$$C_\infty \approx 3.128525 \cdot \sigma^2 \qquad \textit{(for polymers with a silicone-type backbone)} \qquad (12.25)$$

The data points for *trans*-polybutadiene and *trans*-polyisoprene lie slightly above the curve for Equation 12.25, and can therefore also be represented approximately by Equation 12.25.

C_∞ can thus be estimated for most polymers, via Equation 12.24 or Equation 12.25, as appropriate, by using the predicted value of σ and considering the geometry of the chain backbone. For polymers with more unusual chain backbone geometries (such as random or alternating copolymers constructed from repeat units whose homopolymers would follow different relationships between σ and C_∞), the angles between the successive bonds in the shortest path across the chain backbone have to be considered more carefully to estimate the value of the proportionality constant in Equation 12.3.

Table 12.3. Experimental steric hindrance parameters σ and characteristic ratios C_∞ [11] of 51 polymers. Some of the σ values listed below differ from those in Table 12.2, because only the results of those measurements for which both σ and C_∞ were reported are utilized in this table.

Polymer	σ	C_∞	Polymer	σ	C_∞
Poly(1,4-butadiene) *(trans)*	1.23	5.80	Poly(*n*-butyl methacrylate)	2.06	8.50
Polyisoprene *(trans)*	1.30	6.35	Poly(vinyl acetate)	2.08	8.65
Poly(ethylene terephthalate)	1.33	3.70	Polyacrylonitrile	2.13	9.10
Poly[oxy(2-methyl-6-phenyl-1,4-phenylene)]	1.36	3.70	Poly(1-octene)	2.14	9.10
Poly(dimethyl siloxane)	1.39	6.25	Poly(1-pentene)	2.14	9.20
Polyoxyethylene	1.43	4.10	Poly(methyl ethacrylate)	2.15	9.25
Poly(propylene oxide)	1.56	4.85	Poly(cyclohexyl methacrylate)	2.15	9.25
Poly[oxy(methylphenylsilylene)]	1.58	8.35	Poly(*p*-chloro styrene)	2.15	9.25
Poly[oxy(methyl γ-trifluoropropylsilylene)]	1.61	6.30	Poly(2-ethylbutyl methacrylate)	2.16	9.30
Poly(1,4-butadiene) *(cis)*	1.66	5.05	Polystyrene	2.18	9.50
Polyoxyethylethylene	1.66	5.50	Poly(2,5-dichlorostyrene)	2.18	9.50
Polyisoprene *(cis)*	1.67	5.00	Poly(N-vinyl pyrrolidone)	2.21	9.76
Polyoxytetramethylene	1.69	5.73	Poly(α-methyl styrene)	2.25	10.10
Polypropylene	1.76	6.20	Poly(*o*-methoxy styrene)	2.26	10.20
Poly(ε-caprolactam)	1.78	6.35	Poly(*n*-hexyl methacrylate)	2.27	10.30
Poly(hexamethylene adipamide)	1.78	6.37	Poly(methyl butacrylate)	2.28	10.40
Polyisobutylene	1.80	6.50	Poly(*n*-octyl methacrylate)	2.28	10.40
Poly(1-butene)	1.82	6.60	Poly(*p*-methyl styrene)	2.31	10.70
Poly(vinyl bromide)	1.82	6.60	Poly(5-vinyl-2-methylpyridine)	2.35	11.10
Poly(acrylic acid)	1.83	6.70	Poly(*p*-methoxy styrene)	2.37	11.20
Polyethylene	1.87	7.00	Poly(1-methoxycarbonyl-		
Poly[oxy(dipropylsilylene)]	1.89	12.00	1-phenylethylene)	2.42	11.70
Poly(vinyl chloride)	1.92	7.40	Poly(vinyl benzoate)	2.46	12.10
Poly(ethyl methacrylate)	2.00	8.00	Poly(*p*-bromo styrene)	2.50	12.50
Poly(methyl methacrylate)	2.01	8.10	Poly(dodecyl methacrylate)	2.57	13.20
Poly(methyl acrylate)	2.03	8.20	Poly(N-vinyl carbazole)	2.82	15.90

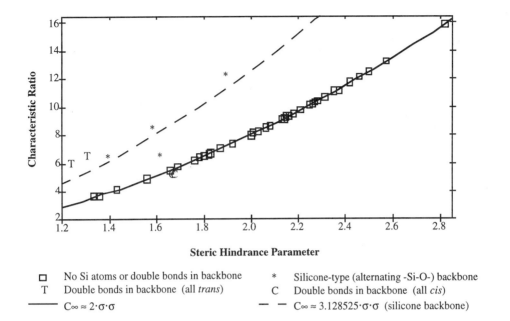

Figure 12.6. Comparison of the experimental values [11] of the steric hindrance parameter σ and characteristic ratio C_∞ for 51 polymers, to show the relation between σ and C_∞ in terms of the geometry of the chain backbone.

12.D.   Correlation for the Molar Stiffness Function

A dataset of 120 polymers was prepared to develop a correlation for K. Group contributions to J were available [12,13] for 94 of these polymers. These group contributions were used with one important change, namely, our own estimate of 24 was used instead of the published value [12,13] of 11 for the contribution of bromine atoms.

Weight factors of 10/J(group) were used in developing the correlation for J. The best fit possible without using any atomic or group correction terms was obtained when the number of hydrogen atoms N_H in the repeat unit and the zeroth-order valence connectivity index $^0\chi^v$ were included in the linear regression. This fit (Equation 12.26) has a standard deviation of only 3.7, a correlation coefficient of 0.9649 for J, and a much larger correlation coefficient of 0.9896 for K when the fixed term $4.2N_{SP}$ is added to J to obtain K.

$$J \approx -1.9546 \cdot N_H + 7.7686 \cdot {}^0\chi^v \qquad (12.26)$$

The correction index N_K, containing both atomic and group correction terms, was defined by Equation 12.27, and used to obtain an improved correlation (Equation 12.23) with a standard deviation of only 0.7 and a correlation coefficient of 0.9990 for J. In Equation 12.28, N is the number of vertices in the hydrogen-suppressed graph.

$$N_K \equiv 5N_{amide} + 7N_{cyanide} + 15N_{carbonate} + 5N_{Cl} + 13N_{Br} + 4N_{hydroxyl} - 3N_{"ether"} - 5N_{C=C}$$
$$+ 3N_{sulfone} - 3N_{(acrylic\ ester)}$$
$$- 5N_{(isolated\ saturated\ aliphatic\ hydrocarbon\ rings,\ i.e.,\ cyclohexyl\ or\ cyclopentyl)} \qquad (12.27)$$

$$J \approx 1.1474 \cdot N - 0.9229 \cdot N_H + 4.1128 \cdot {}^0\chi^v + 0.9778 \cdot N_K \qquad (12.28)$$

Clarifications and reminders concerning some of the terms in Equation 12.27 are as follows:

1. $N_{hydroxyl}$ only includes ordinary (alcohol-type and phenol-type) -OH moieties. It does not include the -OH units in carboxylic acid (-COOH) or sulfonic acid ($-SO_3H$) groups.

2. $N_{"ether"}$ in Equation 12.27 includes both the ordinary types of ether (-O-) linkages flanked by carbon atoms on both sides, and ether linkages flanked by silicon atoms on one or both sides. The word "ether" is enclosed in quotation marks in the subscript, to distinguish $N_{"ether"}$ from N_{ether} encountered in Section 3.C, where only the ether linkages flanked by carbon atoms on both sides were counted.

3. $N_{(acrylic\ ester)}$ includes both acrylic ester groups with substitution on the α-carbon atom of the backbone as in PMMA (Figure 11.1), and acrylic ester groups with no substitution for the hydrogen on the α-carbon atom.

The final expression (Equation 12.29) for K, which is obtained by combining equations 12.8, 12.27 and 12.28, has a correlation coefficient of 0.9997, so that it accounts for 99.94% of the variation of the K(group) values in the dataset. The standard deviation of 0.7 is only 1.6% of the average K(group) value of 44.6. The intrinsic viscosities predicted by substituting Equation 12.29 into Equation 12.7 therefore typically only differ by about $(1.6)^2 \approx 2.6\%$ from the intrinsic viscosities predicted by using group contributions for K. Since the use of group contributions for K only results in a rough estimate of the intrinsic viscosity, Equation 12.29 provides predictions of comparable quality, without the need for group contributions.

$$K \approx 1.1474 \cdot N - 0.9229 \cdot N_H + 4.1128 \cdot {}^0\chi^v + 0.9778 \cdot N_K + 4.2 \cdot N_{SP} \qquad (12.29)$$

The results of these calculations are listed in Table 12.4 and depicted in Figure 12.7. The predictive power of Equation 12.29 is also demonstrated in Table 12.4, where values predicted for K by using Equation 12.29 are listed for the 26 polymers from the dataset for which K(group) could not be calculated because of the lack of the group contributions for J.

Table 12.4. The molar stiffness function K calculated by group contributions, the number N_{SP} of atoms in the shortest path across the backbone of the repeat unit, the number of hydrogen atoms N_H in the repeat unit, the quantity N_K used in the correlation for K, and fitted or predicted (fit/pre) values of K, for 120 polymers. For 26 of these polymers, K cannot be calculated via group contributions, but can be calculated by using the new correlation. N and $^0\chi^v$, which are also used in the correlation, are listed in Table 2.2. K is in units of $grams^{0.25}\cdot cm^{1.5}/mole^{0.75}$.

Polymer	K(group)	N_{SP}	N_H	N_K	K(fit/pre)
Polyoxymethylene	10.85	2	2	-3	10.50
Polyethylene	13.10	2	4	0	12.82
Poly(1,2-butadiene)	--------	2	6	-5	13.13
Polypropylene	15.45	2	6	0	15.70
Poly(vinyl methyl ether)	15.55	2	6	-3	15.59
Poly(vinyl fluoride)	16.15	2	3	0	15.91
Polyoxyethylene	17.40	3	4	-3	16.91
Poly(1-butene)	17.80	2	8	0	17.91
Polyisobutylene	17.85	2	8	0	18.80
Poly(vinyl ethyl ether)	17.90	2	8	-3	17.80
Poly(vinylidene fluoride)	19.15	2	2	0	19.22
Poly(propylene oxide)	19.75	3	6	-3	19.79
Poly(vinyl alcohol)	19.90	2	4	4	19.18
Poly(acrylic acid)	19.90	2	4	0	21.30
Poly(dimethyl siloxane)	20.60	2	6	-3	20.59
Poly(methyl acrylate)	21.85	2	6	-3	21.62
Poly(vinyl cyclohexane)	21.90	2	14	-5	21.97
Poly(1,4-butadiene)	22.00	4	6	-5	21.53
Polytrifluoroethylene	22.20	2	1	0	22.31
Poly(methacrylic acid)	22.30	2	6	0	24.40
Poly(4-methyl-1-pentene)	22.50	2	12	0	23.00
Poly(vinyl *n*-butyl ether)	22.60	2	12	-3	22.22
Poly(vinyl *sec*-butyl ether)	22.60	2	12	-3	22.89
Poly(thiocarbonyl fluoride)	--------	2	0	0	23.19
Poly(ethylene sulfide)	--------	3	4	0	23.20
Polyoxytrimethylene	23.95	4	6	-3	23.32
Poly(vinyl chloride)	24.15	2	3	5	23.91
Poly(ethyl acrylate)	24.20	2	8	-3	23.83
Poly(methyl methacrylate)	24.25	2	8	-3	24.72

☞☞☞   **TABLE 12.4 IS CONTINUED IN THE NEXT PAGE.** ☞☞☞

Table 12.4. CONTINUED FROM THE PREVIOUS PAGE.

Polymer	K(group)	N_{SP}	N_H	N_K	K(fit/pre)
Polyisoprene	24.40	4	8	-5	24.63
Poly(vinyl acetate)	24.45	2	6	0	24.56
Polytetrafluoroethylene	25.20	2	0	0	25.62
Poly(ethyl methacrylate)	26.60	2	10	-3	26.93
Poly(methyl ethacrylate)	26.60	2	10	-3	26.93
Poly(vinyl propionate)	26.80	2	8	0	26.77
Polyacrylonitrile	26.90	2	3	7	26.24
Poly(maleic anhydride)	--------	2	2	0	28.48
Poly(n-butyl acrylate)	28.90	2	12	-3	28.26
Polymethacrylonitrile	29.30	2	5	7	29.34
Poly(vinyl sulfonic acid)	29.90	2	4	3	27.52
Poly(N-vinyl pyrrolidone)	29.90	2	9	0	28.85
Poly(p-vinyl pyridine)	29.90	2	7	0	29.80
Poly[oxy(dipropylsilylene)]	30.00	2	14	-3	29.43
Polystyrene	30.15	2	8	0	29.41
Polyoxytetramethylene	30.50	5	8	-3	29.73
Poly(cyclohexyl methacrylate)	30.70	2	16	-8	30.99
Polyepichlorohydrin	30.80	3	5	2	30.21
Poly(n-butyl methacrylate)	31.30	2	14	-3	31.35
Poly(sec-butyl methacrylate)	31.30	2	14	-3	32.02
Poly(t-butyl methacrylate)	31.35	2	14	-3	32.91
Poly(vinyl pivalate)	31.55	2	12	0	32.74
Poly(propylene sulfone)	31.65	3	6	3	32.15
Poly(p-methyl styrene)	31.75	2	10	0	32.50
Poly(p-methoxy styrene)	31.85	2	10	-3	32.40
Poly(o-methyl styrene)	--------	2	10	0	32.50
Poly(α-methyl styrene)	32.55	2	10	0	32.50
Poly(methyl α-chloroacrylate)	32.95	2	5	2	32.93
Polychloroprene	33.10	4	5	0	32.83
Poly(p-phenylene)	33.10	4	4	0	33.60
Polychlorotrifluoroethylene	33.30	2	0	5	33.61
Poly(2-hydroxyethyl methacrylate)	33.40	2	10	1	32.62
Poly(1-butene sulfone)	34.00	3	8	3	34.36

☞ ☞ ☞   **TABLE 12.4 IS CONTINUED IN THE NEXT PAGE.** ☞ ☞ ☞

Table 12.4. CONTINUED FROM THE PREVIOUS PAGE.

Polymer	K(group)	N_{SP}	N_H	N_K	K(fit/pre)
Poly(β-alanine)	34.10	4	5	5	34.42
Poly(vinylidene chloride)	35.25	2	2	10	35.21
Poly[oxy(methylphenylsilylene)]	35.30	2	8	-3	34.30
Poly(methyl α-cyanoacrylate)	35.70	2	5	4	35.26
Poly(vinyl butyral)	--------	4	14	-6	35.71
Poly(vinyl bromide)	35.90	2	3	13	35.14
Poly(n-hexyl methacrylate)	36.00	2	18	-3	35.77
Poly(2-ethylbutyl methacrylate)	36.00	2	18	-3	36.44
Poly(1-hexene sulfone)	38.70	3	12	3	38.78
Poly(p-t-butyl styrene)	38.85	2	16	0	40.69
Poly(vinyl benzoate)	39.15	2	8	0	38.26
Poly(styrene sulfide)	--------	3	8	0	39.79
Poly(p-chloro styrene)	40.45	2	7	5	40.71
Polyfumaronitrile	40.70	2	2	14	39.67
Poly(n-octyl methacrylate)	40.70	2	22	-3	40.19
Poly(o-chloro styrene)	--------	2	7	5	40.71
Poly(α-vinyl naphthalene)	--------	2	10	0	41.01
Poly(benzyl methacrylate)	41.30	2	12	-3	40.64
Poly[oxy(2,6-dimethyl-1,4-phenylene)]	--------	5	8	-3	43.89
Poly(N-phenyl maleimide)	--------	2	7	0	44.84
Poly(p-xylylene)	46.20	6	8	0	46.42
Poly(2,5-benzoxazole)	--------	5	3	0	47.43
Poly(ε-caprolactone)	50.15	7	10	0	49.31
Poly(N-methyl glutarimide)	--------	4	13	0	50.15
Poly(oxy-2,2-dichloromethyltrimethylene)	50.80	4	8	7	50.14
Poly(oxy-1,1-dichloromethyltrimethylene)	50.80	4	8	7	50.14
Poly(p-bromo styrene)	52.20	2	7	13	51.95
Poly(N-vinyl carbazole)	52.90	2	11	0	49.80
Poly(ε-caprolactam)	53.75	7	11	5	53.65
Poly(vinylidene bromide)	58.75	2	2	26	57.69
Poly(8-aminocaprylic acid)	66.85	9	15	5	66.47
Poly[3,5-(4-phenyl-1,2,4-triazole)-1,4-phenylene]	--------	7	9	0	77.77
Poly(ethylene terephthalate)	81.00	10	8	0	80.94

☞ ☞ ☞   **TABLE 12.4 IS CONTINUED IN THE NEXT PAGE.** ☞ ☞ ☞

Table 12.4. CONTINUED FROM THE PREVIOUS PAGE.

Polymer	K(group)	N$_{SP}$	N$_H$	N$_K$	K(fit/pre)
Poly(tetramethylene terephthalate)	94.10	12	12	0	93.76
Polybenzobisoxazole	--------	11	6	0	99.05
Poly(ethylene-2,6-naphthalenedicarboxylate)	--------	12	10	0	100.94
Polybenzobisthiazole	--------	11	6	0	105.77
Poly(1,4-cyclohexylidene dimethylene terephthalate)	105.80	14	18	-5	106.88
Poly(hexamethylene adipamide)	107.50	14	22	10	107.30
Poly(p-phenylene terephthalamide)	108.20	12	10	10	110.41
Poly(hexamethylene isophthalamide)	---------	13	18	10	111.06
Phenoxy resin	112.55	14	20	-2	113.37
Bisphenol-A polycarbonate	117.60	12	14	15	118.54
Poly(oxy-1,4-phenylene-oxy-1,4-phenylene-carbonyl-1,4-phenylene)	121.10	15	12	-6	119.23
Poly[2,2-propane *bis*{4-(2,6-dimethylphenyl)}carbonate]	---------	12	22	15	130.93
Poly[1,1-(1-phenylethane) *bis*(4-phenyl)carbonate]	132.30	12	16	15	132.25
Poly(hexamethylene sebacamide)	133.70	18	30	10	132.94
Poly[2,2-hexafluoropropane *bis*(4-phenyl)carbonate]	---------	12	8	15	136.18
Poly[1,1-(1-phenyltrifluoroethane) *bis*(4-phenyl)carbonate]	---------	12	13	15	141.07
Poly[diphenylmethane *bis*(4-phenyl)carbonate]	147.00	12	18	15	145.96
Poly[1,1-biphenylethane *bis*(4-phenyl)carbonate]	148.60	12	20	15	149.06
Torlon	---------	16	14	5	148.67
Poly[N,N'-(p,p'-oxydiphenylene)pyromellitimide]	---------	16	10	-3	148.81
Poly[4,4'-diphenoxy di(4-phenylene)sulfone]	157.20	19	16	-3	159.05
Poly[2,2-propane *bis*{4-(2,6-dichlorophenyl)}carbonate]	---------	12	10	35	163.76
Poly[4,4'-isopropylidene diphenoxy di(4-phenylene)sulfone]	168.50	20	22	-3	171.44
Poly[2,2-propane *bis*{4-(2,6-dibromophenyl)}carbonate]	---------	12	10	67	208.71
Ultem	---------	24	24	-6	224.10

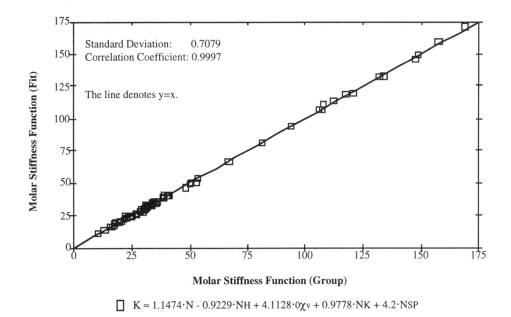

$\square$ $K \approx 1.1474 \cdot N - 0.9229 \cdot NH + 4.1128 \cdot 0\chi v + 0.9778 \cdot NK + 4.2 \cdot NSP$

Figure 12.7. A fit using connectivity indices, to the molar stiffness function K calculated by group contributions, for a set of 94 polymers. K is in units of $grams^{0.25} \cdot cm^{1.5}/mole^{0.75}$.

References and Notes for Chapter 12

1. P. J. Flory, *The Principles of Polymer Chemistry*, Cornell University Press, Ithaca, New York (1953).

2. P. J. Flory, *Statistical Mechanics of Chain Molecules*, Interscience Publishers, New York (1969).

3. P. J. Flory, *Macromolecules*, 7, 381-392 (1974).

4. A. E. Tonelli, article titled "*Conformation and Configuration*", *Encyclopedia of Polymer Science and Engineering*, 4, Wiley-Interscience, New York (1986).

5. P. J. Flory, *Pure & Applied Chemistry*, 56, 305-312 (1984).

6. R. Zallen, *The Physics of Amorphous Solids*, John Wiley & Sons, New York (1983).

7. S. Wu, *Polymer Engineering and Science*, 30, 753-761 (1990).

8. H.-G. Elias, *Macromolecules, Volume 1: Structure and Properties*, second edition, Plenum Press, New York (1984).

9. P.-G. deGennes, *Scaling Concepts in Polymer Physics*, Cornell University Press, Ithaca, New York (1979).

10. J. des Cloizeaux and G. Jannink, *Polymers in Solution: Their Modelling and Structure*, Clarendon Press, Oxford (1990).

11. M. Kurata and Y. Tsunashima, in *Polymer Handbook*, edited by J. Brandrup and E. H. Immergut, third edition, Wiley, New York (1989), VII/1-VII/60.

12. D. W. van Krevelen, *Properties of Polymers*. (a) Second edition, Elsevier, Amsterdam (1976). (b) Third edition, Elsevier, Amsterdam (1990).

13. D. W. van Krevelen, in *Computational Modeling of Polymers*, edited by J. Bicerano, Marcel Dekker, New York (1992), Chapter 1.

14. T. F. Ford, *J. Phys. Chem.*, *64*, 1168-1174 (1960).

15. A. Rudin and G. B. Strahdee, *J. Paint Technology*, *46*, 33-43 (1974).

16. S. Shambayati, J. F. Blake, S. G. Wierschke, W. L. Jorgensen and S. L. Schreiber, *J. Am. Chem. Soc.*, *112*, 697-703 (1990).

CHAPTER 13

POLYMER MELT AND
CONCENTRATED SOLUTION VISCOSITY

13.A. Definitions and General Considerations

The *melt viscosity* η [1-8] is defined by Equation 13.1, as the shear stress (τ) divided by the shear rate ($\dot{\gamma}$). This definition implies that if two resins have different melt viscosities, a larger shear stress must be applied to the resin with the larger η in order to create the same rate of shear.

$$\eta \equiv \frac{\tau}{\dot{\gamma}} \qquad (13.1)$$

It is interesting to note the similarity between the definitions of the shear viscosity (Equation 13.1) and the shear modulus (Equation 13.2), which leads to a relationship (Equation 13.3) between η and G in terms of the ratio of the shear strain (γ) divided by the shear rate ($\dot{\gamma}$).

$$G \equiv \frac{\tau}{\gamma} \qquad (13.2)$$

$$\eta = \frac{\gamma}{\dot{\gamma}} \cdot G \qquad (13.3)$$

Melt processing, by *molding* (i.e., compression molding and injection molding), *extrusion*, or combinations of molding and extrusion, is the most commonly utilized method of manufacturing finished parts and preparing test specimens from polymers. The melt viscosity determines whether a polymer will be melt-processable under a given set of conditions, for the manufacture of parts of desired sizes and shapes. For example, if the melt viscosity of a resin remains very high even at elevated temperatures at which the resin starts to undergo rapid thermal degradation, then other methods must be used to manufacture finished parts from that resin.

Detailed knowledge or prediction of the melt viscosity as a function of important variables is very useful in the design of new polymers. These variables can be classified into two groups, namely, (a) "intrinsic" variables related to the nature of the polymer itself, and (b) "extrinsic" variables related to the external conditions imposed on the polymer.

The following are the most important *intrinsic variables* affecting η:

1. The *structure* of the polymeric repeat unit.
2. The *weight-average molecular weight* (M_w) of the polymer chains.
3. The *distribution of the molecular weights* (polydispersity) of the polymer chains, as described by the ratio of the weight-average and number-average molecular weights (M_w/M_n). For wide

molecular weight distributions, the chains of high molecular weight play a greater role in determining η than is reflected by their contribution to M_w, and the treatment of the molecular weight dependence of η must be modified to take this effect into account.

4. The presence of *additives* (such as plasticizers) *and impurities* (such as residual monomers).

5. *Branching* of the polymer chains.

6. Morphological factors, and especially the presence of *crystallinity*.

The following are the most important *extrinsic variables* affecting η:

1. *Temperature* (T). The melt viscosity decreases with increasing temperature. The only major exceptions to this rule are systems consisting of reactive molecules or oligomers where chain extension (increase of M_w) and/or crosslinking may occur upon heating.

2. *Shear rate* ($\dot{\gamma}$). If η is independent of $\dot{\gamma}$, it is "Newtonian". If η is a function of $\dot{\gamma}$, it is "non-Newtonian". Polymers typically manifest Newtonian behavior at low $\dot{\gamma}$, and non-Newtonian behavior with "shear thinning" (i.e., decrease of the melt viscosity with increasing $\dot{\gamma}$) at high $\dot{\gamma}$.

3. *Hydrostatic pressure* (p). The effects of this variable are usually not as important as those of the temperature and the shear rate, and have therefore not been studied as thoroughly.

The *zero-shear viscosity* η_0 is defined as the melt viscosity in the limit of $\dot{\gamma}=0$, and is a function of T and M_w. It is important to keep in mind, however, that η_0 is very often not measured directly, but extrapolated from measurements at low shear rates. Such extrapolations can introduce an error in the value of η_0 if the range of shear rates used in the extrapolation is sufficiently high for non-Newtonian effects to begin manifesting themselves.

The dependences of η on M_w, T and $\dot{\gamma}$ can all be treated separately, with the exception of polymers of very low M_n where the key material parameters determining the temperature dependence of η become sensitive to the value of M_n. For example, there are no cross-terms between the dependences of η on M_w and T. The M_w dependence of η has the same *functional form* regardless of the value of T. The T dependence of η has the same *functional form* regardless of the value of M_w. Consequently, the dependences of η on M_w and T can be treated separately. The correlations [5-7] for the dependences of η on each of the three variables (M_w, T and $\dot{\gamma}$) can then be combined, to estimate η as a function of all three variables.

The dependences of η for polymer melts on M_w, M_w/M_n, T, $\dot{\gamma}$ and p will be discussed below.

The zero-shear viscosity of a concentrated polymer solution can be treated by a modified version of the method used to calculate the zero-shear viscosity of a polymer melt. The modifications take the two effects of the solvent (plasticization and true dilution of the polymer) into account. Approximations are involved, however, in determining the appropriate mixing rules for the plasticization effect and the magnitude of the true dilution effect. The zero-shear viscosity of concentrated polymer solutions will be discussed briefly at the end of this chapter.

13.B. Dependence of Melt Zero-Shear Viscosity on Average Molecular Weight

13.B.1. Dependence on Critical Molecular Weight

Let M_{cr} denote the *critical molecular weight* of a polymer. M_{cr} is an intrinsic property of a polymer. Let $\eta_{cr}(T)$ (a function of the temperature, but not of the molecular weight) denote the *critical zero-shear viscosity*, i.e., the zero-shear viscosity when $M_w = M_{cr}$. The dependence of η_0 on M_w can be estimated by two power law expressions with crossover from short chain behavior (Equation 13.4) to high polymer behavior (Equation 13.5) at M_{cr}. The crossover usually occurs over a range of M_w rather than precisely at M_{cr}, but is approximated as occurring at M_{cr}. The behavior described by equations 13.4 and 13.5 is illustrated in Figure 13.1.

$$\eta_0(T, M_w) \approx \eta_{cr}(T) \cdot \left(\frac{M_w}{M_{cr}}\right) \qquad \text{(for } M_w \leq M_{cr} \text{ only)} \qquad (13.4)$$

$$\eta_0(T, M_w) \approx \eta_{cr}(T) \cdot \left(\frac{M_w}{M_{cr}}\right)^{3.4} \qquad \text{(for } M_w > M_{cr} \text{ only)} \qquad (13.5)$$

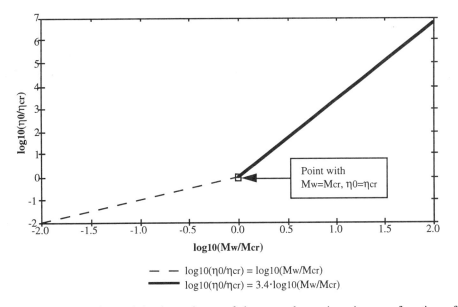

$$-- \quad \log 10(\eta_0/\eta_{cr}) = \log 10(M_w/M_{cr})$$
$$\textbf{———} \quad \log 10(\eta_0/\eta_{cr}) = 3.4 \cdot \log 10(M_w/M_{cr})$$

Figure 13.1. Molecular weight dependence of the zero-shear viscosity as a function of the weight-average molecular weight, described by two power law equations with a crossover at the critical molecular weight.

There has been some controversy concerning whether these two equations are sufficient to describe the molecular weight dependence of η_0 completely, or whether polydispersity effects also need to be taken into account. A review of this issue [5] shows that (unlike the definite effect of polydispersity on the viscosity at high shear rates) there is no consistent effect of polydispersity on η_0, and that some of the apparent evidence to the contrary may have been an artifact of the large uncertainties in the measured values of M_w. We will, hence, use equations 13.4 and 13.5 and not incorporate any corrections related to polydispersity into the calculation of η_0.

13.B.2. A Correlation for the Critical Molecular Weight

The viscosities of polymer solutions and of polymer melts have some very important common features, which are related to the fundamental nature of the motions of polymer chain segments [4,8,9] resulting in the flow of macromolecular chains. At an empirical level, one manifestation of these interrelationships is that M_{cr}, which is the key material parameter determining the molecular weight dependence of the melt viscosity, can be estimated from the intrinsic viscosity of the polymer under Θ conditions, which was discussed in Chapter 12. If $K_\Theta \cdot M_{cr}^{0.5}$ is expressed in units of cc/grams, then M_{cr} can be predicted [7] *(but only to within a factor of two)* by using Equation 13.6.

$$K_\Theta \cdot M_{cr}^{0.5} \approx 13 \tag{13.6}$$

Comparison of equations 12.5 and 13.6 shows that, for narrow molecular weight distributions, Equation 13.6 implies that all polymers have approximately the same intrinsic viscosity under Θ conditions when $M_w = M_{cr}$. M_{cr} depends strongly on the structure of the polymeric repeat unit. Equation 13.6 therefore does not, in any way, contradict the fact that, at a given value of M_w, two polymers can have very different intrinsic viscosities.

Substitution of Equation 12.6 into Equation 13.6 gives Equation 13.7, which is convenient in estimating M_{cr}. K can be calculated by using group contributions if they are available for all of the structural units in the polymer of interest, and by using Equation 12.29 if some of the group contributions are missing.

$$M_{cr} \approx 169 \left(\frac{M}{K} \right)^4 \tag{13.7}$$

Here are two examples of calculations using Equation 13.7:

1. K(group)=30.15 and K(from Equation 12.29)=29.41 for polystyrene (Figure 1.1). Since M=104.15, Equation 13.7 gives M_{cr}=24064 if K=30.15, and M_{cr}=26579 if K=29.41. The experimental value (twice the value of M_e listed in Table 11.4) is M_{cr}≈35702.

2. The group contribution for the imide ring is not available, so that K(group) cannot be calculated, and therefore M_{cr} cannot be predicted by using K(group), for Ultem [see Figure 2.4]. Equation 12.29 gives K=224.10. Since Ultem has M=592.61, it is predicted that M_{cr}≈8264. We are not aware of any measured values of M_{cr} for Ultem, and therefore cannot evaluate the quality of this prediction. The usefulness of being able to predict the M_{cr} of a polymer of arbitrary structure should, however, be obvious.

The relative standard deviation of Equation 12.29 is 1.6%. M_{cr} values predicted by using Equation 12.29 to estimate K thus typically only differ by about $(1.6)^4$≈6.6% from M_{cr} values predicted by using group contributions for K. A difference of 6.6% is much smaller than the typical magnitude of the error of Equation 13.7. On the other hand, since Equation 12.29 is applicable to all polymers, it allows the prediction of M_{cr} for many polymers whose K cannot be calculated by using the existing tables of group contributions. It is often necessary to compare several candidate polymers, some of whose K values can be predicted by using group contributions, and some of whose K values can only be predicted by using Equation 12.29. In such situations, the trends predicted by using Equation 12.29 to calculate M_{cr} for all of these polymers are slightly more reliable than the trends predicted by mixing values of M_{cr} calculated by using two different methods to estimate K.

13.B.3. Alternative Correlation for Critical Molecular Weight

An alternative method of estimating M_{cr} is given by Equation 13.8, where the entanglement molecular weight M_e is calculated by using Equation 11.24. For example, Equation 11.24 (with l_m=2.32Å) gives M_e≈15536 for polystyrene, so that the use of Equation 13.8 results in the prediction of M_{cr}≈31072, in comparison with the experimental value of 35702.

$$M_{cr} \approx 2M_e \qquad (13.8)$$

An especially appealing aspect of the use of equations 11.24 and 13.8 to estimate M_{cr} is that Equation 11.24 was developed [10] by using data on rubbery polymers. It therefore relates more directly to the zero-shear viscosity of a polymer melt than does an extrapolation from data on the viscosities of dilute solutions under theta conditions. We have found that the use of equations 11.24 and 13.8 is generally preferable to the use of Equation 13.7 in calculating M_{cr}. The correlation developed for V_w in Section 3.B has allowed the much more general use of Equation 11.24 than was previously possible by using group contributions.

13.C.   Dependence of Melt Zero-Shear Viscosity on Temperature

13.C.1.   General Relationships

The glass transition temperature (T_g) is the key material parameter in determining the temperature dependence of the melt viscosity. An amorphous polymer will typically be "solidlike" below T_g, "rubberlike" between T_g and a crossover temperature T_x above but not too far from $1.2 \cdot T_g$, and "molten" with viscous flow over the range of temperatures above T_x.

As demonstrated by van Krevelen [7], a universal equation in terms of T_g/T can thus be used for $\eta_{cr}(T)$ for $T_g < T \leq 1.2 \cdot T_g$, while a generalized activated flow expression with a limiting asymptotic activation energy of $E_{\eta\infty}$ is required for $T >> 1.2 \cdot T_g$. The equations needed to perform these calculations are listed below. They provide continuity between the calculated values of $\eta_{cr}(T)$ at the reference temperature of $T=1.2 \cdot T_g$ above and below $T=1.2 \cdot T_g$ while reaching the correct asymptotic limit as $T \rightarrow \infty$. $R=8.31451$ J/(mole·K) is the gas constant. The use of equations 13.9, 13.10 and 13.11, with $E_{\eta\infty}$ expressed in J/mole and all temperatures expressed in degrees Kelvin, results in the prediction of $\eta_{cr}(T)$ in N·sec/m^2. Van Krevelen [7] used a graphical representation instead of equations 13.10 and 13.11. We converted the curves and the data represented graphically in his book into these equations to facilitate practical calculations. On the other hand, Equation 13.9 was taken directly from van Krevelen's book [7].

$$\eta_{cr}(1.2T_g) \approx 3.981 \cdot 10^{-2} \cdot \exp\left[\frac{E_{\eta\infty} \cdot \left(1 - 1.635 \cdot 10^{-3} \cdot T_g\right)}{R \cdot T_g}\right] \tag{13.9}$$

$$\eta_{cr}(T) \approx 9.553 \cdot 10^{-10} \cdot \eta_{cr}(1.2T_g) \cdot \exp\left[60.604 \cdot \left(\frac{T_g}{T}\right)^2 - 25.581 \cdot \left(\frac{T_g}{T}\right)\right] \quad (T_g < T \leq 1.2 \cdot T_g) \tag{13.10}$$

$$\eta_{cr}(T) \approx \eta_{cr}(1.2T_g) \cdot \exp\left(\frac{0.288 \cdot E_{\eta\infty} \cdot T_g^2}{R \cdot T^3}\right) \cdot \exp\left[\frac{(T_g - T) \cdot E_{\eta\infty}}{R \cdot T \cdot T_g}\right] \quad (T > 1.2 \cdot T_g) \tag{13.11}$$

According to Equation 13.10, as a first approximation, $\eta_{cr}(T)/\eta_{cr}(1.2 \cdot T_g)$ is equal for all amorphous polymers at the same ratio of T_g/T for $T_g < T \leq 1.2 \cdot T_g$. For example, if $T=1.1 \cdot T_g$, then $T_g/T \approx 0.90909$, so that Equation 13.10 gives $\log_{10}[\eta_{cr}(T)/\eta_{cr}(1.2 \cdot T_g)] \approx 2.63203$ and thus $\eta_{cr}(1.1 \cdot T_g) \approx 429 \cdot \eta_{cr}(1.2 \cdot T_g)$, for all amorphous polymers.

$E_{\eta\infty}$ is the *activation energy for viscous flow at zero shear rate in the limit of T→∞*. $E_{\eta\infty}$ can be estimated [7] in terms of the *molar viscosity-temperature function* H_η by using Equation 13.12. H_η is expressed in units of $g \cdot J^{1/3} \cdot mole^{-4/3}$ and normally estimated by using group contributions. For example, the use of group contributions gives H_η=4020 $g \cdot J^{1/3} \cdot mole^{-4/3}$ for polystyrene, so that the use of Equation 13.12 with H_η=4020 and M=104.15 results in $E_{\eta\infty} \approx 57504$ J/mole, in comparison with the experimental value [7] of 59000 J/mole.

$$E_{\eta\infty} \approx \left(\frac{H_\eta}{M}\right)^3$$

(13.12)

The value of $\eta_{cr}(T)$ at a given T_g/T varies over a wide range as a function of the structure of the polymer because the parameters which enter equations 13.9 and 13.11 (T_g and $E_{\eta\infty}$) both vary over a very wide range as a function of the structure of the repeat unit.

It is seen from equations 13.9 and 13.11 that $E_{\eta\infty}$ plays a role in determining the optimum melt processing conditions, albeit not as important as the role of T_g. According to Equation 13.9, if two polymers have equal values of T_g, then the polymer with the higher value of $E_{\eta\infty}$ will have a higher η_0 at the reference temperature of $1.2 \cdot T_g$. On the other hand, according to Equation 13.11, the viscosity of the polymer with a larger value of $E_{\eta\infty}$ will decrease faster with increasing temperature for $T>1.2 \cdot T_g$. Since $E_{\eta\infty}$ can be estimated by using Equation 13.12, the temperature dependence of zero-shear viscosity can be estimated if T_g and H_η can be predicted.

Substitution of $\eta_{cr}(T)$ calculated by combining equations 13.9 and 13.12, with Equation 13.10 or Equation 13.11 used as appropriate for the temperature of interest, into equations 13.4 and 13.5, allows the prediction of the zero-shear viscosity of a polymer melt as a function of both the weight-average molecular weight of the polymer chains and the temperature of measurement. This prediction is usually quite approximate in terms of the accuracy of its absolute magnitude. It is, nonetheless, valuable as a quantitative tool, both in evaluating individual candidate polymers and in assessing the trends expected as a function of specific structural changes among variants of a general type of structure.

The methods described in Chapter 6 can be used to estimate T_g, which is the most important input parameter. For polymers of low M_n ($M_n << M_{cr}$), it is important to apply the correction described in Section 6.C for the effects of the molecular weight on T_g.

Because of the limited amount of reliable data for $E_{\eta\infty}$, only a limited number of group contributions is available [7,11] for the molar viscosity-temperature function H_η used to estimate this property via Equation 13.12. It is, therefore, useful to develop a correlation to allow the prediction of this quantity when the required group contributions are not all available. The correlation developed for this purpose will be presented in Section 13.C.2.

13.C.2. Estimation of $E_{\eta\infty}$ Without Using Group Contributions

The molar viscosity-temperature function H_η can be expressed [7] as a sum of two terms:

$$H_\eta \equiv H_{\eta sum} + H_{\eta str} \qquad (13.13)$$

$H_{\eta sum}$ is a simple additive quantity estimated as a sum of the group contributions made by the structural units in the repeat unit. $H_{\eta str}$ is a "structural" term related to side chains, which are a special type of side group, i.e., a side group containing at least one divalent link. (See Figure 2.9 for an example of a polymer with a side chain.)

$$H_{\eta str} = 250 \cdot N_{(side\ chain)} \qquad (13.14)$$

$N_{(side\ chain)}$ is defined [7] as the "extra effect of side chain, per -CH_2- or other bivalent group". Since the $H_{\eta str}$ of any polymer is a constant whose value is exactly determined from the structure of the repeat unit, the only remaining task is to develop a correlation for $H_{\eta sum}$. The total value of H_η can be calculated by adding $H_{\eta str}$ to $H_{\eta sum}$.

A dataset of 140 polymers was prepared. All of the needed group contributions were available [7,11] to calculate $H_{\eta sum}$(group) for 80 of these polymers. A strong correlation was found between $H_{\eta sum}$ and the molar intrinsic viscosity function J, indicating that there is some interrelationship between the molecular weight and temperature dependences of the zero-shear viscosity. This correlation was much stronger than the correlation between $H_{\eta sum}$ and any of the zeroth-order and first-order connectivity indices. Values of J calculated by group contributions were used whenever possible. Equation 12.28 was used to calculate J whenever some of the necessary group contributions were not available. Equation 13.15, which was obtained by using a linear regression procedure with weight factors of $1000/H_{\eta sum}$(group), has a correlation coefficient of 0.9895 and a standard deviation of 531.9.

$$H_{\eta sum} \equiv 182.0 \cdot J \qquad (13.15)$$

The correction term $N_{H\eta}$ was defined by Equation 13.16, to improve the correlation.

$$\begin{aligned}
N_{H\eta} \equiv\ & 5N_{(acrylic\ ester)} - N_{(all\ other\ types\ of\ esters)} - 20N_{carbonate} - 7N_{amide} + 6N_{"ether"} - 3N_{Cl} - N_{Br} \\
& + 5N_{sulfone} - 10N_{Si} + 3N_{(six\text{-}membered\ nonterminal\ aromatic\ rings)} + 8N_{C=C} \\
& + 4N_{(carbon\ atoms\ bonded\ to\ four\ non\text{-}hydrogen\ atoms,\ i.e.,\ with\ \delta=\delta^v=4)} \\
& + 2N_{(carbon\ atoms\ bonded\ to\ three\ non\text{-}hydrogen\ atoms\ and\ one\ H,\ i.e.,\ with\ \delta=\delta^v=3)} \qquad (13.16)
\end{aligned}$$

A "terminal" aromatic ring is attached to the rest of the polymer chain by a single bond, and does not have any non-hydrogen substituents. For example, polystyrene (Figure 1.1) and poly(*p*-vinyl pyridine) [Figure 5.1(e)] both contain terminal aromatic rings. Consequently, all aromatic rings in the chain backbone [as in bisphenol-A polycarbonate, Figure 5.1(b)], and those aromatic rings in side groups which have substituents "sticking out" of them [as in poly(*p*-chloro styrene), Figure 5.1(g)] are defined as "nonterminal" aromatic rings. This correction is made only for six-membered nonterminal aromatic rings, whether they are isolated or in fused rings.

The remaining correction terms in Equation 13.16 have either been encountered earlier or are self-explanatory. For example, see Section 12.D for definitions of $N_{(acrylic\ ester)}$ and $N_{"ether"}$.

Group contributions for H_η were not available for any structural units containing sulfur, fluorine, and/or bromine atoms. The need to use atomic correction terms proportional to N_{Br} and $N_{sulfone}$ in Equation 13.16 was deduced from our own unpublished work.

A fit with a standard deviation of 52.2 and a correlation coefficient of 0.9998 is provided by Equation 13.17. The correlation coefficient indicates that Equation 13.17 accounts for 99.97% of the variation of the 80 $H_{\eta sum}$(group) values in the dataset. The standard deviation is approximately 1.0% of the average $H_{\eta sum}$(group) value of 4982.

$$H_{\eta sum} \approx 177.7 \cdot J + 82.2 \cdot N_{H\eta} \tag{13.17}$$

These calculations are summarized in Table 13.1 and illustrated in Figure 13.2. The results obtained by using Equation 13.17 in a predictive mode for the 60 polymers from our dataset for which $H_{\eta sum}$(group) could not be calculated because of the lack of some of the needed group contributions are also listed in Table 13.1.

The observed values [7] of $E_{\eta\infty}$ are compared with the values predicted by using the correlation developed for H_η in this section, with J values calculated by group contributions (which were available for all of the polymers listed in Table 13.2) inserted into equation 13.17. The correlation coefficient is 0.9920, indicating that the equations accounted for 98.4% of the variation of the $E_{\eta\infty}$ values in the dataset. The standard deviation of $2.8 \cdot 10^3$ J/mole is equal to 6.2% of the average value of $44.9 \cdot 10^3$ J/mole for $E_{\eta\infty}$ in this dataset.

It should be noted that the use of Equation 12.28 instead of group contributions to estimate J for insertion into Equation 13.17 results in slightly different predicted values of $E_{\eta\infty}$ when the group contributions for J are all available and the two methods for estimating J can both be applied. For example, $E_{\eta\infty} \approx 52400$ J/mole is calculated for polystyrene instead of $E_{\eta\infty} \approx 57504$ J/mole when Equation 12.28 is used to estimate J for insertion into Equation 13.17. The important new capability provided by the work described in this section is that the calculation of $E_{\eta\infty}$ is now no longer limited by the lack of group contributions for H_η.

Table 13.1. Additive portion $H_{\eta sum}$ of the molar viscosity-temperature function H_η calculated by using group contributions, molar intrinsic viscosity function J and correction index $N_{H\eta}$ used in the correlation, and fitted or predicted (fit/pre) values of $H_{\eta sum}$, for 140 polymers. For 60 of these polymers, $H_{\eta sum}$(group) cannot be calculated because of the lack of group contributions, but $H_{\eta sum}$(fit/pre) can be calculated by using the new correlation. $H_{\eta sum}$ is in $g \cdot J^{1/3} \cdot mole^{-4/3}$.

Polymer	$H_{\eta sum}$(group)	J	$N_{H\eta}$	$H_{\eta sum}$(fit/pre)
Polyethylene	840	4.70	0	835
Polyoxymethylene	900	2.45	6	929
Polyoxyethylene	1320	4.80	6	1346
Polypropylene	1480	7.05	2	1417
Poly(vinyl fluoride)	------	7.75	2	1542
Poly(1,4-butadiene)	1600	5.20	8	1582
Poly(1,2-butadiene)	------	4.73	10	1663
Polyoxytrimethylene	1740	7.15	6	1764
Poly(dimethyl siloxane)	1830	12.20	-4	1839
Poly(ethylene sulfide)	------	10.60	0	1884
Poly(1-butene)	1900	9.40	2	1835
Poly(propylene oxide)	1960	7.15	8	1928
Poly(vinyl methyl ether)	1960	7.15	8	1928
Polyisoprene	2030	7.60	8	2008
Polyisobutylene	2040	9.45	4	2008
Polyoxytetramethylene	2160	9.50	6	2181
Poly(vinyl alcohol)	-------	11.50	2	2208
Poly(acrylic acid)	-------	11.50	2	2208
Poly(vinylidene fluoride)	-------	10.75	4	2239
Poly(1-pentene)	2320	11.75	2	2252
Poly(vinyl ethyl ether)	2380	9.50	8	2346
Poly(β-alanine)	2490	17.30	-7	2499
Poly(vinyl cyclohexane)	-------	13.50	4	2728
Poly(1-hexene)	2740	14.10	2	2670
Poly(vinyl chloride)	2750	15.75	-1	2717
Poly(methacrylic acid)	-------	13.90	4	2799
Poly(vinyl trimethylsilane)	2830	19.15	-8	2745
Poly(4-methyl-1-pentene)	2880	14.15	4	2843
Poly(vinyl acetate)	2930	16.05	1	2934
Poly(methyl acrylate)	2930	13.45	7	2965

☞ ☞ ☞ **TABLE 13.1 IS CONTINUED IN THE NEXT PAGE.** ☞ ☞ ☞

Table 13.1. CONTINUED FROM THE PREVIOUS PAGE.

Polymer	H$_{\eta}$sum(group)	J	N$_{H\eta}$	H$_{\eta}$sum(fit/pre)
Polytrifluoroethylene	-------	13.80	6	2945
Poly(thiocarbonyl fluoride)	-------	14.79	4	2957
Poly(N-methyl acrylamide)	3130	19.65	-5	3081
Poly(*p*-phenylene)	3200	16.30	3	3143
Poly(vinyl *n*-butyl ether)	3220	14.20	8	3181
Polychloroprene	3300	16.30	5	3308
Poly(vinyl propionate)	3350	18.40	1	3352
Poly(ethyl acrylate)	3350	15.80	7	3383
Poly(vinyl *sec*-butyl ether)	3440	14.20	10	3345
Polyacrylonitrile	-------	18.50	2	3452
Poly(methyl methacrylate)	3490	15.85	9	3556
Poly[oxy(dipropylsilylene)]	3510	21.60	-4	3510
Poly(ε-caprolactone)	3550	20.75	-1	3605
Polytetrafluoroethylene	-------	16.80	8	3643
Poly(2-*t*-butyl-1,4-butadiene)	3650	14.70	12	3599
Polyepichlorohydrin	3650	18.20	5	3645
Poly[oxy(*p*-phenylene)]	3680	16.40	9	3654
Poly(ethylene oxalate)	3740	22.70	-2	3869
Poly(ε-caprolactam)	3750	24.35	-7	3752
Poly(ethylene oxalate)	-------	22.70	-2	3869
Poly(maleic anhydride)	-------	20.08	4	3897
Poly(ethyl methacrylate)	3910	18.20	9	3974
Poly(methyl ethacrylate)	3910	18.20	9	3974
Poly(propylene sulfone)	-------	19.05	7	3961
Poly(*p*-vinyl pyridine)	-------	21.50	2	3985
Poly(N-vinyl pyrrolidone)	-------	21.50	2	3985
Polystyrene	4020	21.75	2	4029
Poly(*p*-xylylene)	4040	21.00	3	3978
Polymethacrylonitrile	-------	20.90	4	4043
Poly(*n*-butyl acrylate)	4190	20.50	7	4218
Poly(octylmethylsilane)	4290	28.55	-10	4251
Poly[oxy(methylphenylsilylene)]	4370	26.90	-4	4451
Poly(1-butene sulfone)	-------	21.40	7	4378

☞☞☞ **TABLE 13.1 IS CONTINUED IN THE NEXT PAGE.** ☞☞☞

Table 13.1. CONTINUED FROM THE PREVIOUS PAGE.

Polymer	H$_{\eta sum}$(group)	J	N$_{H\eta}$	H$_{\eta sum}$(fit/pre)
Poly(vinyl sulfonic acid)	-------	21.50	7	4396
Poly(styrene oxide)	4500	21.85	8	4540
Poly(vinyl pivalate)	4550	23.15	5	4525
Poly(vinylidene chloride)	4580	26.85	-2	4607
Poly(α-methyl styrene)	4580	24.15	4	4620
Poly(8-aminocaprylic acid)	4590	29.05	-7	4587
Poly(p-hydroxybenzoate)	4650	25.30	2	4660
Poly(p-methyl styrene)	4680	23.35	5	4560
Poly(o-methyl styrene)	-------	24.10	5	4694
Poly(n-butyl methacrylate)	4750	22.90	9	4809
Poly(methyl α-chloroacrylate)]	4760	24.55	6	4856
Poly[oxy(2,6-dimethyl-1,4-phenylene)]	-------	22.89	9	4807
Polychlorotrifluoroethylene	-------	24.90	5	4836
Poly(vinyl butyral)	-------	18.91	18	4840
Poly(cyclohexyl methacrylate)	-------	22.30	11	4867
Poly(2,5-benzoxazole)	-------	26.43	3	4943
Poly(vinyl bromide)	-------	27.50	1	4969
Poly(sec-butyl methacrylate)	4970	22.90	11	4974
Poly(styrene sulfide)	-------	27.19	2	4996
Poly(t-butyl methacrylate)	5110	22.95	13	5147
Poly(p-methoxy styrene)	5160	23.45	11	5071
Poly(2-hydroxyethyl methacrylate)	-------	25.00	9	5182
Poly(1-hexene sulfone)	-------	26.10	7	5213
Poly(vinyl benzoate)	5470	30.75	1	5546
Poly(n-hexyl methacrylate)	5590	27.60	9	5644
Poly(methyl α-cyanoacrylate)	-------	27.30	9	5591
Poly(2-ethylbutyl methacrylate)	5810	27.60	11	5809
Poly(o-chloro styrene)	-------	32.31	2	5906
Poly(p-chloro styrene)	5950	32.05	2	5860
Poly(phenyl methacrylate)	6030	30.55	9	6169
Polyfumaronitrile	-------	32.30	4	6069
Poly(p-t-butyl styrene)	6300	30.45	9	6151
Poly(oxy-2,2-dichloromethyltrimethylene)	6320	34.00	4	6371

☞☞☞   **TABLE 13.1 IS CONTINUED IN THE NEXT PAGE.** ☞☞☞

Table 13.1. CONTINUED FROM THE PREVIOUS PAGE.

Polymer	$H_{\eta sum}$(group)	J	$N_{H\eta}$	$H_{\eta sum}$(fit/pre)
Poly(oxy-1,1-dichloromethyltrimethylene)	6320	34.00	4	6371
Poly(α-vinyl naphthalene)	-------	32.61	8	6452
Poly(n-octyl methacrylate)	6430	32.30	9	6480
Poly(benzyl methacrylate)	6450	32.90	9	6586
Poly(N-methyl glutarimide)	-------	33.35	8	6584
Poly(N-phenyl maleimide)	-------	36.44	4	6804
Poly(ethylene terephthalate)	6940	39.00	1	7013
Poly(hexamethylene adipamide)	7500	48.70	-14	7503
Poly(vinyl p-t-butyl benzoate)	7750	39.45	8	7668
Poly(tetramethylene terephthalate)	7780	43.70	1	7848
Poly(p-bromo styrene)	-------	43.80	4	8112
Poly(N-vinyl carbazole)	-------	44.50	8	8565
Poly(1,4-cyclohexylidene dimethylene terephthalate)	-------	47.00	5	8763
Poly[3,5-(4-phenyl-1,2,4-triazole)-1,4-phenylene]	-------	48.37	3	8842
Poly(vinylidene bromide)	-------	50.35	2	9112
Poly(hexamethylene isophthalamide)	-------	56.46	-11	9129
Poly(hexamethylene sebacamide)	9180	58.10	-14	9174
Poly(ethylene-2,6-naphthalenedicarboxylate)	-------	50.54	4	9310
Poly(p-phenylene terephthalamide)	9700	57.80	-8	9613
Polybenzobisoxazole	-------	52.85	6	9885
Poly[methane bis(4-phenyl)carbonate]	9970	62.45	-14	9947
Poly[thio bis(4-phenyl)carbonate]	--------	66.15	-14	10604
Polybenzobisthiazole	--------	59.57	6	11079
Bisphenol-A polycarbonate	11170	67.20	-10	11119
Poly[di(n-octyl) itaconate]	11240	62.70	2	11306
Phenoxy resin	--------	53.75	24	11524
Poly[2,2-pentane bis(4-phenyl)carbonate]	12010	71.90	-10	11955
Poly(ether ether ketone)	--------	58.10	21	12051
Poly[2,2-propane bis{4-(2,6-dimethylphenyl)}carbonate]	--------	80.53	-10	13488
Poly[2,2-propane bis{4-(2,6-difluorophenyl)}carbonate]	--------	81.37	-10	13637
Poly[1,1-(1-phenylethane) bis(4-phenyl)carbonate]	13710	81.90	-10	13732
Poly(bisphenol-A terephthalate)	14120	74.00	11	14054
Torlon	--------	81.47	2	14642

☞ ☞ ☞   TABLE 13.1 IS CONTINUED IN THE NEXT PAGE. ☞ ☞ ☞

Table 13.1. CONTINUED FROM THE PREVIOUS PAGE.

Polymer	$H_{\eta sum}$(group)	J	$N_{H\eta}$	$H_{\eta sum}$(fit/pre)
Poly[2,2-hexafluoropropane *bis*(4-phenyl)carbonate]	--------	85.78	-2	15079
Poly[1,1-(1-phenyltrifluoroethane) *bis*(4-phenyl)carbonate]	--------	90.67	-6	15619
Poly[N,N'-(*p*,*p*'-oxydiphenylene)pyromellitimide]	--------	81.61	15	15735
Poly[4,4'-diphenoxy di(4-phenylene)sulfone]	--------	77.40	29	16138
Poly[diphenylmethane *bis*(4-phenyl)carbonate]	16250	96.60	-10	16344
Poly[1,1-biphenylethane *bis*(4-phenyl)carbonate]	16910	98.20	-7	16875
Poly[4,4'-isopropylidene diphenoxy di(4-phenylene)sulfone]	--------	84.50	33	17728
Poly[2,2-propane *bis*{4-(2,6-dichlorophenyl)}carbonate]	--------	113.36	-22	18336
Poly[4,4'-sulfone diphenoxy di(4-phenylene)sulfone]	--------	89.40	34	18681
Ultem	--------	123.30	31	24459
Poly[2,2-propane *bis*{4-(2,6-dibromophenyl)}carbonate]	--------	158.31	-14	26981

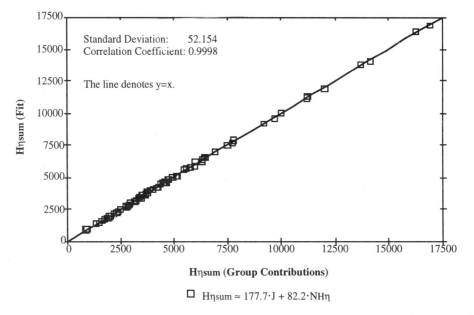

Figure 13.2. A fit using the molar intrinsic viscosity function J, for the simply additive portion $H_{\eta sum}$ of the molar viscosity-temperature function H_η. $H_{\eta sum}$ is in $g \cdot J^{1/3} \cdot mole^{-4/3}$.

Table 13.2. Observed activation energies for viscous flow of polymers at zero shear rate as $T \rightarrow \infty$ ($E_{\eta\infty}$) [7], and the values of $E_{\eta\infty}$ predicted by using the correlation developed for the molar viscosity-temperature function in this section. $E_{\eta\infty}$ is in units of 10^3 J/mole.

Polymer	$E_{\eta\infty}$(exp)	$E_{\eta\infty}$(pre)	Polymer	$E_{\eta\infty}$(exp)	$E_{\eta\infty}$(pre)
Poly(dimethyl siloxane)	15	15	Polypropylene	44	38
Poyisoprene	23	26	Poly(ethylene terephthalate)	45	49
Polyethylene	25	26	Polyisobutylene	48	46
Poly(1,4-butadiene) *(cis)*	26	25	Polystyrene	59	58
Polyoxyethylene	27	28	Poly(methyl methacrylate)	65	67
Poly(decamethylene succinate)	28	31	Poly(vinyl acetate)	67	63
Poly(decamethylene adipate)	29	30	Poly(*n*-butyl methacrylate)	72	77
Poly(decamethylene sebacate)	30	30	Poly(vinyl chloride)	85	82
Poly(ε-caprolactam)	36	36	Bisphenol-A polycarbonate	85	84

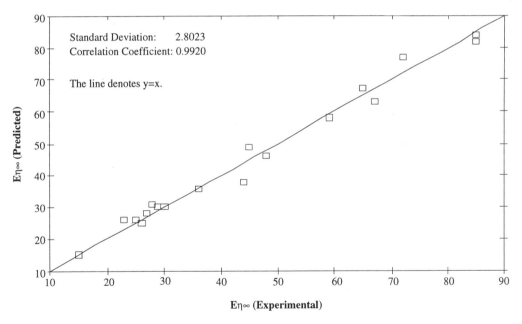

Figure 13.3. Comparison between observed [7] and predicted activation energies of polymers for viscous flow at zero shear rate as $T \rightarrow \infty$ ($E_{\eta\infty}$). $E_{\eta\infty}$ is in units of 10^3 J/mole.

Finally, we must consider the effect of the average molecular weight on $E_{\eta\infty}$ at very low values of M_n ($M_n \ll M_{cr}$). In contrast to the dependence of T_g on M_n, the amount of useful data on the dependence of $E_{\eta\infty}$ on M_n is very limited. In order to be useful for our analysis, the dependence must have been obtained from data on η_0 rather than from data on η at an elevated shear rate, because $E_{\eta\infty}$ is an activation energy for viscous flow in the asymptotic limit of $T\rightarrow\infty$ at zero shear rate. The dependence must have been obtained over a wide range of values of M_n, including both the region of very rapid change of $E_{\eta\infty}$ with M_n ($M_n \ll M_{cr}$) and at least the beginning of the region of asymptotic behavior. The polymers must not have complicating structural or morphological attributes such as crystallinity and/or long-chain branching which can modify the rheological behavior drastically. Finally, data from solution viscosity measurements cannot be used in the analysis, since, as discussed earlier, the mixing rules for the $E_{\eta\omega}$ of a solution in terms of $E_{\eta p}$, $E_{\eta s}$ and Φ_p are not yet firmly established, so that any back-calculated values of $E_{\eta p}$ would have a very large uncertainty. As a result of all of these limitations, only four of the many interesting publications [12-27] found in a literature search on the dependence of the activation energy for viscous flow on the average molecular weight contained data that could be used for the very specific purposes of our analysis.

Data for just four polymers [12-15] are insufficient to develop an independent quantitative structure-property relationship for the dependence of $E_{\eta\infty}$ on M_n. It was, therefore, decided that a reasonable working hypothesis is to assume that the variation of $E_{\eta\infty}$ with M_n follows the variation of T_g with M_n, as described by Equation 13.18, where $E_{\eta\infty}(\infty)$ denotes the value of $E_{\eta\infty}$ at high (more precisely, infinite) M_n, calculated as described above.

$$E_{\eta\infty}(M_n) \approx E_{\eta\infty}(\infty) \cdot \frac{T_g(M_n)}{T_{g\infty}} \qquad (13.18)$$

The results of the use of Equation 13.18 are shown in Figure 13.4, based on experimental data for the $E_{\eta\infty}$ of melts of polystyrene [12], p-nonylphenol formaldehyde novolac epoxy resin [13], linear poly(dimethyl siloxane) [14] and polybutadiene containing 80% cis-groups [15] as functions of M_n. $T_g(M_n)/T_{g\infty}$ was estimated by using Equation 6.6. The following values were used for $T_{g\infty}$: 382K for polystyrene (Table 6.3), 313K for the p-nonylphenol formaldehyde novolac epoxy resin (estimated by using the method introduced in Section 6.B), 150K for poly(dimethyl siloxane) (Table 6.3), and 174K for cis-polybutadiene (Table 6.3). $E_{\eta\infty}$ was estimated by using an equation of the same general form as Equation 6.4 to fit the experimental data for $E_{\eta\infty}$ as a function of M_n. It can be seen that the agreement between the predictions of Equation 13.18 and the experimental results is at best at a semi-quantitative level of accuracy. The ratio $E_{\eta\infty}(M_n)/E_{\eta\infty}(\infty)$ is overestimated significantly in some cases and underestimated significantly in other cases in the range of M_n values over which the transition from high polymer to molecular liquid behavior occurs, when this ratio is assumed to be equal to $T_g(M_n)/T_{g\infty}$.

Consequently, Equation 13.18 must only be utilized as a provisional method for estimating the change of $E_{\eta\infty}$ as a function of M_n, to be replaced in the future if additional and more mutually consistent sets of experimental data allow the development of a better relationship.

Finally, it must be noted that there is evidence that polymer-solvent interactions can cause the $E_{\eta\infty}$ of a polymer in solution to remain significantly dependent on M_n up to higher values of M_n than the E_η of the same polymer in the melt [15,16]. The same evidence [15,16] also suggests that the rules-of-mixture which describe the behavior of the $E_{\eta\infty}$ of a solution as a function of $E_{\eta p}$, $E_{\eta s}$ and Φ_p accurately will probably have to include terms related to the compatibility between the solvent and the polymer instead of having very simple forms such as equations 12.16 or 12.18. There is, presently, no method to predict the effects of such complicating features.

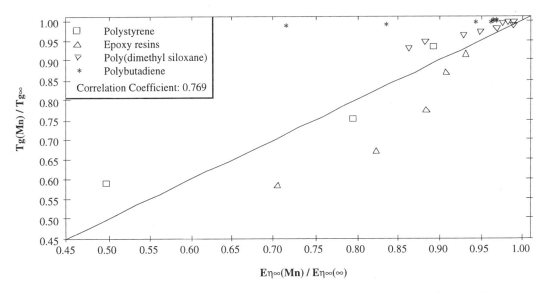

Figure 13.4. Comparison of observed quotient $E_{\eta\infty}(M_n)/E_{\eta\infty}(\infty)$ for melts of four polymers, with results calculated by assuming (Equation 13.18) that this quotient is equal to $T_g(M_n)/T_{g\infty}$.

13.D. Dependence of Melt Zero-Shear Viscosity on Hydrostatic Pressure

The pressure dependence of the melt viscosity (η) can be estimated by using Equation 13.19 (derived from classical thermodynamics to relate the pressure and temperature coefficients of η [7]), where p is the hydrostatic pressure, κ is the isothermal compressibility, α is the coefficient of volumetric thermal expansion, and ∂ is a partial derivative. The sign of the pressure coefficient of the viscosity is opposite to the sign of the temperature coefficient. Consequently, since η

decreases with increasing T, it increases with increasing p. Equation 13.20 is obtained by integrating Equation 13.19.

$$\frac{\partial \eta(T,p)}{\partial p} \approx -\frac{\kappa}{\alpha} \cdot \frac{\partial \eta(T,p)}{\partial T} \tag{13.19}$$

$$\eta(T, p) \approx \eta(T, 0) - \left(\frac{\kappa}{\alpha}\right) \cdot \left[\frac{\partial \eta(T, p)}{\partial T}\right] \cdot p \tag{13.20}$$

Equations 13.19 and 13.20 are approximations since κ and α are defined in terms of the change of the volume with a change of p or T in the limit of very small p. The complete pressure-volume-temperature relationships must be considered to provide a more accurate description of the pressure dependence of η at elevated pressures, at the cost of increased complexity of the calculations. It was shown [7] that Equation 13.21 provides a reasonable first approximation for (κ/α), so that Equation 13.20 can be rewritten in the even simpler form given by Equation 13.22:

$$\frac{\kappa}{\alpha} \approx 4 \cdot 10^{-7} \tag{13.21}$$

$$\eta(T, p) \approx \eta(T, 0) - 4 \cdot 10^{-7} \cdot \left[\frac{\partial \eta(T, p)}{\partial T}\right] \cdot p \tag{13.22}$$

Equation 13.22 is simplified by setting $\eta = \eta_0$ for the zero-shear viscosity η_0. The only remaining task is the estimation of the partial derivative of η_0 with respect to T. Equation 13.23 is obtained by differentiating Equation 13.10 in the temperature regime of $T_g < T \leq 1.2 \cdot T_g$. Equation 13.24 is obtained by differentiating Equation 13.11 in the temperature regime of $T > 1.2 \cdot T_g$. The zero-shear viscosity can then be predicted as a function of the pressure by combining Equation 13.22 with Equation 13.23 or Equation 13.24 based on whether $T_g < T \leq 1.2 \cdot T_g$ or $T > 1.2 \cdot T_g$.

$$\frac{\partial \eta_0}{\partial T} \approx \eta_0 \cdot \left(\frac{T_g}{T^2}\right) \cdot \left(25.581 - 121.208 \cdot \frac{T_g}{T}\right) \qquad \text{(for } T_g < T \leq 1.2 \cdot T_g) \tag{13.23}$$

$$\frac{\partial \eta_0}{\partial T} \approx -\eta_0 \cdot \left(\frac{E_{\eta\infty}}{R \cdot T^2} + \frac{0.864 \cdot E_\eta \cdot T_g^2}{R \cdot T^4}\right) \qquad \text{(for } T > 1.2 \cdot T_g) \tag{13.24}$$

13.E. Melt Zero-Shear Viscosity: Summary, Examples and Possible Refinements

The zero-shear viscosity η_0 of a polymer can be estimated as a function of the weight-average molecular weight (M_w), temperature (T), and hydrostatic pressure (p), as follows:

1. Calculate the critical molecular weight (M_{cr}), the glass transition temperature (T_g), and the activation energy for viscous flow ($E_{\eta\infty}$) of the polymer, by using the methods described in this book. (The values used for T_g and $E_{\eta\infty}$ should be corrected to include the effects of the polymer molecular weight when $M_n << M_{cr}$, as described in Section 6.C for T_g and above for $E_{\eta\infty}$.) If experimental values are available for any of these input parameters (and especially for T_g), these values can be used instead of the calculated values in order to refine the calculations.

2. Calculate $\eta_{cr}(1.2 \cdot T_g)$ (the value of η_0 when $T = 1.2 \cdot T_g$ and $M_w = M_{cr}$) by using Equation 13.9.

3. Calculate $\eta_{cr}(T)$ from Equation 13.10 if $T_g < T \leq 1.2 \cdot T_g$, and from Equation 13.11 if $T > 1.2 \cdot T_g$. (For semicrystalline polymers, equations 13.9, 13.10 and 13.11 are only valid above T_m.)

4. Calculate $\eta_0(T, M_w)$ by using Equation 13.4 if $M_w \leq M_{cr}$ and Equation 13.5 if $M_w > M_{cr}$.

5. If $p > 0$, compute $\eta_0(T, M_w, p)$ from $\eta_0(T, M_w, 0)$ by combining Equation 13.22 (with η replaced by η_0) with Equation 13.23 if $T_g < T \leq 1.2 \cdot T_g$ and with Equation 13.24 if $T > 1.2 \cdot T_g$.

For example, it is estimated, by using the equations developed in this book, that $M_{cr} \approx 31072$ g/mole, $T_g \approx 382K$, and $E_{\eta\infty} \approx 52400$ J/mole for polystyrene. The procedure summarized above can then be used with these values of the input parameters. The calculated and observed results are compared for the temperature and M_w dependence of η_0 at $p=0$ in Figure 13.5, and for the normalized pressure coefficient $(1/\eta_0) \cdot (\partial \eta_0 / \partial p)$ of η_0 as a function of the temperature in Figure 13.6. Experimental data of Cogswell and McGowan [28] were used for both the the temperature and the pressure dependence of η_0. Additional data by Hellwege *et al* [29] were also used for the pressure dependence. The experimental values of the normalized pressure coefficient were calculated from the data [28,29] by subtracting the viscosities at the two lowest pressures used in the measurements, multiplying the measured viscosity at the lowest pressure with the difference between the two lowest pressures, and dividing the first quantity by the second one.

Theory and experiment are in reasonable but rather rough semi-quantitative agreement. This roughness is unavoidable because the theoretical relationships were developed by using data from many different sources for many polymers, showing a very large amount of scatter when a correlation is attempted either with fundamental material properties or with T and p. (An example of the scatter in the data can be seen by comparing data from different sources [28,29] shown in Figure 13.6.) In addition, because of the exponential dependence of η_0 on the fundamental material properties used as input parameters, small errors in the estimated values of these input parameters can be compounded and magnified in the calculated value of the viscosity. The equations for polymer viscosity are much more accurate in reflecting general trends when used in a correlative manner than they are when they are used to make quantitative predictions.

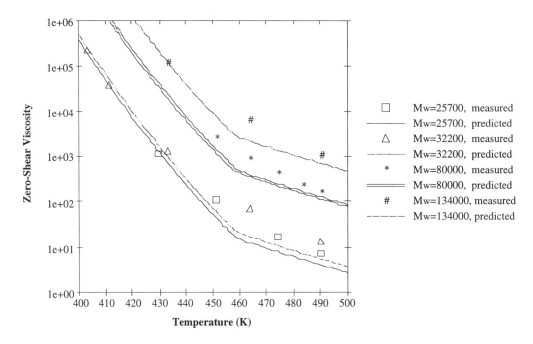

Figure 13.5. Zero-shear viscosity (N·sec/m^2) of polystyrene as a function of T and M$_w$.

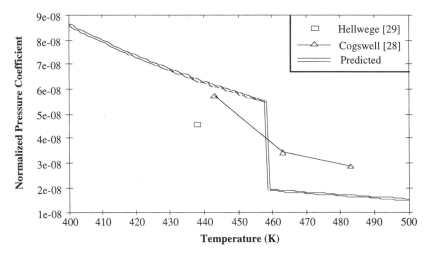

Figure 13.6. Predicted and measured normalized pressure coefficient $(1/\eta_0)\cdot(\partial\eta_0/\partial p)$ of η_0 as a function of the temperature for polystyrene. The units of $(1/\eta_0)\cdot(\partial\eta_0/\partial p)$ are m^2/N.

The very sharp discontinuity in the predicted $(1/\eta_0) \cdot (\partial\eta_0/\partial p)$ curve in Figure 13.6, and the fact that $(1/\eta_0) \cdot (\partial\eta_0/\partial p)$ is overestimated before this discontinuity is reached and underestimated afterwards, are caused by the sudden and unphysical change in the slope of $\eta_0(T)$ as shown in Figure 13.5. These artifacts of the computational procedure can be removed easily.

A spline function $\xi(T)$, utilizing cubic Hermite polynomials [30,31], is used to interpolate $\eta_0(T)$ over an appropriate range of temperature. The spline function passes through all of the specified (T, η_0) data points and maintains monotonicity between successive data points. It is a smooth function, so that the refined $\eta_0(T)$ and $\partial\eta_0(T)/\partial T$ are both continuous over the complete temperature range of the calculation. Equations 13.25 and 13.26 show that η_0 has an essentially exponential dependence on T. Consequently, the best interpolations are obtained by fitting $\xi(T)$ to the natural logarithm of $\eta_0(T)$ instead of fitting $\xi(T)$ to $\eta_0(T)$ itself, making $\eta_0(T)$ equal to the exponential of $\xi(T)$ (Equation 13.25) over region of interpolation. The partial derivative $\partial\eta_0(T)/\partial T$ over the interpolation region is then calculated by using Equation 13.26.

$$\eta_0(T) = \exp[\xi(T)] \qquad \text{(for T in the interpolation range)} \qquad (13.25)$$

$$\frac{\partial\eta_0(T)}{\partial T} = \eta_0(T) \cdot \frac{\partial\,\xi(T)}{\partial T} \qquad \text{(for T in the interpolation range)} \qquad (13.26)$$

Such a spline function can be used to make the derivatives of $\eta_0(T)$ with respect to T and p continuous at all $T > T_g$ in addition to the continuity of $\eta_0(T)$ itself. For example, this refinement can be made by using Equation 13.25 instead of equations 13.9, 13.10 and 13.11 for $1.16 \cdot T_g < T < 1.24 \cdot T_g$, to make $\partial\eta_0(T)/\partial T$ continuous at $T = 1.2 \cdot T_g$. The right-hand-side of Equation 13.9 is then no longer equal to $\eta_{cr}(1.2 \cdot T_g)$, but instead becomes merely a reference value of η_{cr} for use as an input parameter in the calculations at $T_g < T \le 1.16 \cdot T_g$ and at $T \ge 1.24 \cdot T_g$. Equations 13.10 and 13.11 are then used only over these lower and upper temperature regions. The exact form of $\xi(T)$ over the transition region $(1.16 \cdot T_g < T < 1.24 \cdot T_g)$ is determined by using the predicted values of the natural logarithm of $\eta_0(T)$ at $T = 1.14 \cdot T_g$, $1.16 \cdot T_g$, $1.24 \cdot T_g$ and $1.26 \cdot T_g$ as a set of "data points" to obtain the spline fit.

Examples of this refinement are shown for polystyrene in Figure 13.7 for $\eta_0(T)$ and in Figure 13.8 for the normalized pressure coefficient $(1/\eta_0) \cdot (\partial\eta_0/\partial p)$. The refined results do not manifest the mathematical artifacts of a "knee" in $\eta_0(T)$ and a "jump" in $\partial\eta_0/\partial p$ at $T = 1.2 \cdot T_g$, so that they provide a far better qualitative physical picture than Figure 13.5 and Figure 13.6 in addition to better quantitative agreement with experimental data [28,29]. $(1/\eta_0) \cdot (\partial\eta_0/\partial p)$ calculated with the new procedure (Figure 13.8) is less smooth than $\partial\eta_0/\partial p$ (not shown), but much smoother than the results obtained with the basic method.

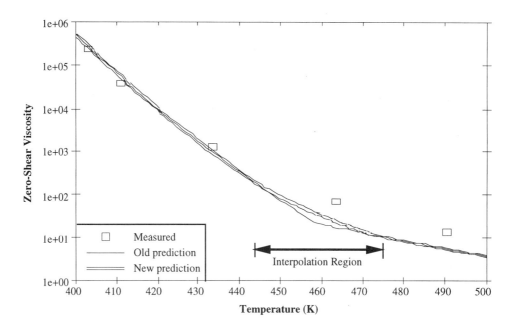

Figure 13.7. Effect of refinement of predicted $\eta_0(T)$ (in units of N·sec/m^2) on comparison between measured [28] and predicted $\eta_0(T)$ of polystyrene with M$_w$=32200.

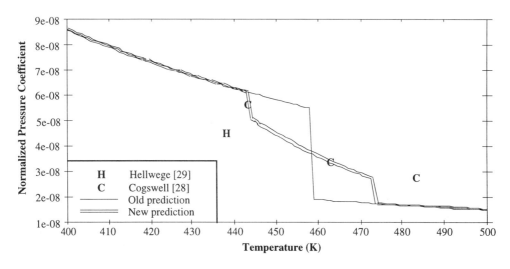

Figure 13.8. Effect of refinement of predicted $\eta_0(T)$ on comparison between measured [28,29] and predicted normalized pressure coefficient $(1/\eta_0)\cdot(\partial\eta_0/\partial p)$ (in units of m^2/N) for polystyrene.

13.F. Combined Effects of Shear Rate and Polydispersity on Melt Viscosity

The effects of the shear rate ($\dot{\gamma}$) and the polydispersity (the ratio M_w/M_n, where M_n denotes the number-average molecular weight) on the melt viscosity have been reviewed by van Krevelen [7] from the viewpoint of fitting experimental data by empirical relationships, and by Kumar [5] from the viewpoint of fundamental theoretical treatments. It is clear from these reviews that the effects of $\dot{\gamma}$ and polydispersity are inextricably intertwined and must be considered simultaneously to provide a useful estimate of η for a polymer with a high polydispersity at high $\dot{\gamma}$.

The quotient (η/η_0), which indicates the magnitude of shear thinning effect with increasing $\dot{\gamma}$, can be estimated quite well as a function of the Weissenberg number (N_{Wg}) and the polydispersity [7]. N_{Wg} is the product of a characteristic time scale for the polymer with $\dot{\gamma}$, and can be approximated by Equation 13.27 where ρ is the polymer bulk density at the temperature of calculation. (The density must be expressed in g/m^3 rather than g/cc in Equation 13.27 for consistency with the units used for the other quantities entering this equation.) Since the polydispersity M_w/M_n enters Equation 13.27, η depends on the polydispersity both indirectly via the dependence of N_{Wg} on M_w/M_n and directly.

$$N_{Wg} \approx \frac{6 \cdot \eta_0 \cdot M_w{}^2 \cdot \dot{\gamma}}{\pi^2 \cdot R \cdot \rho \cdot T \cdot M_n} \tag{13.27}$$

The shear thinning effect increases with increasing N_{Wg}. Since $(M_w{}^2/M_n) = M_w \cdot (M_w/M_n)$, shear thinning generally increases both with increasing M_w at a given polydispersity and with increasing polydispersity at a given M_w. The factor $\eta_0/(\rho \cdot T)$ in Equation 13.27 is a function of the temperature, and decreases very rapidly with increasing temperature. Consequently, the magnitude of the shear thinning effect is also a function of the temperature and decreases with increasing temperature. Van Krevelen showed the dependence of the ratio η/η_0 on N_{Wg} and M_w/M_n graphically [7], but did not attempt to represent this dependence in an algebraic form applicable for all values of these two parameters. Consequently, we extracted data points from van Krevelen's graphical representations, and fitted these data points with a nonlinear regression relationship. The results are summarized by equations 13.28, 13.29 and 13.30, and can be used to calculate the effects of shear rate and polydispersity on melt viscosity.

$$\eta = \eta_0 \cdot \exp\left[f\left(N_{Wg}, \frac{M_w}{M_n}\right) \right] \tag{13.28}$$

$$f\left(N_{Wg}, \frac{M_w}{M_n}\right) = 0 \qquad \textit{[for } 0 \leq N_{Wg} < 0.254761 \textit{ and } 1 \leq (M_w/M_n)] \tag{13.29}$$

$$f\left(N_{Wg}, \frac{M_w}{M_n}\right) = -\left\{0.0509 \cdot \frac{\ln(N_{Wg})^2}{\left[1 + 0.0030 \cdot \ln(N_{Wg})^2\right]} + 0.2252 \cdot \ln(N_{Wg}) + 0.2133\right\} \cdot \left(\frac{M_w}{M_n}\right)^{-0.1371} \quad (13.30)$$

Equation 13.30 holds if $0.254761 \leq N_{Wg} \leq 10^9$ and $1 \leq (M_w/M_n) \leq 20$ [set $(M_w/M_n)=20$ if $20<(M_w/M_n)$]. Here, "ln" is the natural logarithm. Equations 13.28 and 13.29 imply that $\eta \approx \eta_0$ if $N_{Wg}<0.254761$. The upper limit of $N_{Wg}=10^9$ is imposed on Equation 13.30 because the use of equations 13.28 and 13.30 for $N_{Wg}>10^9$ is a potentially dangerous extrapolation far beyond the range of N_{Wg} values of the data considered in developing these equations. Anomalous effects are often seen in the shear rate dependence of η at exceedingly high shear rates [32,33].

The setting of $(M_w/M_n)=20$ if $(M_w/M_n)>20$ in Equation 13.30 implies that the increase of M_w/M_n beyond 20 affects the ratio (η/η_0) only through its effect on N_{Wg} as calculated by using Equation 13.27, and that the separate and direct dependence of (η/η_0) on the polydispersity as quantified by the factor $(M_w/M_n)^{-0.1371}$ in Equation 13.30 levels off at $(M_w/M_n) \approx 20$. Equation 13.22 remains valid as an estimate of the dependence of η on p, but equations 13.23 and 13.24 are replaced by equations 13.31, 13.32 and 13.33 which incorporate the temperature dependence of $f(N_{Wg}, M_w/M_n)$. The value of $\partial \eta_0/\partial T$ is calculated from Equation 13.23 or Equation 13.24 for $T_g < T \leq 1.2 \cdot T_g$ and $T > 1.2 \cdot T_g$, respectively; for use in equations 13.31, 13.32 and 13.33. Equation 13.33 includes terms related to the estimated temperature dependence of $\rho(T)$ for $T > T_g$ based on equations provided Chapter 3. The result of the use of equations 13.22, 13.31, 13.32 and 13.33 is the prediction that $\partial \eta/\partial p$ decreases with further increase of $\dot{\gamma}$ for $N_{Wg} \geq 0.254761$.

$$\frac{\partial \eta}{\partial T} = \frac{\partial \eta_0}{\partial T} \qquad \qquad \textit{[for } 0 \leq N_{Wg} < 0.254761 \textit{ and } 1 \leq (M_w/M_n)] \qquad (13.31)$$

$$\frac{\partial \eta}{\partial T} = -\eta \cdot \left\{\frac{0.1018 \cdot \ln(N_{Wg})}{\left[1 + 0.0030 \cdot \ln(N_{Wg})^2\right]} - \frac{0.0003054 \cdot \ln(N_{Wg})^3}{\left[1 + 0.0030 \cdot \ln(N_{Wg})^2\right]^2} + 0.2252\right\} \cdot \frac{1}{N_{Wg}} \cdot \frac{\partial N_{Wg}}{\partial T} \cdot \left(\frac{M_w}{M_n}\right)^{-0.1371}$$
$$+ \frac{\partial \eta_0}{\partial T} \cdot \exp(f) \qquad (13.32)$$

$$\frac{\partial N_{Wg}}{\partial T} = \frac{6 \cdot M_w^2 \cdot \dot{\gamma}}{\pi^2 \cdot R \cdot M_n} \cdot \frac{\left(T \cdot \rho \cdot \frac{\partial \eta_0}{\partial T} - \rho \cdot \eta_0 + 0.1911 \cdot \frac{\rho \cdot \eta_0 \cdot T}{T_g}\right)}{T^2 \cdot \rho^2} \qquad (13.33)$$

Equation 13.32 holds if $0.254761 \leq N_{Wg} \leq 10^9$ and $1 \leq (M_w/M_n) \leq 20$ [set $(M_w/M_n)=20$ if $20<(M_w/M_n)$]. The only new material property encountered in this section is the bulk density ρ of the polymer, which can be estimated by using the equations presented in Chapter 3.

In summary, the following procedure can be used to estimate polymer melt viscosity as a function of M_w, polydispersity, temperature, hydrostatic pressure, and shear rate:

1. Estimate the zero-shear viscosity by using the procedure summarized in Section 13.E.
2. Calculate the Weissenberg number N_{Wg} at the shear rate of interest via Equation 13.27, with the value of the polymer bulk density estimated by using the equations given in Chapter 3.
3. Use Equation 13.28, combined with Equation 13.29 or Equation 13.30 as appropriate, to estimate the combined effects of the shear rate and the polydispersity on the melt viscosity.
4. Calculate the effect of pressure for p>0 via Equation 13.22, with the partial derivative $\partial\eta/\partial T$ given by Equation 13.31, or by equations 13.32 and 13.33, as appropriate.

For example, Tuminello and Cudré-Mauroux [34] measured η for polystyrene as a function of γ and plotted the results as smoothed master curves at T=433K. We extracted data points from these master curves, for comparison with the results of calculations performed as described above, using $\rho(433K)\approx1.025\cdot10^6$ g/m³, $M_{cr}\approx31072$ g/mole, $T_g\approx382K$, and $E_{\eta\infty}\approx52400$ J/mole (calculated by using the correlations developed in this book) as input values of the key material parameters. The results are shown for one of the samples in Figure 13.9. There is good overall agreement between theory and experiment. The effects of replacing the calculated value of η_0 with the experimental value on the overall agreement with the observations on the shear rate dependence are also shown. This refinement only improves the results of the calculations at low shear rates in this particular example; however, more significant improvements are obtained in calculations on many other polymers, and it is worthwhile to make this refinement when possible.

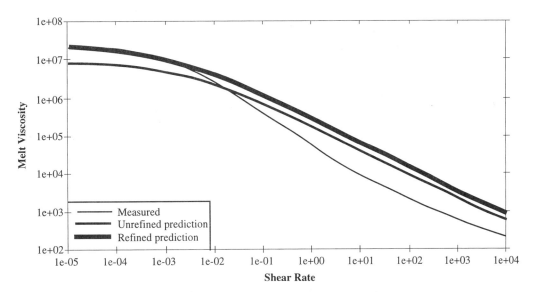

Figure 13.9. Observed [34] and calculated melt viscosity (N·sec/m²) of a polystyrene sample with M_w=435000 and a polydispersity of 6.37 at T=433K as a function of the shear rate (sec⁻¹). The effects of using the observed instead of the predicted value of $\eta_0(433K)$ in the calculations of the shear rate and polydispersity dependence of $\eta(433K)$ are also shown.

13.G.   Zero-Shear Viscosity of Concentrated Polymer Solutions

The effect of polymer concentration on the zero-shear viscosity of a concentrated solution can often be estimated by starting from the viscosity of the polymer melt and assuming that the effect of the addition of solvent can be deconvoluted into two different types of perturbations [7]:

1. The *plasticization effect*, causing a decrease of the viscosity of the pure polymer as a direct result of the decrease of its glass transition temperature because of the addition of the solvent.

2. The *dilution effect*, causing the viscosity of the concentrated solution to be lowered even further, and to fall between the viscosity of the plasticized polymer and that of the pure solvent.

The most important effect of plasticization is the lowering of the T_g of the polymer, which was discussed in Section 6.D. Another effect of plasticization is that the activation energy for viscous flow of the solution at $T>1.2 \cdot T_g$ is usually smaller than the activation energy of the pure polymer since $E_{\eta p} >> E_{\eta s}$ for most polymer-solvent combinations. Our preliminary calculations show that Equation 12.18 may often be preferable to Equation 12.16 for describing the behavior of $E_{\eta ss}$, at least for $\Phi_p << 1$. This issue must, however, be considered in greater detail in order to reach more definitive conclusions. In particular, $E_{\eta ss}$ must be examined as a function of Φ_p in the limit of $\Phi_p \rightarrow 1$. Furthermore, the dependence the behavior of $E_{\eta ss}$ as a function of Φ_p on the strength of the interactions between the polymer and the solvent needs to be considered.

The effect of true dilution is that it becomes increasingly more difficult for the polymer chains to become entangled with each other as the volume fraction of the polymer is reduced. This effect can be described by Equation 13.34 in terms of an increase of M_{cr} with decreasing Φ_p [7], since M_{cr} quantifies the value of M_w at which entanglements become important in limiting viscous flow.

$$M_{cr}(\text{polymer in solution}) \approx \frac{M_{cr}(\text{polymer in melt})}{\Phi_p^{1.5}} \qquad (13.34)$$

It should be noted, however, that the power of $\Phi_p^{1.5}$ in the denominator of Equation 13.34 may be an overestimation of the effect of true dilution. There is evidence that a lower power, such as $\Phi_p^{1.2}$, may be more appropriate. An important portion of this evidence comes from the analysis of the results of dilution on the shear modulus in the rubbery plateau region which suggest that the exponent should be somewhere in the range of 1.0 to 1.3 (average of 1.15 for the range) [35], coupled with the fact that the shear modulus and the shear viscosity are related by Equation 13.3. Another analysis, based completely on theoretical considerations, gives an exponent of 1.25 [36] for the effect of dilution on M_{cr}. (Note, however, that the validity of the analysis provided in [36] has been questioned [37].) Averaging these two estimates [35,36] of 1.15 and 1.25 gives 1.2. The issue is not completely resolved at this time, however, and the power of Φ_p may actually even depend slightly on the strength of polymer-solvent interactions.

If η_{pcr}^* denotes the viscosity of the plasticized but undiluted polymer at the critical molecular weight (calculated by using the lowered values of T_g and $E_{\eta\infty}$ describing the plasticized polymer), then the solution viscosity (η_{ss}) can be calculated by using Equation 13.4 or Equation 13.5, as appropriate, with M_{cr} calculated from Equation 13.34 (or a refined version of it with a slightly smaller power of Φ_p in the denominator) being used instead of the M_{cr} of the polymer melt.

As was the case with the melt viscosity, T_{gp} and $E_{\eta p}$ must also be estimated as functions of M_n for use in the equations presented earlier for polymers with $M_n \ll M_{cr}$ to obtain accurate predictions of the solution viscosity if M_n is low, especially at high polymer concentrations.

In summary, the following procedure can be used to predict, *at a very approximate level of accuracy*, the zero-shear viscosity $\eta_{ss}(M_w,T,\Phi_p)$ for concentrated polymer solutions:

1. Convert polymer concentration to volume fraction if a concentration (rather than volume fraction) has been specified. The variation of the densities of the polymer and the solvent with the temperature can be neglected in this conversion, to a very good approximation.
2. Estimate the T_g of the plasticized polymer as described in Section 6.D.
3. Estimate the $E_{\eta\infty}$ of the plasticized polymer by using Equation 12.16 or Equation 12.18.
4. Estimate the viscosity of the plasticized but undiluted polymer with $M_w = M_{cr}$ at the temperature of interest (η_{pcr}^*) by utilizing equations 13.9, 13.10 and 13.11 with the T_g and the $E_{\eta\infty}$ of the plasticized polymer rather than the T_g and the $E_{\eta\infty}$ of the pure polymer.
5. Describe true dilution by using Equation 13.34 to calculate M_{cr} for the polymer in solution.
6. Describe the dependence of η_{ss} on the M_w of the polymer by utilizing Equation 13.4 or Equation 13.5, as appropriate, with the modified value of M_{cr} given by Equation 13.34.

The viscosities of concentrated solutions of polystyrene in *o*-xylene will now be calculated by using the procedure described above (with Equation 12.18 as the mixing rule for $E_{\eta\infty}$), and compared with experimental data. The properties of many solvents have been tabulated in standard references [38-40], but the needed properties are not always available. This is particularly the case for the glass transition temperature (T_{gs}), which is required to estimate the plasticization effect on T_g by using the procedure of Section 6.D. In this example, T_{gs} was estimated from $T_m \approx 245K$ for *o*-xylene [7], by using Equation 6.14. The quantity X was estimated by using Equation 6.13, since the thermal expansion coefficient of the solvent above and below its unknown T_{gs} could obviously not be assigned directly either. The estimation of $E_{\eta s} \approx 8941$ J/mole was done directly, by fitting an activated flow expression to the zero-shear viscosity of *o*-xylene as a function of the temperature. The observed viscosities of solutions of varying concentration were illustrated graphically along with the viscosities of pure polystyrene and *o*-xylene in van Krevelen's book [7], for a polystyrene resin of $M_w = 3.7 \cdot 10^5$. We extracted data points from this graphical representation, for comparison with the results of the calculations as shown in Figure 13.10. Rough semi-quantitative agreement is seen between the measured and predicted results. The predicted values when multiplied by two agree very well with the data, so that the effects of changing concentration and temperature are both described correctly and the only discrepancy is in the overall scale of the predictions which is off by a factor of roughly two.

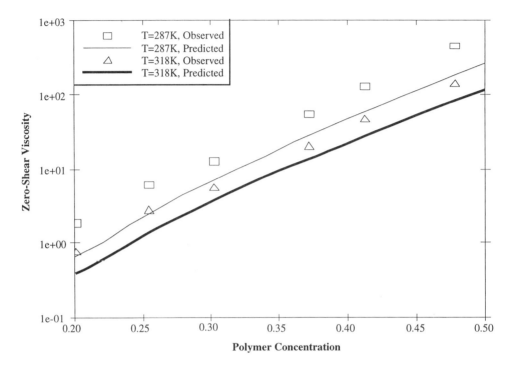

Figure 13.10. Comparison of observed [7] and predicted zero-shear viscosities (N·sec/m^2) of concentrated solutions of polystyrene with $M_w = 3.7 \cdot 10^5$ in xylene, as functions of the polymer concentration (g/cc), at two different temperatures (T=287K and T=318K).

References and Notes for Chapter 13

1. R. B. Bird, R. C. Armstrong and O. Hassager, *Dynamics of Polymeric Liquids, Volume 1: Fluid Mechanics*, second edition, John Wiley & Sons, New York (1987).

2. R. B. Bird, C. F. Curtiss, R. C. Armstrong and O. Hassager, *Dynamics of Polymeric Liquids, Volume 2: Kinetic Theory*, second edition, John Wiley & Sons, New York (1987).

3. J. D. Ferry, *Viscoelastic Properties of Polymers*, John Wiley & Sons, New York (1961).

4. M. Doi and S. F. Edwards, *The Theory of Polymer Dynamics*, Clarendon Press, Oxford (1986).

5. N. G. Kumar, *J. Polym. Sci., Macromol. Rev.*, *15*, 255-325 (1980).

6. C. K. Schoff, article titled "*Rheological Measurements*", *Encyclopedia of Polymer Science and Engineering*, *14*, Wiley-Interscience, New York (1988).

7. D. W. van Krevelen, *Properties of Polymers*, second edition, Elsevier, Amsterdam (1976). Chapter 15 of this book deals with the melt viscosity.

8. R. G. Larson, *Constitutive Equations for Polymer Melts and Solutions*, Butterworth Publishers, Boston (1988).

9. P.-G. deGennes, *Scaling Concepts in Polymer Physics*, Cornell University Press, Ithaca, New York (1979).

10. J. T. Seitz, *J. Appl. Polym. Sci.*, *49*, 1331-1351 (1993).

11. D. W. van Krevelen, in *Computational Modeling of Polymers*, edited by J. Bicerano, Marcel Dekker, New York (1992), Chapter 1.

12. D. R. Terrell and H. Katiofsky, *J. Appl. Polym. Sci.*, *21*, 1311-1322 (1977).

13. C. N. Cascaval, F. Mustata and D. Rosu, *Die Angewandte Makromolekulare Chemie*, *209*, 157-166 (1993).

14. K. Dodgson, D. J. Bannister and J. A. Semlyen, *Polymer*, *21*, 663-667 (1980).

15. G. V. Vinogradov and A. Ya. Malkin, *Rheology of Polymers*, Mir Publishers, Moscow (1980).

16. L. I. Shakhovskaya, L. V. Krayeva and L. I. Mezentseva, *Polymer Science U.S.S.R.*, *22*, 2915-2921 (1980).

17. L. H. Drexler, *J. Appl. Polym. Sci.*, *14*, 1857-1869 (1970).

18. E. A. Collins and A. P. Metzger, *Polymer Engineering and Science*, *10*, 57-65 (1970).

19. J. Miltz and A. Ram, *Polymer Engineering and Science*, *13*, 273-279 (1973).

20. M. Costagliola, R. Greco and E. Martuscelli, *Polymer*, *19*, 860-862 (1978).

21. A. Munari, F. Pilati and G. Pezzin, *Rheologica Acta*, *24*, 534-536 (1985).

22. A. Munari, G. Pezzin and F. Pilati, *Rheologica Acta*, *29*, 469-474 (1990).

23. S. Ottani, G. Pezzin and C. Castellari, *Rheologica Acta*, *27*, 137-144 (1988).

24. K. Kishore and H. K. Shobha, *J. Phys. Chem.*, *96*, 8161-8168 (1992).

25. Y. Oyanagi and M. Matsumoto, *J. Colloid Science*, *17*, 426-438 (1962).

26. T. Yanagida, A. Teramoto and H. Fujita, *J. Physical Chemistry*, *72*, 1265-1271 (1968).

27. A. A. Tager and G. O. Botvinnik, *Polymer Science U.S.S.R.*, *16*, 1483-1488 (1974).

28. F. N. Cogswell and J. C. McGowan, *British Polymer Journal*, *4*, 183-198 (1972).

29. K.-H. Hellwege, W. Knappe, F. Paul and V. Semjonov, *Rheologica Acta*, *6*, 165-170 (1967).

30. F. N. Fritsch and R. E. Carlson, *SIAM J. Numer. Anal.*, *17*, 238-246 (1980).

31. F. N. Fritsch and J. Butland, *An Improved Monotone Piecewise Cubic Interpolation Algorithm*, Lawrence Livermore National Laboratory Preprint UCRL-85104 (1980).

32. H. Takahashi, T. Matsuoka and T. Kurauchi, *J. Appl. Polym. Sci.*, *30*, 4669-4684 (1985).

33. H. Takahashi, T. Matsuoka, T. Ohta, K. Fukumori, T. Kurauchi and O. Kamigaito, *J. Appl. Polym. Sci.*, *37*, 1837-1853 (1989).

34. W. H. Tuminello and N. Cudré-Mauroux, *Polymer Engineering and Science*, *31*, 1496-1507 (1991).

35. W. W. Graessley and S. F. Edwards, *Polymer*, *22*, 1329-1334 (1981).

36. R. P. Wool, *Macromolecules*, *26*, 1564-1569 (1993).

37. L. J. Fetters, D. J. Lohse, D. Richter, T. A. Witten and A. Zirkel, *Macromolecules*, *27*, 4639-4647 (1994).

38. R. C. Reid, J. M. Prausnitz and B. E. Poling, *The Properties of Gases and Liquids*, fourth edition, McGraw-Hill Book Company, New York (1987).

39. N. B. Vargaftik, *Handbook of Physical Properties of Liquids and Gases: Pure Substances and Mixtures*, Hemisphere Publishing Corporation, Washington (1983).

40. J. K. Sears and J. R. Darby, *The Technology of Plasticizers*, Wiley, New York (1982).

CHAPTER 14

THERMAL CONDUCTIVITY
AND THERMAL DIFFUSIVITY

14.A. Background Information

14.A.1. Definition and General Considerations

The *thermal conductivity* [1-8] λ is the amount of heat transported per unit time, through a unit area of a slab of material of unit length, per degree Kelvin of temperature difference between the two faces of the slab, at steady-state heat flow. Let q denote the heat flux in $J/(m^2 \cdot sec)$ where m denotes meters, let dT/dz denote the temperature gradient in the z-direction in (degrees Kelvin)/m, and let λ be expressed in units of $J/(K \cdot m \cdot sec)$. For unidirectional, rectilinear heat flow in the z-direction under steady-state conditions, λ is defined by Equation 14.1:

$$q = -\lambda \cdot \frac{dT}{dz} \tag{14.1}$$

It is useful to be able to estimate the thermal conductivities of polymers, both for optimizing certain processes and for evaluating end-use performance:

1. During *melt processing*, for example by molding or extrusion, a polymer is first heated up, and then cooled, by a considerable amount in a short time period. The thermal conductivity is therefore very important in determining the optimum processing conditions. For three of the many published treatments of heat transfer properties such as the thermal conductivity and the thermal diffusivity within the context of practical polymer processing, see Rauwendaal [9] on extrusion, Menges and Mohren [10] on injection molding, and Throne [11] on thermoforming.
2. If a polymer is to be used in applications where high *thermal insulation* is required, the thermal conductivity is of great importance. The low thermal conductivities of polymers, and of *foams* [12] made by expanding polymers and incorporating pockets of air or other gas molecules in the "cells" resulting from the expansion, when coupled with other desirable properties, have made polymers and polymeric foams the insulating materials of choice in many applications.

The motions of delocalized electrons play the key role in heat transfer in electrically conducting materials, such as metals. On the other hand, in insulators, such as ordinary organic polymers, electrons are localized in inner shell, lone pair and bonding orbitals, and are thus not free to move throughout the material. Heat transfer therefore mainly occurs by atomic vibrations in organic polymers. Consequently, the thermal conductivities of unmodified organic polymers are quite small, and fall into a very narrow portion of the span of several orders of magnitude

393

encompassing all materials. For example, at room temperature, $\lambda \approx 427$ J/(K·m·sec) for silver, $\lambda \approx 237$ J/(K·m·sec) for aluminum, and $\lambda \approx 0.026$ J/(K·m·sec) for air [2], while $0.1 \leq \lambda \leq 0.3$ J/(K·m·sec) for unfilled isotropic amorphous polymers. Even when significant amount of fillers of higher thermal conductivity are used, the λ values of polymers still fall into a rather narrow portion of the complete range of thermal conductivities. For example [2], $\lambda \approx 1.7$ J/(K·m·sec) for 40% graphite-filled polyimide resin, and $\lambda \approx 0.52$ J/(K·m·sec) for high-density polyethylene which has a high percent crystallinity with the crystallites functioning as a "filler" phase of high λ.

Acoustic (sound) waves are also transmitted by atomic vibrations. Atomic vibrations are often called *phonons*, i.e., "particles" of sound. Theories of heat transfer [13] in insulators usually attempt to relate λ to other physical properties which are mainly determined by atomic vibrations, such as the velocity of sound and the heat capacity.

For a given polymeric structure, the morphology (crystallinity and orientation), formulation (additives, fillers and impurities), humidity (especially for polar polymers), temperature, and pressure, are the most important factors which affect the thermal conductivity. References [1-8] review many of these factors. In addition, see Bigg [14] and Ross *et al* [15] for detailed treatments of the effects of fillers and of pressure, respectively, on thermal conductivity.

Thermal conductivities of polymers at temperatures which are not extremely low will be discussed below. See Hust [7] on heat transfer properties of materials at cryogenic temperatures.

Even a small amount of *crystallinity* can cause a very significant increase in λ. The thermal conductivity in the crystalline limit of a semicrystalline polymer can be many times larger than the thermal conductivity in the amorphous limit. The general form of the temperature dependence of λ also changes with crystallinity. For example, λ increases monotonically for amorphous polymers between T=100K and T=T_g. On the other hand, for highly crystalline polymers, λ usually reaches a peak at around 100K and then decreases with increasing T. Consequently, the general form of the temperature dependence of λ for a semicrystalline polymer between T=100K and T=T_g can change qualitatively (from λ increasing with T at low crystallinity to λ decreasing with T at high crystallinity) when specimens of widely differing crystallinity are compared.

The thermal conductivity increases greatly with increasing strength of the bonds located in the direction of heat transport. For example, the covalent chemical bonds along a polymer chain backbone are far more effective in transporting heat, than are nonbonded interactions in directions perpendicular to the backbone which physically link different polymer chains. The alignment of polymer chains via *orientation* therefore results in an increase in λ. This effect can be especially drastic if the polymer being oriented also possesses a significant amount of crystallinity.

Because of the great sensitivity of the thermal conductivity to the crystallinity, orientation and formulation, a wide range of values, measured at the same temperature, are found for many polymers in standard tables, especially when different commercial grades of a given polymer are compared [16]. In fact, the possible range of variation of the thermal conductivity of a given polymer as a result of these factors is larger than the entire range of variation of the thermal conductivities of unmodified isotropic amorphous polymers as a function of the structure of the repeat unit. Furthermore, most of the tests used to measure λ have large margins of error. Predictive methods for the thermal conductivity are inevitably limited by all of these factors, and they can at best provide rough estimates.

The thermal conductivity of a material is defined in terms of the transport of heat under steady-state conditions. On the other hand, one is often interested in the transport of heat when a specimen is not at equilibrium so that the flow of heat is transient. The *thermal diffusivity*, a, which is defined by Equation 14.2, describes these time-dependent, non-steady-state aspects of heat flow. The thermal diffusivity is used to calculate the temperature (T) as a function of the position within the specimen (z) and the time (t) under non-steady-state conditions.

$$\frac{\partial^2 T}{\partial z^2} = \frac{1}{a} \cdot \frac{\partial T}{\partial t} \tag{14.2}$$

The thermal diffusivity is related in a simple way to the thermal conductivity, the density, and the specific heat capacity:

$$a = \frac{\lambda}{c_p \cdot \rho} \tag{14.3}$$

14.A.2. Temperature Dependence of the Thermal Conductivities of Amorphous Polymers

Like many other properties (such as the molar volume, Chapter 3; heat capacity, Chapter 4; and surface tension, Chapter 7), the thermal conductivity of an amorphous polymer can be estimated as a function of the temperature if it has been measured or predicted at some reference temperature, usually taken as room temperature (298±5K). For example, van Krevelen [6] plotted the measured values of $\lambda(T)/\lambda(T_g)$ against the "reduced temperature" T/T_g for amorphous polymers. $\lambda(T)/\lambda(T_g)$ slowly increases for $T < T_g$, has a maximum at $T = T_g$, and very slowly decreases for $T > T_g$. The change of $\lambda(T)/\lambda(T_g)$ with increasing T/T_g is nonlinear (has nonzero curvature) for $T < T_g$, and is almost linear for $T > T_g$. Van Krevelen drew a generalized curve passing through the data points, and showed that $\lambda(T)$ can be estimated graphically from the values of (a) T_g, and (b) λ at a reference temperature (usually room temperature).

We found it convenient to reproduce van Krevelen's generalized curve very closely, but not exactly, by a pair of simple equations (equations 14.4 and 14.5), as an alternative to estimating $\lambda(T)$ graphically. (We later found out that Equation 14.4, expressed in an equivalent but slightly different form, is also recommended by Rauwendaal [9].) The general form of the temperature dependence obtained by using equations 14.4 and 14.5 is shown in Figure 14.1. The rate of change of $\lambda(T)/\lambda(T_g)$ with changing T/T_g is very slow at temperatures of practical interest. $\lambda(T)/\lambda(T_g)$ changes smoothly at $T=T_g$. The apparent cusp at $T=T_g$ in Figure 14.1 is merely an artifact of the use of two different functions for $\lambda(T)/\lambda(T_g)$ below and above T_g.

$$\lambda(T) \approx \lambda(T_g) \cdot \left(\frac{T}{T_g}\right)^{0.22} \qquad \text{(amorphous polymers, } T \leq T_g) \qquad (14.4)$$

$$\lambda(T) \approx \lambda(T_g) \cdot \left[1.2 - 0.2\left(\frac{T}{T_g}\right)\right] \qquad \text{(amorphous polymers, } T > T_g) \qquad (14.5)$$

Equations 14.4 and 14.5 can be used to estimate the thermal conductivity of an amorphous polymer as a function of the temperature, from its thermal conductivity at room temperature and from its known or predicted T_g. For example, assume that a polymer has $T_g=500K$ and $\lambda(298K)=0.200$ J/(K·m·sec), and that we wish to predict $\lambda(600K)$. Since $298K<T_g$, $T=298$ and $T_g=500$ can be substituted into Equation 14.4 to obtain $\lambda(T_g)\approx0.224$ J/(K·m·sec). Since $600K>T_g$, Equation 14.5 with $T=600$ and $T_g=500$ can then be used to estimate that $\lambda(600K)\approx0.215$ J/(K·m·sec). $\lambda(298K)$ is thus the only quantity which still remains to be predicted, in order to predict the $\lambda(T)$ of an arbitrary amorphous polymer.

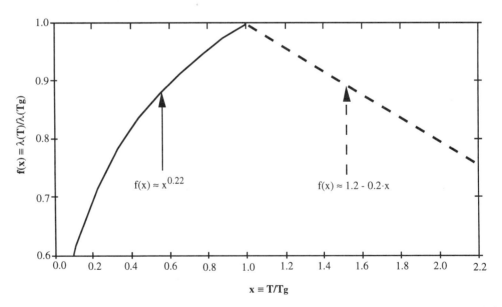

Figure 14.1. Approximate form of the temperature dependence of the thermal conductivity $\lambda(T)$ of amorphous polymers, in terms of the "reduced temperature" $x\equiv(T/T_g)$. The two equations defining these curves can be used to estimate $\lambda(T)$ at any temperature if an estimate is available for T_g and for the value of λ at a reference temperature.

14.A.3. Thermal Conductivities of Amorphous Polymers at Room Temperature

The thermal conductivities of amorphous polymers at room temperature can be estimated by using Equation 14.6 [6]. The heat capacity C_p at constant pressure, Poisson's ratio v, and molar volume V, must all be calculated at room temperature for use in Equation 14.6. U_R is the molar Rao function, encountered in Chapter 11 in the calculation of the bulk modulus. The factor $5 \cdot 10^{-11}$ (in meters) is a typical order of magnitude for the mean free path length of phonons.

$$\lambda(298K) \approx 5 \cdot 10^{-11} \cdot \left[\frac{C_p(298K)}{V(298K)}\right] \cdot \left[\frac{U_R}{V(298K)}\right]^3 \cdot \sqrt{\frac{3 \cdot [1 - v(298K)]}{[1 + v(298K)]}} \qquad (14.6)$$

Equation 14.6 was derived by assuming that there is a direct proportionality between thermal diffusivity and the velocity of sound in an amorphous polymer or polymer melt [6]. See Chapter 4 for the calculation of $C_p(298K)$. $C_p^s(298K)$ is used if $T_g>298K$, while $C_p^l(298K)$ is used if $T_g<298K$. $V(298K)$ is estimated by using equations 3.13 and 3.14. See Section 11.B.2.c for the correlation for U_R. Poisson's ratio at T=298K is estimated by using Equation 11.10 if $T_g>298K$. If $T_g<298K$, $v(298K)\approx0.5$, so that the quantity within the square root in Equation 14.5 becomes equal to 1, and the simpler Equation 14.7 can be used instead of Equation 14.6.

$$\lambda(298K) \approx 5 \cdot 10^{-11} \cdot \left[\frac{C_p^l(298K)}{V(298K)}\right] \cdot \left[\frac{U_R}{V(298K)}\right]^3 \qquad \textit{(for } T_g<298K \textit{ only)} \qquad (14.7)$$

U_R is in units of $cm^{10/3}/(sec^{1/3} \cdot mole)$. $C_p(298K)$ was expressed in J/(mole·K) in Chapter 4, and V(298K) was expressed in cc/mole in Chapter 3. Since the factor $5 \cdot 10^{-11}$ is in meters, the substitution of all quantities in their preferred units into Equation 14.6 or Equation 14.7 gives $\lambda(298K)$ in units of 10^4 J/(K·m·sec). In order to obtain $\lambda(298K)$ in J/(K·m·sec), $\lambda(298K)$ values calculated by using the preferred units for C_p, V and U_R in equations 14.6 and 14.7 must therefore be multiplied by a unit conversion factor of 10^4.

For example, let us estimate $\lambda(298K)$ for polyisoprene, which has $T_g<298K$ so that Equation 14.7 can be used with $C_p^l(298K)$ as the appropriate heat capacity. The correlations presented earlier give $C_p^l(298K)\approx142.0$ J/(mole·K) (Table 4.2), $V(298K)\approx76.6$ cc/mole (Table 3.4), and $U_R\approx4072$ $m^{10/3}/(sec^{1/3} \cdot mole)$ (Table 11.1). Substitution of these values into Equation 14.7, and use of the conversion factor of 10^4 to obtain the correct units, gives the following result: $\lambda(298K) \approx 10^4 \cdot 5 \cdot 10^{-11} \cdot (142.0/76.6) \cdot (4072/76.6)^3 \approx 0.139$ J/(K·m·sec). This result is in reasonable agreement with the commonly quoted experimental value [6] of 0.134 J/(K·m·sec).

14.A.4. Improvements in the Ability to Predict the Thermal
Conductivities of Polymers

An improvement we made in the ability to predict the thermal conductivities of polymers was to develop new correlations for U_R, V(298K), C_p(298K), and for the quantities entering the correlation for v(298K). These correlations do not require group contributions, and can thus be used without being limited by the need for group contributions. The range and the structural diversity of polymers for which Equation 14.6 or Equation 14.7 can be used to predict λ(298K) are therefore significantly expanded by the development of the new correlations.

According to Equation 14.6, λ(298K) is proportional to U_R^3. The relative standard deviation of 1.3% for the correlation for U_R (Section 11.B.2.c) thus implies that there is, typically, a difference of only about 3.95% between λ(298K) calculated by using U_R values obtained from group contributions and λ(298K) calculated by using U_R values obtained from equations 11.16 and 11.17, when the needed group contributions are available to make such a comparison. This difference is much smaller than the uncertainty in the experimental determination of λ(298K).

As another improvement in the ability to predict the thermal conductivities of polymers, a direct correlation was developed between the λ(298K) of amorphous polymers and some of their most important structural features. This correlation will be presented in Section 14.B. While the indirect relationship (Equation 14.6) describes the physics of thermal conduction at a much more fundamental level, the direct correlation reveals some of the connections between the structure of the repeat unit and the magnitude of the thermal conductivity in a more obvious manner.

14.B. Direct Correlation for the Thermal Conductivity at Room Temperature

A dataset of twenty λ(exp) values, all measured at room temperature (298±5K), for amorphous polymers and for the amorphous phases of semicrystalline polymers, was prepared by combining information provided in several sources [2,4,6]. This dataset was used to develop a direct correlation between the structural features of polymers, as encoded into the zeroth-order and first-order connectivity indices, and the thermal conductivity at room temperature. This correlation is a first step towards quantifying the relationships between the structures and thermal conductivities of amorphous polymers. Its value does not lie in its absolute accuracy but in the clues it provides on the structural factors determining thermal conductivity.

The thermal conductivity is an intensive property (Section 2.C). In other words, its value is independent of the size of the system being considered. Consequently, the intensive (ξ-type) connectivity indices, which are defined by Equation 2.8 as the corresponding extensive (χ-type) connectivity indices divided by the number N of vertices in the hydrogen-suppressed graph (i.e., $\xi \equiv \chi/N$), were used in the correlation for $\lambda(298K)$.

Vibrational modes perpendicular to the chain backbone are only weakly coupled across neighboring chains by the various types of nonbonded interchain interactions. Heat transfer via backbone vibrations is therefore much more effective than heat transfer via vibrational modes perpendicular to the backbone. λ is therefore a locally anisotropic property, as defined in Section 2.D. It is necessary to separate the connectivity indices into backbone and side group components because of this local anisotropy.

The correlation coefficient between $\lambda(exp)$ and $^1\xi^{BB}$ is 0.7271. The correlation coefficient between $\lambda(exp)$ and $^1\xi^{SG}$ is -0.6934. There is, thus, a weak but discernible positive correlation [increasing $\lambda(exp)$] with increasing $^1\xi^{BB}$, and a negative correlation [decreasing $\lambda(exp)$] of comparable magnitude with increasing $^1\xi^{SG}$. These two correlations indicate that the observed $\lambda(exp)$ values depend on the relative sizes of the backbone and side group portions of the repeat unit, albeit only weakly because of the overall (global) isotropic morphology of unoriented amorphous polymers. On the other hand, when the backbone and side group connectivity indices are combined into $^1\xi \equiv (^1\xi^{BB} + ^1\xi^{SG})$, the correlation coefficient is only 0.4900. Combination of these two components of the index therefore causes the mutual cancellation of a major portion of the correlations of opposite sign with the two components, masking a qualitatively significant relationship. The same situation also holds if any one of the other three zeroth-order and first-order connectivity indices is used instead of $^1\xi$.

The correction term ξ_{NOH}, which is defined by Equation 14.8, was combined with $^1\xi^{BB}$ in a linear regression relationship, to give the best possible direct fit for the twenty $\lambda(exp)$ values in the dataset. In Equation 14.8, N_N, N_O and N_H denote the total numbers of nitrogen, oxygen and hydrogen atoms, respectively, in the repeat unit.

$$\xi_{NOH} \equiv \frac{(N_N + N_O - 0.125N_H)}{N} \tag{14.8}$$

The final correlation (Equation 14.9) has a standard deviation of 0.0174, which is 10.2% of the average $\lambda(exp)$ of 0.1701 for the dataset. The correlation coefficient is 0.8782, so that Equation 14.9 accounts for 77.1% of the variation of the $\lambda(exp)$ values in the dataset. The results are summarized in Table 14.1 and shown in Figure 14.2.

$$\lambda(298K) \approx 0.135614 + 0.126611 \cdot {}^1\xi^{BB} + 0.108563 \cdot \xi_{NOH} \tag{14.9}$$

As discussed above, the main factor determining $\lambda(298K)$ is the presence of strongly coupled backbone vibrations, which are very roughly quantified by $^1\xi^{BB}$.

Table 14.1. Experimental thermal conductivities λ(exp) of 20 amorphous or nearly amorphous polymers at room temperature (298K), the quantities used in a direct correlation, and the λ(fit) values given by a linear regression with three adjustable parameters. λ is in J/(K·meter·sec).

Polymer	λ(exp)	$^1\xi$	$^1\xi$BB	$^1\xi$SG	ξ_{NOH}	λ(fit)
Poly(p-chloro styrene)	0.116	0.4845	0.0907	0.3937	-0.0972	0.137
Polyisobutylene	0.130	0.4268	0.1768	0.2500	-0.2500	0.131
Poly(vinylidene fluoride)	0.130	0.4268	0.1768	0.2500	-0.0625	0.151
Polychlorotrifluoroethylene	0.132	0.4167	0.0833	0.3333	0.0000	0.146
Polyisoprene	0.134	0.4788	0.3633	0.1155	-0.2000	0.160
Polystyrene	0.142	0.4958	0.1021	0.3937	-0.1250	0.135
Polypropylene	0.146[a]	0.4646	0.2721	0.1925	-0.2500	0.143
Poly(N-vinyl carbazole)	0.155	0.4966	0.0544	0.4422	-0.0250	0.140
Poly(vinyl acetate)	0.159	0.4646	0.1361	0.3285	0.2083	0.175
Poly(dimethyl siloxane)	0.163	0.4268	0.1768	0.2500	0.0625	0.165
Poly(vinyl chloride)	0.168	0.4646	0.2722	0.1925	-0.1250	0.157
Phenoxy resin	0.176	0.4787	0.4036	0.0751	0.0238	0.189
Polychloroprene	0.192	0.4788	0.3633	0.1155	-0.1250	0.168
Poly(methyl methacrylate)	0.193	0.4555	0.1010	0.3545	0.1429	0.164
Bisphenol-A polycarbonate	0.193	0.4765	0.3935	0.0830	0.0658	0.193
Polyoxyethylene	0.205	0.5000	0.5000	0.0000	0.1667	0.217
Poly(2,2,2'-trimethylhexamethylene terephthalamide)	0.210	0.4701	0.3400	0.1301	0.0476	0.184
Amorphous sulfur	0.210	0.5000	0.5000	0.0000	0.0000	0.199
Poly(ethylene terephthalate)	0.218	0.4836	0.4012	0.0825	0.2143	0.210
Polyoxymethylene	0.230[b]	0.5000	0.5000	0.0000	0.3750	0.240

[a] The λ(exp) value (0.146) listed for polypropylene is the average of two published values (0.120 and 0.172).

[b] A much lower alternative value (0.168) is also quoted for polyoxymethylene in the literature.

The positive contribution of N_N and N_O and the negative contribution of N_H to λ(298K) can be tentatively ascribed to the fact that nitrogen and oxygen atoms are usually found in polar moieties, while most hydrogen atoms are attached to carbon atoms. The presence of nitrogen and/or oxygen atoms in the repeat unit usually increases the cohesive interchain interactions, so that some of the vibrational modes perpendicular to the chain backbone become more strongly coupled across the neighboring chains than similar modes in nonpolar polymers.

In concluding this chapter, we must reemphasize that, while quantitative structure-property relationships for the thermal conductivities of amorphous polymers are useful indicators of the general trends as a function of the polymer structure, they must nonetheless be used with some caution in practical applications, since factors such as crystallinity, orientation, additives and/or impurities can modify the thermal conductivity by a very considerable amount, and differences among measurement methods introduce further uncertainties. The demand for and the availability of thermal conductivity data for polymers were discussed in a concise review article [17] which is recommended as an indicator of the need for caution when utilizing either published experimental values or predicted values of this property in practical applications.

References and Notes for Chapter 14

1. *Thermal Conductivity - Nonmetallic Solids*, edited by Y. S. Touloukian, R. W. Powell, C. Y. Ho and P. G. Klemens, *Thermophysical Properties of Matter*, Volume 2, IFI/Plenum Data Corporation, New York (1970).

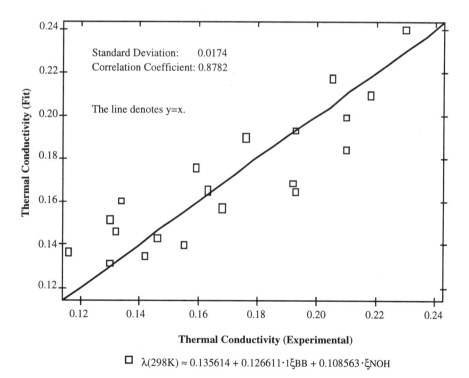

$$\lambda(298K) \approx 0.135614 + 0.126611 \cdot 1\xi_{BB} + 0.108563 \cdot \xi_{NOH}$$

Figure 14.2. Direct correlation with structural features, for the observed thermal conductivities $\lambda(298K)$ of 20 polymers at room temperature. λ is in J/(K·meter·sec).

2. E. V. Thompson, article titled *"Thermal Properties"*, in *Encyclopedia of Polymer Science and Engineering*, *16*, Wiley-Interscience, New York (1989).

3. D. Hands, *Rubber Chemistry and Technology*, *50*, 480-522 (1977).

4. D. Hands, K. Lane and R. P. Sheldon, *J. Polym. Sci., Symp. Ser.*, *42*, 717-726 (1973).

5. C. L. Choy, *Polymer*, *18*, 984-1004 (1977).

6. D. W. van Krevelen, *Properties of Polymers*, second edition, Elsevier, Amsterdam (1976). Thermal conductivity is discussed in Chapter 17, and the effect of orientation is discussed in Chapter 14, of this book.

7. J. G. Hust, in *Materials at Low Temperatures*, edited by R. P. Reed and A. F. Clark, American Society for Metals, Metals Park, Ohio (1983), Chapter 4.

8. Y. K. Godovsky, *Thermophysical Properties of Polymers*, Springer-Verlag, Berlin (1992).

9. C. Rauwendaal, *Polymer Extrusion*, second edition, Hanser Publishers, Munich (1990).

10. G. Menges and P. Mohren, *How to Make Injection Molds*, Hanser Publishers, Munich (1986).

11. J. L. Throne, *Thermoforming*, Hanser Publishers, Munich (1987).

12. L. J. Gibson and M. F. Ashby, *Cellular Solids*, Pergamon Press, New York (1988).

13. S. Saeki, M. Tsubokawa and T. Yamaguchi, *Polymer*, *31*, 1919-1924 (1990).

14. D. M. Bigg, *Polymer Composites*, *7*, 125-140 (1986).

15. R. G. Ross, P. Andersson, B. Sundqvist and G. Baeckstroem, *Rep. Prog. Phys.*, *7*, 1347-1402 (1984).

16. *Plastics: A Desk-Top Data Bank*, International Plastics Selector, Inc., San Diego, California (1980).

17. M. J. Majurey, *Plastics and Rubber International*, *2*, 111-114 (1977).

CHAPTER 15

TRANSPORT OF SMALL PENETRANT MOLECULES

15.A. Background Information

15.A.1. Definitions and Major Industrial Applications

The transport of penetrant molecules through polymers [1,2] is important in many areas of technology. The *permeability* P is the most important quantifier of the transport of penetrant molecules through polymers. The permeability of a polymer to a given type of penetrant molecule is defined as its permittivity to the transport of penetrant molecules of that type, expressed as the quantity Q of penetrant molecules going through a specimen of given surface area (A) and thickness (t) during a fixed amount of time (τ), when there is a partial pressure difference of Δp between the two surfaces of the specimen (see Equation 15.1). There are many different units for permeability [3]. Units of (cc·mil)/[day·(100 inches^2)·atm)], often referred to as "Dow Units" (DU), will be used in this chapter, where "mil" denotes one thousandth of an inch.

$$P \equiv \frac{Q \cdot t}{A \cdot \tau \cdot \Delta p} \tag{15.1}$$

The *barrier performance* [4] of a polymer can be defined as its resistance to the transport of penetrant molecules, i.e., as the inverse of its permeability. A lower value of P therefore indicates better performance as a barrier material.

The *selectivity* of a polymer between two types of molecules is the ratio of its permeabilities to those molecules. For example, P_{O2}/P_{N2} is the selectivity between oxygen and nitrogen. The selectivity is therefore dimensionless. Since the barrier and the selectivity are both defined in terms of the permeability, they can be considered as two sides of the same coin. Both of these properties are determined by the same types of transport phenomena [5,6].

Transport properties play a crucial role in many different types of technological applications of polymers. In particular, the key performance criteria in two types of industrially important polymeric systems are defined directly in terms of the transport properties:

1. *Barrier plastics* [4,7] have a high resistance to the permeation of gases, flavor-aroma molecules and/or liquids, depending on the application. They are used in food and beverage packaging applications, as well as in other applications requiring protective packaging.

2. *Separation membranes* [5,6,8] are used to purify mixtures of gases or liquids flowing through them. A desirable separation membrane material has the following two key characteristics:

    (a) A high selectivity between the components of the mixture flowing through them.

403

(b) A permeability which is sufficiently high to allow the recovery of a product of the desired purity within a reasonable amount of time.

Other areas of technology where the transport of small molecules through polymers plays a key role include foams (where small molecules are used as blowing agents for foam expansion [9-11] and any gas trapped in the cells of a closed-cell foam affects key properties such as the thermal conductivity [12]), plasticization [13,14], and the removal of process solvents, residual monomers or other impurities by techniques such as supercritical fluid extraction [15,16].

15.A.2.   The Solution-Diffusion Mechanism

The permeation of small molecules through polymers usually occurs by the *solution-diffusion mechanism*, which has two key steps. The penetrant molecule is first "sorbed" by the polymer, i.e., dissolves in the polymer. It then crosses the specimen via a succession of diffusive "jumps". P is therefore equal to the product of the *diffusivity* (or *diffusion coefficient*) D and *solubility* S:

$$P = D \cdot S \tag{15.2}$$

According to the *principle of microscopic reversibility*, a given penetrant molecule is equally likely to make a diffusive jump towards either surface of a specimen at any given instant. A net diffusion and thus permeation of molecules from one surface to the other surface only occurs if there is a difference in the amount (expressed in terms of concentration, partial pressure, fugacity, or chemical potential, in different contexts) of the penetrant molecules between the two surfaces. Such a concentration, partial pressure, fugacity or chemical potential gradient causes the net transport of the penetrant molecules towards the surface which is at the low end of the gradient.

We will only deal with the permeability in detail in this chapter, because it is the quantity of most direct interest in applications. In developing a more fundamental theoretical understanding of transport, however, it will be crucial to consider the diffusivity and the solubility separately. Many of the shortcomings of simple structure-property relationships for the permeability and selectivity may possibly be overcome by a more fundamental understanding, which may therefore also be useful in future refinements and practical applications of the correlative schemes.

It is worth keeping in mind that there are many types of systems in which permeation occurs by alternative mechanisms. An obvious example is a beverage container with a large hole, where the beverage leaks rapidly through the hole without having to be sorbed into the walls of the container. At the other extreme, the permeabilities of certain types of extremely densely crosslinked membranes [17,18] cannot be explained in terms of the solution-diffusion mechanism either. See Bitter [6] for a comparative evaluation of transport mechanisms.

When the solution-diffusion mechanism is applicable, permeation can usually be treated as an activated process, as described by Equation 15.3. The *activation energy* E_P changes as the polymer goes through major transitions such as the glass transition or melting. P_0 is a pre-exponential factor which has the same units as $P(T)$.

$$P(T) \approx P_0 \cdot \exp\left(-\frac{E_P}{RT}\right) \tag{15.3}$$

There are many empirical correlations for the activation energy, as well as some theoretical treatments [19-21]. These treatments usually attempt to relate the activation energy to quantities such as the size of the penetrant molecule (usually expressed in terms of a penetrant diameter or the square of this diameter, and sometimes corrected for the penetrant shape if it deviates significantly from spherical), the glass transition temperature of the polymer, and whether the polymer is glassy or rubbery at the temperature of interest. The diffusivity and solubility components of permeability are usually treated separately in such attempts, as described below.

The solubility and diffusivity can each be treated separately as activated processes with relationships similar to Equation 15.3. The activation parameter for solution is called the *Gibbs free energy for sorption* (ΔG_s), and consists of the enthalpy of sorption(ΔH_s) and the entropy of sorption (ΔS_s), combined as shown in Equation 4.7, giving Equation 15.4 as the fundamental thermodynamic relationship for sorption. Note that ΔS_s factors out into the pre-exponential factor (S_0), leaving ΔH_s as the key factor determining the temperature dependence.

$$S(T) = \exp\left(\frac{-\Delta G_s}{RT}\right) \equiv \exp\left(\frac{\Delta S_s}{R}\right) \cdot \exp\left(\frac{-\Delta H_s}{RT}\right) \equiv S_0 \cdot \exp\left(\frac{-\Delta H_s}{RT}\right) \tag{15.4}$$

The thermodynamic functions of the solubilities of many gases in molecular liquids at room temperature have been tabulated [22]. The enthalpy of sorption is negative (exothermic) if the sorption energy exceeds the energy needed to make a hole of molecular size in the polymer or molecular liquid, and positive (endothermic) otherwise. In rough empirical correlations [21], S and ΔH_s are usually related to the boiling temperature, critical temperature, or Lennard-Jones 6-12 potential energy parameter of the gas molecule. ΔH_s can also be modeled atomistically [23-25], and by statistical thermodynamic equation-of-state theories (Section 3.E and Ref. [26]).

The solubility of a gas in a polymer is defined as the sorbed gas concentration (C) divided by the penetrant pressure (p). We have used the capital letter C for gas concentration, to distinguish it from the lower-case letter c used for the concentration of a polymer dissolved in a solvent.

$$S = \frac{C}{p} \tag{15.5}$$

Another quantity which is often encountered in discussions of gas solubility in polymers is Henry's law coefficient of sorption (k_D). For rubbery and molten polymers, k_D is defined very simply as the low-pressure limit ($p \to 0$) of the solubility where C varies almost linearly with p.

For glassy polymers, "dual mode" sorption isotherms, of the general form shown in Equation 15.6, are used in fits to experimental data. In this equation, k_D is the equilibrium liquid limit of the solubility, while the second term represents the effect of the extra unoccupied volume "frozen into" a glassy polymer because of the reduction of chain segment mobility below T_g. C_H' is the Langmuir (frozen-in hole) sorption capacity and can depend strongly on the thermal history for some polymers, while b is the "affinity coefficient" for the Langmuir sorption isotherm.

$$C = k_D \cdot p + \frac{C_H' \cdot b \cdot p}{(1 + b \cdot p)} \qquad (15.6)$$

Figures 15.1 and 15.2 show how the general shapes of the sorption isotherms of polymers are affected by the T_g and the thermal history of the polymer, the measurement temperature, and whether the polymer-penetrant interactions are strong enough for sorption to cause plasticization.

The activation parameter for diffusion (see Equation 15.7) is called the *activation energy for diffusion*. A particularly promising recent advance in modeling diffusion involves the use of atomistic simulations combined with transition state theory [27,28].

$$D(T) = D_0 \cdot \exp\left(-\frac{E_D}{RT}\right) \qquad (15.7)$$

The sizes and shapes of the penetrant molecules, and their solubilities, are the key properties determining their relative permeabilities in a given polymer, i.e., the selectivities of polymers between them, when permeation occurs by the solution-diffusion mechanism. For example, the activation energy for diffusion typically increases with increasing size of penetrant molecules of identical shape permeating through a given polymer.

Three factors play the key role in determining the relative permeabilities of different polymers to a given penetrant molecule permeating by the solution-diffusion mechanism. The "free volume" available for the penetrant molecule to traverse the polymer plays a major role, especially in the diffusivity. The cohesive forces between the polymer chains (i.e., how tightly the chains are held together) are also crucial. Finally, the solubility can be affected very significantly by the strength of the interactions between the penetrant molecule and the structural units in the polymer chains, i.e., by the "compatibility" of the polymer and the penetrant with each other.

Estimation of P(T) at "room temperature" (defined by different authors to have different values in the range of 293K≤T≤303K) is of greatest interest, and we will focus on this problem. For simplicity, P(T) at room temperature will be denoted as P in the remainder of this chapter.

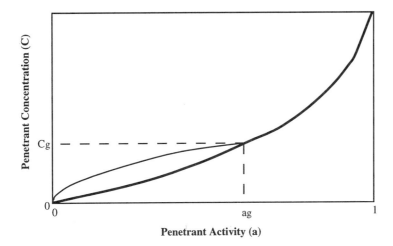

Penetrant Activity (a)

Figure 15.1. Types of sorption isotherms in polymers. The thick line represents sorption into rubbery or molten polymers. The thin line represents sorption into glassy polymers. C_g and a_g are the values of C and a at the plasticization-induced glass transition which may be caused by the sorption of enough CO_2 or another plasticizing penetrant to lower the T_g of a glassy polymer below the measurement temperature. For non-plasticizing penetrants such as O_2 and N_2, and for polymers whose T_g is far above the measurement temperature, the amount of gas sorbed is insufficient to reach C_g and the polymer remains glassy.

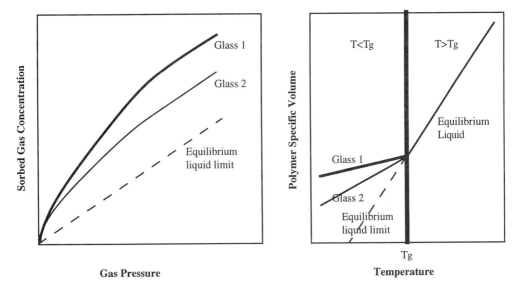

Figure 15.2. Effects of nonequilibrium nature of glass structure on gas sorption behavior (left), and volumetric behavior (right) of polymers below T_g. Glass 1 and Glass 2 are specimens of the same polymer, but subjected to different thermal histories. For example, Glass 1 may have been quenched from the melt very rapidly, while Glass 2 may either have been cooled slowly or subjected to volumetric relaxation via annealing in the glassy state.

15.A.3.   Theories, Simulations and Empirical Correlations

Empirical correlations [21] and computationally-intensive large-scale atomistic simulations [27-37] have been mentioned above as opposite extremes in the spectrum of methods used to treat diffusion phenomena.   In addition, two general types of phenomenological theories (which are much more sophisticated than empirical correlations, but much more coarse-grained than atomistic simulation methods) have been developed to treat diffusion:

1. "Free volume" theories emphasize the amount of "empty space" [38-40] available for diffusion. The theory of Vrentas and Duda [41-48] is an especially elaborate example of such theories.
2. The "statistical mechanical" theories, such as the theory of Pace and Datyner [49-54], consider both E_{coh} and the average interchain distance in idealized models of penetrants traveling through hypothetical "tubular regions" of the polymer.   Interchain cohesion and the amount of "free volume" are of course partially related, since there is usually less empty space between chains held together by very strong interchain interactions.

Recently, Bitter [6] provided an interesting new phenomenological treatment of diffusion, which can be especially useful in calculating mixture diffusivities in polymers.

Simple empirical correlations can be used to provide rough estimates of the solubility [21], while detailed atomistic simulations [23-25] can be used to provide deeper physical insight and hold the potential for greatly improved accuracy in the future.   Flory-Huggins theory and its extensions [6,55] provide the most common phenomenological treatments of solubility.   The use of statistical thermodynamic equations-of-state, which have the advantage of providing correct pressure-volume-temperature relationships for pure components, is also becoming more common [26,56-64] as these methods are improved and refined.

Among the empirical techniques for estimating permeability [21], a correlation based on specific free volume [65] has been found to be quite useful.   Direct correlations with connectivity indices have also been published [66,67].

The most commonly used empirical method for estimating the permeability, however, is the "permachor" method of Salame [21,68-72].   An additive property, which is called the *permachor* and represented by the Greek letter π, is defined, and P is expressed in terms of π.   The values of π are obtained empirically, and are assumed to quantify the key physical factors believed to determine the permeabilities of polymers, namely their cohesive energy densities and their fractional free volumes.   The P's (in DU) of a large number of polymers to O_2, N_2 and CO_2 have been found to correlate fairly well with the following empirical equations in terms of π [71]:

$$P_{O2} \approx 8850 \cdot \exp(-0.112 \cdot \pi) \tag{15.8}$$

$$P_{N2} \approx 3000 \cdot \exp(-0.121 \cdot \pi) \tag{15.9}$$

$$P_{CO2} \approx 55100 \cdot \exp(-0.122 \cdot \pi) \tag{15.10}$$

Equations 15.8-15.10 are useful and widely applied, but suffer from the common weakness of all group contribution methods, i.e., the lack of group contributions for many structural units.

Equations 15.8-15.10 can be used to estimate the selectivities of polymers between oxygen, nitrogen and carbon dioxide, in terms of the value of π. For example, the division of Equation 15.8 by Equation 15.9 gives Equation 15.11 for the selectivity between oxygen and nitrogen.

$$\frac{P_{O2}}{P_{N2}} \approx 2.95 \cdot \exp(0.009 \cdot \pi) \tag{15.11}$$

According to equations 15.8, 15.9 and 15.11, as π increases, the permeability decreases and the selectivity increases simultaneously. Consequently, *as a very rough first approximation,* all polymers with the same value of π should have equal selectivities as well as equal permeabilities.

On the other hand, the experimental selectivities of polymers with equal values π (and thus of P_{O2}) often differ significantly, showing that the use of a single parameter only provides a partial treatment of the transport of small penetrant molecules through polymers. The best separation membranes combine high selectivity and reasonably high permeability, so that polymers deviating significantly from a simple coupling of the permeability and selectivity are often the materials of choice in the development of new membranes. The treatment of transport phenomena must therefore be refined, to allow the predicted selectivity more independence than a simple lock step variation with the permeability in terms of the value of a single parameter.

A new correlation will be developed for P_{O2} in Section 15.B, to enable its prediction at room temperature without having to rely on the availability of group contributions. This correlation will then be used to estimate P_{N2}, P_{CO2}, and hence also the selectivities between O_2, N_2 and CO_2, without being limited by the lack of group contributions. These estimates will, however, be subject to the limitations of using a single parameter to estimate both the permeability and the selectivity, as described above.

An approach to permeability, similar to ours, was published after the completion of our work, by Jia and Xu [73]. These authors correlated the natural logarithm of the permeability to the ratio $[V(T) - V(0)]/E_{coh}$ encompassing the effects of both the expansion volume and cohesive forces.

15.B. Correlations for the Permeability at Room Temperature

Data for 60 polymers, gathered by combining the information provided in several different sources [3,5,68-70,72,74-80], were used to develop the correlation for P_{O2} at room temperature. The structures of some of the more exotic polymers from this dataset are depicted in Figure 15.3.

The permeabilities of polar polymers, and especially of hydrogen-bonding polymers, are often quite sensitive to humidity. Whenever possible, permeabilities measured at the lowest relative humidity were therefore preferred for such polymers.

Data on isotropic amorphous specimens were used whenever available. Some polymers in the dataset are semicrystalline. If an estimate of the crystallinity was available, an empirical correction [70] in terms of the amorphous fraction α, which is given by Equation 15.12, was made for semicrystalline polymers to estimate the permeabilities of their amorphous phases. Crystalline phases of polymers are more dense than their amorphous phases and generally impermeable [81], with very rare exceptions such as poly(4-methyl-1-pentene) [82] whose crystalline regions are packed inefficiently and hence have a lower density than the amorphous regions. Equation 15.12 thus accounts for the usual reduction of permeability with increasing percent crystallinity. The power of 2.1 for α in the denominator is stronger than the power of 2 to be expected from the $1/\alpha$ dependence of both the solubility and the diffusivity because of the exclusion of the gas molecules from the crystalline volume fraction. The dependence of P on α is slightly stronger because of the "tortuosity" caused by the crystallites in the pathway of a penetrant molecule.

$$P_{O2}(\text{amorphous phase}) \approx \frac{P_{O2}(\text{semicrystalline polymer})}{\alpha^{2.1}} \qquad (15.12)$$

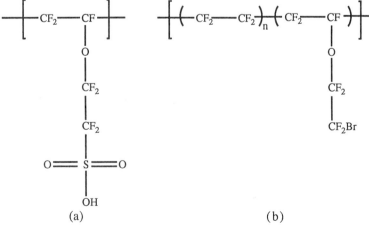

(a)   (b)

Figure 15.3. Schematic illustration of the repeat units of some polymers from the dataset used in developing a correlation for the oxygen permeability. (a) PVPFSA. (b) Poly[TFE(n)/BEVE].

Natural logarithms (ln) of the experimental oxygen permeabilities were taken. Linear correlations were sought for $\ln(P_{O2})$. Linear regressions were performed with weight factors of $abs[\ln(P_{O2})]$, where "abs" is the absolute value. For example, $P_{O2}(exp) \approx 0.004$ DU for poly(vinyl alcohol) in the amorphous limit, $\ln(P_{O2}) = -5.5215$, and the weight factor used in the linear regressions was $abs(-5.5215) = 5.5215$. On the other hand, $P_{O2} \approx 9.0$ DU for amorphous poly(ethylene terephthalate), so that $\ln(P_{O2}) \approx 2.1972$, and the weight factor for this polymer was $abs(2.1972) = 2.1972$.

As a very rough first approximation, the cohesive energy density E_{coh1}/V, where E_{coh1} is the Fedors-type E_{coh} discussed in Section 5.B, has a correlation coefficient of -0.8837 with $\ln(P_{O2})$. There is, therefore, a distinct, albeit not strong, relationship between increasing cohesive energy density and decreasing permeability. When the quantities V/V_w and N_{rot}/N are also included in the regression relationship, a somewhat improved correlation is found. (V_w is the van der Waals volume, N_{rot} is the rotational degrees of freedom parameter defined in Chapter 4, and N is the number of vertices in the hydrogen-suppressed graph of the repeat unit.) Most of the E_{coh1}, V and V_w values used in the calculations were obtained by using correlations in terms of connectivity indices. For a few of the polymers, E_{coh1}, V and V_w values obtained by combining group contributions with connectivity indices, as will be discussed in Chapter 17, were used to expedite the calculations.

Equation 15.13 has a standard deviation of 0.9841 and a correlation coefficient of 0.9091. The most important term in Equation 15.13 is the term proportional to E_{coh1}/V. The small regression coefficient multiplying E_{coh1}/V is an artifact of the large numerical value of E_{coh1}/V when expressed in units of J/cc, compared with the much smaller magnitudes of the dimensionless quantities V/V_w and N_{rot}/N.

$$\ln(P_{O2}) \approx 4.073699 - 0.017905\frac{E_{coh1}}{V} + 6.622610\frac{V}{V_w} - 2.666180\frac{N_{rot}}{N} \qquad (15.13)$$

A correction term N_{per} was defined by equation 15.14 and used to improve the correlation.

$$N_{per} \equiv 2N_{C=C} - 14N_{(BB\ ester)} + 5x_4' - 7N_{hheq\delta} - 6N_{cyanideeq\delta} - 12N_{hb,ar} \qquad (15.14)$$

$N_{C=C}$ was defined in Section 3.B.1.

$N_{(BB\ ester)}$ is the number of ester (-COO-) groups in the backbone of the repeat unit.

The structural parameter x_4' is very similar to the parameter x_4 introduced in Section 6.B.2. One difference between them is that, while x_4 is calculated by taking only substituents attached to rigid backbone rings into account, the x_4' parameter is calculated by taking substituents attached to all backbone rings into account. Another difference is that two or more substituents on a distinct ring can all be located on the same path along the chain backbone, and still count in x_4'. For example, $x_4=0$, but $x_4'=2$, for the ring in Figure 6.9(f).

$N_{hheq\delta}$ is the total number of heavy halogen (Cl and Br) atoms, and $N_{cyanideeq\delta}$ is the total number of cyanide ($-C\equiv N$) groups, attached to carbon atoms which have $\delta=\delta^v$ (i.e., carbon atoms which are singly bonded to all of their neighbors). Cl or Br atoms or cyanide groups attached to other types of carbon atoms which have $\delta\neq\delta^v$, such as carbon atoms in aromatic rings or carbon atoms with a double bond to one of their neighbors, are not counted.

$N_{hb,ar}$ corrects for the effects of strongly hydrogen-bonding moieties, including an apparent synergistic reduction of P_{O2} when aromatic rings are present simultaneously with strongly hydrogen-bonding moieties. It is calculated by using the following set of rules:

1. Ordinary (i.e., alcohol-type) hydroxyl (-OH) groups contribute 1 to $N_{hb,ar}$ if aromatic rings are also present in the repeat unit, but do not contribute anything to $N_{hb,ar}$ if there are no aromatic rings in the repeat unit.

2. The -OH groups in carboxylic acid (-COOH) and sulfonic acid ($-SO_3H$) moieties do not contribute to $N_{hb,ar}$.

3. Each aromatic ring contributes 2 to $N_{hb,ar}$ if the repeat unit also contains amide, urea and/or hydroxyl group(s).

4. Aromatic rings in polymers not containing either amide, urea or hydroxyl groups do not contribute to $N_{hb,ar}$.

The final correlation is expressed in terms of a quantity, the "newchor" v, defined by Equation 15.15. Equation 15.16 has a standard deviation of 0.7159. This standard deviation indicates that approximately two thirds of the fitted P_{O2} values are within a factor of $\exp(0.7159)\approx 2$ of the experimental values, i.e., between $0.5\cdot P_{O2}(exp)$ and $2\cdot P_{O2}(exp)$. The correlation coefficient of Equation 15.16 is 0.9685, indicating that this equation accounts for 93.8% of the variation of $\ln(P_{O2})$ in the dataset.

$$v \equiv \frac{E_{coh1}}{V} - 196\frac{V}{V_w} + 110\frac{N_{rot}}{N} - 57\frac{N_{per}}{N} \tag{15.15}$$

$$\ln(P_{O2}) \approx 8.515520 - 0.017622\cdot v \tag{15.16}$$

The exponential of both sides of Equation 15.16 can be taken, to express P_{O2} by Equation 15.17 which is formally analogous to Equation 15.8.

$$P_{O2} \approx 4991.6\cdot\exp(-0.017622\cdot v) \tag{15.17}$$

The results of these calculations are listed in Table 15.1 and shown in figures 15.4 and 15.5.

As an example of these calculations, consider phenoxy resin [Figure 2.4(h)]. The repeat unit of this polymer simultaneously has a hydroxyl group and two aromatic rings, so that $N_{hb,ar}=5$ according to the rules listed above. Since it does not have any of the other structural features included in Equation 15.14, $N_{per}=-60$. In addition, $N=21$, $N_{rot}=11$, $E_{coh1}\approx120878$ J/mole, $V\approx251.7$ cc/mole and $V_w\approx161.58$ cc/mole. When these quantities are inserted into Equation 15.15, it is found that $v\approx395.4$, so that equations 15.16 and 15.17 give $\ln(P_{O2})\approx1.5477$ and $P_{O2}\approx4.7$ DU, in reasonable agreement (within a factor of two) of $P_{O2}(exp)\approx7.0$ DU.

An advantage of equations 15.14-15.17 is that they can be used to obtain a rough (order of magnitude) estimate of P_{O2} for all polymers built from the elements (C, N, O, H, F, Si, S, Cl and Br) of interest in this work, without being limited by the lack of group contributions.

Comparison of Equation 15.17 with equations 15.8-15.10 leads to two very rough equations:

$$P_{N2} \approx 1692.1 \cdot \exp(-0.019038 \cdot v) \tag{15.18}$$

$$P_{CO2} \approx 31077.9 \cdot \exp(-0.019195 \cdot v) \tag{15.19}$$

According to equations 15.8-15.10, the selectivity of a polymer between a pair of gases is a function of π. Similarly, according to equations 15.17-15.19, the selectivity is a function of v. Equations 15.17-15.19 thus provide a first approximation to the selectivity. For example, the selectivities of polymers between oxygen and nitrogen can be estimated as a function of v by combining equations 15.17 and 15.18. The result (Equation 15.20) is formally analogous to Equation 15.11, and subject to the same general types of limitations as Equation 15.11.

$$\frac{P_{O2}}{P_{N2}} \approx 2.95 \cdot \exp(0.001416 \cdot v) \tag{15.20}$$

The limitations of such a simplistic approach, where the permeability and the selectivity are both treated in terms of a single parameter, have already been discussed above.

Table 15.1. Experimental oxygen permeabilities $P_{O2}(exp)$ at room temperature, and the fitted values of P_{O2}, for 60 polymers. Comments concerning the nature of the specimen and the details of the test conditions are made whenever appropriate. $P_{O2}(exp)$ is in Dow Units.

Polymer	$P_{O2}(exp)$	$P_{O2}(fit)$
Poly(vinyl alcohol) *(dry, amorphous limit)*	0.004	0.005
Polyacrylonitrile *(prior to annealing)*	0.10	0.083
Resin C *(proprietary phenoxy-type thermoplastic)*	0.26	0.54

☞ ☞ ☞   TABLE 15.1 IS CONTINUED IN THE NEXT PAGE. ☞ ☞ ☞

Table 15.1. CONTINUED FROM THE PREVIOUS PAGE.

Polymer	P_{O2}(exp)	P_{O2}(fit)
Polymethacrylonitrile *(unknown percent crystallinity)*	0.50	1.99
Resin G *(proprietary phenoxy-type thermoplastic)*	0.65	0.59
Poly(vinylidene chloride) *(extrapolated to amorphous limit)*	1.3	1.0
Poly(hexamethylene isophthalamide) *(dry, amorphous)*	1.8	1.5
Poly(ethylene isophthalate)	2.5	6.6
Poly(ethylene-2,6-naphthalenedicarboxylate)	3.1	12.9
Poly(hexamethylene adipamide) *(dry)*	5.1	6.7
Phenoxy resin	7.0	4.7
Poly(ε-caprolactam) *(dry, amorphous limit)*	7.5	6.7
Poly(ethylene terephthalate) *(amorphous)*	9.0	6.6
Poly(vinyl chloride) *(unplasticized)*	9.5	8.3
Poly(methoxymethyl hexamethylene adipamide)	9.6	14.7
Poly(oxy-2,2-dichloromethyltrimethylene)	12.0	22.5
Poly[thio(*p*-phenylene)]	16.7	52.7
Poly(methyl methacrylate)	17.0	180.3
Poly(11-aminoundecanoic acid)	26.0	52.5
Poly(ether ether ketone)	38.2	71.3
Poly(1,4-cyclohexylidene dimethylene terephthalate)	40.5	27.4
Equimolar random copolymer of oxy-1,4-phenylene-carbonyloxyethylene		
and oxy-1,3-phenylene-carbonyloxyethylene	41.0	36.8
PVPFSA *(see Figure 15.1)*	60.0	12.1
Poly(vinyl acetate)	60.0	89.8
Polyepichlorohydrin	70.0	39.1
Poly[1-cyano-3,3-butane *bis*{4-(2,6-dibromophenyl)}carbonate]	88.5	75.5
Poly(methyl acrylate)	125	66.2
Poly[3,5-(4-phenyl-1,2,4-triazole)-1,4-phenylene-3,5-(4-phenyl-1,2,4-triazole)-1,3-phenylene]		
(Extrapolated from 338K to 298K, by using Equation 15.3 and the measured value of E_p.		
The polymer is named differently in the source [78].)	144	149
Poly[diphenylmethane *bis*(4-phenyl)carbonate]	146	177
Poly(vinyl benzoate)	149	87
Poly[oxy-1,4-phenylene-2,2-propane-1,4-phenylene-oxy-(2-cyano-1,3-phenylene)]	152	124
Poly(*p*-methyl styrene) *(prior to irradiation)*	170	605
Poly[1,1-cyclohexane *bis*{4-(2,6-dibromophenyl)}carbonate]	180	216
Poly(*p*-xylylene) *(extrapolated to amorphous limit)*	206	414

☞ ☞ ☞   TABLE 15.1 IS CONTINUED IN THE NEXT PAGE. ☞ ☞ ☞

Table 15.1. CONTINUED FROM THE PREVIOUS PAGE.

Polymer	P_{O2}(exp)	P_{O2}(fit)
Poly[4,4'-isopropylidene diphenoxy di(4-phenylene)sulfone]	210	88
Poly[2,2-propane *bis*{4-(2,6-dibromophenyl)}carbonate]		
(measured at a temperature of 308K)	227	286
Poly(ethyl methacrylate)	234	252
Bisphenol-A polycarbonate	250	288
Poly[oxy-(2,6-dimethyl-1,4-phenylene)-ethylene-(2,6-dimethyl-1,4-phenylene)		
-oxy-1,4-phenylene-carbonyl-1,4-phenylene]	334	581
Poly(2,3-dimethyl-1,4-butadiene)	349	3049
Poly[2,2-propane *bis*{4-(2,6-dichlorophenyl)}carbonate]		
(measured at a temperature of 308K)	382	191
Poly(vinyl fluoride) *(extrapolated to amorphous limit)*	418	813
Polystyrene	450	365
Polyoxyethylene *(extrapolated to amorphous limit)*	528	662
Polychloroprene	650	466
Polytetrafluoroethylene *(amorphous limit)*	700	1272
Polypropylene *(extrapolated to amorphous limit)*	780	1308
Poly[2,2-hexafluoropropane *bis*(4-phenyl)carbonate]	842	252
Poly[2,2-propane *bis*{4-(2,6-dimethylphenyl)}carbonate]		
(measured at a temperature of 308K)	934	1596
Equimolar random copolymer of tetrafluoroethylene and hexafluoropropylene	1000	1277
Poly[TFE(9)/BEVE] *(see Figure 15.1)*	1020	954
Poly[TFE(5)/BEVE] *(see Figure 15.1)*	1240	817
Polyisobutylene	1500	2456
Polyethylene *(extrapolated to amorphous limit)*	1560	756
Poly[oxy(2,6-dimethyl-1,4-phenylene)]	3000	2075
Poly(1,4-butadiene)	3340	2101
Cis-polyisoprene *(natural rubber)*	4000	2650
Poly(4-methyl-1-pentene)[a]	4000	1320
Poly(vinyl trimethylsilane)	9500	7690
Poly(dimethyl siloxane) *(measured only approximately)*	16700	9143

[a] Poly(4-methyl-1-pentene) is anomalous. Its crystalline phase has a lower density than its amorphous phase and a nonzero permeability to small gas molecules [82], unlike most other polymers whose crystalline phases are of greater density than the amorphous phase and impermeable to small gas molecules.

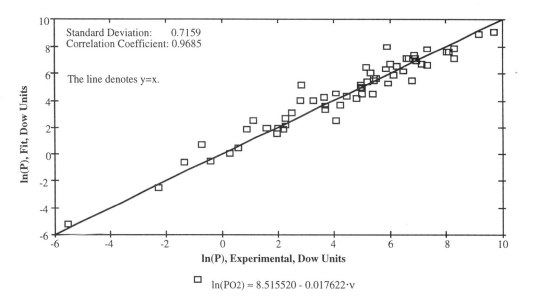

$$\square \quad \ln(PO2) \approx 8.515520 - 0.017622 \cdot v$$

Figure 15.4. Correlation in terms of the "newchor" v, for the natural logarithm (ln) of the oxygen permeabilities of 60 polymers, measured at or near room temperature.

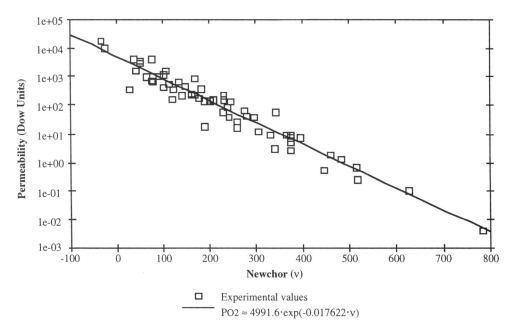

$\square$ Experimental values

—— $PO2 \approx 4991.6 \cdot \exp(-0.017622 \cdot v)$

Figure 15.5. Observed oxygen permeabilities of 60 polymers at or near room temperature, as a function of the "newchor" v. The y-axis is logarithmic.

References and Notes for Chapter 15

1. J. Crank and G. S. Park, *Diffusion in Polymers*, Academic Press, New York (1968).

2. V. Stannett, *J. Membrane Science*, *3*, 97-115 (1978).

3. S. Pauly, in *Polymer Handbook*, edited by J. Brandrup and E. H. Immergut, Wiley, New York, third edition (1989), VI/435-VI/449.

4. *Barrier Polymers and Barrier Structures*, edited by W. J. Koros, *ACS Symposium Series*, *423*, American Chemical Society, Washington, D. C. (1990).

5. W. J. Koros, B. J. Story, S. M. Jordan, K. O'Brien and G. R. Husk, *Polymer Engineering and Science*, *27*, 603-610 (1987).

6. J. G. A. Bitter, *Transport Mechanisms in Membrane Separation Processes*, Plenum Press, New York (1991).

7. L. B. Ryder, *Plastics Engineering*, 41-48 (May 1984).

8. S.-T. Hwang and K. Kammermeyer, *Membranes in Separations*, Robert E. Krieger Publishing Company, Malabar, Florida (1984).

9. W. F. Sullivan and A. K. Thomas, *Cellular Polymers*, *11*, 18-28 (1992).

10. A. Arefmanesh, S. G. Advani and E. E. Michaelides, *Polymer Engineering and Science*, *30*, 1330-1337 (1990).

11. A. Arefmanesh and S. G. Advani, *Rheologica Acta*, *30*, 274-283 (1991).

12. L. J. Gibson and M. F. Ashby, *Cellular Solids*, Pergamon Press, New York (1988).

13. J. K. Sears and J. R. Darby, *The Technology of Plasticizers*, Wiley, New York (1982).

14. A. R. Berens, Ref. [4], Chapter 4.

15. M. A. McHugh and V. J. Krukonis, *Supercritical Fluid Extraction: Principles and Practice*, Butterworth Publishers, Stoneham, Massachusetts (1989).

16. M. A. McHugh and V. J. Krukonis, article titled "*Supercritical Fluids*", *Encyclopedia of Polymer Science and Engineering*, *16*, Wiley-Interscience, New York (1989).

17. R. Y. S. Chen, *ACS Polymer Preprints*, *15 (No. 2)*, 387-394 (1974).

18. H. Yasuda, *J. Membrane Science*, *18*, 273-384 (1984).

19. S. M. Aharoni, *J. Appl. Polym. Sci.*, *23*, 223-228 (1979).

20. A. R. Berens and H. B. Hopfenberg, *J. Membrane Science*, *10*, 283-303 (1982).

21. D. W. van Krevelen, *Properties of Polymers*, third edition, Elsevier, Amsterdam (1990), Chapter 18.

22. E. Wilhelm and R. Battino, *Chemical Reviews*, *73*, 1-9 (1973).

23. J. Bicerano, A. R. K. Ralston and D. J. Moll, ANTEC '94 Preprints, Society of Plastics Engineers, 2105-2109 (1994).

24. A. R. K. Ralston, J. Bicerano and D. J. Moll, to be published.

25. A. A. Gusev and U. W. Suter, *Phys. Rev. A*, *43*, 6488-6494 (1991).

26. J. Bicerano, *Computational Polymer Science*, *2*, 177-201 (1992).

27. A. A. Gusev, S. M. P. Arizzi, U. W. Suter and D. J. Moll, *J. Chem. Phys.*, *99*, 2221-2227 (1993).

28. A. A. Gusev and U. W. Suter, *J. Chem. Phys.*, *99*, 2228-2234 (1993).

29. J. Bicerano, A. F. Burmester, P. T. DeLassus and R. A. Wessling, Ref. [4], Chapter 6.

30. S. Trohalaki, D. Rigby, A. Kloczkowski, J. E. Mark and R. J. Roe, *ACS Polymer Preprints*, *30 (No. 2)*, 23-24 (1989).

31. S. Arizzi and U. W. Suter, *Proceedings of the ACS Division of Polymeric Materials: Science and Engineering*, *61*, 481-485 (1989).

32. E. Smit, M. H. V. Mulder, C. A. Smolders, H. Karrenbeld, J. van Eerden and D. Feil, *J. Membrane Science*, *73*, 247-257 (1992).

33. K. Matsumoto, Y. Minamizaki and P. Xu, *Membrane*, *17*, 395-402 (1992).

34. P. V. Krishna Pant and R. H. Boyd, *Macromolecules*, *25*, 494-495 (1992).

35. P. V. Krishna Pant and R. H. Boyd, *Macromolecules*, *26*, 679-686 (1993).

36. F. Müller-Plathe, S. C. Rogers and W. F. van Gunsteren, *Macromolecules*, *25*, 6722-6724 (1992).

37. S. Trohalaki, A. Kloczkowski, J. E. Mark, R. J. Roe and D. Rigby, *Computational Polymer Science*, *2*, 147-151 (1992).

38. V. V. Volkov, A. V. Gol'danskii, G. G. Durgar'yan, V. A. Onischuk, V. P. Shantorovich and Yu. P. Yampol'skii, *Polymer Science USSR*, *29*, 217-224 (1987).

39. J. G. Victor and J. M. Torkelson, *Macromolecules*, *20*, 2241-2250 (1987).

40. J. G. Victor and J. M. Torkelson, *Macromolecules*, *20*, 2951-2954 (1987).

41. J. S. Vrentas and J. L. Duda, *J. Appl. Polym. Sci.*, *21*, 1715-1728 (1977).

42. J. S. Vrentas, H. T. Liu and J. L. Duda, *J. Appl. Polym. Sci.*, *25*, 1297-1310 (1980).

43. S. T. Ju, J. L. Duda and J. S. Vrentas, *Ind. Eng. Chem. Prod. Res. Dev.*, *20*, 330-335 (1981).

44. J. S. Vrentas, J. L. Duda and H.-C. Ling, *J. Polym. Sci., Polym. Phys. Ed.*, *23*, 275-288 (1985).

45. J. S. Vrentas, J. L. Duda, H.-C. Ling and A.-C. Hou, *J. Polym. Sci., Polym. Phys. Ed.*, *23*, 289-304 (1985).

46. J. S. Vrentas, J. L. Duda and A.-C. Hou, *J. Polym. Sci., Polym. Phys. Ed.*, *23*, 2469-2475 (1985).

47. J. S. Vrentas, J. L. Duda and A.-C. Hou, *J. Appl. Polym. Sci.*, *33*, 2581-2586 (1987).

48. K. Ganesh, R. Nagarajan and J. L. Duda, *Ind. Eng. Chem. Res.*, *31*, 746-755 (1992).

49. R. J. Pace and A. Datyner, *J. Polym. Sci., Polym. Phys. Ed.*, *17*, 437-451 (1979).

50. R. J. Pace and A. Datyner, *J. Polym. Sci., Polym. Phys. Ed.*, *17*, 453-464 (1979).

51. R. J. Pace and A. Datyner, *J. Polym. Sci., Polym. Phys. Ed.*, *17*, 465-476 (1979).

52. R. J. Pace and A. Datyner, *J. Polym. Sci., Polym. Phys. Ed.*, *17*, 1675-1692 (1979).

53. R. J. Pace and A. Datyner, *J. Polym. Sci., Polym. Phys. Ed.*, *17*, 1693-1708 (1979).

54. R. J. Pace and A. Datyner, *J. Polym. Sci., Polym. Phys. Ed.*, *18*, 1103-1124 (1980).

55. P. J. Flory, *Principles of Polymer Chemistry*, Cornell University Press, Ithaca, New York (1953).

56. D. C. Bonner, *Polymer Engineering and Science*, *17*, 65-72 (1977).

57. D. D. Liu and J. M. Prausnitz, *J. Appl. Polym. Sci.*, *24*, 725-733 (1979).

58. M. Ohzono, Y. Iwai and Y. Arai, *J. Chem. Eng. Japan*, *17*, 550-553 (1984).

59. Y. L. Cheng and D. C. Bonner, *J. Polym. Sci., Polym. Phys. Ed.*, *16*, 319-333 (1978).

60. C. G. Panayiotou, *Makromol. Chem.*, *187*, 2867-2882 (1986).

61. E. J. Beckman, R. Konigsveld and R. S. Porter, *Macromolecules*, *23*, 2321-2329 (1990).

62. M. B. Kiszka, M. A. Meilchen and M. A. McHugh, *J. Appl. Polym. Sci.*, 36, 583-597 (1988).

63. D. S. Pope, I. C. Sanchez, W. J. Koros and G. K. Fleming, *Macromolecules*, *24*, 1779-1783 (1991).

64. R. G. Wissinger and M. E. Paulaitis, *Ind. Eng. Chem. Res.*, *30*, 842-851 (1991).

65. W. M. Lee, *Polymer Engineering and Science*, *20*, 65-69 (1980).

66. M. R. Surgi, A. J. Polak and R. C. Sundahl, *J. Polym. Sci., Polym. Chem. Ed.*, *27*, 2761-2776 (1989).

67. G. S. Patil, M. Bora and N. N. Dutta, *J. Membrane Science*, *101*, 145-152 (1995).

68. M. Salame, *ACS Polymer Preprints*, *8*, 137-144 (1967).

69. M. Salame, *Proceedings of Future-PAK '83* (November 28-30), 233-259 (1983).

70. M. Salame, *J. Plastic Film & Sheeting*, *2*, 321-334 (1986).

71. M. Salame, *Polymer Engineering and Science*, *26*, 1543-1546 (1986).

72. M. Salame, in the *Encyclopedia of Packaging Technology*, edited by M. Bakkar, Wiley, New York (1986), 48-54.

73. L. Jia and J. Xu, *Polymer Journal*, *23*, 417-425 (1991).

74. R. R. Light and R. W. Seymour, *Polymer Engineering and Science*, *22*, 857-864 (1982).

75. N. Muruganandam, W. J. Koros and D. R. Paul, *J. Polym. Sci., Polym. Phys. Ed.*, *25*, 1999-2026 (1987).

76. Unpublished data, measured at Dow by various researchers, and provided to the author by A. F. Burmester and D. J. Moll.

77. Compilation of literature data on the permeabilities of polymers, provided by D. J. Moll.

78. B. Gebben, M. H. V. Mulder and C. A. Smolders, *J. Membrane Science*, *46*, 29-41 (1989).

79. R. Greenwood and N. Weir, *Die Macromoleculare Chemie*, *176*, 2041-2051 (1975).

80. Measured by J. E. White, J. W. Ringer, S. L. Brick, C. N. Brown, H. C. Silvis and S. L. Kram (1989).

81. D. H. Weinkauf and D. R. Paul, Ref. [4], Chapter 3.

82. A. C. Puleo, D. R. Paul and P. K. Wong, *Polymer*, *30*, 1357-1366 (1989).

CHAPTER 16

THERMAL STABILITY

16.A. Background Information

16.A.1. Definitions

Polymers are often exposed to high temperatures during processing and/or use. Thermal and thermooxidative stability (TTOS) are therefore among the most important properties of polymers for a wide range of applications. *Thermal stability* is stability against degradation upon exposure to elevated temperatures in an inert environment. Typical inert testing environments include the vacuum and unreactive gases such as nitrogen, helium and argon. *Thermooxidative stability* is stability against degradation upon exposure to elevated temperatures in an oxidizing environment. Air, which contains oxygen molecules, is a typical example of an oxidizing environment.

16.A.2. Measurement of Thermal and Thermooxidative Stability

There are numerous techniques for measuring TTOS. Some techniques use the resistance to deterioration of certain key physical properties at elevated temperatures as their criterion for stability. Other techniques use the resistance to weight loss at elevated temperatures as their major criterion. A material can degrade significantly without losing much weight. For example, chemical bonds could break, and other bonds could form, resulting in a very different material of comparable weight. A substantial amount of loss of the critical physical properties of a material can therefore occur without a concomitant loss of weight. It is, therefore, more desirable to measure the retention of critical physical properties rather than the retention of weight whenever possible, in measuring TTOS.

The most important testing protocol for resistance to deterioration of critical physical properties upon exposure to heat, especially over prolonged periods of time, was designed by the Underwriters Laboratories (UL) [1]. The *UL temperature index* is used to qualify materials for applications (such as use in electrical equipment) regulated by UL. The main purpose of these tests is to ensure that a polymer can be used safely for a prolonged period of time in these applications. A larger value of the UL temperature index implies greater stability, defined in terms of the retention of the key properties which are tested. There are UL temperature indices [1] for the retention of electrical properties, the retention of mechanical properties, and the simultaneous retention of electrical and mechanical properties. These indices are excellent indicators of the maximum continuous use temperature under very low continuous or low intermittent stresses in applications requiring maximum retention of the properties being tested. On the other hand, the tests that need to be performed to determine these indices are extremely laborious and expensive, as well as requiring large amounts of material. These tests therefore demand resources exceeding what is available on a routine basis in industrial research and development. They are thus normally only performed by UL itself, on commercial materials whose certification for specific applications is desired by their manufacturers. A listing of the UL temperature indices of a large number of commercial materials is available [1].

Measuring the retention of weight is not as reliable a method of testing for TTOS as determining the retention of the desired physical properties directly. On the other hand, weight retention measurements are much easier, faster, and less expensive to perform, than direct methods such as the measurement of the UL temperature indices. When used cautiously, the results of weight retention experiments can provide some useful indications of TTOS. A material which loses large amounts of weight under test conditions resembling the requisite processing or operating conditions will definitely not be useful for a given application. A material which loses little or no weight under the same conditions has a better chance of having sufficient TTOS; however, little weight loss does not automatically guarantee high TTOS with regard to the retention of the critical physical properties.

The principal method for measuring weight loss is *thermogravimetric analysis* (TGA) [2]:

1. In *dynamic TGA*, the temperature is increased at a specific (usually constant) rate, normally a few degrees Kelvin per minute, and the weight of the specimen is measured at regular intervals. The American Society for Testing and Materials (ASTM) has published protocols [3] for standardized dynamic TGA testing.

2. In *isothermal TGA*, the specimen is kept at a constant elevated temperature, and its weight is measured at regular intervals, to provide data on the weight loss as a function of time at the temperature of measurement.

Dynamic TGA provides a better indication of short-term TTOS. Isothermal TGA provides a better indication of long-term TTOS. Dynamic TGA is much less time-consuming, and thus more routinely performed, than isothermal TGA. It was once believed that the results of isothermal TGA could be correlated in a simple manner with the results of dynamic TGA, and that isothermal TGA behavior could thus be predicted from dynamic TGA behavior. Such extrapolations have, however, been found to have only very crude qualitative validity, and to fail at a quantitative level. It is possible for two polymers to have similar weight retention curves in dynamic TGA scans, but for one of them to lose weight considerably more rapidly in an isothermal TGA test [4,5].

Kinetic parameters, such as rate (frequency) factors, activation energies, and orders of reaction, can be deduced from TGA measurements [2]. There are many contradictions between different sets of such data, leading some workers to question the meaning and usefulness of these kinetic parameters [6,7]. This state of affairs is partially caused by the specimen dependence of the results of TGA experiments, and the consequent difficulty in interpreting the data and/or comparing data obtained in different laboratories or under different test conditions. For example, the size and shape of a specimen, the number of structural defects in the chains of the resin from which the specimen was made, and the presence of additives (such as plasticizers or stabilizers) and impurities (such as residual monomer molecules or catalysts) can all affect the TGA results. A well-known effect is the increase in the apparent time for thermal degradation in isothermal TGA with increasing specimen thickness, and has been explained [8] by using a transport model which accounts for the observed scaling behavior of degradation time with specimen thickness by incorporating the change in diffusivity resulting from material volatilization during degradation. In addition, the nominal kinetic parameters extrapolated from the results of TGA measurements often fail to take the numerous types of chemical reactions that may be competing during degradation into account in a proper manner. Finally, the use of such kinetic parameters to extrapolate the maximum use temperature or the useful lifetime of a polymer at elevated temperatures is fraught with difficulties of a fundamental nature [9,10].

16.A.3. Mechanisms of Weight Loss During Degradation

Weight loss by the thermal degradation or thermooxidative degradation of a polymer itself (as opposed to the volatilization of small molecules which might have been trapped in the polymeric structure) invariably requires the breakage of chemical bonds. Once chemical bonds start to break, reactive chain ends and other free radicals are created, and degradation can proceed either by *depolymerization* or by *random chain scission* [11,12].

The dominant degradation mechanism may depend significantly on the details of structure and composition, including the types of end groups terminating the polymer chains, and the presence of structural defects, additives or impurities. In any case, loss of weight can only start after the first bond breakage event. The first bond to break will usually be the weakest bond in the polymer. The strength of the weakest bond should therefore play an important role in determining TTOS. This statement is especially accurate for ordinary thermal stability measured in inert (non-oxidizing) environments, where the complications caused by reactions with oxygen do not exist.

Depolymerization (sometimes also referred to as *unzipping*) is the reverse of the building up of polymer chains by polymerization. A stepwise breakdown starting at the chain ends yields monomer molecules. For example, PMMA normally undergoes substantial amounts of depolymerization during thermal degradation.

In *random chain scission*, thermal degradation occurs by the random scission of bonds along the chain backbone. The shorter chain segments resulting from random chain scission continue to degrade by additional occurrences of random scission. Most aromatic heterocyclic polymers, or polymers formed by step-growth polymerization, undergo degradation by random scission.

Oxygen reacts with a polymer, while the vacuum or an inert atmosphere does not react with it. Weight loss is therefore usually (but not always) both faster and greater in an oxidative environment than in vacuum or in an inert atmosphere. A variety of chemical reactions are possible between oxygen and different types of structural units. Significant differences are therefore often found between the orderings of different types of polymers according to relative thermal stability and relative thermooxidative stability [5]. The detailed study of the effects of oxygen on different polymeric structures requires extensive quantum mechanical calculations of the degradation mechanism, by using model molecules. The difficulty of such studies is compounded by the fact that various types of free radical species are likely to be present with significant abundances during degradation.

16.A.4.   Effects of Structure on Thermal and Thermooxidative Stability

16.A.4.a.   Qualitative Summary of Trends

There are many books and review articles which provide extensive discussions of TTOS. For example, Arnold [5], Hergenrother [12], and Wright [13] have provided excellent qualitative analyses of the relationships between the structures and the thermal stabilities of polymers. The following general trends are found:

1. The strength of the covalent chemical bonds in the polymer, and especially of the weakest backbone bond, is the single most important factor in determining the TTOS.

2. Certain types of secondary chemical interactions, such as van der Waals interactions, hydrogen bonding, and the resonance stabilization of aromatic rings, increase the average effective bond strength and/or the intermolecular attraction, and thus stabilize the polymer against thermal and thermooxidative degradation.

3. Chemical crosslinks increase the number of bonds that must be broken in order for the material to exhibit weight loss or deterioration of important properties. Crosslinking hence increases the thermal stability.

4. High average chain molecular weight favors greater TTOS for the same reason as crosslinking.

5. Structural defects and/or irregularities are weak points. They may function as initiation sites for degradation, and thus usually decrease the TTOS. Such weak points include chain ends, branches, chain backbone unsaturation (such as double bonds involving at least one atom of the chain backbone), and carbonyl groups incorporated in polymers (such as polyethylene) where they are not a part of the normal structure.

6. Ordering of polymer chains, and especially crystallinity, usually increases the TTOS.

7. Molecular symmetry, and the resulting regularity of the chemical structure, also usually increases the TTOS.

8. Aromatic functionalities incorporated into the chain backbone increase the TTOS.

9. Aromatic functionalities with fused rings often increase thermal stability more than individual aromatic rings.

The following observations can be made on the basis of these general trends:

1. Strong secondary (nonbonded physical) interactions (i.e., a high cohesive energy density), the presence of aromatic rings and/or fused heterocyclic aromatic rings in the chain backbone, and the presence of crosslinks, both increase T_g and result in a high TTOS. There are, therefore, many similarities between the trends observed for T_g and for TTOS.

2. These similarities of rather limited scope, however, should not be used to reach incorrect generalizations. It is also possible for a polymer to have a very high TTOS and a very low T_g. For example, many high-performance and high-TTOS elastomers and lubricants, such as the silicones [14,15] and many of their carborane-containing variants [16,17], perfluorinated elastomers [18,19], and polyphosphazenes [20], have strong primary chemical bonds and thus have high TTOS. On the other hand, they have weak physical interchain interactions and/or low chain stiffness, imparting them with a T_g well below room temperature, and therefore with elastomeric or lubricating properties instead of structural rigidity at the use temperature.

3. Outstanding TTOS can often only be achieved at the expense of low processability and great difficulty of fabrication into finished parts. Tradeoffs between TTOS and processability (whether in the melt phase or in solution) and ease of fabrication are therefore often required to design a material with an optimum balance of properties. Here are two examples:

(a) Many high-TTOS polymers containing stiff fused aromatic heterocyclic ring moieties are extremely difficult or even impossible to process in the melt phase, and soluble in very few solvents. Consequently, special techniques are required to process them, and they are only used in specialized applications requiring certain extraordinary properties.

(b) *Para*-phenylene moieties located in the chain backbone, which generally increase the TTOS considerably more than *meta*-phenylene or *ortho*-phenylene moieties, also usually result in the lowest solubility.

16.A.4.b.   Quantitative Structure-Property Relationships

Van Krevelen [11] developed a correlation for the *temperature of half decomposition* $T_{d,1/2}$ which he defined as "the temperature at which the loss of weight during pyrolysis (at a constant rate of temperature rise) reaches 50% of its final value". In this correlation, which is given by Equation 16.1, $T_{d,1/2}$ is approximated in terms of the ratio of the *molar thermal decomposition function* $Y_{d,1/2}$ divided by the molecular weight M per repeat unit. The observed values of $T_{d,1/2}$ [11] are listed in Table 16.1 for a number of polymers.

$$T_{d,1/2} \approx \frac{Y_{d,1/2}}{M} \tag{16.1}$$

Table 16.1. Temperatures of half decomposition ($T_{d,1/2}$) of polymers [11], in degrees Kelvin.

Polymer	$T_{d,1/2}$	Polymer	$T_{d,1/2}$
Linear polyethylene	687	Polybutadiene	680
Branched polyethylene	677	Polyisoprene	596
Polypropylene	660	Poly(p-phenylene)	925
Polyisobutylene	621	Polybenzyl	703
Polystyrene	637	Poly(p-xylylene)	715
Poly(m-methyl styrene)	631	Polyoxyethylene	618
Poly(α-methyl styrene)	559	Poly(propylene oxide)	586
Poly(vinyl fluoride)	663	Poly[oxy(2,6-dimethyl-1,4-phenylene)]	753
Poly(vinyl chloride)	543	Poly(ethylene terephthalate)	723
Polytrifluoroethylene	682	Bisphenol-A terephthalate	750
Polychlorotrifluoroethylene	653	Bisphenol-A polycarbonate	750
Polytetrafluoroethylene	782	Poly(hexamethylene adipamide)	693
Poly(vinyl cyclohexane)	642	Poly(ε-caprolactam)	703
Poly(vinyl alcohol)	547	Poly(p-phenylene terephthalamide)	800
Poly(vinyl acetate)	542	Kapton	840
Polyacrylonitrile	723	Poly(m-phenylene-2,5-oxadiazole)	800
Poly(methyl acrylate)	601	Cellulose	600
Poly(methyl methacrylate)	610		

The formal analogy between Equation 16.1 and van Krevelen's expression for T_g (Equation 1.1) should be obvious. $Y_{d,1/2}$ is estimated [11] by using group contributions. These group contributions are not strictly additive, since they incorporate some of the environmental effects which play a role in determining the thermal stability, such as whether or not a phenyl ring is adjacent to another group with which it can have resonance interactions.

$Y_{d,1/2}$ is expressed in K·kg/mole. Since M is in g/mole, a conversion factor of 1000 must be used to obtain $T_{d,1/2}$ in degrees Kelvin when $Y_{d,1/2}$ and M are substituted in Equation 16.1. For example, $Y_{d,1/2}$=66 K·kg/mole and M=104.15 g/mole for polystyrene (Figure 1.1). Equation 16.1, with a unit conversion factor of 1000, then gives $T_{d,1/2} \approx 1000 \cdot 66/104.15 = 634$K, in excellent agreement with the experimental value [10] of 637K.

Van Krevelen [11] also showed that the *temperature of initial decomposition* ($T_{d,0}$, where the weight loss during heating becomes just measurable), the *temperature of maximum decomposition*

($T_{d,max}$, where the rate of weight loss reaches its maximum), and the *average activation energy of decomposition* ($E_{act,d}$, determined from the temperature dependence of the weight loss rate), all correlate with $T_{d,1/2}$. The correlation is weakest for $E_{act,d}$, partly because $E_{act,d}$ sometimes varies during pyrolysis. Equations 16.2, 16.3 and 16.4 (with $T_{d,0}$, $T_{d,1/2}$ and $T_{d,max}$ in degrees Kelvin, and $E_{act,d}$ in 10^3 J/mole), imply that $T_{d,1/2}$ can be used as the key indicator of thermal stability, with the other indicators being roughly estimated from it.

$$T_{d,0} \approx 0.9 \cdot T_{d,1/2} \tag{16.2}$$

$$T_{d,max} \approx T_{d,1/2} \tag{16.3}$$

$$E_{act,d} \approx T_{d,1/2} - 423 \tag{16.4}$$

The contribution of our work to the prediction of the thermal stabilities of polymers is the development a new correlation for $Y_{d,1/2}$, to allow the estimation of this quantity, and hence also of $T_{d,1/2}$, even when the required group contributions are not all available for a polymer. This correlation will be presented in Section 16.B.

16.B.   Correlation for the Molar Thermal Decomposition Function

A dataset of 180 polymers was prepared. The necessary group contributions were available [11,21] to calculate $Y_{d,1/2}$(group) for 140 of these polymers. The $Y_{d,1/2}$(group) values correlated fairly well with the connectivity indices. The best correlation was with the first-order simple connectivity index $^1\chi$. The correlation with the simpler variable N (the number of vertices in the hydrogen-suppressed graph of the repeat unit) was slightly stronger, with a correlation coefficient of 0.9794. The use of the number of hydrogen atoms N_H and the first-order valence connectivity index $^1\chi^v$ along with N resulted in the best possible regression relationship while remaining completely within the framework of the zeroth-order and first-order connectivity indices and not using correction terms. Weight factors of $100/Y_{d,1/2}$(group) were used in the linear regression procedure. Equation 16.5 has a standard deviation of 9.6, and a correlation coefficient of 0.9875, between $Y_{d,1/2}$(group) and $Y_{d,1/2}$(fit).

$$Y_{d,1/2} \approx 7.17 \cdot N - 2.31 \cdot N_H + 12.52 \cdot {}^1\chi^v \tag{16.5}$$

Correction terms were introduced to improve the correlation. These terms were formulated in as general a manner as possible, by considering the trends summarized in Section 16.A.4.a. This procedure allows many of the improvements made in the correlation to carry over to types of structural units not included in the dataset of polymers for which $Y_{d,1/2}$(group) was calculated.

The correction term N_{Yd} was defined by Equation 16.6, where BB denotes "backbone" and SG denotes "side group". The terms entering this equation will be described and examples will be given below. The remarks made in Chapter 6 concerning the structural terms entering the correlation for T_g are also germane for these terms:

1. The large number of terms used below is a result of the complexity of thermal degradation.

2. With a little practice, these terms become very easy to evaluate by inspection of the polymeric repeat unit. (However, this is unnecessary, since the correlations developed in this book have been automated in a software package.)

3. The resulting correlation for $Y_{d,1/2}$ enables the use of Equation 16.1 to estimate $T_{d,1/2}$ for all polymers constructed from the nine elements (C, N, O, H, F, Si, S, Cl and Br) which are of interest in this book.

$$
\begin{aligned}
N_{Yd} \equiv {} & 14N_{(-S-)} + 21N_{(BB\ sulfone)} + 5N_{(BB\ amide)} + 4N_{(BB\ Si-O\ bonds)} \\
& + 12N_{(perhalogenated\ carbon\ atoms)} + 41N_{(BB\ aromatic\ imide\ groups)} \\
& + 17N_{(BB\ heterocyclic\ fused\ aromatic\ rings\ except\ imides)} + 9N_{(oxygen\ atoms\ in\ aromatic\ BB\ rings)} \\
& - 7N_{(side\ group\ correction)} + 20N_{(Br\ attached\ to\ SG\ atoms\ or\ to\ non-aromatic\ BB\ atoms)} \\
& + 25N_{(Br\ attached\ to\ aromatic\ BB\ atoms)} - 10N_{(single,\ unsubstituted\ meta\ aromatic\ BB\ rings)} \\
& - 30N_{(single,\ unsubstituted\ ortho\ aromatic\ BB\ rings)} \\
& + 5N_{(hydrogen-containing\ substituents,\ attached\ to\ single\ aromatic\ BB\ rings)} \\
& + 10N_{(substituents\ not\ containing\ hydrogen,\ attached\ to\ single\ aromatic\ BB\ rings)} \\
& + 8N_{(resonance\ around\ six-membered\ single\ unsubstituted\ aromatic\ BB\ rings)} \\
& - 5N_{(BB\ ester\ with\ aliphatic\ groups\ on\ both\ sides)} - 5N_{(acetate-type\ ester)}
\end{aligned}
\tag{16.6}
$$

$N_{(-S-)}$ is the number of divalent sulfur atoms located anywhere in the repeat unit, i.e., the number of sulfur atoms in the R-S-R' bonding configuration, including those located in the chain backbone, in side groups, and in rings. The effect of divalent sulfur atoms in side groups on the value $T_{d,1/2}$ may be overestimated when this term is used; however, the available group contributions and experimental data are insufficient to make a finer distinction between the magnitudes of the correction terms for divalent sulfur atoms in the backbone and in side groups.

The correction terms for sulfone ($-SO_2-$) groups, amide ($-CONH-$) groups, silicon-oxygen (Si-O) bonds, imide rings which are parts of aromatic moieties, other types of heterocyclic fused rings, and oxygen atoms in aromatic rings, are only used if the structural feature in question is a part of the backbone of the repeat unit. [Urea and urethane groups are not counted as amide groups in determining $N_{(BB\ amide)}$ for use in calculating N_{Yd}.]

Note that polymers with a simple silicone-type backbone, i.e., with a backbone consisting solely of alternating silicon and oxygen atoms, such as poly(dimethyl siloxane) (Figure 5.1) and poly[oxy(methylphenylsilylene)] (Figure 6.11), have *two* Si-O backbone bonds per repeat unit.

The correction for aromatic imide groups (+41) is much larger than the correction (+17) for other types of aromatic heterocyclic fused rings, such as benzimidazole, benzoxazole, benzothiazole and quinoxaline rings.

Imide groups which are not connected to an aromatic ring are not counted in determining when +41 should be added to N_{Yd}. For example, the imide rings in poly(N-methyl glutarimide) and poly(N-phenyl maleimide) (see Figure 2.4) do not contribute to N_{Yd}.

Furthermore, +41 or +17, as appropriate, is not added if the imide group or other type of heterocyclic ring moiety contains any non-aromatic rings.

Imide groups involving aromatic rings contribute the same amount (+41) to N_{Yd} whether two rings are fused to each other (as in the phthalimide groups in Torlon and Ultem, Figure 2.4), or three rings are fused to each other (as in the pyromellitimide group in Kapton, Figure 6.10). Similarly, non-imide heterocyclic groups, such as the benzoxazole, benzobisoxazole and benzobisthiazole units (see Figure 11.4), all contribute equally (+17) to N_{Yd}.

A "perhalogenated carbon atom" is an aliphatic carbon atom in the bonding configuration -CXY- or -CXYZ. X, Y and Z are halogen (fluorine, chlorine or bromine) atoms, which may or may not be identical in -CXY- or -CXYZ. For example, this correction term is applied for -CF$_2$-, -CFCl-, -CCl$_2$-, -CBr$_2$- and -CF$_3$ groups found in various polymers in our dataset. The use of a single term for all of these groups as a simplification leads to a slight overestimation of the effects of perchlorination and a slight underestimation of the effects of perfluorination.

The correction of +25 for a bromine atom directly attached to an aromatic backbone carbon atom is larger than the correction of +20 for a bromine atom attached to an aliphatic backbone atom or to any type of side group atom.

By far the most common example of a single (i.e., unfused) and unsubstituted six-membered aromatic backbone ring is a simple phenyl (-C$_6$H$_4$-) ring located along the backbone. The value of $Y_{d,1/2}$, and thus also of $T_{d,1/2}$, typically decreases significantly in the order *para>meta>ortho* for the bonding configuration of such rings, leading to no correction for the *para* configuration, -10 for the *meta* configuration, and -30 for the *ortho* configuration.

The correction for the contribution of a substituent on a single aromatic backbone ring to N_{Yd} differs depending upon whether or not the substituent contains hydrogen atoms. For example, the chlorine substituent in poly(chloro-*p*-xylylene) (Figure 5.1) contributes +10 to N_{Yd}, while the methyl (-CH$_3$) substituent at the same location in poly(methyl-*p*-xylylene) contributes only +5.

The amount, if any, of extended resonance effects between a single unsubstituted backbone aromatic ring and the structural units by which it is flanked, is quantified by the number of sides (zero, one or two) of the ring where other groups with electron delocalization (such as another aromatic ring, a carbonyl or sulfone group, or a C=C double bond) are directly bonded to it. For example, if the ring is flanked by a resonating group on one side only, +8 is added to N_{Yd}. On the other hand, if it is flanked by resonating groups on both sides, +16 is added to N_{Yd}.

$N_{(side\ group\ correction)}$, which is multiplied by -7 in Equation 16.6, is counted as follows:

1. A contribution of +1 is made to $N_{(side\ group\ correction)}$ per backbone atom carrying one or more of the following types of structural units in its side group(s):

    (a) A pendant aromatic unit of any type (i.e., whether carbocyclic or heterocyclic, single or fused). Some examples are the phenyl ring in the side groups of polystyrene (Figure 1.1), poly(N-phenyl glutarimide) (Figure 2.4) and poly[oxy(methylphenylsilylene)] (Figure 6.11).

    (b) A pendant non-aromatic simple hydrocarbon ring (such as a cyclohexyl or a cyclopentyl ring). [A pendant ring has all of its atoms in the side group. Rings containing a backbone atom at which the side group portions of the ring come together (see Figure 6.2 for three examples) are not counted as pendant rings.]

2. An *additional* contribution of +1 is made if one or both of the following circumstances hold:

    (a) If the backbone atom carrying such a ring or rings is a tertiary carbon atom, with no heteroatoms elsewhere (i.e., other than those, if any, in the aromatic side groups) in its two side groups. *Examples:* (i) There is no additional +1 for poly[oxy(methylphenylsilylene)] [Figure 6.11(a)], because the tertiary backbone atom is a silicon atom instead of being a carbon atom. (ii) There is an additional +1 for a structural variant of poly[oxy(methylphenylsilylene)] where the backbone silicon atom is replaced by a carbon atom. (iii) If an ester (-COO-) moiety is inserted between the backbone atom and the phenyl ring, the additional contribution is not made regardless of whether the backbone atom is a carbon or a silicon atom.

    (b) If the backbone atom carrying the ring is a tertiary atom carrying another such ring in its *other* side group. For example, there is an additional +1 for the polymer in Figure 6.11(b), where a tertiary backbone silicon atom has two phenyl rings as its two separate side groups.

Poly(glycolic acid) (Figure 2.4) is an example of a polymer which contains a backbone ester linkage flanked by aliphatic groups on both sides.

An acetate-type ester group is an ester linkage in a side group which is attached to the chain backbone via its divalent oxygen atom, as in poly(vinyl acetate) whose repeat unit is shown in Figure 16.1. The difference between an acetate-type and an acrylate-type ester linkage in a side group should become obvious by comparing poly(vinyl acetate) with poly(methyl methacrylate) (Figure 11.1) whose acrylic ester group is attached to the backbone via its carbon atom.

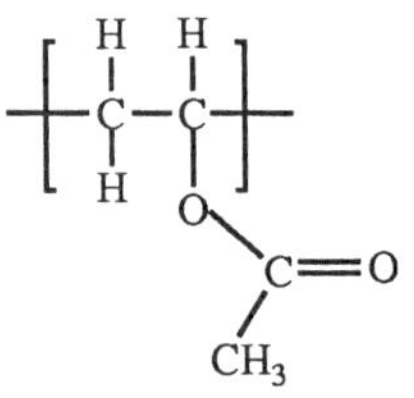

Figure 16.1. Schematic illustration of the repeat unit of poly(vinyl acetate).

It should be kept in mind, in calculating N_{Yd}, that the contribution of *each* structural term must be considered independently of all others, for each structural unit or feature of the polymer. For example, the benzobisoxazole and benzobisthiazole groups (Figure 11.4), when incorporated into the backbone, both contribute +17 to N_{Yd} because they are non-imide heterocyclic backbone rings. Benzobisoxazole contributes an *additional* +18 because of its two aromatic backbone O atoms, to give a total contribution of +35. Benzobisthiazole contributes an *additional* +28, to give a total contribution of +45, because its two aromatic S atoms can be formally represented as divalent atoms in the bonding configuration of -S- in a simple valence bond diagram.

The final correlation given by Equation 16.7 has a standard deviation of 2.4, and a correlation coefficient of 0.9991, so that it accounts for 99.8% of the variation of the $Y_{d,1/2}$(group) values in the dataset. The standard deviation is 2.2% of the average $Y_{d,1/2}$(group) of 109.9.

$$Y_{d,1/2} \approx 6.50 \cdot N + 0.99 \cdot (10^{-1}\chi^v + N_{Yd} - N_H) \tag{16.7}$$

The results of these calculations are shown in Figure 16.2 and listed in Table 16.2. $Y_{d,1/2}$ values estimated by using equations 16.6 and 16.7, for the 40 polymers in the dataset for which $Y_{d,1/2}$(group) could not be calculated because of the lack of some of the group contributions, are also listed in Table 16.2. The quantity ($N_{Yd} - N_H$) is listed instead of N_{Yd} and N_H separately, to avoid overcrowding Table 16.2. A further check of the quality of Equation 16.7 is provided in Table 16.3. The $T_{d,1/2}$ values of several commercial polymers, whose $Y_{d,1/2}$ cannot be calculated by using the available group contribution tables because of the lack of some of the needed group contributions, were estimated by utilizing Equation 16.7 to calculate $Y_{d,1/2}$. The predicted values of $T_{d,1/2}$ were then compared with the results of dynamic TGA scans [22] obtained in an inert (nitrogen) atmosphere at a rate of 10K/minute on a DuPont TGA module. The agreement between the calculated and the experimental $T_{d,1/2}$ values is quite good. In conclusion, the use of Equation 16.7 to predict $Y_{d,1/2}$ allows prediction of $T_{d,1/2}$, which provides very valuable information on thermal stability, at an accuracy comparable to (but not better than) the use of group contributions to predict $Y_{d,1/2}$. Equation 16.7 has an important advantage, however, since it can be used to predict $Y_{d,1/2}$ (and hence $T_{d,1/2}$) without being limited by the lack of group contributions.

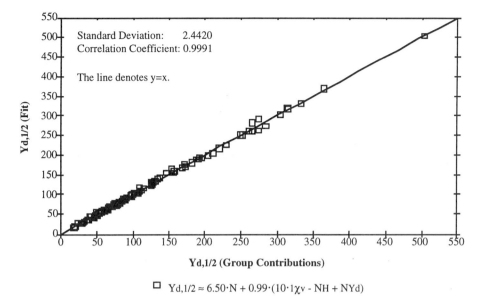

$$Y_{d,1/2} \approx 6.50 \cdot N + 0.99 \cdot (10 \cdot {}^1\chi v - NH + NYd)$$

Figure 16.2. A fit using connectivity indices, to the molar thermal decomposition function $Y_{d,1/2}$ calculated by group contributions, for 140 polymers. $Y_{d,1/2}$ is in units of K·kg/mole.

Table 16.2. Molar thermal decomposition function $Y_{d,1/2}$ calculated by group contributions, the quantity $(N_{Yd} - N_H)$ entering the correlation equation, and the fitted or predicted (fit/pre) values of $Y_{d,1/2}$, for 180 polymers. N_{Yd} is a correction index. N_H is the number of hydrogen atoms in the repeat unit. For 40 of these polymers, $Y_{d,1/2}$(group) cannot be calculated because of the lack of some of the group contributions, but $Y_{d,1/2}$(fit/pre) can be calculated by using the new correlation. The values of N and ${}^1\chi v$, which are also used in the correlation equation, are listed in Table 2.2. $Y_{d,1/2}$ is in units of K·kg/mole.

Polymer	$Y_{d,1/2}$(group)	$(N_{Yd}\text{-}N_H)$	$Y_{d,1/2}$(fit/pre)
Polyoxymethylene	17.5	-2	16.7
Polyethylene	19.0	-4	18.9
Poly(vinyl alcohol)	23.5	-4	26.2
Polyoxyethylene	27.0	-4	26.2
Poly(vinyl fluoride)	27.5	-3	26.8
Polypropylene	28.0	-6	27.4
Poly(glycolic acid)	29.5	-7	29.5
Poly(vinyl chloride)	33.0	-3	31.1
Polyisobutylene	35.0	-8	35.0

☞ ☞ ☞ TABLE 16.2 IS CONTINUED IN THE NEXT PAGE. ☞ ☞ ☞

Table 16.2. CONTINUED FROM THE PREVIOUS PAGE.

Polymer	$Y_{d,1/2}$(group)	$(N_{Yd}\text{-}N_H)$	$Y_{d,1/2}$(fit/pre)
Poly(1,2-butadiene)	-----	-6	35.5
Poly(vinyl methyl ether)	36.0	-6	34.5
Poly(propylene oxide)	36.0	-6	35.0
Polyoxytrimethylene	36.5	-6	35.7
Poly(1,4-butadiene)	37.0	-6	36.4
Polyacrylonitrile	37.5	-3	36.2
Poly(1-butene)	37.5	-8	37.2
Polyisoprene	40.5	-8	44.9
Poly(vinyl methyl ketone)	42.0	-6	45.3
Polyacrylamide	-----	-5	43.4
Poly(acrylic acid)	-----	-4	43.7
Poly(vinyl ethyl ether)	45.5	-8	44.9
Polyoxytetramethylene	46.0	-8	45.1
Polymethacrylonitrile	46.5	-5	44.2
Poly(1-pentene)	47.0	-10	46.7
Poly(vinyl acetate)	47.0	-11	47.5
Poly(vinylidene fluoride)	48.0	10	46.6
Polychloroprene	-----	-5	48.5
Poly(vinylidene chloride)	48.5	10	54.1
Poly(β-alanine)	49.0	0	48.9
Polyepichlorohydrin	50.5	-5	48.8
Poly(methacrylic acid)	-----	-6	51.7
Poly(ethylene sulfide)	52.0	10	51.5
Poly(methyl acrylate)	52.0	-6	52.1
Poly[oxy(m-phenylene)]	52.0	-14	53.7
Poly(vinyl sulfonic acid)	-----	-4	54.3
Poly(vinyl bromide)	-----	17	55.6
Polyfumaronitrile	56.0	-2	53.8
Poly(4-methyl-1-pentene)	56.0	-12	54.7
Polytrifluoroethylene	56.5	11	55.0
Poly(1-hexene)	56.5	-12	56.1
Poly(vinyl propionate)	56.5	-13	57.6
Poly(ethylene oxalate)	59.0	-14	59.4

☞ ☞ ☞   **TABLE 16.2 IS CONTINUED IN THE NEXT PAGE.** ☞ ☞ ☞

Table 16.2. CONTINUED FROM THE PREVIOUS PAGE.

Polymer	$Y_{d,1/2}(group)$	$(N_{Yd}-N_H)$	$Y_{d,1/2}(fit/pre)$
Poly(methyl methacrylate)	60.5	-8	60.1
Poly(propylene sulfide)	61.0	8	59.3
Poly(vinyl cyclopentane)	-----	-19	61.0
Poly(vinyl methyl sulfide)	61.0	8	61.1
Poly(ethyl acrylate)	61.5	-8	62.4
Poly[oxy(p-phenylene)]	62.0	-4	63.6
Poly(m-xylylene)	63.0	-18	64.2
Poly(maleic anhydride)	-----	-2	63.9
Poly(vinyl sec-butyl ether)	64.0	-12	63.1
Poly(vinyl n-butyl ether)	64.5	-12	63.8
Poly(m-phenylene)	65.0	2	61.5
Poly(methyl α-chloroacrylate)	65.5	-5	63.7
Poly(α-methyl styrene)	65.5	-24	68.1
Polystyrene	66.0	-15	67.0
Poly(o-vinyl pyridine)	-----	-14	66.6
Poly(ε-caprolactone)	67.5	-15	67.3
Poly(dimethyl siloxane)	68.0	2	69.8
Poly(vinyl cyclohexane)	69.5	-21	70.5
Poly(methyl α-cyanoacrylate)	70.0	-5	69.3
Poly(methyl ethacrylate)	70.0	-10	70.2
Poly(ethyl methacrylate)	70.0	-10	70.4
Poly(thiocarbonyl fluoride)	71.5	26	67.6
Poly(vinyl pivalate)	72.5	-17	73.5
Poly(p-xylylene)	73.0	-8	74.1
Poly(styrene oxide)	74.0	-15	74.7
Poly(N-vinyl pyrrolidone)	-----	-9	74.5
Poly(p-phenylene)	75.0	12	71.4
Poly(2-hydroxyethyl methacrylate)	75.0	-10	78.0
Poly(m-methyl styrene)	75.5	-17	75.6
Poly(p-methyl styrene)	75.5	-17	75.6
Poly(o-methyl styrene)	-----	-17	75.7
Polytetrafluoroethylene	77.0	24	75.2
Poly[thio(m-phenylene)]	77.0	0	75.7

☞ ☞ ☞   TABLE 16.2 IS CONTINUED IN THE NEXT PAGE. ☞ ☞ ☞

Table 16.2. CONTINUED FROM THE PREVIOUS PAGE.

Polymer	$Y_{d,1/2}$(group)	$(N_{Yd}-N_H)$	$Y_{d,1/2}$(fit/pre)
Poly(m-hydroxybenzoate)	77.0	-6	79.1
Poly(ε-caprolactam)	77.5	-6	77.4
Polychlorotrifluoroethylene	77.5	24	78.9
Poly(propylene sulfone)	78.0	15	76.3
Poly(ethylene succinate)	78.0	-18	77.9
Poly(o-chloro styrene)	-----	-14	79.3
Poly(2-vinylthiophene)	-----	1	80.0
Poly(p-chloro styrene)	80.5	-14	79.2
Poly(oxy-2,2-dichloromethyltrimethylene)	80.5	-8	79.7
Poly(oxy-1,1-dichloromethyltrimethylene)	80.5	-8	80.3
Poly(n-butyl acrylate)	80.5	-12	81.4
Poly(p-methoxy styrene)	83.5	-17	83.2
Poly(vinyl benzoate)	85.0	-20	87.5
Poly(t-butyl methacrylate)	86.0	-14	86.5
Poly[thio(p-phenylene)]	87.0	10	85.6
Poly(p-hydroxybenzoate)	87.0	4	89.0
Poly(1-butene sulfone)	87.5	13	86.2
Poly(vinyl trimethylsilane)	88.0	-12	88.3
Poly(sec-butyl methacrylate)	88.5	-14	88.7
Poly(n-butyl methacrylate)	89.0	-14	89.4
Poly(vinyl n-butyl sulfide)	89.5	2	88.0
Poly[oxy(2,6-dimethyl-1,4-phenylene)]	90.0	2	90.8
Poly(vinyl butyral)	-----	-14	92.6
Poly(chloro-p-xylylene)	-----	3	96.2
Poly(8-aminocaprylic acid)	96.5	-10	96.3
Poly(ethylene adipate)	97.0	-22	96.8
Poly(phenylmethylsilane)	98.0	-15	100.6
Poly(phenyl methacrylate)	98.5	-17	100.6
Poly(styrene sulfide)	99.0	-1	98.9
Poly(p-t-butyl styrene)	101.0	-23	101.5
Poly(cyclohexylmethylsilane)	101.5	-21	104.8
Poly(vinylidene bromide)	------	50	101.9
Poly(cyclohexyl methacrylate)	102.0	-23	103.9

☞ ☞ ☞   **TABLE 16.2 IS CONTINUED IN THE NEXT PAGE.** ☞ ☞ ☞

Table 16.2. CONTINUED FROM THE PREVIOUS PAGE.

Polymer	$Y_{d,1/2}(group)$	$(N_{Yd}-N_H)$	$Y_{d,1/2}(fit/pre)$
Poly(p-bromo styrene)	------	6	103.1
Poly(α-vinyl naphthalene)	------	-17	105.0
Poly[oxy(dipropylsilylene)]	106.0	-6	103.1
Poly[oxy(methylphenylsilylene)]	106.0	-7	104.9
Poly(1-hexene sulfone)	106.5	9	105.1
Poly(N-methyl glutarimide)	------	-13	106.9
Poly(2-ethylbutyl methacrylate)	107.5	-18	107.6
Poly(n-hexyl methacrylate)	108.0	-18	108.3
Poly(benzyl methacrylate)	108.0	-19	109.4
Polyoxynaphthoate	108.0	-6	119.0
Poly(2,5-benzoxazole)	110.0	23	109.4
Poly(N-phenyl maleimide)	------	-14	112.6
Poly(N-vinyl phthalimide)	------	-14	112.8
Poly(acenaphthylene)	------	-8	114.5
Poly(m-phenylene-2,5-oxadiazole)	115.0	11	115.8
Poly[oxy(methyl γ-trifluoropropylsilylene)]	------	13	122.9
Poly(2,5-benzothiazole)	125.0	28	122.5
Poly(p-phenylene-2,5-oxadiazole)	125.0	21	125.7
Poly(m-phenylene-2,5-thiodiazole)	125.0	16	128.9
Poly(octylmethylsilane)	126.5	-20	129.5
Poly(n-octyl methacrylate)	127.0	-22	127.2
Poly(ethylene isophthalate)	128.0	-2	130.8
Poly(α,α,α',α'-tetrafluoro-p-xylylene)	131.0	20	134.7
Poly(N-vinyl carbazole)	------	-18	133.4
Poly(p-phenylene-2,5-thiodiazole)	135.0	26	138.8
Poly(ethylene terephthalate)	138.0	8	140.7
Poly(α-naphthyl methacrylate)	------	-19	138.6
Poly[methane bis(4-phenyl)carbonate]	147.5	-10	153.8
Poly(ethylene-2,6-naphthalenedicarboxylate)	152.0	-10	162.7
Poly(hexamethylene adipamide)	155.0	-12	154.7
Poly(tetramethylene terephthalate)	157.0	4	159.6
Poly(o-phenylene terepthalamide)	------	-14	158.2
Poly(pentachlorophenyl methacrylate)	------	-12	162.0

☞☞☞   TABLE 16.2 IS CONTINUED IN THE NEXT PAGE. ☞☞☞

Table 16.2. CONTINUED FROM THE PREVIOUS PAGE.

Polymer	$Y_{d,1/2}$(group)	(N_{Yd}-N_H)	$Y_{d,1/2}$(fit/pre)
Poly(m-phenylene isophthalamide)	169.0	-4	168.0
Poly[3,5-(4-phenyl-1,2,4-triazole)-1,4-phenylene]	------	5	170.4
Poly[oxy(2,6-diphenyl-1,4-phenylene)]	------	-16	170.9
Poly[thio bis(4-phenyl)carbonate]	171.0	6	174.7
Bisphenol-A polycarbonate	173.0	-14	170.7
Poly(hexamethylene isophthalamide)	182.0	-2	178.7
Poly(p-phenylene terephthalamide)	189.0	16	187.8
Phenoxy resin	192.0	-20	188.0
Poly(hexamethylene terephthalamide)	192.0	8	188.6
Poly(hexamethylene sebacamide)	193.0	-20	192.6
Poly(m-phenylene diimidazobenzene)	195.0	15	190.0
Poly(1,4-cyclohexylidene dimethylene terephthalate)	------	-2	199.0
Poly[1,1-cyclohexane bis(4-phenyl)carbonate]	------	-18	203.1
Poly(p-phenylene diimidazobenzene)	205.0	25	199.9
Poly[1,1-(1-phenylethane) bis(4-phenyl)carbonate]	211.0	-30	203.8
Polybenzobisoxazole	------	45	217.9
Poly(dicyclooctyl itaconate)	------	-48	219.8
Poly(ether ether ketone)	220.0	4	216.1
Poly[2,2-propane bis{4-(2,6-dimethylphenyl)}carbonate]	229.0	-2	225.0
Poly[1,1-(1-phenyltrifluoroethane) bis(4-phenyl)carbonate]	------	-15	241.2
Polybenzobisthiazole	------	55	243.9
Poly[2,2-hexafluoropropane bis(4-phenyl)carbonate]	------	16	245.6
Poly[diphenylmethane bis(4-phenyl)carbonate]	249.0	-32	250.7
Bisphenol-A isophthalate	252.0	-12	249.6
Poly[2,2-propane bis{4-(2,6-dichlorophenyl)}carbonate]	------	30	259.4
Bisphenol-A terephthalate	262.0	-2	259.5
Poly[1,1-biphenylethane bis(4-phenyl)carbonate]	265.0	-34	259.3
Poly(quinoxaline-quinoxaline-m-phenylene)	265.0	28	280.7
Poly[2,2'-(m-phenylene)-5,5'-bibenzimidazole]	275.0	28	262.3
Poly(quinoxaline-quinoxaline-p-phenylene)	275.0	38	290.6
Poly(pentabromophenyl methacrylate)	------	88	281.6
Poly[2,2'-(p-phenylene)-5,5'-bibenzimidazole]	285.0	38	272.2
Torlon	304.5	40	300.4

☞ ☞ ☞   **TABLE 16.2 IS CONTINUED IN THE NEXT PAGE.** ☞ ☞ ☞

Table 16.2. CONTINUED FROM THE PREVIOUS PAGE.

Polymer	$Y_{d,1/2}(group)$	$(N_{Yd}\text{-}N_H)$	$Y_{d,1/2}(fit/pre)$
Poly[4,4'-diphenoxy di(4-phenylene)sulfone]	314.0	37	318.8
Poly[N,N'-(p,p'-oxydiphenylene)pyromellitimide]	316.0	47	322.8
Poly[4,4'-isopropylidene diphenoxy di(4-phenylene)sulfone]	333.0	15	328.9
Poly[4,4'-sulfone diphenoxy di(4-phenylene)sulfone]	364.0	58	367.7
Poly[2,2-propane bis{4-(2,6-dibromophenyl)}carbonate]	------	130	374.8
Ultem	503.0	64	498.6

Table 16.3. Comparison of the predicted and observed values of $T_{d,1/2}$, for several polymers. The new correlation developed in this section was used to calculate $Y_{d,1/2}$, since these polymers contain structural units whose group contributions to $Y_{d,1/2}$ are not available. $Y_{d,1/2}$ is in units of K·kg/mole. $T_{d,1/2}$ is in degrees Kelvin.

Polymer	M	$Y_{d,1/2}$	$T_{d,1/2}(pre)$	$T_{d,1/2}(exp)$
Poly(o-vinyl pyridine)	105.14	66.6	633	644
Poly(vinyl butyral)	142.20	92.6	651	645
Poly(α-vinyl naphthalene)	154.21	105.0	681	657
Poly(1,4-cyclohexylidene dimethylene terephthalate)	274.31	199.0	725	691
Poly(N-vinyl pyrrolidone)	111.14	74.5	670	692
Poly(N-vinyl carbazole)	193.25	133.4	690	699

References and Notes for Chapter 16

1. *Modern Plastics Encyclopedia*, McGraw-Hill, New York (1989), 412-413 and 659-699.

2. J. H. Flynn, in *Aspects of Degradation and Stabilization of Polymers*, edited by H. H. G. Jellinek, Elsevier, Amsterdam (1978), Chapter 12.

3. *Annual Book of ASTM Standards*: (a) D3850-84, Vol. 10.02 (1985). (b) E472-79, Vol. 14.02 (1985). (c) E473-82, Vol. 14.02 (1985). (d) E914-83, Vol. 14.02 (1985).

4. C. D. Doyle, *J. Appl. Polym. Sci.*, 6, 639-642 (1962).

5. C. Arnold, Jr., *J. Polym. Sci., Macromol. Rev.*, 14, 265-378 (1979).

6. J. R. MacCallum, in *Comprehensive Polymer Science, Volume 1: Polymer Characterization*, edited by G. Allen, Pergamon Press, Oxford (1988).

7. J. H. Flynn, *J. Thermal Analysis*, 34, 367-381 (1988).

8. A. Cohen, C. J. Carriere and A. J. Pasztor, Jr., *Polymer Engineering and Science*, *33*, 317-321 (1993).

9. P. E. Cassidy, *Thermally Stable Polymers: Syntheses and Properties*, Marcel Dekker, New York (1980), Chapter 1.

10. J. H. Flynn, *ANTEC '88 Preprints*, 930-932 (April 1988).

11. D. W. van Krevelen, *Properties of Polymers*, third edition, Elsevier, Amsterdam (1990), Chapter 21.

12. P. M. Hergenrother, article titled "*Heat-Resistant Polymers*", in *Encyclopedia of Polymer Science and Engineering*, *7*, Wiley-Interscience, New York (1987).

13. W. W. Wright, in *Degradation and Stabilization of Polymers*, edited by G. Geuskens, John Wiley & Sons, New York (1975), Chapter 3.

14. W. Lynch, *Handbook of Silicone Rubber Fabrication*, Van Nostrand Reinhold Company, New York (1978).

15. *Silicon-Based Polymer Science: A Comprehensive Resource*, edited by J. M. Zeigler and F. W. Gordon Fearon, *Advances in Chemistry Series*, *224*, American Chemical Society, Washington, D.C. (1990).

16. K. O. Knollmueller, R. N. Scott, H. Kwasnik and J. F. Sieckhaus, *J. Polym. Sci., A-1*, *9*, 1071-1088 (1971).

17. E. N. Peters, *Ind. Eng. Chem. Prod. Res. Dev.*, *23*, 28-32 (1984).

18. H. Schroeder, in *High Performance Polymers: Their Origin and Development*, edited by R. B. Seymour and G. S. Kirshenbaum, Elsevier Science Publishing, New York (1986), 389-399.

19. J. A. Vaccari, *Design Engineering*, *51 (No. 3)*, 35-38 (1980).

20. H. R. Allcock, *Phosphorus-Nitrogen Compounds*, Academic Press, New York (1972), Part III.

21. D. W. van Krevelen, in *Computational Modeling of Polymers*, edited by J. Bicerano, Marcel Dekker, New York (1992), Chapter 1.

22. Measured by T. Sangster and C. A. Wedelstaedt (1990).

CHAPTER 17

EXTENSIONS, GENERALIZATIONS, SHORTCUTS, AND POSSIBLE DIRECTIONS FOR FUTURE WORK

17.A. Introduction

Simple extensions, generalizations and shortcuts, which facilitate the calculations and/or increase their accuracy, will now be discussed. These methods can all be used to calculate many other properties and to solve many other problems besides the ones chosen as examples.

Specialized "designer correlations" [1] can be developed for structurally or compositionally related polymers. The resulting correlations of limited applicability can sometimes be utilized to calculate the properties of selected types of polymers with greater accuracy than is possible with a general-purpose correlation, while using fewer adjustable parameters. Designer correlations are only useful if the general-purpose correlation for a property utilizes many atomic and/or group correction terms. For example, the correlations for the cohesive energy (Chapter 5) and the glass transition temperature (Chapter 6) require many correction terms, and can be replaced by several designer correlations. On the other hand, the correlations for the heat capacity (Chapter 4) only utilize general parameters (connectivity indices and rotational degrees of freedom), and their separation into designer correlations does not either reduce the number of adjustable parameters or increase the accuracy. Three examples of designer correlations will be given, for the glass transition temperatures of polyphthalimides constructed completely from C, N, O, S and H atoms (Section 17.B.1) and of polyesters (Section 17.B.2), and for the Fedors-type cohesive energies (E_{coh1}) of polymers containing only carbon and hydrogen atoms (Section 17.B.3).

Many correlations developed in this book can be used to calculate group contributions and extend existing group contribution tables. The new group contributions can then be added to the tables of group contributions, and used to calculate the properties of other polymers containing the appropriate structural units. An example of this procedure will be given in Section 17.C, where the contribution of the α-naphthyl group to the Fedors-type cohesive energy will be calculated and used to augment the existing group contribution tables. In addition, it will be shown how group contributions can often be combined with the new correlations to speed up the calculations.

The correlations developed in this book can also be combined with experimental data to provide predictions of greater accuracy for many polymers. Two general methods by which the accuracy can be improved by using experimental data to refine theoretical predictions have been discussed by van Krevelen [2]. The *method of standard properties* allows greater precision in the estimation of a physical property from accurate knowledge of the experimental value of a related property which can be used as a standard value. The *method of standard substances* involves the use of data on a polymer, whose property of interest is known accurately, as a standard for structurally related polymers, i.e., other polymers obtained from the reference polymer by a relatively small structural and/or compositional modification. Such refinements can be very useful when used judiciously. An example of the use of the method of standard properties was given in Chapter 13, where the use of the predicted and the observed zero-shear rate viscosity as input for the calculation of the shear rate dependence of the viscosity were compared for polystyrene. An example of the use of the method of standard substances in a "corresponding states" model is the refinement of the predicted zero-shear viscosity of poly(α-methyl styrene) by multiplying it with the ratio of the observed divided by the predicted values for polystyrene at the same T/T_g.

Most of the calculations in earlier chapters dealt with homopolymers. The techniques developed in this book can also be used to calculate the physical properties of alternating and random copolymers. The properties of block copolymers, on the other hand, cannot be calculated without additional information concerning the block sizes, and whether or not the different blocks aggregate into domains. The prediction of the properties of alternating copolymers will be discussed in Section 17.D. It will also be shown how the prediction of the properties of some homopolymers can be simplified by formally breaking their repeat units down into the smaller alternating repeat units of homopolymers whose properties have already been calculated. The calculation of the properties of random copolymers will be discussed in Section 17.E.

A software package has been developed based on the correlations presented in this book. This software package will be briefly discussed in Section 17.F.

The results of calculations made by using the methods developed in this book can be inserted into composite models to predict some of the key properties (such as the moduli, coefficients of thermal expansion, thermal conductivities and gas permeabilities) of multiphase polymeric systems such as blends and block copolymers of immiscible polymers, semicrystalline polymers, and polymers containing various types of fillers. Some of the composite models which we have found to be useful in such applications will be briefly reviewed in Section 17.G.

The method developed in this book can also provide input parameters (either calculated material properties or a set of simple and unambiguous structural descriptors) into many other types of models. Such models range from phenomenological theories of polymer properties to software tools based on artificial intelligence. Some of the work stimulated in these directions as a result of the publication of the first edition of this book will be discussed in Section 17.H.

Some possible directions for future work will be discussed in Section 17.I.

17.B.   Examples of Designer Correlations

17.B.1.   Glass Transition Temperatures of a Family of Polyimides

The "master correlation" for T_g (Equation 6.3) consistently overestimates the T_g's of the polyphthalimides completely built from C, N, O, S and H atoms. (See Torlon and Ultem in Figure 2.4, and the polymers in figures 6.6 and 6.7.) A specialized "designer correlation" will now be developed for the T_g's of this important class of polymers. Fourteen of the polymers in the dataset (Table 6.2) used to develop the master correlation for T_g fall into this family of polymers. The same two parameters as in Equation 6.3, namely (a) the solubility parameter δ calculated by using E_{coh2}, and (b) the combination of structural parameters represented by N_{Tg} (Equation 6.2), were used to correlate the T_g's of these 14 polymers. Weight factors of $1000/T_g(exp)$ were used in the linear regression. The resulting correlation (Equation 17.1) has a standard deviation of 12.34K, which equals 2.4% of the average $T_g(exp)$ value of 509.3K for the 14 polymers in the dataset. The correlation coefficient of 0.9767 indicates that Equation 17.1 accounts for 95.4% of the variation of the 14 $T_g(exp)$ values in this specialized dataset. The results of these calculations are summarized in Table 17.1 and shown in Figure 17.1.

$$T_g \approx 18.94 \cdot \delta + 23.28 \cdot \frac{N_{Tg}}{N} \qquad (17.1)$$

Table 17.1. Experimental glass transition temperatures T_g in degrees Kelvin, the quantities (δ, N and N_{Tg}) used in the specialized correlation equation for T_g, and the fitted values of T_g, for 14 polymers containing phthalimide groups, and completely built from C, N, O, S and H atoms.

Polymer	$T_g(exp)$	δ	N	N_{Tg}	$T_g(fit)$	Polymer	$T_g(exp)$	δ	N	N_{Tg}	$T_g(fit)$
Polyetherimide 1	401	23.5	43	-50	418	Polyetherimide 3	500	24.3	43	55	491
Polyetherimide 6	473	24.3	51	12	466	Polyetherimide 4	512	24.7	44	101	521
Polyetherimide 2	482	24.3	43	55	490	Polyimide 10	513	26.1	38	60	531
Polyetherimide 7	485	24.4	51	3	463	Polyetherimide 5	520	24.2	42	109	520
Polyetherimide 8	486	24.4	51	31	476	Torlon	550	25.7	27	78	555
Ultem	493	23.7	45	93	497	Polyimide 11	606	27.1	30	122	609
Polyetherimide 9	494	24.4	51	50	485	Polyetherimide 12	615	26.8	29	122	605

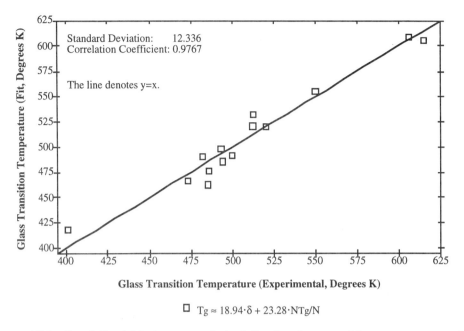

Figure 17.1. Specialized "designer correlation" for the glass transition temperatures of 14 polyphthalimides containing only C, N, O, S and H atoms.

17.B.2.   Glass Transition Temperatures of Polyesters

A designer correlation was developed by Mumby [3] for the glass transition temperatures of the 21 polyesters included in Table 6.2. In developing this correlation, which is given by Equation 17.2, the solubility parameter was consistently calculated by the method described in this book. This is in contrast to the procedure of Chapter 6, where the shortcut of using group contributions whenever available and supplementing them with the results of calculations by the new method had been employed in work performed prior to the computer implementation of the new method. Equation 17.2 has a standard deviation of 25.73K, which equals 7.3% of the average T_g(exp) value of 351.8K for the 21 polymers in the dataset. The correlation coefficient of 0.9765 indicates that Equation 17.2 accounts for 95.4% of the variation of T_g(exp) in this dataset. The results of these calculations are summarized in Table 17.2 and shown in Figure 17.2.

$$T_g \approx 467.00 + 0.298 \cdot \delta + 35.88 \cdot \frac{N_{Tg}}{N} \qquad (17.2)$$

A further attempt was also made to obtain an improved fit, by treating the linear regression coefficient of each of the structural parameters in the equation separately instead of using them in the combination designated as N_{Tg}. It was found, however, that this further attempt at refinement resulted in the overfitting of the data instead of producing significant additional improvements.

Table 17.2. Experimental glass transition temperatures T_g in degrees Kelvin, the quantities (δ, N and N_{Tg}) used in the specialized correlation equation developed by Mumby [3] for T_g, and the fitted values of T_g, for 21 polyesters.

Polymer	$T_g(exp)$	δ	N	N_{Tg}	$T_g(fit)$
Poly(tetramethylene adipate)	205	17.95	14	-96	226
Poly(decamethylene adipate)	217	17.55	20	-144	214
P(oxycarbonyl-3-methylpentamethylene)	220	17.36	9	-60	233
Poly(ethylene azelate)	228	17.86	15	-104	224
Poly(ethylene adipate)	233	18.20	12	-80	233
Poly(oxycarbonyl-1,5-dimethylpentamethylene)	240	16.90	10	-64	242
Poly(ethylene sebacate)	243	17.78	16	-112	221
Poly(ethylene succinate)	272	18.59	10	-64	243
Poly(ethylene isophthalate)	324	19.77	14	-60	319
Poly(ethylene-1,4-naphthalenedicarboxylate)	337	20.38	18	-41	391
Poly(ethylene-1,5-naphthalenedicarboxylate)	344	20.38	18	-73	328
Poly(ethylene terephthalate)	345	19.77	14	-41	368
Poly(1,4-cychexylidene dimethylene terephthalate)	368	20.04	20	-49	385
Poly(ethylene-2,6-naphthalenedicarboxylate)	397	20.31	18	-26	421
Poly(*m*-phenylene isophthalate)	411	20.27	18	-56	361
Poly(*p*-hydroxybenzoate)	420	20.27	9	-9	437
Poly(oxyterephthaloyloxy-2-methyl-1,4-phenyleneisopropylidene-3-methyl-1,4-phenylene)	444	19.61	29	-35	430
Poly(bisphenol-A terephthalate)	478	19.53	27	-27	437
Poly(oxyterephthaloyloxy-2,6-dimethyl-1,4-phenylene-isopropylidene-3,5-dimethyl-1,4-phenylene)	498	19.67	31	5	479
Polyphenolphthalein 2	580	20.48	40	71	537
Polyphenolphthalein 3	583	20.70	50	146	578

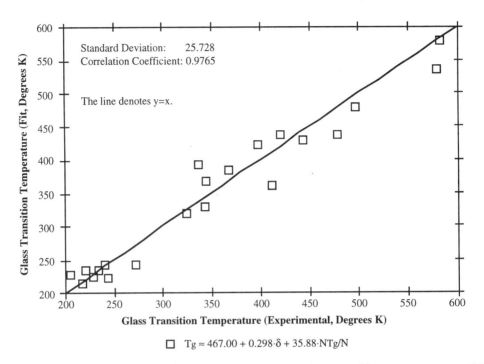

$$\square \quad Tg \approx 467.00 + 0.298 \cdot \delta + 35.88 \cdot NTg/N$$

Figure 17.2. Specialized "designer correlation" for the glass transition temperatures of 21 polyesters.

17.B.3.   Fedors-Type Cohesive Energies of Hydrocarbon Polymers

The cohesive energies of 25 "hydrocarbon" polymers (i.e., polymers containing only carbon and hydrogen atoms), as calculated by using the group contributions of Fedors [4,5], were correlated with connectivity indices. Weight factors of $10000/E_{coh}(group)$ were used in the linear regression procedure. The resulting correlation for E_{coh1} is given by Equation 17.2. It has a standard deviation of only 976 J/mole and a correlation coefficient of 0.9990. The standard deviation is only 2.9% of the average $E_{coh}(group)$ value of 34204 J/mole for this dataset. The results of these calculations are summarized in Table 17.3 and depicted in Figure 17.3.

$$E_{coh1} \approx 12482.1 \cdot {}^1\chi - 3005.7 \cdot {}^1\chi^v \tag{17.3}$$

Equation 17.3 uses only two adjustable parameters, does not require any atomic or group correction terms, and gives a better fit, for a limited class of polymers only, than the "general-purpose" correlation developed in Section 5.B. On the other hand, equations 5.8-5.10 have the advantages of much greater generality and applicability.

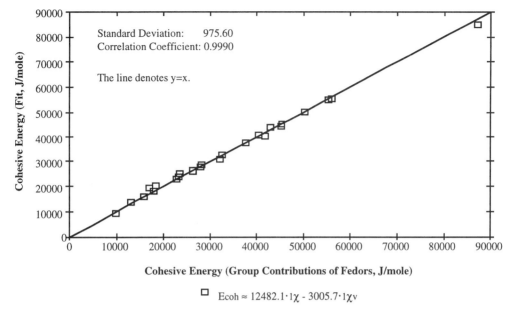

Figure 17.3. Designer correlation for Fedors-type cohesive energies of 25 hydrocarbon polymers.

In the early stages of the work presented in this book, E_{coh}(Fedors) was the first property for which correlations were developed in terms of connectivity indices. Specialized correlations, each one of which was applicable to a different class of polymers, were developed for this property. This approach was abandoned later in favor of the development of general-purpose correlations, for the reasons discussed below.

The polymers shown in Figure 17.4 each contain a structural unit whose contribution to E_{coh}(Fedors) is not listed in the group contribution tables [4,5], so that E_{coh}(Fedors) cannot be calculated for these polymers by using group contributions. Calculations of E_{coh1} for these seven polymers by using the new correlations (equations 5.8-5.10 for the general-purpose correlation, and Equation 17.3 for the designer correlation) are summarized in Table 17.4.

Four of the polymers shown in Figure 17.4 contain oxygen atoms. Their E_{coh1} therefore cannot be calculated by using the designer correlation developed for polymers containing only carbon and hydrogen atoms. This example therefore also illustrates the limited applicability of designer correlations compared with general-purpose correlations.

E_{coh1} values predicted for the three hydrocarbon polymers by using the designer correlation differ by less than 4% from E_{coh1} values predicted by using the general-purpose correlation. These differences are smaller than the differences between E_{coh1} values deduced for most polymers from the results of different experiments. The advantages of using designer correlations for the cohesive energies of the different types of polymers within the formalism developed in this

Table 17.3. Cohesive energies calculated by using the group contributions of Fedors, the indices used in the correlation, and the fitted values of E_{coh1}, for 25 "hydrocarbon" polymers.

Polymer	$E_{coh1}(group)$	$^1\chi$	$^1\chi^v$	$E_{coh1}(fit)$
Polyethylene	9880	1.0000	1.0000	9476
Polypropylene	13080	1.3938	1.3938	13208
Polyisobutylene	15830	1.7071	1.7071	16177
Poly(1,2-butadiene)	16990	1.9319	1.5581	19431
Poly(1-butene)	18020	1.9319	1.9319	18307
Poly(1,4-butadiene)	18500	2.0000	1.6498	20005
Poly(1-pentene)	22960	2.4319	2.4319	23046
Polyisoprene	23210	2.3938	2.0505	23716
Poly(1,4-pentadiene)	23440	2.5000	2.1498	24744
Poly(4-methyl-1-pentene)	26160	2.7877	2.7877	26417
Poly(1-hexene)	27900	2.9319	2.9319	27784
Poly(2-methyl-1,4-pentadiene)	28150	2.8938	2.5505	28455
Poly(p-phenylene)	31940	2.9663	2.0714	30800
Poly(vinyl cyclopentane)	32610	3.4663	3.4663	32848
Poly(vinyl cyclohexane)	37550	3.9663	3.9663	37586
Polystyrene	40310	3.9663	3.0159	40443
Poly(p-xylylene)	41820	3.9495	3.0285	40195
Poly(α-methyl styrene)	43060	4.3123	3.3678	43704
Poly(o-methyl styrene)	45020	4.3770	3.4325	44317
Poly(p-methyl styrene)	45020	4.3602	3.4265	44125
Poly(3-phenyl-1-propene)	45250	4.4495	3.4890	45052
Poly(4-phenyl-1-butene)	50190	4.9495	3.9890	49790
Poly(5-phenyl-1-pentene)	55130	5.4495	4.4890	54529
Poly(p-t-butyl styrene)	55910	5.5715	4.6765	55488
Poly(1-octadecene)	87180	8.9319	8.9319	84642

book are thus at best marginal. The same situation also holds for many of the other properties discussed in earlier chapters. In general, it was therefore found to be much preferable to develop general-purpose correlations containing structural parameters and atomic and/or group correction terms in addition to connectivity indices, instead of sets of specialized correlations, providing comparable accuracy without requiring the development of several correlations for each property.

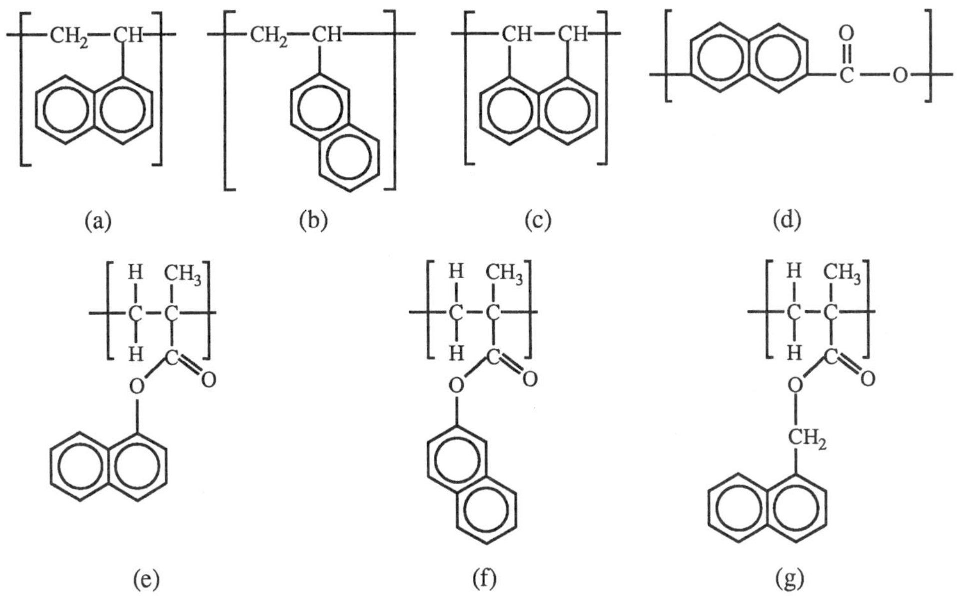

Figure 17.4. Repeat units of several polymers. (a) Poly(α-vinyl naphthalene). (b) Poly(β-vinyl naphthalene). (c) Poly(acenaphthylene). (d) Polyoxynaphthoate. (e) Poly(α-naphthyl methacrylate). (f) Poly(β-naphthyl methacrylate). (g) Poly(α-naphthyl carbinyl methacrylate).

Table 17.4. Comparison of the predictions made by using the general-purpose correlation (Equations 5.8-5.10), and the designer correlation (Equation 17.3), for the Fedors-type cohesive energies E_{coh1} of seven polymers. Each one of these polymers contains a structural unit whose group contribution to E_{coh}(Fedors) is not listed. The group contribution table of Fedors cannot, therefore, itself be used to calculate E_{coh}(Fedors) for any of these polymers.

Polymer	Fedors-Type Cohesive Energy E_{coh1} (J/mole)	
	General-Purpose Correlation	"Designer" Correlation
Poly(α-vinyl naphthalene)	58796	60958
Poly(β-vinyl naphthalene)	58630	60766
Poly(acenaphthylene)	58962	61000
Polyoxynaphthoate	66109	--------
Poly(α-naphthyl methacrylate)	79577	--------
Poly(β-naphthyl methacrylate)	79410	--------
Poly(α-naphthyl carbinyl methacrylate)	84518	--------

17.C. Combination of New Correlations and Group Contributions

17.C.1. Calculation of Group Contributions from the New Correlations

It is usually preferable to use the correlations developed in terms of connectivity indices consistently, instead of mixing values calculated for polymeric properties via connectivity indices and group contributions. This section should, nonetheless, help readers accustomed to thinking about the properties of polymers in terms of group contributions, as well as pointing out some possible additional uses of the formalism developed in this book.

Group contributions can be calculated from many of the correlations developed in this book. The contribution of the α-naphthyl group to the Fedors-type cohesive energy will be calculated as an example. This contribution is not completely independent of the environment of the α-naphthyl group, but its dependence on the environment is small. This point will be illustrated by calculating the group contribution in both poly(α-vinyl naphthalene) and poly(α-naphthyl methacrylate) (Figure 17.4). The general-purpose correlation (equations 5.8-5.10) will be utilized since the designer correlation (Equation 17.3) cannot be used for poly(α-naphthyl methacrylate). The α-naphthyl group does not contribute to either N_{atomic} (Equation 5.9) or N_{group} (Equation 5.10). Its contribution to E_{coh1} can thus be estimated by using Equation 17.4, which is obtained by setting $N_{atomic}=N_{group}=0$ in Equation 5.8:

$$E_{coh1}(\alpha\text{-naphthyl group}) \approx 9882.5 \cdot (\text{contribution of the } \alpha\text{-naphthyl group to } {}^1\chi) \qquad (17.4)$$

See Figure 17.5. The contribution of the α-naphthyl group to ${}^1\chi$ (or ${}^1\chi^v$) is calculated by drawing its hydrogen-suppressed graph, and adding up the inverse square roots of the numbers along the bonds. This calculation differs from the calculation illustrated for a complete repeat unit in Figure 2.3 because the α-naphthyl group has only one bond connecting it to the rest of the polymer. The reciprocal square roots of all numbers at the vertices inside the boxes shown in Figure 17.5 must be added to calculate the contribution of this group to ${}^0\chi$. The reciprocal square roots of all numbers along the bonds inside these boxes have to be added to calculate the contribution to ${}^1\chi$. There is a slight difference between the contributions of the α-naphthyl groups in poly(α-vinyl naphthalene) and poly(α-naphthyl methacrylate). The indices at the vertices outside the boxes are different, resulting in different contributions to ${}^1\chi$ from the bonds connecting the α-naphthyl group to the rest of the polymer chain.

The α-naphthyl group in poly(α-vinyl naphthalene) contributes 5.1330 to ${}^1\chi$ and thus 50727 J/mole to E_{coh1}. The α-naphthyl group in poly(α-naphthyl methacrylate) contributes 5.2079 to ${}^1\chi$ and thus 51467 J/mole to E_{coh1}. These two contributions only differ by 1.5%. Their average, which is 51097 J/mole, can be accepted as the typical contribution of an α-naphthyl group to E_{coh}(Fedors).

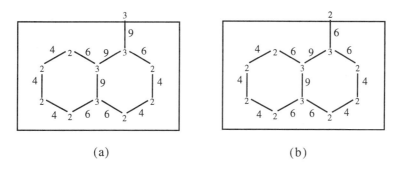

(a)   (b)

Figure 17.5. Hydrogen-suppressed graph of the α-naphthyl group in (a) poly(α-vinyl naphthalene) and (b) poly(α-naphthyl methacrylate). The simple atomic index δ (see Chapter 2) is shown at the vertices. The products of pairs of δ values are shown along the edges. The two graphs differ slightly because the vertex outside the box has $\delta=3$ in (a) and $\delta=2$ in (b), resulting in a small difference between the contributions of the α-naphthyl unit to the first-order connectivity index $^1\chi$. Both graphs make the same contribution to the zeroth-order index $^0\chi$.

If a very accurate estimate of a group contribution is required, the subunit of interest can be placed in many types of bonding environments, and the resulting group contributions can all be averaged. Such an approach is only useful for small groups ranging in size up to about six atoms. The importance of the increment made to a group contribution by the bond or bonds connecting a subunit to the rest of the polymer rapidly decreases with increasing size of the subunit, and is negligible for all practical purposes when large groups such as α-naphthyl subunits are considered. Finally, if a correlation depends only on the zeroth-order connectivity indices $^0\chi$ and/or $^0\chi^v$, the environment need not be considered at all in estimating the group contributions of the subunits.

The group contribution computed above can be used to augment the group contribution table of Fedors [4,5], and to calculate E_{coh}(Fedors) for polymers containing α-naphthyl groups by using group contributions. The results are 59467 J/mole for poly(α-vinyl naphthalene), 80217 J/mole for poly(α-naphthyl methacrylate), and 85157 J/mole for poly(α-naphthyl carbinyl methacrylate). These values only differ by about 1% from the cohesive energies calculated for these three polymers by using the general-purpose correlation and listed in Table 17.4.

It was shown in Figure 17.5 that the contribution of the bond connecting the subunit to the rest of the polymer must be included in calculating the contribution of a monovalent group to a first-order connectivity index. It was shown in Figure 2.3 that only one of the two bonds connecting a repeat unit to the rest of the polymer is counted when calculating $^1\chi$ or $^1\chi^v$. On the other hand, the contribution of the bonds connecting a divalent group smaller than the entire repeat unit to the rest of the repeat unit is half of the sum of the contributions of these two bonds. Because of the lack of periodicity at the boundaries when only a portion of the repeat unit is considered, the calculation must include both bonds and must use the average of their total contribution, instead of moving the square brackets by half a bond and only using the contribution of one of the bonds. See Figure 17.6 for an example. In the same manner, the contributions of the n bonds connecting an n-valent group with n≥3 (for example, trivalent for n=3) to the rest of the repeat unit must all be added up and then divided by n (for example, by 3 for n=3) to obtain the contribution of the n bonds connecting the n-valent group to the rest of the repeat unit.

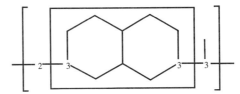

Figure 17.6. Hydrogen-suppressed graph of polyoxynaphthoate, with the value of the simple atomic index δ indicated at selected vertices. The lack of periodicity at the boundaries of the box, which contains a subunit smaller than the full repeat unit, causes different values to be calculated for the contribution of the group enclosed in the box to $^1\chi$ if only one of the two bonds linking it to the rest of the repeat unit is considered. One of these bonds contributes $(1/\sqrt{6}) \approx 0.4082$, and the other bond contributes $(1/\sqrt{9}) \approx 0.3333$. The correct contribution of these two bonds to $^1\chi$ is the average of these two numbers, i.e., $0.5 \cdot [(1/\sqrt{6}) + (1/\sqrt{9})] \approx 0.3708$.

17.C.2. Combined Use of New Correlations and Group Contributions

Many of the existing tables of group contributions were initially developed for ordinary molecules rather than for macromolecules. The augmentation of these tables as described above is therefore especially useful in calculating the key physical properties of ordinary molecules.

Group contributions can also often be combined with results calculated for simpler polymers by using correlations in terms of connectivity indices, to predict the properties of more complicated polymers. This shortcut can be used whether the group contributions were found in a standard reference or estimated by utilizing the procedure described above. *When used judiciously*, this shortcut can expedite the calculations on polymers with large repeat units, and result in major savings in time if the calculations are being performed manually, i.e., on a hand calculator instead of a computer. The connectivity indices are calculated for the polymer with smallest repeat unit containing the group(s) whose contributions are unknown. Correlations in terms of connectivity indices are then used to estimate the properties of this "reference" polymer. The extensive properties of polymers which differ from the reference polymer only by having additional and/or fewer numbers of structural units whose group contributions are known, can then be quickly predicted by adding and/or subtracting the appropriate group contributions, without carrying out all of the calculations in terms of connectivity indices. The defining equations in terms of extensive properties are then used to predict the intensive properties. (See Section 2.C for more detailed discussions on extensive and intensive properties.)

For example, the correlation using connectivity indices for the Fedors-type cohesive energy gives $E_{coh1} \approx 79577$ J/mole for poly(α-naphthyl methacrylate) and 84518 J/mole for poly(α-naphthyl carbinyl methacrylate). The only difference between the structures of poly(α-naphthyl carbinyl methacrylate) [Figure 17.4(g)] and poly(α-naphthyl methacrylate) [Figure 17.4(e)] is the extra methylene (-CH_2-) group in poly(α-naphthyl carbinyl methacrylate). The E_{coh1} of poly(α-naphthyl carbinyl methacrylate) can therefore also be estimated by adding the group contribution of a methylene unit (4940 J/mole) to the E_{coh1} calculated for poly(α-naphthyl methacrylate). This procedure gives $E_{coh1} \approx (79577 + 4940) = 84517$ J/mole, which is almost identical to the result (84518 J/mole) of using the correlation in terms of connectivity indices.

17.D.   Calculation of the Properties of Alternating Copolymers

The calculation of the properties of alternating copolymers does not require the extension of the methods developed thus far in this book. If the properties of the alternating copolymer of repeat units A and B are of interest, they can be predicted simply by treating the alternating copolymer as the homopolymer (-A-B)$_n$ of repeat unit (-A-B).

The calculation of the properties of many homopolymers with large repeat units can be simplified by treating such polymers formally as alternating copolymers of the smaller repeat units of polymers whose properties have already been calculated. (Whether a polymer was, or will be, synthesized by copolymerizing two different monomers, or simply by homopolymerizing a single larger monomer, is irrelevant for structure-property calculations which only consider the structure of the final repeat unit.)   Simple additivity can then be assumed to hold for the extensive properties of the alternating copolymer, such as its connectivity indices, cohesive energy, and molar volume. All of the extensive properties can thus be calculated. The intensive properties, such as the solubility parameter, are defined in terms of the extensive properties. Their prediction therefore does not require any detailed calculations either.

This procedure, at worst, introduces a very small error. It will now be shown that this error is usually completely negligible, and that it generally should become increasingly more insignificant with increasing size of the repeat unit.

See Figure 17.7 for two simple examples of the utilization of the shortcut described above. Polyoxytrimethylene can be treated formally as the alternating copolymer of ethylene and oxymethylene. Similarly, poly(*p*-xylylene) can be treated formally as the alternating copolymer of ethylene and p-phenylene. The major topological quantities and the predicted values of a few physical properties are listed in Table 17.5 for these polymers.

The predicted properties of polyoxytrimethylene are identical to the results calculated by adding up the predicted values of the extensive properties of polyethylene and polyoxymethylene, and using the resulting extensive properties to calculate the intensive properties.

The predicted properties of poly(p-xylylene) differ by a negligible amount from the results calculated by adding up the predicted values of the extensive properties of polyethylene and poly(p-phenylene). The difference is caused by the bond connecting a pair of successive repeat units of poly(p-xylylene), which makes a different contribution to $^1\chi$ and $^1\chi^v$ than the connecting bonds in polyethylene and poly(p-phenylene). The differences in $^1\chi$ and $^1\chi^v$ introduce a small difference in the predicted values of all properties whose calculation utilizes these indices.

Regardless of the size of the repeat unit, there can be no more than one bond in the repeat unit of the alternating copolymer which could introduce small differences of this type. Such effects therefore usually decrease in importance with increasing size of the repeat unit. The shortcut of adding the predicted extensive properties of the homopolymers to calculate the properties of the alternating homopolymer is therefore acceptable as a general method.

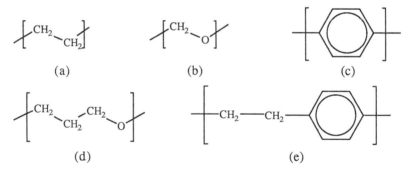

Figure 17.7. Repeat units of several polymers. (a) Polyethylene. (b) Polyoxymethylene. (c) Poly(p-phenylene). (d) Polyoxytrimethylene. (e) Poly(p-xylylene).

Table 17.5. Number of vertices N in the hydrogen-suppressed graph, connectivity indices $^0\chi$, $^0\chi^v$, $^1\chi$ and $^1\chi^v$, and predicted cohesive energies E_{coh1} in J/mole (Equations 5.8-5.10), amorphous molar volumes V at room temperature in cc/mole (Equations 3.13 and 3.14), and solubility parameters δ in $(J/cc)^{0.5}$.

Polymer	N	$^0\chi$	$^0\chi^v$	$^1\chi$	$^1\chi^v$	E_{coh1}	V	δ
Polyethylene	2	1.4142	1.4142	1.0000	1.0000	9883	32.2	17.5
Polyoxymethylene	2	1.4142	1.1154	1.0000	0.5774	8089	23.0	18.8
Poly(p-phenylene)	6	3.9831	3.3094	2.9663	2.0714	29314	66.5	21.0
Polyoxytrimethylene	4	2.8284	2.5296	2.0000	1.5774	17972	55.2	18.0
(Polyethylene+Polyoxymethylene)	4	2.8284	2.5296	2.0000	1.5774	17972	55.2	18.0
Poly(p-xylylene)	8	5.3973	4.7236	3.9495	3.0285	39031	97.9	20.0
[Polyethylene+Poly(p-phenylene)]	8	5.3973	4.7236	3.9663	3.0714	39197	98.7	19.9

17.E. Calculation of the Properties of Random Copolymers

The properties of random copolymers can be estimated by using weighted averages for all extensive properties, and the appropriate definitions for the intensive properties in terms of the extensive properties. Let m_1, m_2, ..., m_n denote the mole fractions of n different types of repeat units in a random copolymer. [The most common random copolymers have n=2. Terpolymers (n=3) are also often encountered.] The n mole fractions then add up to one, and the extensive properties of a random copolymer can be estimated by using the mole fractions as weight factors:

$$\sum_{i=1}^{n} m_i = 1 \tag{17.5}$$

$$\text{(Extensive copolymer property)} \approx \sum_{i=1}^{n} \left[m_i \cdot \text{(value for homopolymer of repeat unit i)} \right] \tag{17.6}$$

For example, the cohesive energy and the molar volume (extensive properties) and the solubility parameter (an intensive property) of a random copolymer containing two different types of repeat units with mole fractions of m_1 and m_2 can be estimated by using equations 17.7-17.10:

$$m_1 + m_2 = 1 \tag{17.7}$$

$$E_{coh1}(\text{copolymer}) \approx m_1 \cdot E_{coh1}(\text{homopolymer of Repeat Unit 1})$$
$$+ m_2 \cdot E_{coh1}(\text{homopolymer of Repeat Unit 2}) \tag{17.8}$$

V(copolymer) ≈ m_1·V(homopolymer of Repeat Unit 1)

+ m_2·V(homopolymer of Repeat Unit 2)   (17.9)

$$\delta(\text{copolymer}) \approx \sqrt{\frac{E_{coh1}(\text{copolymer})}{V(\text{copolymer})}}$$ (17.10)

The dielectric constant of the low-hydroxyl grade of poly(vinyl butyral) (see Figure 2.9) will be calculated in detail as an example. The method used to synthesize poly(vinyl butyral) typically results in the presence of a substantial fraction of unreacted hydroxyl (-OH) groups randomly located on polymer chains. The amount of these hydroxyl groups is normally described in terms of the weight fraction of residual poly(vinyl alcohol) (see Figure 4.1). Even the most completely reacted grade of poly(vinyl butyral) contains approximately twelve weight percent of poly(vinyl alcohol) [6]. As discussed in Chapter 9, the value of the dielectric constant is very sensitive to the presence of such a considerable amount of exposed hydroxyl groups, which therefore have to be considered in order to calculate the dielectric constant of poly(vinyl butyral) correctly.

The weight fractions w_2=0.12 of poly(vinyl alcohol) and w_1=0.88 of poly(vinyl butyral) must be converted into mole fractions. The molecular weight of a repeat unit is 142.2 for poly(vinyl butyral), and 44.1 for poly(vinyl alcohol). The mole fractions m_1 and m_2 of poly(vinyl butyral) and poly(vinyl alcohol) can be determined by solving equations 17.7 and 17.11 as a pair of simultaneous equations with two unknown variables.

44.1·m_2 = 0.12·(142.2·m_1 + 44.1·m_2)   *[for 12 weight percent poly(vinyl alcohol)]*   (17.11)

The left-hand-side of Equation 17.11 is the contribution of the repeat units of poly(vinyl alcohol) to the average molecular weight per repeat unit of the copolymer. The quantity in parentheses in the right-hand-side is the average molecular weight per repeat unit of the copolymer. The factor of 0.12 indicates that only 12% of this average molecular weight comes from the repeat units of poly(vinyl alcohol). The simultaneous solution of equations 17.7 and 17.11 gives m_1=0.694 and m_2=0.306. Since the molecular weight of one repeat unit of poly(vinyl alcohol) is much smaller than the molecular weight of one repeat unit of poly(vinyl butyral), the mole fraction of poly(vinyl alcohol) is much larger than its weight fraction.

Equations 9.11 and 9.12 will be used to calculate the dielectric constant at room temperature. Three extensive variables are needed, namely E_{coh1} as calculated by equations 5.8-5.10, the van der Waals volume V_w, and the correction term N_{dc}, to express ε(298K) as an intensive property. The results of calculations using Equation 17.6 to estimate E_{coh1}, V_w and N_{dc}, and equations 9.11 and 9.12 to estimate ε(298K), for the low-hydroxyl grade of poly(vinyl butyral) containing twelve weight percent of poly(vinyl alcohol), are summarized in Table 17.6.

Since the calculations assume that the specimens are completely dry, and do not consider the effects of the humidity, the predicted value of ε(298K) in the limit of pure poly(vinyl alcohol) is much lower than the extremely high reported typical values [7] of 8 to 14 for poly(vinyl alcohol).

The calculated value of ε(298K)$\approx$2.67 for poly(vinyl butyral) as a random copolymer containing 12% by weight of poly(vinyl alcohol), is in close agreement with the experimental value [7] of 2.69 for the low-hydroxyl grade of poly(vinyl butyral). If the presence of hydroxyl groups is neglected and the calculations are carried out for the homopolymer of poly(vinyl butyral), a dielectric constant of 2.42, which is much lower than the observed value, is calculated.

The software implementation of this method allows the user the convenience of specifying the copolymer composition either in mole fractions or in weight fractions. The method outlined above is used in either case in the calculations, with the composition being first converted into mole fractions by the program if it was supplied by the user in weight fractions.

The main limitation of the weighted averaging scheme outlined above is that each intensive property varies monotonically with composition. Deviations from such behavior, which may occur because of specific interactions between different repeat units, and manifest themselves as a maximum or a minimum in a property at an intermediate composition, are not taken into account.

Table 17.6. Cohesive energy E_{coh1} (Equations 5.8-5.10), van der Waals volume V_w, the term N_{dc} used in predicting the dielectric constant ε, and predicted values of ε at room temperature, for the "low-hydroxyl" grade of poly(vinyl butyral) containing roughly 12% by weight of poly(vinyl alcohol). The subscript 1 denotes poly(vinyl butyral), while the subscript 2 denotes poly(vinyl alcohol), for the weight fractions and mole fractions which are denoted by w and m, respectively.

Polymer	w_1	w_2	m_1	m_2	E_{coh1}	V_w	N_{dc}	ε
Poly(vinyl butyral) *(100%)*	1.00	0.00	1.000	0.000	44652	83.68	0.0	2.42
Poly(vinyl alcohol) *(100%, dry)*	0.00	1.00	0.000	1.000	35296	25.01	12.0	4.55
Poly(vinyl butyral) *(low-hydroxyl grade)*	0.88	0.12	0.694	0.306	41789	65.73	3.7	2.67

17.F. A Software Package Implementing the Key Correlations

A flow chart describing calculations using a software package [8], SYNTHIA, which automates the use of key correlations from this book, is shown in Figure 17.8. This program enables the prediction of the properties of uncrosslinked polymers constructed from the nine elements (C, N, O, H, F, Si, S, Cl and Br) considered in developing the correlations. For silicon and sulfur atoms, only the bonding configurations listed in Table 2.1 are allowed.

The execution of this program begins with the interactive construction or retrieval of a *coordinate file* and a *molecular data file* for the polymeric repeat unit by using a molecular modeling program. These files contain the coordinates of the atoms as well as a connectivity table describing the bonding pattern. The program then uses this information to predict the material properties of the polymer, by using a subset of the correlations presented in this monograph.

The software package is modular in structure. It can be easily modified to run on different types of computers, and to accept molecular data files written in different file formats. In its *original version* as developed at Dow, it had three modes of execution:

1. The *automatic homopolymer mode* was convenient when calculations on many homopolymers (each represented by its own molecular data file containing both the atomic coordinates and the connectivity table), all under identical conditions, were needed. The directory containing the molecular data files was specified by the user. The program then automatically performed calculations on all requested molecular data files in that directory, in alphabetical order by their names. The output for each structure listed the calculated properties.

2. In the *interactive homopolymer mode*, the calculations were performed one homopolymer (i.e., one molecular data file) at a time. The interactive homopolymer mode allowed greater flexibility than the automatic homopolymer mode in the conditions of calculation and the output options. For example, if the user has a good independent estimate of the T_g and/or the density of the polymer, or an estimate of the repeat unit length obtained by using an accurate molecular modeling program, these values could be supplied to the program, and used by the program instead of the values calculated internally, when required as parameters for the prediction of other properties. The interactive mode allowed the user to plot several properties as a function of the temperature, in addition to providing an output table listing the calculated properties.

3. The *interactive copolymer mode* allowed the prediction, and presentation in tables and graphs, of the properties of alternating or random copolymers, as discussed in sections 17.D and 17.E. The input consisted of the names of the molecular data files of the repeat units incorporated into the copolymer, and the mole or weight fraction (selected by the user) of each repeat unit.

The user interface of this program has been rewritten completely by Biosym Technologies (renamed Molecular Simulations, Inc., after a recent merger) since its commercialization, to provide an extremely flexible and fully interactive user interface. The capabilities of this interface include the options for the user to provide designer correlations for any property of interest, to supply experimental values for three important properties (glass transition temperature, density and solubility parameter), to plot any calculated property against any other with a variety of display options, to select subsets of properties for calculation, and to obtain both the key structural descriptors and the predicted properties in a spreadsheet format (in addition to the usual output text file) to facilitate any further desired data analysis.

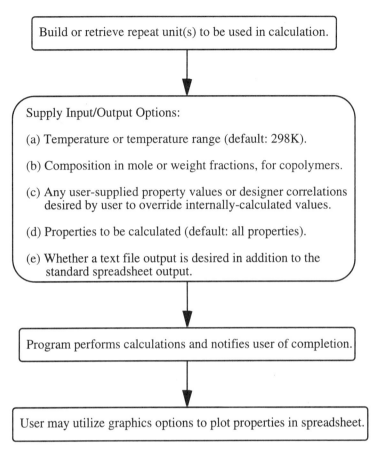

Figure 17.8. Flow chart of calculations using the SYNTHIA software package.

In a few cases, the automation and generalization of the rules to calculate the structural parameters required extensions to cover very specialized types of structural features. Whenever possible, experimental data generated at Dow were used in making such generalizations, and/or in evaluating their quality.

For example, for complicated assemblies of fused rings located along the chain backbone, rules were developed to determine in a consistent and systematic manner whether an assembly is closest to being in a *para*, *meta*, or *ortho* configuration in terms of the effect of its motions on the rest of the chain backbone. These rules involved the detailed consideration of both the location of the two bonds connecting the assembly to the rest of the repeat unit, and the angle between these two bonds.

Another generalization involved situations where a fused ring assembly with more than two rings, some of which are aromatic and some of which are not aromatic, is located along the chain backbone. The following expression was programmed to determine the contributions of these types of units to the structural parameter x_1 (see Section 6.B.2) in calculating the glass transition temperatures of such polymers:

x_1 = (number of aromatic rings in assembly)

  + 0.75·(number of non-aromatic rings adjacent to aromatic rings in same assembly)

  + 0.5·(number of non-aromatic rings not adjacent to any aromatic rings in same assembly)

  + [(number of aromatic rings in assembly) + 0.5·(number of non-aromatic rings adjacent

      to aromatic rings in assembly)] / (total number of rings in assembly)   (17.12)

The rules generalizing the determination of whether a complicated fused ring assembly is in a *para*, *meta*, or *ortho* configuration, and the calculation of its contribution to x_1 when it contains both aromatic and non-aromatic rings, were found to be useful in making predictions on a number of polymers in our practical work.

Another extension of the rules, which was made for the sake of generality but could not be tested in practice, was to assume that the correction term $N_{C=C}$ used for several properties also encompasses carbon-carbon triple bonds ($C\equiv C$).

For copolymers built from a mixture of repeat units whose homopolymers would be described by different relationships for the characteristic ratio (Equation 12.24 or Equation 12.25) and/or the glass transition temperature (Equation 6.3, or the specialized Equation 17.1), weighting rules were developed to express these properties by combining the correlations. These rules also could not be tested in practice, and were considered to be merely reasonable mathematical interpolations.

A simple scheme for predicting the physical properties of polymers from their structures cannot obviate the user's need to have an understanding of materials, and of the relationships between the important material properties and the performance requirements of specific technological applications. A software package implementing a set of simple quantitative structure-property relationships is most useful if the user has some knowledge of polymer properties, and exercises good judgment in deciding which ones among the many properties automatically calculated by the program are relevant or even meaningful for a particular problem or a particular type of polymer. When used judiciously, the program implementing our correlations is especially useful in evaluating the trends in the properties, expected to be caused by structural variations. See Chapter 18 for detailed examples of such calculations.

17.G. Prediction of Properties of Multiphase Polymeric Systems

17.G.1. Types of Materials and Morphologies

Many materials of industrial interest are multiphase materials. As the demands for the required balance of properties become increasingly stringent for novel materials, the technological importance of multiphase materials, which may manifest synergistic combinations of desirable properties not simultaneously attainable with simpler single-phase materials, will also be increasing [9].

Examples of multiphase polymeric materials [10] include:

1. *Composites,* where a matrix material is "filled" with fibers, platelets or particulates [11]. The quality of the adhesion between the phases is a very important factor in determining the mechanical properties of composites, since adhesion determines the effectiveness of the interface in transferring an applied load from the matrix to the filler phase.

2. *Blends* of immiscible components which undergo phase separation [12]. Blends usually have rather irregular morphologies, and the shapes of their phases are difficult to predict in advance.

3. *Block copolymers* of immiscible components which undergo phase separation [13].

    (a) *Conventional block copolymers*, which usually consist of a thermoplastic "hard phase" such as polystyrene, and a rubbery "soft phase" such as polybutadiene or polyisoprene, in a few long blocks (as in diblock copolymers where each chain has two blocks and triblock copolymers where each chain has three blocks). Unlike blends, conventional block copolymers of immiscible components usually manifest a series of well-defined morphologies (such as spherical, cylindrical, ordered bicontinuous double diamond, and lamellar) as a function of the component volume fractions [14]. These more regular morphologies arise from the fact that the phases are interconnected by chemical bonds between the blocks on each chain. For the same reason, the rheological properties of immiscible blends and block copolymers also differ since the chemically distinct parts of a block copolymer remain interlinked along the chains even after melting.

    (b) *Segmented block copolymers*, such as polyurethanes, where the average chain contains many relatively short blocks. The possibility of mixing between the hard phase and the soft phase increases as the segmented blocks are shortened. The quantitative understanding and prediction of the morphologies of segmented block copolymers is at a more primitive stage of development than those of the morphologies of conventional block copolymers.

4. *Semicrystalline polymers* [15]. The crystalline and amorphous phases are either chemically identical or very similar (as in cases where the amorphous phase may contain a slightly higher mole fraction of a comonomer which does not pack effectively into the crystallites). A variety of crystalline morphologies are possible, such as extended-chain crystallites, fringed micelles, folded chain lamellae, and spherulites consisting of nearly-spherical aggregations of lamellae with some amorphous material between the lamellae. Some properties of the crystalline and amorphous phases are similar when the two phases are in similar physical states. For example

(Chapter 4), C_p is similar for the amorphous and crystalline phases when $T<T_g$ and both phases are "solids", and equal when $T>T_m$ and both phases are "liquids". When $T_g<T<T_m$, the crystalline phase is "solid" and the amorphous phase is "liquid", so that C_p is much larger for the amorphous phase. The crystalline phase is generally stiffer than the amorphous phase, and its modulus can be higher by several orders of magnitude when $T_g<T<T_m$.

5. *Laminates* (multilayer structures), where each layer itself may be either a single-phase material or a multiphase material [16].

6. *Foams* (cellular structures made by expanding a material by growing bubbles in it) [17].

(a) *Closed-cell foams* have polymeric walls separating adjacent cells. The polymer is the only continuous phase. A gas phase is trapped as a "filler" inside the cells. Gas molecules can move slowly between the cells, and between the foam and its environment, by the solution-diffusion mechanism. As a result, in most closed-cell foams, the gas inside the cells eventually becomes replaced with air. This may, however, take an extremely long time if the polymer has very good barrier properties and/or the blowing agent (the gas used to expand the foam during fabrication) consists of very large molecules.

(b) *Open-cell foams* where the rupture of the cell walls results in a network of interconnected cells through which air molecules can move very easily. Since there are no cell walls to slow down the exit of the blowing agent from the foam, it is rapidly replaced by air after fabrication. The polymer and air are co-continuous phases in an open-cell foam.

It is obvious from this discussion that the properties of the individual components, their volume fractions, and the morphology at the microscopic scale, are the key factors in determining the properties of multiphase materials. Very loosely, the *microscopic scale* of morphology can be defined as the features, such as the shapes, sizes, and extents of spatial continuity of the components, which can be investigated by microscopy. Readers interested in understanding multiphase materials are urged to browse through the book by Woodward [18] where micrographs are presented for a wide variety of such materials with informative explanations.

An important morphological distinction involves the difference between systems which can and which cannot manifest phase inversion with changing composition. For example, the binary blends and conventional block copolymers of immiscible components often manifest phase inversion as the volume fraction of one of the components is varied from 0% to 100%, with the "filler" phase first becoming co-continuous with the matrix phase, and eventually becoming the matrix phase itself. An example of this phenomenon is found in the morphology of injection-molded blends of bisphenol-A polycarbonate with copolymers of styrene and acrylonitrile, where phase inversion occurs roughly in the volume fraction range of 0.31 to 0.69 of the components and can be explained in terms of percolation theory [19]. On the other hand, polymers containing inorganic fillers such as glass spheres or fibers, carbon fibers, or talc or mica particles, remain the matrix (sometimes called the *binder*) regardless of the volume fraction of the filler, which can go as high as the maximum packing fraction of fillers of that particular shape (or shape distribution) and size (or size distribution).

17.G.2. Prediction of Properties

Most of the methods developed in this book are, by themselves, only applicable to amorphous polymers and amorphous polymeric phases. Their combination with other types of methods, which have been developed over several decades to calculate the properties of multiphase materials from component properties and multiphase system morphology, enables us to expand their applications to include the prediction of selected properties of multiphase polymeric systems where one or more of the phases are amorphous polymers. In other words, the methods developed in this book are used to predict the properties of the amorphous polymeric phases of the multiphase system, and these predicted properties are then inserted into the equations of the composite models (along with the material parameters of the other components, obtained from other sources such as literature tabulations) to predict the properties of the multiphase system.

Many books and review articles are available for general treatments of composite theory. Most notable are the book of Nemat-Nasser and Hori [20] for its mathematical thoroughness, the book of Christensen [11] for its emphasis on the engineering aspects, and an article by Tucker *et al* [21] for its emphasis on the internally consistent combination of a set of judiciously chosen techniques to predict the thermoelastic properties of a wide variety of multiphase polymeric systems.

There are two types of properties for which relatively simple closed-form expressions based on composite theory have been found to be especially useful:

1. *Thermoelastic properties* (moduli and thermal expansion coefficients). This is the area where the largest amount of successful past method development has taken place. See Halpin and Kardos [22,23], and Chow [24], for examples of equations which are particularly useful and versatile.
3. *Transport properties* (dielectric constant, electrical conductivity, magnetic permeability, thermal conductivity, and gas diffusivity and permeability). Models for these properties often utilize mathematical treatments [25] similar to those used for the thermoelastic properties, once the appropriate mathematical analogies [26] are made.

Our own experience (which has been very encouraging) has mainly involved the use of composite theory coupled with SYNTHIA calculations to predict the thermoelastic properties of multiphase polymeric systems. In particular, with the exception of foams, the procedure of Tucker *et al* [21,27] can be used to predict the thermoelastic properties of all types of multiphase systems containing any number of components, of any shape (with the shape represented by the *aspect ratio* which can be defined as the ratio of the largest dimension of a filler phase divided by its smallest dimension), in any type of average orientation state, and of any number of layers:

1. The properties of a two-phase system (a continuous "matrix" and a discontinuous "filler") are calculated in terms of the properties and volume fractions of the components, by *micromechanical methods* of which the equations of Halpin and Kardos (historically referred to as the Halpin-Tsai equations [22,23]) and Chow [24], are examples. If the filler is anisotropic (such as a fiber or a platelet), it is oriented uniaxially. Perfect adhesion is assumed between the phases.

2. If another average orientation state besides uniaxial alignment is of interest for an anisotropic filler, orientation tensors are used to perform *orientation averaging* [28].

3. If there is more than one type of filler phase, *aggregate averaging* is performed in terms of the volume fractions to calculate the properties of the hybrid composite system. (We have found that phase inversion is also best incorporated during aggregate averaging into the calculations.)

4. If the system of interest is multilayered, the results of the calculations for the individual layers are combined within the context of *lamination theory* to calculate its properties.

Foams are a special case in terms of the prediction of their mechanical properties. Their unique morphologies have allowed the development of a very simple alternative set of equations, to provide first estimates of both their thermoelastic properties and their mechanical strengths in terms of the modulus and the strengh of the bulk material, the relative density (density of foam divided by density of bulk material), and whether the cells are closed or open [17].

17.H.   Utilization to Provide Input Parameters for Other Types of Methods

The methods developed in this book can be used to provide the key input parameters for other methods intended to predict polymer properties, in addition to their use, as discussed above, to provide input parameters for composite models of multiphase materials. Two recent examples of predictive methods which use the computational tools developed in this book will now be given.

A method named *group interaction modeling* was developed by Porter [29] for predicting the engineering properties of polymers. This method is based on a new equation-of-state including both energy terms resulting from interchain interactions in the polymer and those resulting from the imposition of external fields. This equation-of-state is then used as the basis of calculations of the thermal and mechanical properties. A few key characteristics of the polymeric repeat unit must be assigned input values in order to use this method. These characteristics include the van der Waals volume and the cohesive energy. The development of our new correlations has enabled the calculation of V_w and E_{coh} for a much wider range of polymeric repeat units, significantly expanding the range of applicability of Porter's model [29] whose main value appears to be in providing some thought-provoking physical insights into the behavior of polymers.

An approach to predicting polymer properties based on some of the concepts and tools of artificial intelligence has been proposed by Sumpter and Noid [30]. These authors have found that *artificial neural networks* can be used to predict polymer properties, with the structural descriptors developed in this book providing the best choice of descriptors based on a comparison of several types of descriptors. These structural descriptors are used as inputs into an artificial neural network and a set of desired properties are used as outputs. After this "training" phase is completed, the neural network can be given the values of the descriptors for polymers which were not used in the training phase and will output a set of "predicted" properties using the internal correlations learned during the training phase. The neural network procedure has been applied to nine properties (V, C_p^s, ΔC_p, E_{coh1}, δ, T_g, n, ε and λ) discussed in this book, with satisfactory results. The main disadvantage of a neural network is that it is a "black box" whose internal workings are hidden from the user, so that no direct means is provided to interpret or understand how an answer was estimated. The main significance of the work of Sumpter and Noid [30] is that it may begin the integration of the methods developed in this book with the powerful tools of artificial intelligence, perhaps leading to the development of an expert system for materials design.

17.I. Possible Directions for Future Work

A new method allowing the prediction of many important physical properties of polymers prior to synthesis was presented in this book. The quantitative structure-property relationships developed based on this method enable the prediction of the properties of uncrosslinked isotropic amorphous polymers constructed from nine key elements (carbon, nitrogen, oxygen, hydrogen, fluorine, silicon, sulfur, chlorine and bromine) from which most of the technologically important synthetic polymers are built. Some properties of crosslinked polymers can also be predicted.

The correlations developed in this book allow the polymer designer to transcend the limitations of traditional group contribution methods, and to predict the properties of polymers containing completely novel types of structural units without being limited by the need for group contributions. The method presented in this book is therefore of much wider applicability than group contribution techniques, while being of comparable accuracy.

The methods developed in this book also provide input parameters for calculations using composite theory for the thermoelastic and transport properties of many types of multiphase polymeric systems in terms of material properties and phase morphology. Material properties calculated by the correlations presented in this book can also be used as input parameters in computationally-intensive continuum mechanical simulations (for example, by finite element analysis) for the properties of composite materials and/or finished parts of diverse sizes, shapes and configurations. The work presented in this book therefore constitutes a "bridge" from the molecular structure and fundamental properties of materials to the performance of finished parts.

Despite of the broad scope of this book, many important improvements in the techniques for the prediction of the properties of polymers from their structures have been left to the future. It is hoped that this new edition of the book will continue to stimulate additional research, both in the industrial and the academic communities, on the development of structure-property relationships to predict the properties of polymers. The following are some possible directions for such work:

1. Straightforward refinements and extensions of the correlations presented in this book. These refinements and extensions will be made more easily in the future as a result of the flexibility of the user interface of the SYNTHIA program implementing the method, which allows users to obtain "molecular spreadsheets" of the structural descriptors and the predicted properties, to analyze data statistically within the program, and to enter their own correlations.

   (a) Addition of new data as they become available into data tables, to refine correlations.

   (b) Further analysis of data, to find ways of simplifying or combining the correlations.

   (c) Simplification and/or generalization of some of the correlations in terms of connectivity indices, for example by using δ^v as an adjustable parameter to eliminate the atomic correction terms and/or by using higher-order indices to eliminate many of the group correction terms.

   (d) Development of additional "designer correlations".

   (e) Use of the new correlations to extend the existing tables of group contributions.

2. There is still room for major improvements in the capabilities for predicting the mechanical and rheological properties. Areas open to such improvements include prediction of rate dependence of mechanical properties (including fatigue, creep and stress relaxation), prediction of crazing stress (including stress required for craze propagation, as distinguished from the lower craze inception stress), and prediction of temperature dependence of the zero-shear viscosity.

3. Development of additional correlations to treat the effects of crosslinking on the properties.

4. Extension of the correlations to polymers containing other elements, such as inorganic and organometallic polymers [31] containing boron [32,33] or phosphorus [34]. Such an extension appears feasible since the properties of such solids as amorphous selenium, and complicated chalcogenide-based alloys used in electrical and optical applications, can often be calculated by techniques resembling the methods used to study ordinary polymers [35,36].

5. The coupling of quantitative structure-property relationship schemes with software based on artificial intelligence tools [30,37], such as expert systems with reverse engineering capabilities [38], to deduce the optimum size, shape and layout of finished parts, and the optimum composition and molecular structure of the material to be used in a given application, from the performance requirements by using quantitative design criteria for material selection [39], is an important long-term goal. Reverse engineering would provide an efficient computerized method to search for plausible structures, and the structure-property relationships would then be used to evaluate these candidates. The easiest way to make progress in this area may be to

use the new method to extend and augment the existing tables of group contributions, and then use these vastly expanded tables of group contributions in the pattern searches required by the expert system for reverse engineering. A key stumbling block to overcome is the need to build updatable rules into the expert system to eliminate candidates which cannot be made by using any of the available techniques of synthetic chemistry. Otherwise, the user may end up with the frustration of having the expert system provide a very long list of candidate materials which may have desirable properties but cannot be synthesized.

In addition to their practical utility in industrial research and development, the correlations and extensive tables of data presented in this book should help researchers who are pursuing more fundamental theoretical approaches to the structure-property relationships in polymers. The correlations may point out interrelationships which have not been noticed in previous work. The tables of data may provide many test cases to evaluate new predictive theories. Empirical (correlative) types of approaches, and their embodiments in predictive schemes, often serve as catalysts for conceptual advances. In many areas of science and technology, phenomenological and empirical treatments have preceded significant theoretical and conceptual advances. Such treatments have provided foundations of critically evaluated data and carefully quantified sets of interrelationships, upon which later workers have erected theoretical edifices transcending the systematic correlation of available data. We therefore hope that theoretically inclined polymer scientists will find some of the information presented in this book helpful in their efforts.

References and Notes for Chapter 17

1. The term "designer correlation" is modified from A. J. Hopfinger, M. G. Koehler, R. A. Pearlstein and S. K. Tripathy, *J. Polym. Sci., Polym. Phys. Ed.*, 26, 2007-2028 (1988), who referred to "designer equations" for calculating the glass transition temperatures of structurally related polymers.

2. D. W. van Krevelen, *Properties of Polymers*, second edition, Elsevier, Amsterdam (1976). See Chapter 3, Section E, of this book, for discussions of the methods of standard properties and standard substances.

3. The designer correlation for the glass transition temperatures of polyesters was developed by S. J. Mumby at Molecular Simulations, Inc.

4. R. F. Fedors, *Polymer Engineering and Science*, *14*, 147-154 (1974).

5. R. F. Fedors, *Polymer Engineering and Science*, *14*, 472 (1974).

6. T. P. Blomstrom, article titled "*Vinyl Acetal Polymers*", *Encyclopedia of Polymer Science and Engineering*, *17*, Wiley-Interscience, New York (1989).

7. *Tables of Dielectric Materials*, Volume IV: Technical Report No. 57, Laboratory for Insulation Research: Massachusetts Institute of Technology (January 1953).

8. The original software package was developed by J. Bicerano and E. R. Eidsmoe at The Dow Chemical Company (1991). It was then licensed to Biosym Technologies, Inc. (renamed to Molecular Simulations, Inc., after a recent merger), for global commercialization. Its first commercial version became available in 1993.

9. *Materials Science and Engineering for the 1990's: Maintaining Competitiveness in the Age of Materials*, National Academy Press, Washington, D.C. (1989).

10. *Multicomponent Polymer Systems*, edited by I. S. Miles and S. Rostami, Longman Scientific & Technical, Essex, England (1992).

11. R. M. Christensen, *Mechanics of Composite Materials*, John Wiley & Sons, New York (1979).

12. J. A. Manson and L. H. Sperling, *Polymer Blends and Composites*, Plenum Press, New York (1976).

13. *Thermoplastic Elastomers: A Comprehensive Review*, edited by N. R. Legge, G. Holden and H. E. Schroeder, Hanser Publishers, Munich (1987).

14. D. J. Meier, Ref. [13], Chapter 11.

15. B. Wunderlich, *Macromolecular Physics*, Academic Press, New York (1973).

16. K. Nichols, J. Solc, M. Barger, D. Pawlowski and J. Bicerano, *ANTEC '94 Preprints*, 1433-1437 (May 1994).

17. L. J. Gibson and M. F. Ashby, *Cellular Solids*, Pergamon Press, New York (1988).

18. A. E. Woodward, *Atlas of Polymer Morphology*, Hanser Publishers, Munich (1989).

19. C. B. Arends, *Polymer Engineering and Science*, *32*, 841-844 (1992).

20. S. Nemat-Nasser and M. Hori, *Micromechanics: Overall Properties of Heterogeneous Materials*, North-Holland, New York (1993).

21. C. W. Camacho, C. L. Tucker III, S. Yalvac and R. L. McGee, *Polymer Composites*, *11*, 229-239 (1990).

22. J. C. Halpin and J. L. Kardos, *J. Appl. Phys.*, *43*, 2235-2241 (1972).

23. J. C. Halpin and J. L. Kardos, *Polymer Engineering and Science*, *16*, 344-352 (1976).

24. T. S. Chow, *J. Materials Science*, *15*, 1873-1888 (1980).

25. J. T. Mottram, in *Computer Aided Design in Composite Material Technology III*, edited by S. G. Advani, W. R. Blain, W. P. de Wilde, J. W. Gillespie, Jr., and O. H. Griffin, Jr., Computational Mechanics Publications, Southampton (1992), 615-626.

26. Z. Hashin, *J. Applied Mechanics*, *46*, 543-550 (1979).

27. R. S. Bay and C. L. Tucker III, unpublished calculations.

28. S. G. Advani and C. L. Tucker III, *J. Rheology*, *31*, 751-784 (1987).

29. D. Porter, *Group Interaction Modelling*, Marcel Dekker, New York (in press).

30. B. G. Sumpter and D. W. Noid, *Macromol. Theory Simul.*, *3*, 363-378 (1994).

31. C. E. Carraher, *J. Macromol. Sci.-Chem.*, *A17*, 1293-1356 (1982).

32. K. O. Knollmueller, R. N. Scott, H. Kwasnik and J. F. Sieckhaus, *J. Polym. Sci., A-1*, *9*, 1071-1088 (1971).

33. E. N. Peters, *Ind. Eng. Chem. Prod. Res. Dev.*, *23*, 28-32 (1984).

34. H. R. Allcock, *Phosphorus-Nitrogen Compounds*, Academic Press, New York (1972), Part III.

35. J. Bicerano and D. Adler, *Pure & Applied Chemistry*, *59*, 101-144 (1987).

36. J. Bicerano and S. R. Ovshinsky, in *Applied Quantum Chemistry*, edited by V. H. Smith, H. F. Schaefer III and K. Morokuma, D. Reidel Publishing Company, Holland (1986), 325-345.

37. P. Sargent, *Materials Information for CAD/CAM*, Butterworth-Heinemann, Oxford (1991).

38. K. G. Joback, *Designing Molecules Possessing Desired Physical Property Values*, digest of doctoral thesis, Department of Chemical Engineering, Massachusetts Institute of Technology (June 17, 1989).

39. M. F. Ashby, *Materials Selection in Mechanical Design*, Pergamon Press, Oxford (1992).

CHAPTER 18

DETAILED EXAMPLES

18.A. Introductory Remarks

It is instructive to calculate the key properties of specific polymers by using the methodology and the correlations developed in this book, to complement the earlier chapters where individual properties were calculated for large sets of polymers. Polystyrene, and random copolymers of styrene and oxytrimethylene, will be used as the examples. Only the correlations which provide the preferred embodiment of our work will be used in this chapter.

In previous chapters, information provided by group contributions, and available experimental data, were incorporated in the calculation of "best estimates" of some properties. In this chapter, the properties will be calculated by using the new correlations consistently in estimating all of the parameters encountered in intermediate steps of the calculations. Some values predicted below for the properties therefore differ from the results listed in the previous chapters. The steps involved in calculations of physical properties will be listed below. See earlier chapters for comparisons of the results with experimental data and with results of calculations by using group contributions.

The calculations summarized below can be performed automatically by the software package, implementing the methodology, for difunctional repeat units of any degree of complexity. Users of this computer program do not have to perform any calculations manually. *Any details provided below concerning this software package only refer to its original version developed at Dow, and not to any of its commercial versions which are subject to changes at the discretion of the vendor.*

18.B. Polystyrene

Step 1. **Draw the structure of the polymeric repeat unit.** [See Figure 1.1(a).]

Step 2. **Calculate the molecular weight M per repeat unit.** A repeat unit of polystyrene has 8 C and 8 H atoms ($N_C=N_H=8$). The atomic weights are 12.011 atomic mass units (amu) for C and 1.00794 amu for H, so that $M = 8 \cdot (12.011 + 1.00794) \approx 104.15$ g/mole.

Step 3. **Calculate the number of backbone rotational degrees of freedom N_{BBrot}, the number of side group rotational degrees of freedom N_{SGrot}, and the total number of rotational degrees of freedom N_{rot},** by using the rules listed in Section 4.C. For polystyrene [Figure 1.1(a)], $N_{BBrot}=2$, $N_{SGrot}=1$, and therefore $N_{rot}=3$.

Step 4. **Calculate the length l_m of the polymeric repeat unit** *in its fully extended conformation.* For most polymers, l_m is the only quantity used in the calculations which is not uniquely determined from the connectivity table of the repeat unit (i.e., from topological input), and instead calculated from the atomic coordinates (i.e., from geometrical input). It can be estimated, as illustrated for PMMA in Figure 11.1 since PMMA and polystyrene have identical (vinylic) chain backbones. The result is $l_m \approx 2.52$ Å $= 2.52 \cdot 10^{-8}$ cm. As pointed out in Chapter 11, more sophisticated molecular modeling methods give shorter repeat unit lengths for vinylic polymers ($l_m \approx 2.32$ Å according to Biosym's Polymer Version 6.0), and this refinement can make a significant difference in some calculated properties. We will use the l_m value estimated "manually" in most of the calculations below, indicating the effects of using the refined l_m value when the refinement makes a large difference in the predicted value of a property.

Step 5. **Build the hydrogen-suppressed graph of the repeat unit,** assign the simple atomic indices δ and the valence atomic indices δ^v, **and calculate the zeroth-order connectivity indices $^0\chi$ and $^0\chi^v$ and the first-order connectivity indices $^1\chi$ and $^1\chi^v$** by using equations 2.2-2.7. The values of δ and δ^v used for various atomic configurations are listed in Table 2.1. The values used for polystyrene are shown in Figue 18.1. The number N of non-hydrogen atoms (equivalently, the number of vertices in the hydrogen-suppressed graph) is 8 for polystyrene, which has $^0\chi \approx 5.3973$, $^1\chi \approx 3.9663$, $^0\chi^v \approx 4.6712$ and $^1\chi^v \approx 3.0159$.

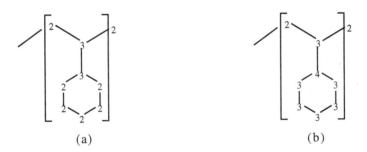

         (a)   (b)

Figure 18.1. Hydrogen-suppressed graph of polystyrene. (a) The simple atomic indices δ are shown at the vertices. (b) The valence atomic indices δ^v are shown at the vertices.

Step 6. **Calculate the van der Waals volume V_w** by using equations 3.10 and 3.11. The result of this calculation is $V_w \approx 64.04$ cc/mole for polystyrene. These equations contain various types of structural parameters, atomic correction indices, and group correction indices, in addition to the connectivity indices, as do many of the correlation equations to be used later. All such parameters are determined automatically by the software package from the connectivity table.

Step 7. **Calculate the amorphous molar volume at room temperature** (298K), i.e., V(298K), by using equations 3.13 and 3.14; **and the amorphous density at room temperature**, i.e., ρ(298K), by using Equation 3.2. The results are V(298K)$\approx$97.0 cc/mole and ρ(298K)$\approx$1.074 grams/cc for polystyrene. (At this step, we calculate V and ρ only at room temperature. Prediction of these properties at other temperatures requires calculation of the glass transition temperature T_g.) The observed values are V(298K)$\approx$99.1 cc/mole and ρ(298K)$\approx$1.05 grams/cc, respectively. The "interactive homopolymer" and "interactive copolymer" modes of execution of the software package both allow the user to supply the density at room temperature as an input parameter, if a better value is available than the internally calculated value for a given polymer. We will, however, use the "internally calculated" values in all calculations below.

Step 8. **Calculate the cohesive energy E_{coh}.** Two correlations are useful for E_{coh}, each one being preferred in providing E_{coh} as an input parameter for the correlations for different "derived" properties. E_{coh1} (equations 5.8-5.10) is an extension and generalization of E_{coh} calculated by the group contributions of Fedors. E_{coh2} (equations 5.12 and 5.13) is an extension and generalization of E_{coh} calculated by the group contributions of van Krevelen and Hoftyzer. The calculated values for polystyrene are $E_{coh1}\approx$39197 J/mole and $E_{coh2}\approx$36933 J/mole.

Step 9. **Calculate the solubility parameter** at room temperature, defined by Equation 5.4, which gives $\delta\approx$20.1 (J/cc)$^{0.5}$ with E_{coh1} and $\delta\approx$19.5 (J/cc)$^{0.5}$ with E_{coh2} for polystyrene with the "internally calculated" V(298K)$\approx$97.0 cc/mole, so that we obtain two estimates of δ.

Step 10. **Calculate the glass transition temperature T_g,** by using equations 6.2 and 6.3 for most polymers (including polystyrene), and the specialized correlations given by Equation 17.1 or Equation 17.2 for certain types of polymers. (The software package identifies automatically, from the connectivity table, when Equation 17.1 needs to be used.) $T_g\approx$382K is calculated for polystyrene. The experimental value is $T_g\approx$373K. The "interactive homopolymer" and "interactive copolymer" modes of execution of the software package both allow the user to supply T_g as an input parameter, if a better value is available than the internally calculated value. We will, however, use the "internally calculated" value in all of the calculations below.

Step 11. **Calculate the coefficient of volumetric thermal expansion at room temperature**, by substituting $T_g\approx$382K into Equation 3.6. The result is 255$\cdot$10^{-6}/(degree K). Equation 3.6 is only valid if $T_g\geq$298K. See Section 3.D for a complete listing of the equations used to estimate the temperature dependences of the volumetric properties, as a function of whether $T_g\geq$298K or $T_g<$298K, and of where T is located relative to T_g and 298K.

***Step 12*.** **Calculate the molar volume as a function of temperature**, for $T \leq T_g$, by substituting $V(298K) \approx 97.0$ cc/mole and $T_g \approx 382K$ into Equation 3.15. **The density as a function of the temperature can then be calculated** by inserting the value of $V(T)$ into Equation 3.2. For example, let us predict these properties at absolute zero temperature (0K) and at $T_g \approx 382K$ for polystyrene. The results are $V(0K) \approx 89.6$ cc/mole, $\rho(0K) \approx 1.162$ grams/cc, $V(T_g) \approx 99.1$ cc/mole, and $\rho(T_g) \approx 1.05$ grams/cc. See Section 3.D for a complete listing of the equations used to estimate $V(T)$ and $\rho(T)$. For example, Equation 3.16 is used to predict that $V(412K) \approx 100.57$ cc/mole and $\rho(412K) \approx 1.036$ grams/cc for polystyrene.

Calculated values of $V(T)$ and $\rho(T)$ over the range of $0K \leq T \leq 500K$, which includes both the glassy and rubbery temperature regimes, are shown in Figure 18.2. The limitations, at very low temperatures, of the equations used to calculate $V(T)$ and $\rho(T)$, were summarized in Section 3.A.

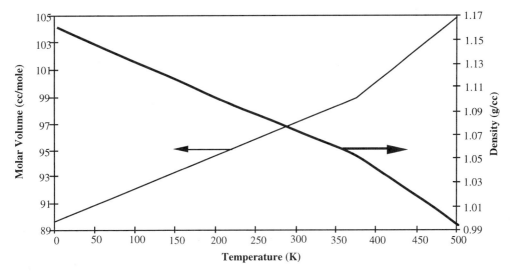

Figure 18.2. Calculated molar volume (thin line, left-hand y-axis) and density (thick line, right-hand y-axis) of polystyrene as functions of the temperature.

***Step 13*.** **Calculate the molar heat capacity at constant pressure (C_p).** Polystyrene has $T_g \approx 382K$ (Step 10), and is a "solid" at room temperature (298K). Equation 4.13 gives $C_p^s(298K) \approx 133.5$ J/(mole·K). Equation 4.14 gives $C_p^l(298K) \approx 174.8$ J/(mole·K) for the extrapolation of the C_p above T_g to room temperature. When $T \neq 298K$, equations 4.10 and 4.15 can be used to estimate $C_p^s(T)$ and $C_p^l(T)$, respectively, for $T \geq 100K$. The predicted $C_p^l(T_g)$ and $C_p^s(T_g)$ with $T_g = 382K$ can then be inserted into Equation 4.9, to estimate that the C_p of polystyrene should increase by approximately 26.8 J/(mole·K) at T_g. Equations 4.10 and 4.15 are invalid for $T < 100K$, because they both predict an unphysical nonzero residual C_p at $T = 0K$.

Equations 18.1 and 18.2, which give $C_p=0$ at T=0K, are preferred for use at $0K \leq T \leq 100K$, with $C_p^s(100K)$ and $C_p^l(100K)$ calculated from equations 4.10 and 4.15, respectively. The results of the calculation of $C_p(T)$ for $0K \leq T \leq 500K$ are shown in Figure 18.3.

$$C_p^s(T) \approx 0.01T \cdot C_p^s(100K) \qquad \text{(for T<100K only)} \qquad (18.1)$$

$$C_p^l(T) \approx 0.01T \cdot C_p^l(100K) \qquad \text{(for T<100K only)} \qquad (18.2)$$

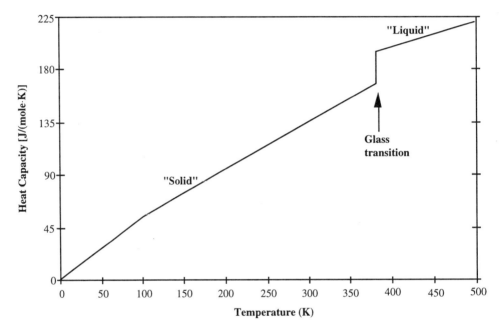

Figure 18.3. Calculated molar heat capacity of polystyrene at constant pressure, as a function of the temperature. $C_p^s(T)$ is used up to the calculated $T_g \approx 382K$, and $C_p^l(T)$ above T_g, with a jump at T_g indicating the magnitude of the estimated value of $\Delta C_p(T_g)$.

Step 14. **Calculate the surface tension γ.** Three predictions of the surface tension at room temperature, namely γ(298K), can be obtained. Equation 7.1, with E_{coh} and V(298K) calculated above, provides two estimates. The use of E_{coh1} gives γ(298K)≈41.0 dyn/cm, and the use of E_{coh2} gives γ(298K)≈39.4 dyn/cm, with V(298K)≈97.0 cc/mole, for polystyrene. The third prediction (45.9 dyn/cm) is obtained by using Equation 7.2, with V(298K)≈97.0 cc/mole inserted for V(T), and the "molar parachor" $P_S \approx 252.5$ (cc/mole)·(dyn/cm)$^{1/4}$ estimated by utilizing equations 7.10 and 7.11 with $V_w \approx 64.04$ cc/mole.

The value of γ(T) at T≠298K can be predicted by substituting P_S (which is independent of the temperature), and V(T) at the temperature of interest, into Equation 7.2. The results of this calculation are shown in Figure 18.4.

Alternatively, the values of γ(298K) estimated by using Equation 7.1 can be scaled by the inverse fourth power of V(T) as in Equation 7.4, to provide two alternative estimates of γ(T), with exactly the same dependence on the temperature but displaced by a constant factor from the curve calculated by using Equation 7.2.

Step 15. **Calculate the key optical properties.** The **refractive index** n at room temperature (298K) is calculated by using equations 8.12 and 8.13. The predicted n(289K) for polystyrene is 1.6037. The predicted values of V(298K)≈97.0 cc/mole and n(298K)≈1.6037 are inserted into Equation 8.7 to predict the **molar refraction** according to Lorentz and Lorenz to be R_{LL}≈33.351. To estimate n(T) at temperatures other than 298K, V(T) and R_{LL} are substituted into Equation 8.5. For example, R_{LL}≈33.351 cc/mole and V(T_g)≈99.1 (Step 12) for polystyrene, so that Equation 8.5 predicts that n(T_g)≈1.5880.

The predicted n(T) of polystyrene is shown in Figure 18.4. Note that the shapes of the predicted surface tension and refractive index curves are very similar, although the magnitudes of the two properties and their relative changes over the temperature range are quite different.

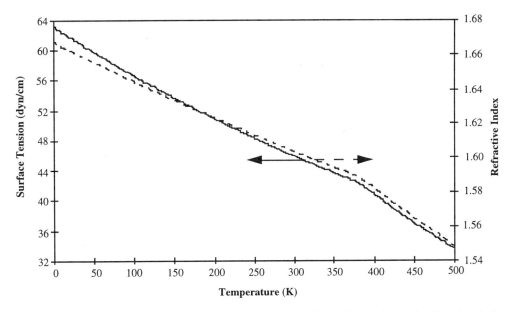

Figure 18.4. Calculated surface tension (solid line, left-hand y-axis) and refractive index (dashed line, right-hand y-axis) of polystyrene as functions of the temperature.

Step 16. **Calculate the key electrical properties.** The **dielectric constant** ε at room temperature can be calculated from equations 9.11 and 9.12. For polystyrene, ε(298K)=2.57 is predicted. The predicted ε(298K) can be inserted into Equation 9.7, to predict the **volume resistivity** at room temperature. The prediction for polystyrene is 7.2·10^{17} ohm·cm. For polar polymers, these two properties can only be predicted at T=298K with the present correlations.

Step 17. **Calculate the magnetic susceptibility** by using equations 10.4, 10.5 and 10.2, which give 0.681·10^{-6} cc/g for polystyrene.

Step 18. Predict the key **mechanical properties at small strains** for 0≤T≤(T$_g$-20K) and at a "typical" rate of measurement. V$_w$≈64.04 cc/mole and l$_m$≈2.52·10^{-8} cm (for polystyrene) are inserted into Equation 11.10 to predict **Poisson's ratio** ν(298K)≈0.3590 at T=298K. ν(298K) and T$_g$ are inserted into equations 11.11 and 11.12 to predict ν(T) for T≤(T$_g$-20K) when T≠298K.

E$_{coh1}$≈39197 J/mole, V(0K)≈89.6 cc/mole and V(T) (97.0 cc/mole if T=298K) are inserted into Equation 11.13 to predict the **bulk modulus** B(T). For polystyrene at room temperature, it is estimated that B(298K)≈3594 MPa.

Equation 11.7 is used to predict the **Young's modulus** E(T) and **shear modulus** G(T). For polystyrene at T=298K, it is calculated that E(298K)≈3041 MPa and G(298K)≈1119 MPa.

The moduli and Poisson's ratio calculated for polystyrene are shown as functions of temperature in Figure 18.5 and Figure 18.6. (See Step 20 for the calculations above T$_g$.)

The calculation of ν(T) via equations 11.10-11.12 [and thus of E(T) and G(T) via Equation 11.7] is only possible if T$_g$>298K, i.e., if the polymer is glassy at room temperature. Only B(T) can be calculated below (T$_g$+30K) with our present structure-property relationships if T$_g$<298K.

Step 19. Calculate the **failure stresses** under mechanical loading, for temperatures up to T$_g$, again at some "typical" rate of measurement. The **brittle fracture stress** σ$_f$ is predicted from Equation 11.28 which gives 59.5 MPa, and the **yield stress under uniaxial tension** σ$_y$ is predicted from Equation 11.30 which gives 85.1 MPa, for polystyrene at room temperature.

The calculated σ$_y$ and σ$_f$ of polystyrene are shown as functions of the temperature in Figure 18.7. Unlike the other figures showing calculated properties, the temperature axis in Figure 18.7 only goes up to the calculated T$_g$≈382K instead of 500K. This is done because the mechanisms of failure are different for glassy and rubbery polymers, and σ$_y$ and σ$_f$ therefore only provide a meaningful description of failure below T$_g$. Note, from Figure 18.7, that unmodified glassy polystyrene is expected to be quite brittle under uniaxial tension, all the way up to T$_g$.

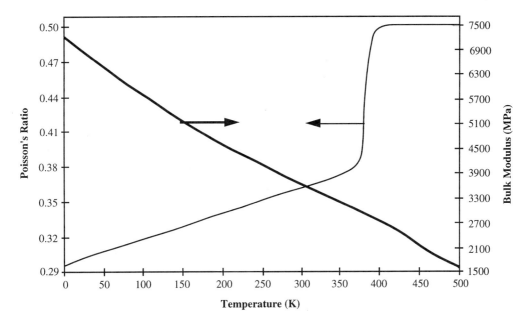

Figure 18.5. Calculated Poisson's ratio (thin line, left-hand y-axis) and bulk modulus (thick line, right-hand y-axis) of polystyrene as functions of the temperature.

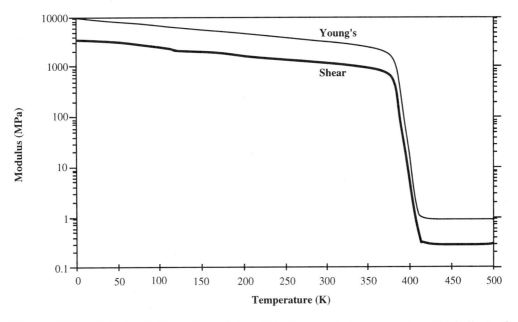

Figure 18.6. Calculated Young's modulus (thin line) and shear modulus (thick line) of polystyrene as functions of the temperature. A logarithmic y-axis has been used for clarity.

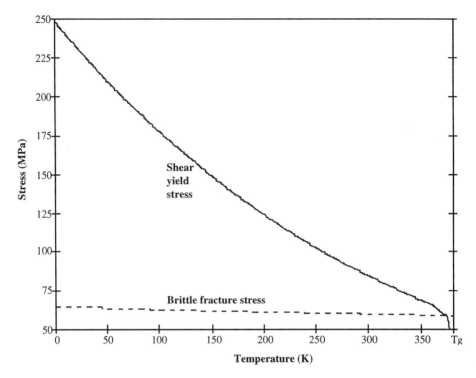

Figure 18.7. Calculated shear yield stress under uniaxial tension (solid line) and brittle fracture stress (dashed line) of polystyrene below the glass transition temperature. Note that the shear yield stress has a much stronger temperature dependence than the brittle fracture stress.

The preferred failure mechanism, and the relative ductility of a specimen, also depend on the mode of deformation. A standard yield criterion (such as modified von Mises or modified Tresca) can be used to estimate the yield stress under other modes of deformation from σ_y.

***Step 20.* Calculate the mechanical properties in the "rubbery plateau" region above T_g.**

The **bulk modulus** B(T) for $T\geq(T_g+30K)$, which is shown in Figure 18.5, is calculated by using Equation 11.25. For example, for polystyrene, $T_g\approx382K$, $V(412K)\approx100.57$ cc/mole, $V_w\approx64.04$ cc/mole, and therefore B(412K)≈2623 MPa.

The **entanglement molecular weight M_e** is calculated by using Equation 11.24. Use of $N_{BBrot}=2$, M=104.15 g/mole, $V_w\approx64.04$ cc/mole and $l_m\approx2.52\cdot10^{-8}$ cm gives $M_e\approx12410$ g/mole. Use of the refined value of $l_m\approx2.32\cdot10^{-8}$ cm from molecular modeling gives $M_e\approx15536$ g/mole.

The **shear modulus in the rubbery plateau regime** is estimated by using Equation 11.22 at $T=(T_g+30K)$, assumed to be the inception of the rubbery plateau. For example, for rubbery polystyrene, Equation 11.22, with a predicted $\rho(T_g+30K)\approx\rho(412K)\approx1.036$ grams/cc and $M_e \approx 15536$ g/mole gives a rubbery plateau shear modulus of 0.219 MPa. Since we have no general method to estimate how the shear modulus changes with the temperature in the rubbery plateau regime, the inception value at $T=(T_g+30K)$ is used in Figure 18.6 for $T>(T_g+30K)$.

Poisson's ratio for a rubbery polymer can be calculated by inserting the shear and bulk moduli into Equation 11.26. For polystyrene, $G(412K) \approx 0.286$ MPa and $B(412K) \approx 2623$ MPa, so that $\nu(412K) \approx 0.4999455$ is predicted. For most practical purposes, $\nu(T) \approx 0.5$ for a rubbery polymer, as can be seen from Figure 18.5.

Equation 11.7 gives a **Young's modulus** (Figure 18.6) of 0.858 MPa for rubbery polystyrene at $T=412K$, which is assumed to be the inception temperature of the rubbery plateau.

Mathematical interpolations are used to predict $B(T)$, $G(T)$, $E(T)$ and $\nu(T)$ in the temperature range of (T_g-20K) to (T_g+30K), where they change rapidly and drastically. The margins of error for the predicted mechanical properties in this temperature regime are therefore quite large.

At temperatures far above T_g, i.e., in the "terminal zone", $G(T)$ decreases further by a large amount, and becomes much lower than it is in the rubbery plateau regime. At present, there is no general method to predict the temperature of onset of the terminal zone of a polymer.

Step 21. **Calculate the key properties of polymers in dilute solutions.**

The **steric hindrance parameter** σ is predicted by using equations 12.19-12.23. For polystyrene, the result is $\sigma \approx 2.22$. The **characteristic ratio** C_∞ is predicted from the value of σ, by using Equation 12.24 *or* Equation 12.25, as appropriate for the type of backbone of the polymer chain. (The software package identifies automatically, from the connectivity table, which correlation needs to be used.) For most polymers, including polystyrene, Equation 12.24 is appropriate. This equation, with $\sigma=2.22$ inserted in it, gives $C_\infty \approx 9.86$ for polystyrene.

The **intrinsic viscosity** in dilute solution **under theta conditions** is given by Equation 12.7, where K is the **molar stiffness function**. K can be estimated by using equations 12.27 and 12.29. For, example, $K \approx 29.41$ is calculated for polystyrene. Methods to estimate the intrinsic viscosity under non-theta conditions, and a method to estimate the solution viscosity at small but finite concentrations, have also been presented in Chapter 12.

Step 22. **Calculate the viscosity of a polymer melt and/or the zero-shear viscosity of a concentrated polymer solution.** The molecular weight dependence of the zero-shear viscosity is given by equations 13.2 and 13.3, where the **critical molecular weight** M_{cr} (Equation 13.6) is the key parameter. $M_{cr} \approx 31072$ g/mole is calculated for polystyrene. The **activation energy for viscous flow** at zero shear rate and in the limit of $T \rightarrow \infty$ ($E_{\eta \infty}$) is predicted by combining Equation 13.8 with equations 13.11, 13.12, 13.14 and 13.15. $E_\eta \approx 52421$ J/mole is calculated for polystyrene. The dependence of the melt viscosity of polystyrene on the average molecular weight, polydispersity, temperature, shear rate, and hydrostatic pressure, can then be calculated as described in great detail in Chapter 13. As also described in Chapter 13, the zero-shear viscosity of a concentrated solution of polystyrene can be calculated as a function of M_w, T, and the properties of the solvent.

Step 23. **Calculate the thermal conductivity** λ**.** The thermal conductivity at T=298K is predicted for isotropic amorphous polymers by using equations 14.8 and 14.9, resulting in $\lambda(298K) \approx 0.135$ J/(K·meter·sec) for polystyrene. $\lambda(T)$ for T$\neq$298K is predicted by using equations 14.4 and 14.5, with T_g and $\lambda(298K)$ as input parameters. For polystyrene, $T_g \approx 382K$ and $\lambda(298K) \approx 0.135$ J/(K·meter·sec) were calculated. Equation 14.4 therefore gives $\lambda(T_g) \approx 0.143$ J/(K·meter·sec), and Equation 14.5 gives $\lambda(500K) \approx 0.134$ J/(K·meter·sec).

See Figure 18.8 for the calculated $\lambda(T)$ of polystyrene. The lower end of the temperature axis in Figure 18.8 is at T=100K instead of T=0K because the calculation of $\lambda(T)$ by the present correlations is expected to have the largest margin of error at very low temperatures.

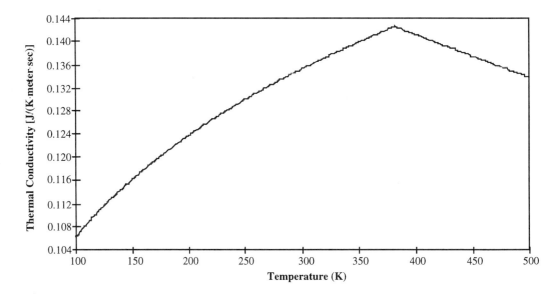

Figure 18.8. Calculated thermal conductivity of polystyrene as a function of the temperature.

Step 24. **Calculate the permeability to small gas molecules at room temperature.** A rough estimate for the permeability to oxygen is given by equations 15.14, 15.15 and 15.17. The permeabilities to nitrogen and carbon dioxide can be predicted (even more roughly) by using equations 15.18 and 15.19, respectively. The use of $E_{coh1} \approx 39197$ J/mole, $V(298K) \approx 97.0$ cc/mole, $V_w \approx 64.04$ cc/mole, $N_{rot}=3$ and N=8 for polystyrene gives predicted permeabilities of 365 Dow Units (DU) to O_2, 100 DU to N_2, and 1798 DU to CO_2.

Step 25. **Calculate the temperature of half decomposition $T_{d,1/2}$ in dynamic thermogravimetry**, at a "typical" rate of temperature rise, in an inert atmosphere (such as N_2, He or Ar), by using equations 16.1, 16.6 and 16.7. $T_{d,1/2} \approx 643K$ is calculated for polystyrene. In addition, it can be estimated by using equations 16.2, 16.3 and 16.4 that $T_{d,0} \approx 578K$, $T_{d,max} \approx 643K$, and $E_{act,d} \approx 220 \cdot 10^3$ J/mole.

18.C. Random Copolymers of Styrene and Oxytrimethylene

The properties of random copolymers are also predicted by following the steps listed above. The weighted averaging scheme outlined in Section 17.E is used, so that intensive properties at a given composition are calculated from extensive properties which are weighted averages of the extensive properties of the constituent repeat units. As a result of this simple weighted averaging scheme, each intensive property either monotonically increases or monotonically decreases with changing composition. Deviations from such monotonic behavior, which occur in rare instances because of specific interactions between different repeat units, and which manifest themselves as a maximum or a minimum in a property at an intermediate composition, are not taken into account.

The types of information which can be obtained from copolymer calculations by using our methodology will now be illustrated, with amorphous random copolymers of styrene (Figure 1.1) and oxytrimethylene [Figure 17.7(d)] at room temperature (T=298K) as the example.

The calculated glass transition temperature T_g and temperature of half decomposition $T_{d,1/2}$ are shown in Figure 18.9, as functions of the weight fraction of polystyrene repeat units (w_{sty}). It is predicted that both properties will increase with increasing w_{sty}, and that T_g will vary over a much wider range than $T_{d,1/2}$. The copolymer is predicted to be rubbery for $w_{sty} < 0.51$, and glassy for $w_{sty} \geq 0.51$, at the calculation temperature (T=298K).

The calculated coefficient of volumetric thermal expansion α and density are shown in Figure 18.10. The change in the nature of the copolymer from rubbery to glassy at $w_{sty} \approx 0.51$ is reflected in the drastic decrease of α at $w_{sty} \approx 0.51$. Both in the rubbery and glassy composition regimes, α decreases slowly with increasing w_{sty}, as a result of the gradual increase of T_g.

The specific heat capacity c_p, which is defined as the quotient of the molar heat capacity C_p and the molecular weight M per repeat unit (Equation 4.8), is shown in Figure 18.11. The weight of one mole of the average repeat unit of a copolymer is an extensive property which changes with w_{sty} (see Section 17.E). It is, therefore, far more instructive to examine the intensive property c_p rather than the extensive property C_p as a function of w_{sty}. Note that c_p decreases slowly with increasing w_{sty} for $w_{sty} < 0.51$ (where C_p^l is the appropriate type of heat capacity), drops drastically at $w_{sty} \approx 0.51$ as the copolymer becomes glassy, and continues to decrease slowly with increasing w_{sty} for $w_{sty} > 0.51$ (where C_p^s is the appropriate type of heat capacity). This behavior is very similar to the behavior calculated for the coefficient of thermal expansion.

The solubility parameter δ calculated by using E_{coh1}, and the surface tension γ calculated by using the generalized molar parachor, are shown in Figure 18.12. Both properties are predicted to increase with increasing w_{sty}. The δ values calculated by using E_{coh2}, and the γ values calculated by using E_{coh1} or E_{coh2}, manifest the same trend with w_{sty}, and therefore are not shown. There exist, however, some combinations of types of repeat units for which the different estimates of δ and γ give opposite trends as a function of changing copolymer composition.

The calculated refractive index n and dielectric constant ε are shown in Figure 18.13. It is predicted that, with increasing w_{sty}, n will increase because of the increasing amount of the highly refractive phenyl rings, while ε will decrease slightly because of the decreasing amount of the polar ether oxygen moieties.

The calculated small-strain mechanical properties are shown in Figure 18.14 (Poisson's ratio and bulk modulus) and Figure 18.15 (Young's modulus and shear modulus). These properties all manifest the effect of the change from rubbery to glassy behavior with increasing w_{sty}. Unlike the results for α (Figure 18.10) and c_p (Figure 18.11), the drastic changes in the mechanical properties are calculated to occur more gradually. This is because the mechanical properties are assumed to change gradually from the glassy regime to the rubbery regime over the wide temperature range of $(T_g-20K) \leq T \leq (T_g+30K)$ in the correlations summarized in Chapter 11.

The calculated large-strain mechanical properties (shear yield stress σ_y under uniaxial tension and brittle fracture stress σ_f) are shown in Figure 18.16. Since these two properties are not relevant to the failure mechanism above T_g, they are only calculated for $w_{sty} \geq 0.51$ where the copolymer is predicted to be glassy. Relative toughness (very roughly quantified by σ_f/σ_y) is predicted to decrease rapidly with increasing w_{sty} because of increasing σ_y and decreasing σ_f. The σ_y and σ_f curves cross near $w_{sty} \approx 0.9$. On the other hand, if these calculations were being performed to select a candidate copolymer manifesting "toughness" at room temperature, we would predict the copolymer to become brittle at a somewhat lower w_{sty}, when $(\sigma_f/\sigma_y) \approx 1.2$. The incorporation of a safety margin of at least this magnitude (and in some cases even greater) into the estimation of "toughness", is crucial for the reasons summarized in Section 11.C.3.

The calculated characteristic ratio (Figure 18.17), critical molecular weight (Figure 18.17, with the refined value of $l_m \approx 2.32$ Å used for the repeat unit of polystyrene) and activation energy for viscous flow (Figure 18.18) increase significantly, and thermal conductivity (Figure 18.18) decreases significantly, with increasing w_{sty}. These trends can be ascribed mainly (but not solely) to large changes in the backbone and side group portions of the polymer chains, and especially to the drastic increase in the amount of bulky phenyl side groups.

Finally, it can be seen from Figure 18.19 that the oxygen permeability is predicted to decrease gradually with increasing w_{sty}.

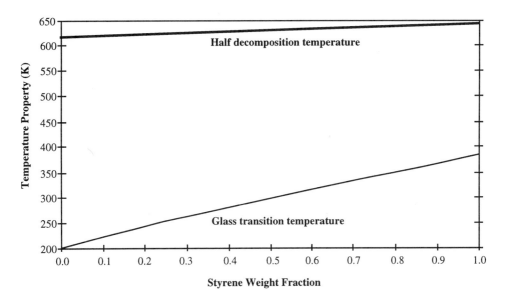

Figure 18.9. Predicted glass transition temperature (thin line) and half decomposition temperature (thick line) of amorphous random copolymers of styrene and oxytrimethylene, as functions of the composition.

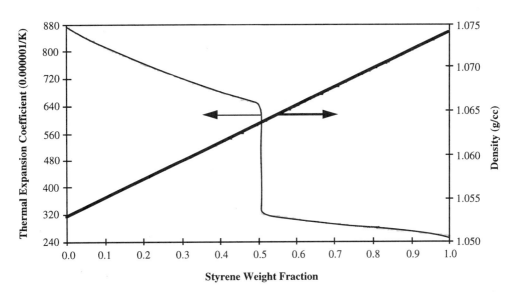

Figure 18.10. Predicted coefficient of volumetric thermal expansion (thin line, left-hand y-axis) and density (thick line, right-hand y-axis) of amorphous random copolymers of styrene and oxytrimethylene at room temperature, as functions of the composition.

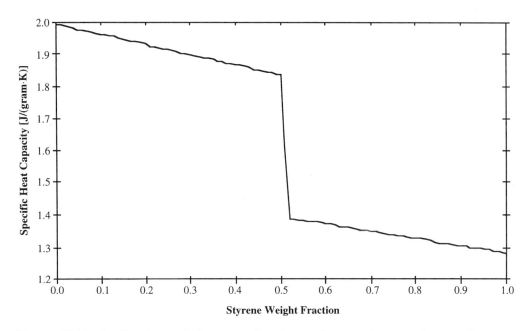

Figure 18.11. Predicted specific heat capacity of amorphous random copolymers of styrene and oxytrimethylene at room temperature, as a function of the composition.

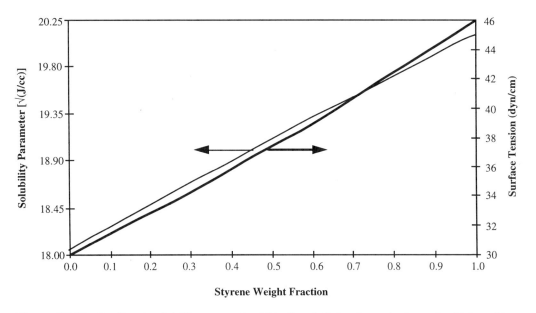

Figure 18.12. Predicted solubility parameter (thin line, left-hand y-axis, from the Fedors-like cohesive energy E_{coh1}) and surface tension (thick line, right-hand y-axis, from generalized molar parachor) of amorphous random copolymers of styrene and oxytrimethylene at room temperature, as functions of the composition.

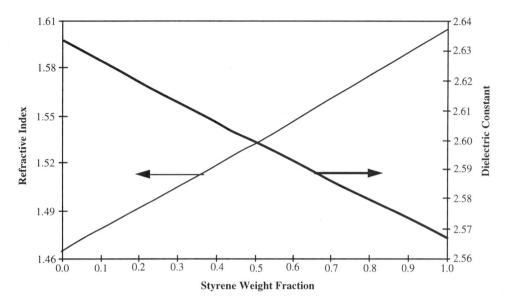

Figure 18.13. Predicted refractive index (thin line, left-hand y-axis) and dielectric constant (thick line, right-hand y-axis) of amorphous random copolymers of styrene and oxytrimethylene at room temperature, as functions of the composition.

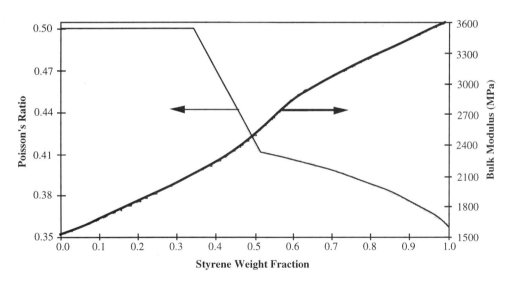

Figure 18.14. Predicted Poisson's ratio (thin line, left-hand y-axis) and bulk modulus (thick line, right-hand y-axis) of amorphous random copolymers of styrene and oxytrimethylene at room temperature, as functions of the composition.

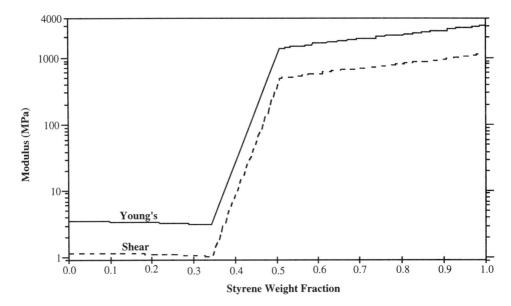

Figure 18.15. Predicted Young's modulus (solid line) and shear modulus (dashed line) of amorphous random copolymers of styrene and oxytrimethylene at room temperature, as functions of the composition. The y-axis is logarithmic.

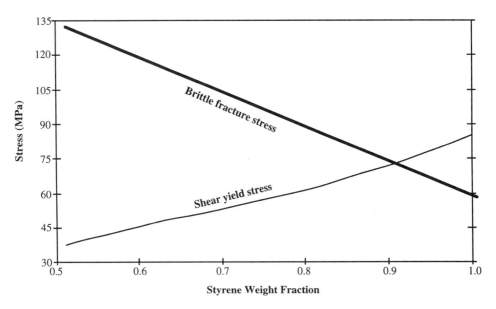

Figure 18.16. Predicted shear yield stress under uniaxial tension (thin line) and brittle fracture stress (thick line) of amorphous random copolymers of styrene and oxytrimethylene at room temperature, as functions of the composition, in the composition range where the copolymers are predicted to be below the glass transition temperature at room temperature.

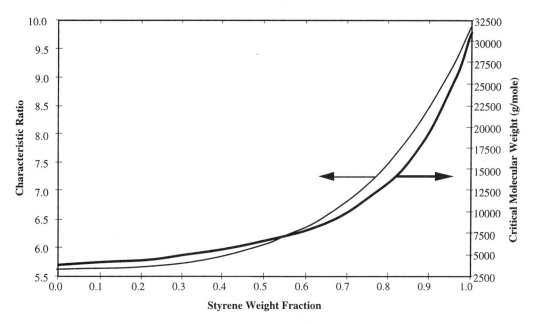

Figure 18.17. Predicted characteristic ratio (thin line, left-hand y-axis) and critical molecular weight (thick line, right-hand y-axis) of amorphous random copolymers of styrene and oxytrimethylene, as functions of the composition.

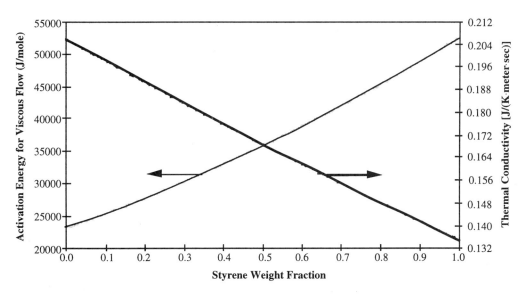

Figure 18.18. Predicted activation energy for viscous flow (solid line, left-hand y-axis, this property is only relevant and meaningful for $T \geq 1.2T_g$) and thermal conductivity (dashed line, right-hand y-axis, at room temperature) of amorphous random copolymers of styrene and oxytrimethylene, as functions of the composition.

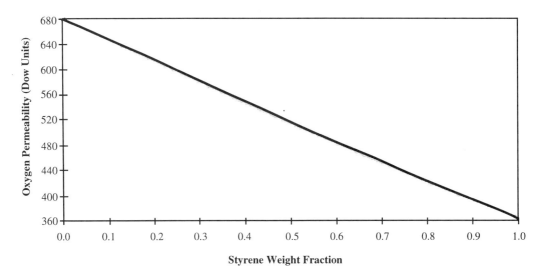

Figure 18.19. Predicted oxygen permeability of amorphous random copolymers of styrene and oxytrimethylene at room temperature, as a function of the composition.

GLOSSARY

SYMBOLS AND ABBREVIATIONS

A. Terms Starting with a Lower-Case Letter of the Latin Alphabet

a	(a) Mark-Houwink exponent.
	(b) Thermal diffusivity.
	(c) "Activity" of a penetrant molecule.
	(d) Generic symbol for a constant or an adjustable parameter in a regression equation.
abs	Absolute value function: $abs(x)=x$ if $x>0$, and $abs(x)=-x$ if $x<0$.
a_g	Penetrant "activity" at which plasticization-induced glass transition occurs.
amu	Atomic mass units.
atm	Atmospheres (a unit of pressure).
b	(a) Generic symbol for a constant or an adjustable parameter in a regression equation.
	(b) Affinity coefficient for the Langmuir sorption isotherm.
c	(a) Generic symbol for a constant or an adjustable parameter in a regression equation.
	(b) Concentration of a polymer dissolved in a solvent.
cal	Calorie (a unit of energy, equal to 4.184 J).
cc	Cubic centimeter (a unit of volume, equal to cm^3).
cm	Centimeters.
coth	Hyperbolic cotangent function.
c_p	Specific heat capacity (i.e., heat capacity per unit weight) at constant pressure.
cS	Centistokes (a unit of viscosity).
d	The total derivative of a function with respect to a variable.
d_{Cpl}	Temperature coefficient of C_p^l.
dn/dc	Specific refractive index increment of a polymer in solution.
dT/dz	Temperature gradient in the z-direction.
dV	Change in volume as a result of deformation.
$d\varepsilon_x$	Deformation (strain) in the x-direction.
$d\varepsilon_y$	Deformation (strain) in the y-direction.
$d\varepsilon_z$	Deformation (strain) in the z-direction.
∂	Partial derivative of a function with respect to one of its variables.
e_{coh}	Cohesive energy density, i.e., E_{coh}/V.
exp	(a) Abbreviation for "experimental".
	(b) Exponential function.
f	Generic symbol for a function. For example, $f(x,y)$ denotes a function of x and y.
g	Grams (a unit of weight).

k	Boltzmann's constant.
kcal	One thousand calories.
k_D	Henry's law solubility coefficient.
kg	Kilograms (one thousand grams).
kHz	One kilohertz, i.e., one thousand Hertz (a unit of frequency).
l	Bond length.
l_i	Length of the i'th type of bond along the shortest path across the chain backbone.
l_k	Length of a statistical chain segment (Kuhn segment).
lim	Limit, in the mathematical sense of a variable approaching a certain value.
l_m	Length (end-to-end distance) of a single repeat unit of a polymer.
ln	Natural logarithm.
$\log_{10}$	Logarithm to the base ten.
m	(a) Meters.
	(b) Mole fraction.
	(c) Number of repeat units in one of the blocks of a block copolymer.
m-	Abbreviation for *meta* substitution.
mil	One thousandth of an inch.
n	(a) Degree of polymerization, i.e., average number of repeat units per polymer chain.
	(b) Number of repcat units in one of the blocks of a block copolymer.
	(c) Order of a connectivity index.
	(d) Refractive index.
	(e) Number of bonds along the shortest path across the chain backbone.
	(f) Average number of repeat units between chemical crosslinks.
	(g) Average number of statistical chain segments between entanglement junctions.
	(h) Generic symbol for the number of occurrences of something.
n^*	Complex refractive index.
n-	Abbreviation for a linear (unbranched) structural unit, as in *n*-butyl.
n_C	Refractive index measured at a wavelength of 6563Å.
n_D	Refractive index measured at a wavelength of 5890Å.
n_{data}	Number of data points.
n_F	Refractive index measured at a wavelength of 4861Å.
n_P	Refractive index of the polymer component of a polymer/solvent system.
n_S	Refractive index of the solvent component of a polymer/solvent system.
o-	Abbreviation for *ortho* substitution.
p	(a) Pressure.
	(b) "p" electrons, i.e., electrons with an angular momentum quantum number of 1.
p^*	Reducing pressure parameter in an equation-of-state.
$\tilde{p}$	Reduced pressure in an equation-of-state.

p-	Abbreviation for *para* substitution.
$\tilde{p}_{mix}$	Reduced pressure of a mixture in an equation-of-state.
pre	Abbreviation for "predicted".
q	Heat flux.
r	Atomic radius.
$<r^2>_0$	Mean-square end-to-end distance of an unperturbed linear chain molecule in solution.
$<r^2>_{0f}$	Mean-square end-to-end distance of a hypothetical "freely rotating" polymer chain.
s	"s" electrons, i.e., electrons with an angular momentum quantum number of 0.
sec	Seconds.
sec-	Abbreviation for a unit attached from its second carbon atom, as in *sec*-butyl.
t	Thickness.
t-	Abbreviation for a unit attached from a tertiary carbon atom, as in *t*-butyl.
tan δ	Dissipation factor (also known as "power factor" or "loss tangent") in electrical tests.
tan δ_E	Mechanical loss tangent under the uniaxial tension mode of deformation.
v	(a) Valence (used as a superscript to indicate a valence connectivity index).
	(b) Specific volume, i.e., volume per unit weight.
v_s	Activation volume for shear yielding.
w	Weight fraction.
x	(a) A "reduced temperature" parameter, defined as T/T_g.
	(b) A generic symbol for a variable of a function, as in f(x,y).
x_i	The i'th one among structural parameters x_1, x_2, ..., x_{13} defined for correlating T_g.
y	A generic symbol for a variable of a function, as in f(x,y).
z	The Cartesian z-coordinate.

B. Terms Starting with a Capital Letter of the Latin Alphabet

A	(a) Molecular cross-sectional area of a polymer chain.
	(b) Surface area of a macroscopic specimen.
	(c) Abbé number.
Å	Ångstroms (a unit of length, equal to 10^{-8} cm).
ACS	American Chemical Society.
ASTM	American Society for Testing and Materials.
B	Bulk modulus.
B^*	Complex bulk modulus, with real (elastic) and imaginary (viscous) components.
B'	Real (elastic) component of B^*, equivalent to B.
BB	Polymer chain backbone.

BCB   Benzocyclobutene.

BPAC-BCB$_2$ Poly(bisphenol-A carbonate - *bis* benzocyclobutene).

Br   Symbol for bromine.

C   (a) Symbol for carbon.

             (b) Concentration of gas sorbed in a polymer.

C_g Sorbed concentration of penetrant at which plasticization induces the glass transition.

C_H' Langmuir (frozen-in hole) sorption capacity.

Cl   Symbol for chlorine.

C_p Molar heat capacity at constant pressure.

C_p^l Molar heat capacity of "liquid" (molten or rubbery) polymers at constant pressure.

C_p^s Molar heat capacity of "solid" (glassy or crystalline) polymers at constant pressure.

C_v Molar heat capacity at constant volume.

C_∞ Characteristic ratio of a polymer chain, usually measured in dilute solutions.

D   (a) Diffusion coefficient.

             (b) Tensile compliance (reciprocal of Young's modulus E).

D_0 Pre-exponential factor in equation for temperature dependence of diffusivity.

DSC   Differential scanning calorimetry.

DK-BCB$_2$ Poly(diketone - *bis* benzocyclobutene).

DVS-BCB$_2$ Poly(divinylsiloxane *bis*-benzocyclobutene).

DU   Dow Units (for permeability), more formally (cc·mil)/[day·(100 inches^2)·atm)].

E   Young's Modulus.

E^* Complex Young's modulus, with real (elastic) and imaginary (viscous) components.

E' Real (elastic) component of E^*, equivalent to E.

E'' Imaginary (viscous) component of E^*.

$E_{act,d}$ Average activation energy for decomposition (weight loss).

E_{coh} Cohesive energy per mole.

E_{coh1} Fedors-like E_{coh} calculated by using the correlation presented in Section 5.B.

E_{coh2} Van Krevelen-like E_{coh} calculated by using the correlation presented in Section 5.C.

E_d Dispersion component of the cohesive energy.

E_D Activation energy for diffusion.

E_h Hydrogen bonding component of the cohesive energy.

E_p Polar component of the cohesive energy.

E_P Activation energy for permeation of small penetrant molecules through polymers.

$E_{\eta p}$ Activation energy for viscous flow of polymer.

$E_{\eta s}$ Activation energy for viscous flow of solvent.

$E_{\eta ss}$ Activation energy for viscous flow of solution.

$E_{\eta\infty}$ Activation energy for viscous flow of polymer at zero shear rate as T→∞.

ESR   Electron spin resonance spectroscopy.

F	(a) Molar attraction constant.
	(b) Fluorine.
	(c) Generic symbol for a function. For example, F(x,y) denotes a function of x and y.
F_d	Dispersion component of the molar attraction constant.
F_h	Hydrogen-bonding component of the molar attraction constant.
Fig.	Abbreviation for "Figure".
F_p	Polar component of the molar attraction constant.
G	(a) Gibbs free energy.
	(b) Shear modulus.
G^*	Complex shear modulus, with real (elastic) and imaginary (viscous) components.
G'	Real (elastic) component of G^*, equivalent to the second definition of G.
G_E^0	Equilibrium shear modulus for $T>T_g$ in a polymer crosslinked beyond its gel point.
G_N^0	Shear modulus of an uncrosslinked polymer in the rubbery plateau region, i.e., for $T \geq (T_g+30K)$ but not so high as to be in the terminal zone.
H	(a) Enthalpy.
	(b) Hydrogen.
Hyb	Valence shell hybridization.
Hz	Hertz (a unit of frequency).
H_η	Molar viscosity-temperature function.
$H_{\eta str}$	Non-additive "structural" portion of H_η.
$H_{\eta sum}$	Simply additive portion of H_η.
IEEE	Institute of Electrical and Electronics Engineers.
J	(a) Joule (a unit of energy).
	(b) Shear compliance (reciprocal of the shear modulus G).
	(c) Molar intrinsic viscosity function.
K	(a) Abbreviation for "degrees Kelvin".
	(b) Molar stiffness function (also called "molar limiting viscosity number function").
	(c) Absorption index.
K_g	An empirical parameter used in the equation of Fox and Flory.
K_g'	An empirical parameter used in the equation of Fox and Loshaek.
K_g''	An empirical parameter used in the equation of Fox and Loshaek.
K_Θ	The quantity $(K/M)^2$, where K denotes the molar stiffness function.
$\pounds^{-1}$	Inverse Langevin function.
M	Molecular weight per repeat unit.
M_c	Average molecular weight between chemical crosslinks in a polymer.
M_{cr}	Critical molecular weight of a polymer.
MDI	4,4'-Methylenediisocyanate.
M_e	Entanglement molecular weight of a polymer.

MHz	One megahertz, i.e., one million Hertz (a unit of frequency).
MPa	Megapascals (a unit with dimensions of pressure, used for stresses and moduli).
M_n	Number-average molecular weight of a polymer.
M_v	Viscosity-average molecular weight of a polymer.
M_w	Weight-average molecular weight of a polymer.
N	(a) Number of non-hydrogen atoms in a molecule or a polymeric repeat unit.
	(b) Equivalently, number of vertices in a hydrogen-suppressed graph.
	(c) Symbol for nitrogen.
N_A	Avogadro's number ($6.022169 \cdot 10^{23}$).
N_{BB}	Total number of atoms in the backbone of a polymeric repeat unit.
N_{BBrot}	Number of rotational degrees of freedom in the backbone of a polymeric repeat unit.
N_H	Number of hydrogen atoms bonded to a given non-hydrogen atom.
N_{item}	Number (or weighted sum of numbers) of item(s) identified by subscript.
NMR	Nuclear magnetic resonance spectroscopy.
$N_{property}$	"Correction index" used in the correlation for the property identified by the subscript.
N_{rot}	Total number of rotational degrees of freedom ($N_{BBrot} + N_{SGrot}$) in a repeat unit.
N_{row}	Row of non-hydrogen atom in periodic table (1 for C, N, O, F; 2 for Si, S, Cl; 3 for Br).
N_{SG}	Total number of non-hydrogen atoms in the side groups of a polymeric repeat unit.
N_{SGrot}	Total number of rotational degrees of freedom in the side groups of a repeat unit.
N_{SP}	Number of atoms in the shortest path across the backbone of a polymeric repeat unit.
N_{Wg}	Weissenberg number.
O	Symbol for oxygen.
P	(a) Molar dielectric polarization, often used in predicting the dielectric constant ε.
	(b) Permeability.
P_0	Pre-exponential factor in equation for temperature dependence of permeability.
P_{CO2}	Permeability of a polymer to carbon dioxide.
PET	Poly(ethylene terephthalate).
P_{LL}	Molar dielectric polarization according to Lorentz and Lorenz.
PMET	A polythiocarbonate.
PMBiT	A polythiocarbonate.
PMMA	Poly(methyl methacrylate).
PMMT	A polythiocarbonate.
PMPhT	A polythiocarbonate.
P_{N2}	Permeability of a polymer to nitrogen.
P_{O2}	Permeability of a polymer to oxygen.
P_S	Molar parachor, which is used in calculations of the surface tension.
PVF	Poly(vinyl fluoride).
PVT	Pressure-volume-temperature.

Q	Quantity of penetrant molecules going through a specimen.
QSAR	Quantitative structure-activity relationships (term mainly used for ordinary molecules).
QSPR	Quantitative structure-property relationships (term mainly used for polymers).
R	(a) Generic symbol for a structural unit in a molecule or a polymer.
	(b) Gas constant, equal to 1.9872 cal/(mole·K) or equivalently 8.31451 J/(mole·K).
	(c) Molar refraction, often used in the prediction of the refractive index n.
	(d) Volume resistivity, in ohm·cm.
	(e) Correlation coefficient.
R'	Generic symbol for a structural unit in a molecule or a polymer.
R"	Generic symbol for a structural unit in a molecule or a polymer.
Ref.	Abbreviation for "Reference".
R_{GD}	Molar refraction according to Gladstone and Dale.
R_{LL}	Molar refraction according to Lorentz and Lorenz.
RMS	Root mean square.
S	(a) Solubility.
	(b) Entropy.
	(c) Sulfur.
S_0	Pre-exponential factor in equation for temperature dependence of solubility.
SC	A "side chain", i.e., side group consisting of a series of divalent links attached to the backbone at one end and terminated by a univalent (dangling) link at the other end.
SG	"Side groups" of a polymer, i.e., all atoms which are not in its backbone.
Si	Symbol for silicon.
SP	Shortest path across the backbone of a polymeric repeat unit.
T	Absolute temperature, in degrees Kelvin.
T^*	Reducing temperature parameter in an equation-of-state.
$\widetilde{T}$	Reduced temperature in an equation-of-state.
TA	Thermal analyzer.
T_b	Boiling temperature of a simple liquid.
TBBPAPC	Tetrabromobisphenol-A polycarbonate.
$T_{d,max}$	Temperature of maximum rate of decomposition (weight loss), in degrees Kelvin.
$T_{d,0}$	Temperature of initial decomposition (weight loss), in degrees Kelvin.
$T_{d,1/2}$	Temperature of half decomposition (weight loss), in degrees Kelvin.
T_g	Glass transition temperature, in degrees Kelvin.
T_{gp}	Glass transition temperature of polymer, in degrees Kelvin.
T_{gs}	Glass transition temperature of solvent, in degrees Kelvin.
$T_{g\infty}$	The limiting value of T_g for $M_n \rightarrow \infty$.
TGA	Thermogravimetric analysis.
T_m	Melting temperature, in degrees Kelvin.

TM	Trademark.
$\widetilde{T}_{mix}$	Reduced temperature of a mixture in an equation-of-state.
T_{ref}	A reference temperature, in degrees Kelvin.
TTOS	Thermal and thermooxidative stability.
T_x	A crossover temperature, in degrees Kelvin.
T_α	Main relaxation (glass transition or melting) temperature.
T_β	A secondary relaxation temperature.
T_γ	A secondary relaxation temperature.
U	Internal energy of a thermodynamic system.
U_H	Molar Hartmann function.
UL	Underwriters Laboratories.
U_R	Molar Rao function (also known as the "molar sound velocity function").
V	Molar volume.
V_i	Contribution of the i'th structural unit to V, in a calculation by an additive scheme.
V_{ref}	Reference volume used in expression for the Flory-Huggins interaction parameter.
V_w	van der Waals volume.
W_{adh}	Reversible work of adhesion.
X	(a) Generic symbol for a structural unit in a molecule or a polymer.
	(b) An intermediate quantity used in calculating the effect of plasticization on T_g.
X'	Generic symbol for a structural unit in a molecule or a polymer.
Y	Generic symbol for a structural unit in a molecule or a polymer.
$Y_{d,1/2}$	Molar thermal decomposition function.
Y_g	Molar glass transition function.
Y_{gi}	Contribution of the i'th structural unit to Y_g.
Y_m	Molar melt transition function.
Z	(a) Atomic number.
	(b) Generic symbol for a structural unit in a molecule or a polymer.
Z^v	Number of valence electrons of an atom.

C. Terms Starting with a Lower-Case Letter of the Greek Alphabet

In alphabetical order of the corresponding letters of Latin alphabet: a→α, b→β, c→χ, d→δ, e→ε, g→γ, h→η, k→κ, l→λ, m→μ, n→ν, p→π, q→θ, r→ρ, s→σ, t→τ, x→ξ *and* z→ζ.

α	(a) Coefficient of volumetric thermal expansion.
	(b) Amorphous fraction of a semicrystalline material.

(c) Optical loss (used with various subscripts and superscripts, as appropriate).

α_g The coefficient of volumetric thermal expansion of a "glassy" polymer ($T<T_g$).

α_{gp} The coefficient of volumetric thermal expansion of a polymer for $T<T_{gp}$ (same as α_g).

α_{gs} The coefficient of volumetric thermal expansion of a solvent for $T<T_{gs}$.

α_{lp} The coefficient of volumetric thermal expansion of a polymer for $T>T_{gp}$.

α_{ls} The coefficient of volumetric thermal expansion of a solvent for $T>T_{gs}$.

α_r Volumetric thermal expansion coefficient of a "rubbery" polymer ($T>T_g$).

α_{elec} Electronic component of intrinsic optical loss.

$\alpha_{extrinsic}$ Extrinsic optical loss.

$\alpha_{intrinsic}$ Intrinsic optical loss

α_{scat} Scattering component of intrinsic optical loss.

α_{total} Total optical loss.

α_{vibr} Vibrational component of intrinsic optical loss.

β (a) Bond simple connectivity index, defined in terms of δ values.

     (b) Coefficient of linear thermal expansion.

β^v Bond valence connectivity index, defined in terms of δ^v values.

χ (a) A connectivity index of a molecule or a polymeric repeat unit. For example, $^n\chi$ and $^n\chi^v$ are n'th-order simple and valence connectivity indices, respectively.

     (b) Flory-Huggins interaction parameter between two polymers. For example, χ_{AB} denotes the Flory-Huggins interaction parameter between polymers A and B.

δ (a) Atomic simple connectivity index.

     (b) Solubility parameter.

δ_1 Solubility parameter calculated by using E_{coh1} for the cohesive energy per mole.

δ_2 Solubility parameter calculated by using E_{coh2} for the cohesive energy per mole.

δ_d Dispersion component of the solubility parameter.

δ_h Hydrogen bonding component of the solubility parameter.

δ_p Polar component of the solubility parameter.

$\delta_{polymer}$ Solubility parameter of polymer.

$\delta_{solvent}$ Solubility parameter of solvent.

δ^v Atomic valence connectivity index.

$\delta^{v'}$ A modified version of δ^v, used in some of the calculations.

ε (a) Dielectric constant.

     (b) Amount of strain a specimen is subjected to, as the fractional change of its length.

ε_{ci} Strain required for the inception of crazing.

ε^* Complex dielectric constant, with real (storage) and imaginary (lossy) components.

ε' Real (storage) component of ε^*, equivalent to the static dielectric constant ε.

ε'' Imaginary (lossy) component of ε^*.

$\dot{\varepsilon}$ Strain rate, defined as (fractional change in length of specimen)/seconds.

$\dot{\varepsilon}_0$	A parameter in equation for the shear yield stress as a function of temperature and $\dot{\varepsilon}$.
γ	(a) Surface tension.
	(b) Elastic shear strain.
$\dot{\gamma}$	Shear rate.
γ_{12}	Interfacial tension between Surface 1 and Surface 2.
γ_{cr}	Critical surface tension of wetting.
γ_i	Surface tension of i'th material.
γ_d	Dispersion component of surface tension.
γ_{di}	Dispersion component of surface tension of i'th material.
γ_x	All non-dispersion components of surface tension lumped together.
γ_{xi}	All non-dispersion components of surface tension of i'th material lumped together.
η	Melt viscosity.
η_0	Zero-shear viscosity of a melt.
η_{cr}	Critical viscosity of a melt, i.e., the value of η_0 at the critical molecular weight.
$\eta_{pcr}{}^*$	Viscosity of a plasticized but undiluted polymer at the critical molecular weight.
η_s	Viscosity of a solvent.
η_{ss}	Viscosity of a solution consisting of a polymer dissolved in a solvent.
$[\eta]$	Intrinsic viscosity (or "limiting viscosity number") of a polymer in dilute solution.
$[\eta]_\Theta$	Intrinsic viscosity of a polymer in dilute solution under "theta conditions".
κ	Compressibility (defined as the bulk compliance, i.e., reciprocal of bulk modulus B).
λ	(a) Thermal conductivity.
	(b) Draw ratio.
μ	Dipole moment.
ν	(a) Poisson's ratio, which relates the bulk, shear and Young's moduli to each other.
	(b) Frequency of measurement in a dielectric relaxation experiment.
	(c) The "newchor", defined for use in permeability calculations.
ν^*	Complex Poisson's ratio, with real (elastic) and imaginary (viscous) components.
ν'	Real component of ν^*, equivalent to the static Poisson's ratio ν.
π	(a) The number 3.141593.
	(b) Permachor.
	(c) Pi electrons and/or orbitals.
θ	Angle between successive C-C backbone bonds in a polymer with a vinyl backbone.
ρ	Density.
ρ^*	Reducing density parameter in an equation-of-state.
$\tilde{\rho}$	Reduced density in an equation-of-state.
$\tilde{\rho}_{mix}$	Reduced density of a mixture in an equation-of-state.
ρ_P	Density of the polymer component of a polymer/solvent system.

σ (a) Steric hindrance parameter (usually measured in dilute solutions).

(b) Sigma electrons and/or orbitals.

(c) Stress.

(d) Standard deviation.

σ_c Stress required to cause failure of a specimen by crazing.

σ_{ci} Stress required for the inception of crazing.

σ_f Stress required to cause failure of a specimen by brittle fracture.

σ_y Stress required to cause failure of a specimen by shear yielding.

τ (a) Angle between two successive bonds in shortest path across the chain backbone.

(b) Shear stress.

(c) Time.

τ_y Stress required to cause failure of a specimen by shear yielding under simple shear.

ξ (a) An "intensive" connectivity index, defined as χ/N, with the same subscripts and superscripts as the corresponding "extensive" (χ-type) connectivity index.

(b) A spline function used in making interpolations.

ζ Magnetic susceptibility.

ζ_m Molar diamagnetic susceptibility.

D. Terms Starting with a Capital Letter of the Greek Alphabet

In alphabetical order of the corresponding letters of Latin alphabet: $D \rightarrow \Delta$, $F \rightarrow \Phi$, $P \rightarrow \Pi$, $Q \rightarrow \Theta$ and $S \rightarrow \Sigma$.

$\Delta C_p(T_g)$ Change in the molar heat capacity of a polymer upon undergoing the glass transition.

ΔG_m Gibbs free energy of fusion (melting).

ΔG_s Gibbs free energy of sorption.

ΔH_m Enthalpy (heat) of fusion (melting).

ΔH_s Enthalpy (heat) of sorption.

ΔS_m Entropy of fusion (melting).

ΔS_s Entropy of sorption.

Δp Partial pressure difference of penetrant molecules between two surfaces of a specimen.

Φ_p Volume fraction of polymer in solution.

Φ_{solv} Solvated volume fraction of polymer in solution (generally different from Φ_p).

Π Product sign.

Θ Indicator that a polymer in solution is under the so-called "theta conditions".

Σ Summation sign.

APPENDIX

REPEAT UNIT MOLECULAR WEIGHTS

The molecular weight per repeat unit (M) is useful in calculating many properties. For example, it can be used to obtain densities and specific volumes from the molar volumes listed in Chapter 3, specific heat capacities from the molar heat capacities listed in Chapter 4, the numbers of repeat units between crosslinks from M_c values, and the number of repeat units between entanglements from M_e values. Table A.1 lists the values of M for many polymers, in units of g/mole, and in alphabetical order by the polymer name.

In searching for a polymer in this table, note that the sorter of the word processing software used in its generation sorts all characters (and not just letters) in a particular order. A parenthesis precedes a square bracket, a square bracket precedes a number, the numbers are ordered in an ascending sequence, and the letters are ordered in the usual alphabetical order after the numbers. That is, the following "alphabetical" order is used: (, [,1,2,...,8,9,10,11,...,a,b,...,y,z.

Table A.1. Repeat unit molecular weights (M), in g/mole.

Polymer	M	Polymer	M
Bisphenol-A polycarbonate	254.29	Poly(2,2,2'-trimethylhexamethylene	
Nylon 6,12	310.45	terephthalamide)	288.39
Poly(1,2,2-trimethylpropyl methacrylate)	170.25	Poly(2,2,2-trifluoro-1-methylethyl	
Poly(1,2-diphenylethyl methacrylate)	266.34	methacrylate)	182.14
Poly(1,3-dimethylbutyl methacrylate)	170.25	Poly(2-chloroethyl methacrylate)	148.59
Poly(1,4-cyclohexylidene dimethylene terephthalate)	274.32	Poly(2-ethylbutyl methacrylate)	170.25
Poly(1,4-pentadiene)	68.12	Poly(2-methyl-1,4-pentadiene)	82.15
Poly(1-butene sulfone)	120.17	Poly(3,3,3-trifluoropropylene)	96.05
Poly(1-butene)	56.11	Poly(3,3-dimethylbutyl methacrylate)	170.25
Poly(1-hexene sulfone)	148.23	Poly(3-chloropropylene oxide)	92.53
Poly(1-hexene)	84.16	Poly(3-phenyl-1-propene)	118.18
Poly(1-methylbutyl methacrylate)	156.22	Poly(4-fluoro-2-trifluoromethylstyrene)	190.14
Poly(1-methylpentyl methacrylate)	170.25	Poly(4-methyl-1-pentene)	84.16
Poly(1-octadecene)	252.48	Poly(4-phenyl-1-butene)	132.21
Poly(1-pentene)	70.13	Poly(5-phenyl-1-pentene)	146.23
Poly(1-phenylethyl methacrylate)	190.24	Poly(8-aminocaprylic acid)	141.20

☞☞☞   **TABLE A.1 IS CONTINUED IN THE NEXT PAGE.** ☞☞☞

499

Table A.1. CONTINUED FROM THE PREVIOUS PAGE.

Polymer	M	Polymer	M
Poly(11-aminoundecanoic acid)	183.29	Poly(glycolic acid)	58.04
Poly(12-aminododecanoic acid)	197.32	Poly(hexamethylene adipamide)	226.32
Poly(α,α-dimethylpropiolactone)	100.12	Poly(hexamethylene azelamide)	268.37
Poly(α-methyl acrylamide)	85.11	Poly(hexamethylene isophthalamide)	246.31
Poly(α-methyl styrene)	118.18	Poly(hexamethylene sebacamide)	282.40
Poly(α-vinyl naphthalene)	154.21	Poly(hexamethylene sebacate)	284.40
Poly(β-propiolactone)	72.06	Poly(isobutyl acrylate)	128.17
Poly(benzyl methacrylate)	176.22	Poly(isobutyl methacrylate)	142.20
Poly(cyclohexyl α-chloroacrylate)	188.65	Poly(isopentyl methacrylate)	156.22
Poly(cyclohexyl methacrylate)	168.24	Poly(isopropyl α-chloroacrylate)	148.59
Poly(δ-valerolactone)	100.12	Poly(isopropyl methacrylate)	128.17
Poly(dicyclooctyl itaconate)	350.50	Poly(m-phenylene isophthalamide)	238.25
Poly(diethnyl diphenylsilylene)	230.34	Poly(m-phenylene terephthalamide)	238.25
Poly(dimethyl itaconate)	158.15	Poly(m-trifluoromethylstyrene)	172.15
Poly(dimethyl siloxane)	74.16	Poly(methacrylic acid)	86.09
Poly(diphenylmethyl methacrylate)	252.31	Poly(methyl α-cyanoacrylate)	110.10
Poly(dodecyl methacrylate)	254.41	Poly(methyl acrylate)	86.09
Poly(ϵ-caprolactam)	113.16	Poly(methyl methacrylate)	100.12
Poly(ϵ-caprolactone)	114.14	Poly(n-butyl α-chloroacrylate)	162.62
Poly(ether ether ketone)	288.30	Poly(n-butyl acrylate)	128.17
Poly(ethyl α-chloroacrylate)	134.56	Poly(n-butyl methacrylate)	142.20
Poly(ethyl acrylate)	100.12	Poly(n-hexyl methacrylate)	170.25
Poly(ethyl methacrylate)	114.14	Poly(n-octyl methacrylate)	198.31
Poly(ethylene adipate)	172.18	Poly(n-propyl α-chloroacrylate)	148.59
Poly(ethylene isophthalate)	192.17	Poly(n-propyl methacrylate)	128.17
Poly(ethylene oxalate)	116.07	Poly(N-vinyl carbazole)	193.25
Poly(ethylene phthalate)	192.17	Poly(N-vinyl pyrrolidone)	111.14
Poly(ethylene sebacate)	228.29	Poly(neopentyl methacrylate)	156.22
Poly(ethylene suberate)	200.23	Poly(o-methyl styrene)	118.18
Poly(ethylene succinate)	144.13	Poly(o-phenylene isophthalamide)	238.25
Poly(ethylene terephthalate)	192.17	Poly(o-phenylene terephthalamide)	238.25
Poly(ethylene-2,6-naphthalenedicarboxylate)	242.23	Poly(octadecyl methacrylate)	338.57
Poly(γ-butyrolactone)	86.09	Poly(oxy-2,2-dichloromethyltrimethylene)	155.02

☞☞☞   **TABLE A.1 IS CONTINUED IN THE NEXT PAGE.** ☞☞☞

Table A.1. CONTINUED FROM THE PREVIOUS PAGE.

Polymer	M	Polymer	M
Poly(oxymethyleneoxyethylene)	74.08	Poly(vinyl dimethylbenzylsilane)	176.33
Poly(oxymethyleneoxytetramethylene)	102.13	Poly(vinyl dimethylphenylsilane)	162.30
Poly(p-bromo styrene)	183.05	Poly(vinyl ethyl ether)	72.11
Poly(p-chloro styrene)	138.60	Poly(vinyl fluoride)	46.04
Poly(p-cyclohexylphenyl methacrylate)	244.34	Poly(vinyl isobutyl ether)	100.16
Poly(p-fluoro styrene)	122.14	Poly(vinyl isopropyl ether)	86.13
Poly(p-hydroxybenzoate)	120.11	Poly(vinyl methyl ketone)	70.09
Poly(p-methacryloxy benzoic acid)	206.20	Poly(vinyl methyl sulfide)	74.15
Poly(p-methyl styrene)	118.18	Poly(vinyl n-butyl ether)	100.16
Poly(p-phenylene isophthalamide)	238.25	Poly(vinyl n-butyl sulfide)	116.23
Poly(p-phenylene)	76.10	Poly(vinyl n-decyl ether)	184.32
Poly(p-t-butyl styrene)	160.26	Poly(vinyl n-dodecyl ether)	212.38
Poly(p-xylylene)	104.15	Poly(vinyl n-hexyl ether)	128.21
Poly(phenyl methacrylate)	162.19	Poly(vinyl n-octyl ether)	156.27
Poly(propylene oxide)	58.08	Poly(vinyl n-pentyl ether)	114.19
Poly(propylene sulfone)	106.15	Poly(vinyl p-ethylbenzoate)	176.22
Poly(sec-butyl α-chloroacrylate)	162.62	Poly(vinyl p-isopropylbenzoate)	190.24
Poly(sec-butyl methacrylate)	142.20	Poly(vinyl p-t-butylbenzoate)	204.27
Poly(styrene oxide)	120.15	Poly(vinyl propionate)	100.12
Poly(t-butyl acrylate)	128.17	Poly(vinyl sec-butyl ether)	100.16
Poly(t-butyl methacrylate)	142.20	Poly(vinyl trimethylsilane)	100.24
Poly(tetramethylene adipate)	200.23	Poly(vinylene diphenylsilylene)	208.34
Poly(tetramethylene isophthalate)	220.22	Poly(vinylidene chloride)	96.94
Poly(tetramethylene terephthalate)	220.22	Poly(vinylidene fluoride)	64.03
Poly(trimethylene adipate)	186.21	Poly[1,1-(1-phenylethane)	
Poly(trimethylene succinate)	158.15	bis(4-phenyl)carbonate]	316.35
Poly(vinyl 2-ethylhexyl ether)	156.27	Poly[1-(o-chlorophenyl)ethyl methacrylate]	224.69
Poly(vinyl acetate)	86.09	Poly[2,2'-(m-phenylene)-	
Poly(vinyl alcohol)	44.05	5,5'-bibenzimidazole]	308.34
Poly(vinyl benzoate)	148.16	Poly[2,2-hexafluoropropane	
Poly(vinyl butyral)	142.20	bis{4-(2,6-dibromophenyl)}carbonate]	677.81
Poly(vinyl chloride)	62.50	Poly[2,2-propane	
Poly(vinyl chloroacetate)	120.54	bis{4-(2,6-dibromophenyl)}carbonate]	569.87
Poly(vinyl cyclohexane)	110.20		

☞ ☞ ☞ TABLE A.1 IS CONTINUED IN THE NEXT PAGE. ☞ ☞ ☞

Table A.1. CONTINUED FROM THE PREVIOUS PAGE.

Polymer	M	Polymer	M
Poly[2,2-propane *bis*{4-(2,6-dichlorophenyl)}carbonate]	392.07	Poly[oxy(vinylmethylsilylene)]	86.17
Poly[2,2-propane *bis*{4-(2,6-dimethylphenyl)}carbonate]	310.39	Poly[thio *bis*(4-phenyl)carbonate]	244.27
Poly[3,5-(4-phenyl-1,2,4-triazole)-1,4-phenylene		Poly[thio(*p*-phenylene)]	108.16
-3,5-(4-phenyl-1,2,4-triazole)-1,3-phenylene]	438.50	Polyacrylamide	71.08
Poly[4,4'-diphenoxy di(4-phenylene)sulfone]	400.45	Polyacrylonitrile	53.06
Poly[4,4'-isopropylidene diphenoxy		Polybutadiene	54.09
di(4-phenylene)sulfone]	442.53	Polychlorotrifluoroethylene	116.47
Polychloroprene	88.54	Polyethylene	28.05
Poly[4,4'-sulfone diphenoxy di(4-phenylene)sulfone]	464.51	Polyheteroarylene IX	524.54
Poly[di(*n*-heptyl) itaconate]	326.48	Polyheteroarylene VIII	488.51
Poly[di(*n*-nonyl) itaconate]	382.59	Polyheteroarylene X	324.32
Poly[di(*n*-octyl) itaconate]	354.53	Polyheteroarylene XII	504.50
Poly[di(*n*-propyl itaconate)]	214.26	Polyheteroarylene XIII	528.53
Poly[ethylene-N-(β-trimethylsilylethyl)imine]	143.31	Polyheteroarylene XIV	496.49
Poly[methane *bis*(4-phenyl)carbonate]	226.23	Polyheteroarylene XV	576.66
Poly[N,N'-(*p,p*'-oxydiphenylene)pyromellitimide]	382.33	Polyisobutylene	56.11
Poly[oxy(2,6-diisopropyl-1,4-phenylene)]	176.26	Polyisoprene	68.12
Poly[oxy(2,6-dimethyl-1,4-phenylene)]	120.15	Polymethacrylonitrile	67.09
Poly[oxy(2,6-dimethyl-5-bromo-1,4-phenylene)]	185.02	Polyoxyethylene	44.05
Poly[oxy(2,6-diphenyl-1,4-phenylene)]	244.29	Polyoxyhexamethylene	100.16
Poly[oxy(acryloxypropylmethylsilylene)]	172.26	Polyoxymethylene	30.03
Poly[oxy(diethylsilylene)]	102.21	Polyoxynaphthoate	170.17
Poly[oxy(mercaptopropylmethylsilylene)]	134.27	Polyoxyoctamethylene	128.21
Poly[oxy(methyl γ-trifluoropropylsilylene)]	156.18	Polyoxytetramethylene	72.11
Poly[oxy(methyl *m*-chlorophenylethylsilylene)]	198.72	Polyoxytrimethylene	58.08
Poly[oxy(methyl *m*-chlorophenylsilylene)]	170.67	Polypentadecanolactone	240.39
Poly[oxy(methyl *n*-hexadecylsilylene)]	284.56	Polyphenylenediamide	360.37
Poly[oxy(methyl *n*-hexylsilylene)]	144.29	Polypropylene	42.08
Poly[oxy(methyl *n*-octadecylsilylene)]	312.62	Polystyrene	104.15
Poly[oxy(methyl *n*-octylsilylene)]	172.34	Polytetrafluoroethylene	100.02
Poly[oxy(methyl *n*-tetradecyl silylene)]	256.51	Polytridecanolactone	212.33
Poly[oxy(methylhydrosilylene)]	60.13	Polytrifluoroethylene	82.03
Poly[oxy(methylphenylsilylene)]	136.23	Torlon	354.37
Poly[oxy(*p*-phenylene)]	92.10	Ultem	592.61
Polyundecanolactone	184.28		

INDEX

Abbé number, 216

Absorption, 2,213,217,268

Acceptor, 11

Achromatic lenses, 216

Acoustics, 288,296

Acoustic wave (*see* Sound)

Activation energy, 52,311

    for degradation, 421,426,479

    for diffusion, 405,406

    for permeation, 405,414

    for sorption, 405

    for viscous flow, 345,368,369,371,
        377-379,381,387-389,477,480,
        485

    table of values, 377

Activation volume, 318,319

Activity, 407

Additive, 77,112,139,211,240,241,244,
    257,364,394,401,422

Additive techniques (*see* Group contributions)

Adhesion, 193,208,210,211,459,462

Adhesive failure, 211

Affinity coefficient, 406

Aggregate, 440,459,462

Aging (*see* Physical aging;
    see also Thermal history)

Agricultural chemistry, 3

Alignment (*see* Orientation)

Alloy, 464 (*see also* Copolymer)

Alternating copolymer (*see* Copolymer)

American Society for Testing and Materials,
    421

Amorphous fraction, 140,410

Amorphous phase *or* material, 4,5,11,52,
    53,75,82,84,110,137,138,140,182,185,
    197,208,215,218,219,228,233,241,
    280,285-287,307,308,315,316,322,
    325,336,368,394-400,410,411,
    413-415,459-461,463,464,470,478,479

Anharmonicity, 51

Anisotropy, 2,8,11,45,47,50,51,217,399,
    462

    (*see also* Orientation)

Annealing, 2,139,407,413

Area (*see* Cross-sectional area;
    see also Surface area)

Aroma molecules, 403

Aromatic ring (*see under* Ring)

Artificial intelligence, 440,463,464

Artificial neural network, 463

Aspect ratio, 461

Atactic polymers (*see under* Tacticity)

Atom, 9,44,45,54,80,83,84,120,201,214,
    243,468 (*see also* Vertex)

Atomic charge, 7

Atomic configuration
    (*see* Electronic configuration)

Atomic corrections (*see* Correction terms)

Atomic number, 19

Atomic radius, 55

Atomic size, 2,51

Atomistic group vibrations, 83

Atomistic simulations, 77,405,406,408

Attractiveness to customers, 213

Automation (*see* Software)

Avogadro's number, 50